# DIATOMIC INTERACTION POTENTIAL THEORY

*Volume 2*

**Applications**

This is Volume 31-2 of
PHYSICAL CHEMISTRY
A series of monographs
Edited by ERNEST M. LOEBL, *Polytechnic Institute of Brooklyn*

A complete list of the books in this series appears at the end of the volume.

# DIATOMIC INTERACTION
# POTENTIAL THEORY

*Jerry Goodisman*

DEPARTMENT OF CHEMISTRY
SYRACUSE UNIVERSITY
SYRACUSE, NEW YORK

*Volume 2*

**Applications**

QD462
G65
V. 2
1973

ACADEMIC PRESS   New York and London   1973

*A Subsidiary of Harcourt Brace Jovanovich, Publishers*

ACADEMIC PRESS, INC.
111 Fifth Avenue, New York, New York 10003

*United Kingdom Edition published by*
ACADEMIC PRESS, INC. (LONDON) LTD.
24/28 Oval Road, London NW1

Library of Congress Cataloging in Publication Data

Goodisman, Jerry.
    Diatomic interaction potential theory.

    (Physical chemistry, a series of monographs)
    Includes bibliographies.
    CONTENTS: v.  1.  Fundamentals.–v.   2.  Applications.
    1.  Quantum chemistry.   I.  Title.   II.  Series.
QD462.G65          541'.28          72-9985
ISBN 0–12–290202–5  (v.  2)

PRINTED IN THE UNITED STATES OF AMERICA

# Contents

## Chapter I  Large and Small Distances

## Chapter II   Intermediate $R$

## Chapter III   Semiempirical Calculations and Simple Models

# Preface

The calculation of the energy of a diatomic system as a function of internuclear separation is a problem which has a long history and has generated an enormous amount of literature. Due in part to advances in computational hardware and software in the past few years, quantum chemists can now produce reliable interaction potentials for diatomic systems in their ground states. The situation for excited states and for polyatomic systems is less satisfactory, but there is hope that it will shortly improve.

These two volumes cover the theoretical material involved in calculations for diatomic systems in their ground states, with attention given to the variety of the approaches one may use. The first volume contains mostly basic and general material; the second includes more in the way of specific descriptions of modern calculations.

The problem is defined in Chap. I, Vol. 1. A discussion is given of the nature of an interatomic interaction potential or potential energy curve, including its relation to reality (experiment). Chapter II presents a general discussion of its shape. Chapter III treats the main approaches to schemes of calculation: variation theory, perturbation theory, the virial and Hellmann–Feynman theorems, local energy principles, and quantum statistical theories. In Chapter I of Volume 2, the calculation of the interaction potential for large and small values of the internuclear distance $R$ (separated and united atom limits) is considered. Chapter II treats the methods used for intermediate values of $R$, which in principle means *any* values of $R$. The Hartree–Fock and configuration interaction schemes described here have been the most used of all the methods. Semiempirical theories and methods constitute the subject of the last Chapter of Volume 2.

The level of treatment throughout, it is hoped, is sufficiently elementary for the material to be understood after an introductory quantum mechanics course. By means of this book, the reader should be able to go from that degree of preparation to the current literature.

Work on this book started about five years ago; its proximate cause was my participation in a special topics graduate course, with Prof. D. Secrest and Prof. J. P. Toennies, at the University of Illinois. The course largely dealt with scattering experiments and their relation to potential energy curves. At that time I was struck by the fact that there was much material which was common knowledge among those involved in quantum chemical calculations but unfamiliar to students, even those with good course backgrounds. Thus, one goal of the present book is to make that material conveniently available to students and others interested in the subject, and to introduce them to the current literature.

For those interested in the theory of quantum chemical calculations, I want to provide in one place as much information as I can on the varied methods which are available. For those interested in potential curves or in quantum chemistry, but not particularly interested in calculating potential curves, I hope this book will be a guide to what has been going on, as well as an aid in reading the literature.

The subject has been limited to diatomic interactions, and, still further, to diatomic ground-state interactions. Of course, the limitation on the calculations discussed does not mean that the methods of calculation have no other applicability. I hope that the general discussions, particularly in the first volume, will be of interest to those who care about other systems. Some of the methods may even find their greatest applicability to those other systems. However, only by severely limiting my subject could I hope to attain some measure of completeness of coverage. Even so, I have had to give very limited space to certain topics. I have not discussed calculations specifically applicable to one- and two-electron systems (another book should be written on these, as was done for the corresponding atoms); I have slighted relativistic and magnetic effects; I have given unjustly brief coverage to many-body theory. There are undoubtedly other sins of omission.

Nevertheless, I believe that this book gives a balanced picture of the enormous amount of work that has been done on ground state diatomic potential curves. While limitations of certain methods have been discussed, I hope the main point is still clear: after many years of effort, reliable potential curves can now be generated for most systems of interest.

# Notes on Notation and Coordinate Systems

As an aid in keeping formulas more legible, a Dirac-like bracket notation is employed frequently, without necessarily implying notions of states, representations, and so on. The triangular bracket $\langle \Phi_i/\Phi_k \rangle$ means the product of $\Phi_i^*$ and $\Phi_k$, integrated over the entire configuration space, which must be the same for the two functions. This means integration over all spatial coordinates and sums over spin coordinates. The arguments of $\Phi_i$ and $\Phi_k$ need not be stated, although sometimes they are. The "factors" in this "scalar product" may be considered separately. Thus $| \Phi \rangle$ and $\Phi$ are equivalent, both being wave functions.

The adjoint wavefunction is written $\langle \Phi |$, but this is also used to denote an operator in the following sense: $\langle \Phi |$ multiplying $| \Psi \rangle$ gives the bracket $\langle \Phi | \Psi \rangle$, a number which is computed by multiplying $\Phi^*$ by $\Psi$ and integrating over the configuration space. Thus, $\langle \Phi |$ may be interpreted as the operation of multiplication by $\Phi^*$, followed by integration. If the functions $\psi_i$ form a complete set,

$$\left[ \sum_i | \psi_i \rangle \langle \psi_i | \right] \phi \rangle = | \phi \rangle$$

so that the operator in the square bracket is the identity operator.

Italics are used for operators, e.g., $F$ or $h$. We use $\langle \Psi | F | X \rangle$ equivalently to $\langle \Psi | FX \rangle$: thus $F$ operates on the function $X$ and the scalar product of the result with $\Psi$ is then taken. Operators are assumed to operate to the right in all bracket expressions. Thus,

$$\langle \Phi | P | \Psi \rangle = \langle P^\dagger \Phi | \Psi \rangle = \langle \Psi | P^\dagger \Phi \rangle^*$$

where $P^\dagger$ is the adjoint of $P$.

In general, we use bold face for matrices, e.g., $\mathbf{M}$ is the matrix with elements $M_{ij}$. The determinant of the matrix $\mathbf{M}$ is written $| \mathbf{M} |$ or $\det[\mathbf{M}]$

and the trace is written Tr[**M**]. The dagger (†) is used to indicate the adjoint of matrices, so $(\mathbf{M}^\dagger)_{ij}$, the $(i, j)$ element of the matrix $\mathbf{M}^\dagger$, is $M_{ji}^*$. The transpose is similarly denoted by $T$: $(\mathbf{M}^T)_{ij} = M_{ji}$. The transpose or adjoint of a column vector is a row vector and vice versa.

The expression "matrix element of the operator $P$ between states (or functions) $\Psi_i$ and $X$" means the integral

$$\int \Psi_i^* P X \, d\tau = \langle \Psi_i \mid P \mid X \rangle.$$

The "scalar product" of $X_i$ and $X_j$, $\langle X_i \mid X_j \rangle$, is sometimes referred to as an element of the overlap matrix. In these integrals or brackets an integration, or summation where appropriate, is implied over all coordinates, unless there is a specific remark to the contrary.

"Real part of" is denoted by Re, e.g., Re($\phi$). Curly brackets are used to refer to all the members of a set of functions, as in: "we orthonormalize the $\{\phi_i\}$ among themselves."

Several different coordinate systems are used in our discussions. Consider the two nuclei separated by a distance $R$, and a Cartesian coordinate system located at the midpoint, with the $Z$ axis along the internuclear axis.

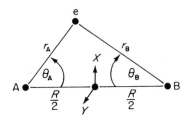

The position of an electron can be specified by its distances from the nuclei, plus the angle $\varphi$ between the plane containing the ABe triangle and a reference plane containing the $Z$ axis. Alternatively, one can use the angles $\theta_A$ and $\theta_B$ together with $\varphi$, or one of the sets $(r_A, \theta_A, \varphi)$ and $(r_B, \theta_B, \varphi)$. The last two are of course just spherical polar coordinates centered on one nucleus or the other.

It is convenient for many purposes to use coordinates involving $r_A$ and $r_B$, together with $\varphi$. Most often, one uses rather than $r_A$ and $r_B$ themselves, the confocal ellipsoidal coordinates

$$\xi = (r_A + r_B)/R$$
$$\eta = (r_A - r_B)/R.$$

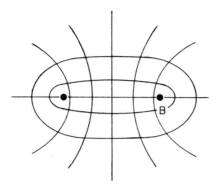

$\xi$ and $\eta$ are dimensionless. The surfaces of constant $\xi$ are ellipsoids of revolution with foci at the nuclei. The value of $\xi$ can run from 1 (corresponding to the line between the nuclei) to $\infty$. The surfaces of constant $\eta$ are hyperboloids of revolution with foci at the nuclei. The value of $\eta$ is between $-1$ (corresponding to the extension of the internuclear axis to the left of A) and $+1$ (corresponding to the extension of the axis to the right of B). $\eta = 0$ is the plane bisecting the internuclear axis.

The volume element for this coordinate system is

$$d\tau = \left(\frac{R}{2}\right)^3 (\xi^2 - \eta^2)\, d\xi\, d\eta\, d\varphi$$

and the Laplacian is

$$\Delta = \left(\frac{2}{R}\right)^2 (\xi^2 - \eta^2)^{-1} \left[ \frac{\partial}{\partial \xi} (\xi^2 - 1) \frac{\partial}{\partial \xi} + \frac{\partial}{\partial \eta} (1 - \eta^2) \frac{\partial}{\partial \eta} \right.$$
$$\left. + \frac{\xi^2 - \eta^2}{(\xi^2 - 1)(1 - \eta^2)} \frac{\partial^2}{\partial \varphi^2} \right].$$

All distances are proportional to $R$ for given values of the coordinates $\xi$ and $\eta$. Thus

$$r_A = \frac{R}{2} (\xi + \eta), \qquad x = \frac{R}{2} [(\xi^2 - 1)(1 - \eta^2)]^{1/2} \cos \varphi,$$

$$r_B = \frac{R}{2} (\xi - \eta); \qquad y = \frac{R}{2} [(\xi^2 - 1)(1 - \eta^2)]^{1/2} \sin \varphi,$$

$$z = \frac{R}{2} \xi\eta.$$

and the distance between two points $(\xi_1, \eta_1)$ and $(\xi_2, \eta_2)$ is

$$r_{12}^2 = \left(\frac{R}{2}\right)^2 \{([(\xi_1^2 - 1)(1 - \eta_1^2)]^{1/2} - [(\xi_2^2 - 1)(1 - \eta_2^2)]^{1/2})^2$$
$$+ (\xi_1\eta_1 - \xi_2\eta_2)^2\}.$$

# Contents of Volume 1

# DIATOMIC INTERACTION POTENTIAL THEORY

*Volume 2*

**Applications**

# Chapter I    Large and Small Distances

---

## A. Very Large $R$

In most of this chapter, we will be discussing the interaction of two atoms A and B at internuclear distances large enough so that their electron distributions may be considered nonoverlapping. This will mean (see the discussion of differential overlap in Section A of Chapter II, Volume I, and Sections 1 and 2 of this chapter) that exchange between atoms can be neglected and that the wavefunction may be written as a product:

$$\Psi = \Psi_A(1 \cdots n_A)\Psi_B(n_A + 1, \ldots, n_A + n_B) \tag{1}$$

The $n_A$ electrons on A are described by the wave function $\Psi_A$, which is antisymmetric with respect to interchanges between these electrons, and the $n_B$ electrons on B are described by $\Psi_B$, which is antisymmetric to interchanges between the electrons on B. $\Psi_A$ is an eigenfunction of an atomic Hamiltonian $H_A$ with eigenvalue $E_A$:

$$H_A\Psi_A = E_A\Psi_A \tag{2}$$

$H_A$ includes electronic kinetic energy operators for electrons $1 \cdots n_A$, as well as Coulombic interactions between these electrons and between these electrons and nucleus A. In a similar way, we have

$$H_B\Psi_B = E_B\Psi_B \tag{3}$$

The correct wave function is an eigenfunction of

$$H = H_A + H_B + V \tag{4}$$

where $V$ includes all potential energy terms not appearing in $H_A$ or $H_B$.

$$V = \sum_{i > n_A} (-e^2/r_{iA}) + \sum_{i < n_A + 1} (-e^2/r_{iB}) + Z_A Z_B e^2/R$$

$$+ \sum_{i > n_A} \sum_{j < n_A + 1} (e^2/r_{ij}) \tag{5}$$

In Subsection 1, we will consider treatments of the effect of $V$ by perturbation theory. Although $V$ may be interpreted as the interaction between the atoms, it is not possible in general to treat it simply as a perturbation on the separated atom wave functions, since the molecular wave function must be antisymmetric with respect to exchanges of electrons between the atoms. The resulting complications will be discussed in Subsection 2, and in Section B. If the internuclear distance is very large, terms in the energy due to interatomic exchange may be neglected. Consider the limit of vanishing interaction. The exact energy of the system, i.e., the sum of the energies of the two atoms, is exactly equal to the expectation value of the Hamiltonian (4) over the product wavefunction (1). In the absence of degeneracy, this implies that (1) is the eigenfunction of (4). Thus, antisymmetrization to exchanges of electrons between the atoms becomes of negligible importance as the interaction approaches zero. In Subsections 3 *et seq.*, the interactions are calculated by performing perturbation theory with $V$, Eq. (5), as the perturbation, starting with the wave function of Eq. (1).

The interaction energy is often divided into "electrostatic" and "dispersive" parts, the former being treated in Subsection 3. The remainder of this section A discusses calculations of dispersive forces or van der Waals constants, a subject on which the literature is rapidly growing. Recent reviews of calculation of long-range forces are given by Dalgarno [1], Dalgarno and Davison [2], and by Margenau and Kestner [3].

### 1. *Perturbation Theory*

In order to provide a more general framework for discussion of the interaction between atoms, we do not invoke the assumption of non-overlapping distributions immediately. To consider the treatment of the interaction by perturbation theory, a convenient basis set would seem to be the set of product functions like that of Eq. (1).

$$\Psi_i = \Psi_{Ai'}(1 \cdots n_A)\Psi_{Bi''}(n_A + 1, \ldots, n_A + n_B), \qquad i', i'' = 0, 1, \ldots \quad (6)$$

We would like to write

$$\Psi = \sum_i c_i \Psi_i$$

with

$$c_i = c_i^{(0)} + c_i^{(1)} + \cdots \tag{7}$$

A problem with this approach is that the $\Psi_i$ are not correctly antisymmetric with respect to interchanges of electrons between the atoms. A set of functions which are, is obtained by writing

$$\tilde{\Psi}_i = \sum_{\mathscr{R}} (-1)^R \mathscr{R} \Psi_i = P\Psi_i$$

where $\mathscr{R}$ is a permutation, and the sum is over all permutations of the $n_A + n_B$ electrons. The operator $P$, defined by this equation, is referred to as a projection operator, since it produces (projects), from an arbitrary function, a function having the desired behavior on electronic exchange. Now $\tilde{\Psi}_i$ is not an eigenfunction of $H_A + H_B$ for any particular assignment of electrons to A and B. Correspondingly, it is impossible to separate out $V$ as in Eq. (4) and consider it as a perturbation, since this presupposes a particular assignment of certain electrons to $A$ and certain others to $B$.

Furthermore, the set of $\tilde{\Psi}_i$ is overcomplete or linearly dependent, as Eisenschitz and London showed in their basic paper [4]. Therefore an expansion of $\Psi$ in the $\tilde{\Psi}_i$ would have coefficients which were not uniquely defined. Eisenschitz and London also showed how to circumvent this problem, and we present a summary of their treatment here, leaving out details of the manipulations. We will mention the nature of the linear dependence and indicate how the expansion coefficients can be uniquely defined. Then we will give the perturbation formalism of Eisenschitz and London, which will be exploited in Subsection 2 and in Section B.

Perhaps it should be pointed out first how projection of a linearly independent set (by antisymmetrization) can yield a linearly dependent set. Suppose $\{f_i\}$ are linearly independent, so that $\sum_j c_j f_j$ can vanish only if all $c_j$ vanish. Let the projected functions be $\{Pf_i\}$. Some of these may vanish. Then $\sum_j c_j Pf_j$ can be zero without all $\{c_j\}$ being zero: The nonzero $c_j$ go with the $f_j$ which are annihilated by $P$. $\{Pf_i\}$ is thus a linearly dependent set.

Suppose we have a complete set of functions in the space of $n_A$ electronic coordinates $A_i(1 \cdots n_A)$, and a complete set $B_j(n_A + 1, \ldots, n_A + n_B)$ in the space of electronic coordinates $n_A + 1, \ldots, n_A + n_B$. The set of products

$$u_k = A_{k'}B_{k''} \tag{8}$$

forms a complete set of functions in $(n_A + n_B)$-electron space. We form the antisymmetric projections of the $u_k$:

$$\Phi_k(I) = \sum_R (-1)^R \mathscr{R} u_k$$

where the sum is over the $(n_A + n_B)!$ permutations of all the electrons. Here, $I$ refers to the identity permutation. To conform somewhat to Eisenschitz and London's notation (their treatment is more general than ours) we write $\chi_R$ for $(-1)^R$, $\chi_R$ being the character of the matrix corresponding to $R$ in some chosen representation. We are concerned with the completely antisymmetric representation. In addition to the above functions we define

$$\Phi_k(S) = \sum_R \chi_R \mathscr{R} \mathscr{S} u_k \tag{9}$$

Since the $\{u_k\}$ form a complete set we may expand $\mathscr{R} u_k$ on them

$$\mathscr{R} u_k = \sum_l \Delta_{kl}^R u_l$$

The coefficients $\Delta_{lk}^R$ form a real unitary representation of the permutation group. Putting $\mathscr{S}^{-1}\mathscr{P}$ for $\mathscr{R}$ leads to

$$\mathscr{P} u_k = \sum_l \Delta_{kl}^{S^{-1}P} \mathscr{S} u_l \tag{10}$$

The relations of linear dependence among the $\Phi_k(Q)$ were elucidated by Eisenschitz and London [4]. First, from Schur's lemma and the group properties (EYRING, Chap. X, PILAR, Chap. 14)

$$\sum_S \chi_{RS^{-1}}^* \chi_{TS^{-1}} = \sum_S \sum_{ijkl} \Gamma_{ij}^*(R)\Gamma_{ji}^*(S^{-1})\Gamma_{kl}(T)\Gamma_{lk}(S^{-1})$$

$$= (g/f)\sum_{ij} \Gamma_{ij}^*(R)\Gamma_{ij}(T)$$

$$= (g/f)\chi(R^{-1}T) = (g/f)\chi(TR^{-1}) \tag{11}$$

Here, $g$ is the order of the group and $f$ is the dimension of the representation.

Using (11),

$$\sum_S \chi_{SR^{-1}}\Phi_k(S) = \sum_S \sum_P \chi_{SR^{-1}}\chi_P \,\mathscr{P}\mathscr{S}u_k$$

$$= \sum_S \sum_T \chi^*_{RS^{-1}}\chi_{TS^{-1}}\,\mathscr{C}\,u_k$$

$$= (g/f)\sum_T \chi_{TR^{-1}}\mathscr{C}u_k \qquad (12)$$

Now, for any set of constants $\alpha_i^s$, arbitrary except that

$$\sum_S \alpha_i^s = 1 \qquad \text{for each} \quad i,$$

let the four-dimensional matrix

$$[T(\alpha)]_{ik}^{RQ} = (f/g)\sum_S \alpha_i^s \,\Delta_{ik}^{S^{-1}}\,\chi^*_{SRQ^{-1}}$$

be defined. Using the relations (10)–(12),

$$\sum_{k,Q}[T(\alpha)]_{ik}^{RQ}\Phi_k(Q) = \sum_{Q,S}(f/g)\alpha_i^s\chi^*_{RSQ^{-1}}\Phi_i(QS^{-1})$$

$$= \sum_{S,T}(f/g)\alpha_i^s\chi_{TR^{-1}}\Phi_i(T) = \Phi_i(R), \qquad (13)$$

which shows the linear dependence between the $\Phi_k(Q)$. Now if one expanded an antisymmetric function in the $\Phi_i(R)$, say

$$G(1,\ldots,n_A+n_B) = \sum_{i,R}c_{iR}\Phi_i(R), \qquad (14)$$

the relations (13) would give

$$G(1,\ldots,n_A+n_B) = \sum_{i,R}\sum_{k,Q}c_{iR}[T(\alpha)]_{ik}^{RQ}\Phi_k(Q).$$

Defining

$$\tilde{c}_{kQ} = \sum_{i,R}[T(\alpha)]_{ik}^{RQ}c_{iR} \qquad (15)$$

we have

$$G(1,\ldots,n_A+n_B) = \sum_{k,Q}\tilde{c}_{kQ}\Phi_k(Q)$$

which is another expansion of $G$ in the same functions. Thus the expansion (14) is not unique. Eisenschitz and London [4] showed that limiting the sum in (14) to one value of $R$ only does not suffice to give a unique expansion, and that the different expansions (15) corresponding to different choices of the $\alpha_i$ are the only ones possible. Then the nonuniqueness problem associated with the linear dependence in the basis set would be solved if a choice among these expansions were made. If no choice were specified, the problem of finding the expansion coefficients (7) and the energies by perturbation theory would not be defined.

We denote by **T** (suppressing the arguments $\alpha$) the particular choice of the matrix $[T(\alpha)]_{ik}^{RQ}$ in which, for each $i$, all the $\alpha_i{}^s$ are taken equal.

$$T_{ik}^{RQ} = (f/g^2) \sum_S \Delta_{ik}^{S^{-1}N^{-1}} \chi^*_{3RQ^{-1}}$$

The matrix **T** is shown to be idempotent, $\mathbf{T}^2 = \mathbf{T}$, as follows:

$$\sum_{i,R} T_{ki}^{QR} T_{ij}^{RS} = (f^2/g^4) \sum_R \sum_{M,N} \Delta_{kj}^{M^{-1}N^{-1}} \chi_{RQ^{-1}M^{-1}} \chi^*_{RS^{-1}N}$$

$$= (f/g^3) \sum_{M,P} \Delta_{kj}^{P^{-1}} \chi^*_{S^{-1}PQ} = T_{kj}^{QS}$$

We have used (11) as well as relations like $\chi_{ABC} = \chi_{BCA} = \chi^*_{A^{-1}B^{-1}C^{-1}}$. Similarly, we can show that

$$\sum_{i,R} [T(\alpha)]_{ki}^{QR} T_{ij}^{RS} = (f^2/g^3) \sum_R \sum_{M,N} \Delta_{kj}^{M^{-1}N^{-1}} \alpha_k{}^M \chi_{RQ^{-1}M^{-1}} \chi_{RS^{-1}N}$$

$$= \frac{f}{g^2} \sum_{M,P} \Delta_{kj}^{P^{-1}} \chi^*_{S^{-1}PQ} \alpha_k{}^M = T_{kj}^{QS}. \tag{16}$$

Now, starting from an expansion like (14) we can transform to new coefficients using **T**,

$$c_{kQ}^0 = \sum_{i,R} T_{ik}^{RQ} c_{iR}. \tag{17}$$

We now show that $c_{kQ}^0$ will be invariant to the starting set of coefficients, so that the set $c_{kQ}^0$ is uniquely defined. Suppose we have two expansions like (14), with the two sets of coefficients differing by a transformation like (15):

$$c_{kQ}^{(1)} = \sum_{iR} [T(\alpha)]_{ik}^{RQ} c_{iR}^{(2)}.$$

Then

$$\sum_{iR} T_{ik}^{RQ} c_{iR}^{(1)} = \sum_{iR} T_{ik}^{RQ} c_{iR}^{(2)}$$

follows from (16), so that both sets lead to the same $\{c^0_{kQ}\}$. It was also shown by Eisenschitz and London [4] that the set $\{c^0_{kQ}\}$ has the property that the sum of the absolute squares of the coefficients is minimized.

The functions $G$ we want to expand will always be antisymmetric, which means $c^0_{iR}$ can be related to $c^0_{il}$. Let $G$ be expanded in the unsymmetrized $\mathscr{R}u_k$ for some fixed $\mathscr{R}$, the set $\{\mathscr{R}u_k\}$ being complete but not overcomplete:

$$G = \sum_j \mathscr{R}u_j \langle (\mathscr{R}u_j) \mid G \rangle$$

The expansion assumes that $\langle \mathscr{R}u_j \mid \mathscr{R}u_k \rangle = \delta_{jk}$. Eisenschitz and London [4] showed that $c^0_{iR}$ is just $(f/g^2)\langle (\mathscr{R}u_i) \mid G \rangle$. Thus, we will always write expansions in the $\Phi_i(R)$, but choose the coefficients $c_{iR}$ as:

$$c_{iR} = (f/g^2)\langle (\mathscr{R}u_i) \mid G \rangle \tag{18}$$

since we are interested in the one-dimensional completely antisymmetric representation.

We can get some idea of how this special choice of the coefficients solves the linear dependence problem. We mentioned earlier that antisymmetrization can produce a linearly dependent set of functions from a linearly independent set if some of the starting functions are annihilated by the antisymmetrization. Suppose $u_k$ is such a function. Then $\langle (\mathscr{R}u_k) \mid G \rangle$ will vanish because of the antisymmetry of $G$. Our recipe, Eq. (18), means that we choose the coefficients of all $\Phi_k(S)$ as zero. The situation will usually be not quite this simple: It will be a linear combination of the starting functions, say $\sum_k \alpha_k u_k$, which is annihilated by the antisymmetrization. Then our recipe means that the coefficient of $\sum_k \alpha_k \Phi_k(S)$, for each $S$, is taken as zero in the expansion of $G$.

Having specified uniquely the expansion of the exact function in the functions (10), Eisenschitz and London [4] developed, by perturbation theory, equations for the coefficients. The functions of equation (8) are atomic wave functions, as in (1):

$$A_{k'} = \Psi_{Ak'}(1 \cdots n_A) \quad \text{and} \quad B_{k''} = \Psi_{Bk''}(n_A + 1, \ldots, n_A + n_B).$$

Here, $\Psi_{Ak'}(1 \cdots n_A)$ is normalized and antisymmetric in its $n_A$ electronic coordinates, and $\Psi_{Bk''}(n_A + 1, \ldots, n_B)$ is normalized and antisymmetric in its $n_B$ electronic coordinates. To ensure antisymmetry for permutations of electrons between the atoms, the exact molecular wave function $\Psi$ is expanded in the set

$$\Psi_i(S) = \sum_R \chi_R \mathscr{R} \mathscr{S} u_i$$

$$= \sum_R \chi_R \mathscr{R} \mathscr{S} \{\Psi_{Ai'}(1 \cdots n_A) \Psi_{Bi''}(n_A + 1, \ldots, n_A + n_B)\}. \quad (19)$$

By taking the coefficients according to (18),

$$c_{iS} = g^{-2} \langle \mathscr{S} u_i) \mid G \rangle \chi_S,$$

we avoid the nonuniqueness of the expansion caused by the linear dependence. In the perturbation theory, the orders of smallness follow from the fact that $V u_k$ is small. It is small because the conditions that make $u_k$ large (small values of $r_{1A}$, $r_{2A}$, etc.) are just the conditions which make $V$ small [see Eq. (5)] and vice versa. Similarly, $\mathscr{R}(V u_k) = (\mathscr{R} V)(\mathscr{R} u_k)$ is a small quantity. We now derive the equations for the expansion coefficients.

The $\{c_{iS}\}$ and the energy are written as sums of zero-order, first-order, etc. terms. Superscripts are used to identify orders of perturbation theory. Substituting into the eigenvalue equation, we have

$$(c_{iS}^{(0)} + c_{iS}^{(1)} + \cdots) H \Psi_{iS} = (E^{(0)} + E^{(1)} + \cdots) \sum_{iS} (c_{iS}^{(0)} + c_{iS}^{(1)} + \cdots) \Psi_{iS}. \quad (20)$$

Since the full Hamiltonian commutes with all permutations,

$$H \mathscr{R} u_i = \mathscr{R}(H_A + H_B + V) u_i = (E_{Ai'} + E_{Bi''}) \mathscr{R} u_i + \mathscr{R} V u_i.$$

The last term is first order. We introduce the abbreviations

$$\varepsilon_i = E_{Ai'} + E_{Bi''}$$

(remembering $i$ means the pair of indices, $i'\ i''$) and

$$\mathscr{A} = \sum_R \chi_R \mathscr{R}$$

for the antisymmetrizer. Note that $\mathscr{A}$ is Hermitian so that

$$\langle F \mid \mathscr{A} \mid G \rangle = \langle \mathscr{A} F \mid G \rangle.$$

Now we write

$$H \Psi_i(S) = \varepsilon_i \Psi_i(S) + \mathscr{A} \mathscr{S} V u_i$$

and substitute into (20). With the specification (18), we have a procedure for finding the eigenfunctions of (4), in the presence of exchange, by perturbation theory.

Separating off terms of various orders, we have

$$\sum_{i,S} c_{iS}^{(0)}(\varepsilon_i - E^{(0)})\Psi_i(S) = 0 \qquad (21)$$

$$\sum_{i,S} c_{iS}^{(1)}(\varepsilon_i - E^{(0)})\Psi_i(S) + \sum_{i,S} c_{iS}^{(0)}(\bar{\mathscr{A}}\mathscr{S}Vu_i - E^{(1)}\Psi_i(S)) = 0 \qquad (22)$$

$$\sum_{i,S} c_{iS}^{(2)}(\varepsilon_i - E^{(0)})\Psi_i(S)$$
$$+ \sum_{i,S} c_{iS}^{(1)}(\bar{\mathscr{A}}\mathscr{S}Vu_i - E^{(1)}\Psi_i(S)) + \sum_{i,S} c_{iS}^{(0)}(-E^{(2)}\Psi_i(S)) = 0 \qquad (23)$$

and so on. A solution to (21) is clearly

$$E^{(0)} = \varepsilon_0$$

and $c_{iS}^{(0)} = 0$ except for states having $\varepsilon_i = E^{(0)}$. If we assume no degeneracy in the ground state, we may put

$$c_{iS}^{(0)} = \delta_{i0}\chi_S.$$

The subsequent equations are inhomogeneous equations for $c_{iS}^{(1)}$, $c_{iS}^{(2)}$, etc. In each case, we use the results of previous equations and expand the inhomogeneity in the $\Psi_i(S)$ according to our recipe, Eq. (18). Thus (22) becomes

$$\sum_{iS} c_{iS}^{(1)}(\varepsilon_i - \varepsilon_0)\Psi_i(S) = \sum_S \sum_{jR} g^{-2}\chi_S\Psi_j(R)\langle \mathscr{R}u_j \mid E^{(1)}\Psi_0(S) - \bar{\mathscr{A}}\mathscr{S}Vu_0\rangle.$$

From the coefficient of $\Psi_0(R)$ we obtain

$$E^{(1)} = \frac{\langle \mathscr{R}u_0 \mid \bar{\mathscr{A}}Vu_0\rangle}{\langle \mathscr{R}u_0 \mid \bar{\mathscr{A}}u_0\rangle} = \frac{\sum_P \chi_P\langle u_0 \mid \mathscr{P}V \mid u_0\rangle}{\sum_P \chi_P\langle u_0 \mid \mathscr{P} \mid u_0\rangle}. \qquad (24)$$

From the coefficient of $\Psi_j(T)$ we obtain

$$c_{jT}^{(1)} = \frac{\sum_S \chi_S\langle \mathscr{C}u_j \mid E^{(1)}\Psi_0(S) - \bar{\mathscr{A}}\mathscr{S}Vu_0\rangle}{(\varepsilon_i - \varepsilon_0)g^2} = \frac{\chi_T\langle u_j \mid E^{(1)}\Psi_0 - \bar{\mathscr{A}}Vu_0\rangle}{(\varepsilon_i - \varepsilon_0)g}.$$

The second-order equation is

$$\sum_{iS} c_{iS}^{(2)}(\varepsilon_i - \varepsilon_0)\Psi_i(S)$$
$$+ \sum_{iS} c_{iS}^{(1)}\sum_{jR} \Psi_j(R)\langle \mathscr{R}u_j \mid \bar{\mathscr{A}}\mathscr{S}Vu_i - E^{(1)}\Psi_i(S)\rangle/g^2$$
$$- \sum_S \chi_S E^{(2)}\sum_{jR} \Psi_j(R)\langle \mathscr{R}u_j \mid \Psi_0(S)\rangle/g^2 = 0.$$

The second order energy comes from the coefficient of $\Psi_0(R)$:

$$E^{(2)} = \frac{\sum_{iS} c_{iS}^{(1)} \langle \mathscr{P} u_0 \mid \mathscr{A} \mathscr{S} V u_i - E^{(1)} \Psi_i(S) \rangle}{\sum_S \chi_S \langle \mathscr{P} u_0 \mid \Psi_0(S) \rangle}$$

$$= \frac{\sum_i \langle u_0 \mid \mathscr{A} V u_i - E^{(1)} \Psi_i \rangle \langle u_i \mid E^{(1)} \Psi_0 - \mathscr{A} V u_0 \rangle}{\sum_S \chi_S \langle u_0 \mid \mathscr{S} u_0 \rangle (\varepsilon_i - \varepsilon_0) g} \tag{25}$$

We emphasize the difference between formulas (24) and (25) and those of the Rayleigh–Schrödinger perturbation theory. Whereas the zero-order wave function is $\tilde{\Psi}_0 = \mathscr{A}\{\Psi_{A0}\Psi_{B0}\}$, the first-order energy is not $\langle \tilde{\Psi}_0 \mid V \mid \tilde{\Psi}_0 \rangle$, which may remain finite as $R \to \infty$ and thus does not resemble an interatomic potential. Equation (24) is a possible potential, being exactly equal to the difference between the expectation value of $H$ over $\tilde{\Psi}_0$ and the energies of the separated atoms; other choices are possible [3].

## 2. Neglect of Exchange

In this subsection we will be interested in the energy for very large internuclear distance. In this case a product of functions $u_i(\mathscr{P} u_j)$ vanishes if the permutation $\mathscr{P}$ mixes electrons between the atoms. Since the wave functions die off exponentially with distance, either $u_i$ or $\mathscr{P} u_j$ is exponentially small in any region of configuration space. If $\mathscr{P}$ does not mix electrons between the atoms, $\mathscr{P} u_j = u_j$. We now have

$$E^{(1)} = \langle u_0 \mid V \mid u_0 \rangle \tag{26}$$

(since the $u_i$ are normalized) and

$$E^{(2)} = \sum_j^{(j \neq 0)} \langle u_0 \mid V - E^{(1)} \mid u_j \rangle \langle u_j \mid V - E^{(1)} \mid u_0 \rangle / (\varepsilon_0 - \varepsilon_j) \tag{27}$$

[the $E^{(1)}$'s may be removed in (27) because the $\{u_i\}$ are orthogonal]. Equations (26) and (27) are just the formulas of first- and second-order perturbation theory, with unsymmetrized (product) functions. We can use, for large distances, wave functions which have not been antisymmetrized with respect to permutations of electrons between the atoms. The effect of neglected permutations corresponds to exchange, and becomes of dominant importance (valence bond theory) for smaller $R$. Returning to the indexing of individual atom states, we write Eq. (27) as

$$E^{(2)} = -\sum_{k,l}' \frac{|\langle 00 \mid V \mid kl \rangle|^2}{E_{Ak} + E_{Bl} - E_{A0} - E_{B0}} \tag{28}$$

where the prime means that the case $k = l = 0$ is not included. Here, the wave functions are products of atomic wave functions, and $\Psi_{Ak}\Psi_{Bl}$ is abbreviated by specifying the indices $k$ and $l$.

Since at large distances the charge distributions are nonoverlapping, the perturbation may be expanded in powers of $R^{-1}$, as in Hirschfelder *et al.* [5, Sect. 12.1], or Margenau and Kestner [3, Section 2.1], where different expressions are compared. We have

$$V = \sum_i^{(A)} \sum_j^{(B)} q_i q_j / r_{ij} \tag{29}$$

where $i$ sums over the electrons and nucleus of atom A and $j$ over the electrons and nucleus of atom B, $q_i$ being the charge on particle $i$. Now

$$r_{ij}^2 = |\, \mathbf{r}_{iA} - (\mathbf{R} + \mathbf{r}_{jB})\,|^2$$

where $\mathbf{r}_{iA}$ is the vector from nucleus A to particle $i$, $\mathbf{r}_{jB}$ the vector from nucleus B to particle $j$, and $\mathbf{R}$ the internuclear vector, from A to B. Since by hypothesis $\mathbf{r}_{iA}$ and $\mathbf{r}_{jB}$ are small compared to $\mathbf{R}$, $r_{ij}^{-1}$ may be expanded in a Taylor series in $r_{iA}/R$ and $r_{jB}/R$. the expansion coefficients involving angular factors. Then on grouping terms together, $V$ becomes a series in $R^{-1}$. One convenient way of writing the series is that of Hirschfelder *et al.* [5]

$$V = \sum_{l_A=0}^{\infty} \sum_{l_B=0}^{\infty} \sum_{m=-l_<}^{l_<} \frac{(-1)^{l_B+m}(l_A + l_B)!}{(l_A + |\,m\,|)!(l_B + |\,m\,|)!} \frac{(Q_{l_A}^m)^*(Q_{l_B}^m)}{R^{l_A+l_B+1}} \tag{30}$$

where $l_<$ is the smaller of $l_A$ and $l_B$. The $Q_l^m$ correspond to multipoles:

$$Q_l^m = \sum_i q_i r_i^l P_l^m(\cos\theta)e^{im\phi_i} \tag{31}$$

where the sum is over all the particles for one atom and $P_l^m$ is the associated Legendre function. The coordinates used in (31) are spherical polar coordinates centered at the nucleus of the atom, with the polar axes for both atoms parallel to the vector from A to B. The expectation value of $Q_l^m$ is a component of a permanent moment. Note that the nuclei contribute only to $Q_0^0$. Another expression for $V$ is given in Eq. (32), for the case of neutral atoms.

More generally, $r_{ij}^{-1}$ may be expanded [6] in products of spherical harmonics, one factor for the orientation of the vector from A to $i$, one factor for the orientation of the vector from B to $j$, and one factor for the orientation of the internuclear axis. Correspondingly, formulas like (30) and (31) may be derived in terms of irreducible tensors. as was done by Rose [7]. He introduced a "combined multipole moment" tensor whose components are sums of products of components of the multipole moment tensors of the two charge distributions. When the interaction energy is written in terms of this new tensor, and spherical harmonics of the angles specifying the direction of the internuclear axis, the invariance of this energy to change of coordinate system is obvious. The general formulation of the interaction in terms of irreducible tensors (spherical harmonics) may also be applied to the first-order energy, which is the expectation value of $V$ and represents the interaction between permanent moments. This has been done by Fontana [8] for the interaction between two alkali atoms, considering the uncoupled and coupled representations. The first ignores the effect of electron spin and takes the state function as a simple product; the second is diagonal with respect to spin–orbit coupling.

The expansion of $V$ in powers of $R^{-1}$ actually is valid only for a limited region of configuration space, as may be seen from its derivation: It is not correct where the charge distributions on the two centers overlap. This was emphasized by Brooks [9], who considered interacting harmonic oscillators so that the matrix elements could be explicitly evaluated. It appeared that the series in $R^{-1}$ for the second-order energy was divergent (in fact, asymptotic), which Brooks ascribed to the incorrect use of the $R^{-1}$ expansion for $V$. He proposed to correct the situation by neglecting the contribution of $V$ in the region of configuration space where the $R^{-1}$ expansion was invalid. He tested this idea for $H_2^+$ and obtained a convergent series for the energy. The energy was in better agreement with that from an accurate $H_2^+$ calculation than was the energy from the unmodified series. Corrections to the energy were then derived for $He_2$ on the basis of the $H_2^+$ results. But Roe [10] argued that the improvement for $H_2^+$ was a coincidence. Dalgarno and Lewis [11] showed that the divergence for the $H_2^+$ problem remains even if the perturbation potential $V$ is handled correctly. But since there is only one series in $R^{-1}$ for the energy, the usual series, even if not convergent, should be retained. The series is asymptotic, and may be truncated to give an error of the size of the last term used. Furthermore, the neglect of exchange is likely to be a more serious error than the divergence of the series. For further discussion of the asymptotic nature of the series, see Margenau and Kestner [3, Sect. 2.2.3(b)].

Dalgarno and Lewis [11] also emphasized the inconsistency in considering higher terms of the series in $R^{-1}$ arising in the second order of perturbation theory while neglecting higher orders of perturbation theory. Indeed, while the second order gives a series in even powers of $R^{-1}$, odd powers appear in the third-order terms.

It should be noted that, at least in simple cases, it is not necessary to resort to the $R^{-1}$ expansion of $V$. Eisenschitz and London [4] treated the H–H interaction using the full $V$ in $E^{(1)}$ and $E^{(2)}$. They assumed that the contributions of the excited states were such that the energy denominator could be taken as a constant, so that the sum was evaluated using a "mean energy denominator" [Volume I Chapter III, Section B, Eq. (160)]:

$$E^{(2)} = -\sum_{k,l}{}' \frac{|\langle 00 | V | kl \rangle|^2}{\beta} = -\beta^{-1}(\langle 00 | V^2 | 00 \rangle - \langle 00 | V | 00 \rangle^2).$$

$\langle 00 | V | 00 \rangle$ vanishes because the ground states have no permanent moments, and $\langle V^2 \rangle_0$ was evaluated in closed form. Contributions to this integral vanishing exponentially with $R$ were dropped and the exponential integral functions which arose were replaced by asymptotic series in $R^{-1}$. The leading term is $-6/\beta$, which is, of course, what we get as the leading term in the energy when the series expansion of $V$ is used (see page 21). The choice of $\beta$ is also discussed later. Fukui and Yanabe [12] performed a similar calculation for H–H and He–He, and derived an approximate formula for the difference, $\langle 00 | V^2 | 00 \rangle - \langle 00 | V | 00 \rangle^2$, for more complex spherically symmetric atoms. But this is not particularly useful unless a method for fixing $\beta$ is given.

Longuet-Higgins [13] emphasized that one could obtain a satisfactory theory of the long-range interatomic interaction, in which electron exchange between the atoms is neglected, but not overlap of the atomic charge distributions, and other workers have investigated this point further (see Kreek and Meath [14] for some bibliography). The neglect of exchange means that each electron is assigned to one or the other atom in the zero-order function, so that the perturbation $V$, Eq. (29), is unambiguously defined. Formulation of a theory when this cannot be done is difficult, as we discuss in Section B. Then $V$ may be taken as the perturbation and the second order energy may be evaluated. The expansion of $V$ in powers of $R^{-1}$ assumes in addition nonoverlapping charge distributions, i.e., that $R$ is always greater than the sum of the distance from A of an electron assigned to A and the distance from B of an electron assigned to B. It is possible to write $V$ as a two-center expansion [7, 14] without making this assumption.

Kreek and Meath [14] used $V$ in this form to calculate the second and third-order perturbation energies for the H–H$^+$ and H–H systems. They pointed out that it is not possible, in such calculations, to express the energies in terms of properties of the individual atoms. Kreek and Meath expanded their energies in powers of $R^{-1}$ to investigate the behavior of the expansion and confirmed its asymptotic character. The errors accompanying the use of truncated expansions in $R^{-1}$, and use of perturbation theory to second order only, were considered as a function of $R$ and compared with the error due to neglect of exchange. At distances of 10 $a_0$ and less, neglect of exchange gives errors of 10% or so for $H_2$, while neglect of energies higher than second order only introduces an error of 1%. Thus the important contribution neglected in Eqs. (27) and (28) is exchange, except at very large $R$. Treatments of the long-range interaction of two atoms which include exchange can be obtained from the Eisenschitz–London formalism and are discussed in Section B. They provide the basis for the valence bond theory.

### 3. *Electrostatic and Induction Forces*

While all forces in quantum mechanics may be considered in terms of classical electrostatics according to the Hellmann–Feynman theorem (Section D of Chapter III, Volume I), the term "electrostatic" as applied to interatomic forces has a more specific meaning. As we shall see later, the first-order energy (26) may be written in terms of the permanent moments of the individual atoms, corresponding to the expansion (30) of $V$ as a sum of products of operators representing atomic multipole moments. We refer to these terms as "electrostatic forces."

The leading term in $V$ [Eq. (30)] is the interaction of the total charges, $(\sum_i^{(A)} q_i)(\sum_j^{(B)} q_j)R^{-1}$. This vanishes for neutral molecules, since all known molecules in their ground states dissociate adiabatically to neutral atoms. Explicitly, the first few terms in $V$ for the interaction of neutral atoms (as written in parallel Cartesian coordinate systems centered on A for electrons $i$ and on B for electrons $j$ and with the internuclear axis as $z$-axis) are:

$$V = (e^2/R^3) \sum_i^{(A)} \sum_j^{(B)} [x_i x_j + y_i y_j - 2z_i z_j]$$

$$+ (3e^2/2R^4) \sum_i^{(A)} \sum_j^{(B)} [r_i^2 z_j - z_i r_j^2 + (2x_i x_j + 2y_i y_j - 3z_i z_j)(z_i - z_j)]$$

$$+ (3e^2/4R^5) \sum_i^{(A)} \sum_j^{(B)} [r_i^2 r_j^2 - 5r_j^2 z_i^2 - 5r_i^2 z_j^2 - 15z_i^2 z_j^2$$

$$+ 2(4z_i z_j - x_i x_j - y_i y_j)^2]$$

$$+ (e^2/2R^5) \sum_i^{(A)} \sum_j^{(B)} [5(z_i^2 + z_j^2)(3x_i x_j + 3y_i y_j - 4z_i z_j)$$

$$+ 3(r_i^2 + r_j^2)(4z_i z_j - x_i x_j - y_i y_j)] \tag{32}$$

The terms in (32) may be referred to as the dipole–dipole, dipole–quadrupole, quadrupole–quadrupole, and dipole–octupole terms. Obviously, use of an expansion like (32) in the formulas for $E^{(1)}$, $E^{(2)}$, and corresponding higher-order energies leads to a series in $R^{-1}$ for the energy.

The first-order energy, the expectation value of $V$, is easily interpreted as the electrostatic interaction between the permanent moments of the atomic systems. If either or both atoms has no permanent moments, $E^{(1)}$ vanishes. The second-order energy involves induced moments as well, and does not vanish even if neither atomic system has permanent moments. In this case, both $k$ and $l$ in (28) refer to excited state wave functions for the two atoms A and B. Inserting in $\langle 00 \mid V \mid kl \rangle$ the expansion (30) for $V$, we see that the contribution of the terms for $l_A = 0$ or $l_B = 0$ vanishes because of the orthogonality of the wave functions. Thus the leading term in $R^{-1}$ in $V$ is for $l_A = l_B = 1$, i.e., it is proportional to $R^{-3}$, and the leading term in $E^{(2)}$ goes as $R^{-6}$. If one atom has a permanent moment, $\Psi_{kl}$ may be "singly excited," i.e., either $k$ or $l$ is zero. Classically, this corresponds to an inductive force, the charge distribution of one atom being perturbed by the field set up by the permanent moment of the other atom, and the resulting induced moments interacting with the permanent moments. For example, the following terms could occur in Eq. (28):

$$-\sum_{k} \frac{\mid \langle \Psi_{B0} \mid (Q_{n_B}^m) \mid \Psi_{B0} \rangle \mid^2 \mid \langle \Psi_{A0} \mid (Q_{n_A}^m)^* \mid \Psi_{Ak} \rangle \mid^2}{E_{Ak} - E_{A0}}$$
$$= \langle Q_{n_B}^m \rangle^2 A_{n_A}^m \qquad (33)$$

Here, $\langle Q_{n_B}^m \rangle$ is a permanent moment (of order $2^{n_B}$) on atom B (expectation value of $Q_{n_B}^m$ over the ground state of atom B), while $A_{n_A}^m$ is a component of the $n_A$-pole polarizability of atom A. This polarizability determines the effect on the energy of atom A of an electric field of $n_A$-pole symmetry: $n_A = 1$ means a dipole (uniform field), $n_A = 2$ means a quadrupole (field gradient), etc.

The distinction between the "singly excited" and the "doubly excited" cases is important, since the latter (dispersion force) has no simple classical analog. Induction forces can be handled within the SCF–MO framework, simply by putting the proper additional term into the potential for the atom which is to be polarized. This cannot be done for the dispersion force. So far, no calculation of this kind has given an attractive interaction between closed-shell atoms. A general proof of this statement is not available, although Margenau and Kestner have showed [3, Appendix 8] that, when the interaction of two one-electron atoms with like spins is treated by the molecular orbital theory, the energy (for $R$ large enough to neglect overlap) is always greater than the sum of the unperturbed atomic energies. They stated that the proof can be extended to the He–He case. Here, we would argue that, since the source of the dispersion or van der Waals attraction is in correlated dipoles (see Volume I, Chapter II), a one-electron model such as the MO cannot give the van der Waals attraction. Configuration interaction wavefunctions, which include such correlations by way of simultaneous excitation on the two atoms (interatomic correlation) can give the dispersion as well as the induction energies [15, 16].

Induction forces are important for molecule-ions which dissociate to a neutral atom and an ion. Let us assume for simplicity that the neutral atom has no permanent moments. Let $Q_0{}^0 = Ze$ for system B and consider the terms for $l_B = 0$ in (30).

$$\bar{V} = \sum_{l_A=1}^{\infty} (Q_{l_A}^0)^* Ze/R^{n_A+1} \tag{34}$$

Because system A has no permanent moments, $E^{(1)}$ vanishes. For $E^{(2)}$, Eq. (28) becomes

$$-\sum_{k>0} \frac{|\langle 00 | \bar{V} | k0 \rangle|^2}{E_{Ak} - E_{A0}} = -Z^2 e^2 \sum_{k>0} \frac{|\langle 0 | (Q_{l_A}^0)^* / R^{n_A+1} | k \rangle|^2}{E_{Ak} - E_{A0}}$$

We will give the leading terms in powers of $R^{-1}$. Since $Q_1{}^0$ and $Q_2{}^0$ have different parities, $\langle 0 | Q_1{}^0 | k \rangle$ and $\langle 0 | Q_2{}^0 | k \rangle$ are never simultaneously nonzero [17]. There are thus no cross-terms between them in Eq. (28), and the first two terms in the sum are

$$-Z^2 \left[ R^{-4} \sum_{k>0} \frac{|\langle 0 | \mu | k \rangle|^2}{E_{Ak} - E_{A0}} + R^{-6} \frac{|\langle 0 | q | k \rangle|^2}{E_{Ak} - E_{A0}} \right]$$

$$= \frac{-Z^2 e^2}{2} \left[ \frac{\alpha}{R^4} + \frac{\gamma}{R^6} \right], \tag{35}$$

where $\mu = \sum_i q_i z_i$ is the dipole moment along the $z$ (internuclear) axis and $q = \sum_i q_i (3z_i^2 - r_i^2)/2$ is the quadrupole moment operator. In the second member of (35) we have introduced the dipole polarizability $\alpha$ [cf. Eq. (50)] and the quadrupole polarizability $\gamma$ of atom A. The dipole polarizability can only be approximately calculated. Calculations of static polarizabilities involve the second-order energy of perturbation theory, when the perturbation is the interaction of the dipole of the system with an electric field, or the quadrupole with an electric field gradient. The techniques of variation–perturbation theory may be used to compute polarizabilities [18]. Their calculation is discussed in the books of Davies [19], and Hirschfelder *et al.* [5], and in Subsection 7. The van der Waals energy must be added in to (35) if the term in $\gamma$ is used.

Dalgarno and Kingston [20] have considered several cases. For some, the van der Waals term is relatively unimportant (as H–He$^+$); for others (e.g., H–H$^-$), the van der Waals term is more important than the quadrupole induction term. Tani and Inokuti [21] have discussed the bound vibrational states of a diatomic system interacting with a potential consisting of a hard core repulsion and an attraction going as $R^{-4}$. This model appeared to be a good one for the Li$^+$–He system, but was not expected to be as useful for interacting partners having open-shell structures. The one-electron molecule–ions have been discussed [22] to elucidate the origin of a maximum in $U(R)$ for molecule–ions.

We conclude this section with discussion of calculation of forces due to permanent moments on each atom (electrostatic forces). Since the electric dipole operator is of odd parity, the lowest permanent moment of a neutral system is the quadrupole, and the leading electrostatic interac'ion is the quadrupole–quadrupole term, which is proportional to $R^{-5}$.

The calculation of this interaction was discussed by Knipp [23]. He found it convenient to rewrite the quadrupole–quadrupole term in $V$ in terms of the five-dimensional irreducible representation of the three-dimensional rotation group. Knipp defined the quantities:

$$f^{(2)} = \tfrac{1}{2}\sqrt{\tfrac{3}{2}}\,(x + iy)^2 \tag{36a}$$

$$f^{(1)} = -\sqrt{\tfrac{3}{2}}\,(x + iy)z \tag{36b}$$

$$f^{(0)} = \tfrac{1}{2}(3z^2 - r^2) \tag{36c}$$

$$f^{(-1)} = \sqrt{\tfrac{3}{2}}\,(x - iy)z \tag{36d}$$

$$f^{(-2)} = \tfrac{1}{2}\sqrt{\tfrac{3}{2}}\,(x - iy)^2 \tag{36e}$$

so that the term in $V$ becomes

$$W = (e^2/R^5) \sum_i^{(A)} \sum_j^{(B)} [f_i^{(2)}f_j^{(-2)} + 4f_i^{(1)}f_j^{(-1)} + 6f_i^{(0)}f_j^{(0)} + 4f_i^{(-1)}f_j^{(1)} + f_i^{(-2)}f_j^{(2)}] \tag{37}$$

For the quadrupole–quadrupole interaction, we are dealing necessarily with degenerate states. This would lead to difficulties in Eqs. (20) *et seq.* If exchange is neglected at the outset, the zero-order wave function is written: $\Sigma c_{ij}^{(0)}\Psi_{ij}$, where $\Psi_{ij} = \Psi_{Ai}\Psi_{Bj}$, and $\Psi_{Ai}$ is one of the degenerate wave functions for the lowest energy term of atom A, $\Psi_{Bj}$ one of the degenerate wave functions for the lowest term of atom B. The properties of angular momenta may be used to form linear combinations of $\Psi_{ij}$ which transform according to representations of the atomic and molecular rotation groups, thus simplifying the secular equation and reducing the number of integrals one must calculate. Knipp [23] worked out a number of important cases.

We will consider only one simple example, that of Russell–Saunders coupling with negligible spin–orbit interaction for both atoms. We label the spin and orbital angular momenta for the lowest energy terms as $S_A$ and $L_A$ for atom A, and $S_B$ and $L_B$ for atom B. The atomic wave functions for atom A may be labeled by $M_{SA}(-S_A \leq M_{SA} \leq S_A)$ and $M_{LA}(-L_A \leq M_{LA} \leq L_A)$ and similarly for atom B. We write them as:

$$\Psi_A(M_{SA}, M_{LA}); \qquad \Psi_B(M_{SB}, M_{LB}).$$

The states of the diatomic system formed from A and B are characterized by the total spin $S$, its component along the internuclear axis $\Sigma$, and the component of orbital angular momentum along this axis $\Lambda$. For the case of like atoms, there is also the gerade–ungerade symmetry to consider. Knipp first formed normalized wavefunctions of correct spin symmetry for the diatomic system (see Section B,1) by standard vector coupling methods:

$$\Phi(S\Sigma M_{LA}M_{LB}) = \sum_{M_S} \Psi_A(M_S, M_{LA})\Psi_B(\Sigma - M_S, M_{LB})$$
$$\times (S_A S_B M_S, \Sigma - M_S \mid S_A S_B S\Sigma) \qquad (38)$$

The coefficients on the right are known. For unlike atoms and $\Lambda \neq 0$, $\Phi(S\Sigma M_{LA}, \Lambda - M_{LA})$ and $k\Phi(S\Sigma, -M_{LA}, -\Lambda + M_{LA})$ transform according to the two rows of the $\Lambda$th representation of the two-dimensional rotation–reflection group, with $k$ equal to $+1$ or $-1$, depending on parity and orbital angular momentum [23]. For $\Lambda = 0$, linear combinations of these two wave functions must be taken to yield wave functions of $\Sigma^+$ and $\Sigma^-$ symmetry. Now we have to consider only matrix elements of $W$ [Eq. (37)] between wave functions with the same values of $\Lambda$, $\Sigma$, and $S$. Each of the two rows of the $\Lambda$-representation may be considered separately, as may the $\Sigma^+$ and $\Sigma^-$ states, because $W$ is totally symmetric under the symmetry operators. Thus one requires only matrix elements

$$\langle \Phi(S\Sigma M_{LA}, \Lambda - M_{LA}) \mid W \mid \Phi(S\Sigma M'_{LA}, \Lambda - M'_{LA}) \rangle, \qquad (39)$$

Because $W$ is spin-independent, substitution of Eq. (38) into Eq. (39) gives a sum of terms like

$$G = \langle \Psi_A(M_S, M_{LA}) \Psi_B(\Sigma - M_S, M_{LB}) \mid W \mid \Psi_A(M_S, M'_{LA}) \Psi_B(\Sigma - M_S, M'_{LB}) \rangle$$

where $M_{LA} + M_{LB} = M'_{LA} + M'_{LB} = \Lambda$. The value of such a term is independent of $M_S$. Since the transformation represented by (38) is unitary, the coefficient of $G$ [a sum of squares of the coefficients in (38)] is unity.

We now write $W$ in the form of Eq. (37), so that the matrix element is a sum of products of terms from atom A and atom B [23]. In computing

$$\left\langle \Psi_A(M_S, M_{LA}) \left| \sum_i f_i^{(k)} \right| \Psi_A(M_S, M'_{LA}) \right\rangle$$

one can take advantage of the fact that the operator and the two wave functions transform according to representations of the three-dimensional rotation group. All such matrix elements are related by vector coupling coefficients. Hence all the matrix elements, Eq. (39), have as a common factor the product of two atomic quantities. Therefore the secular equation corresponding to the representation specified by $S$, $\Sigma$, $\Lambda$, and the row (for $\Lambda = 0$, one must specify the $\Sigma^+$ or $\Sigma^-$ term) can be solved without evaluating any integrals. The eigenvalues of the secular equation are proportional to this atomic factor. The atomic quantity, if $\Psi_A$ is constructed from atomic orbitals, is a sum of one-electron contributions, all of which are $\langle r^2 \rangle$ times an angular function which can be evaluated exactly. Closed shells do not contribute. Then there is only one one-electron integral for each open shell.

For the case in which spin–orbit coupling is large, the atomic wave functions are characterized by $S$, $L$, $J$, and $M$ ($J$ is the total angular momentum and $M$ is its component along the figure axis) instead of by $S$, $L$, $M_S$, and $M_L$. Only the $J$ value corresponding to the lowest energy state need be considered. This situation, as well as the one in which spin–orbit coupling is large for one atom and not the other, is discussed in Knipp's article [23]. The elements in the secular equation may always be reduced to terms proportional to a product of atomic contributions. The atomic contributions may be calculated, for a wave function made up of atomic orbitals, from the radial parts of their wave functions.

Recently, Burke *et al.* [24] extended Knipp's calculations. They considered in particular two atoms having one open shell each. Chang [25] gave an extensive discussion of the long-range interaction for atoms with nonvanishing angular momenta. He considered cases where the atomic spin–orbit interaction was small, large, and comparable to the interaction energy. For the last case, almost-degenerate perturbation theory (see Volume I, Chapter III, Section B.1) was used, considering a zero-order Hamiltonian which included spin-orbit coupling as well as intramolecular Coulombic interactions. Russell–Saunders coupling was assumed for the atoms in all cases. Chang showed that, even for atoms in degenerate states, the first-order energy is an infinite series of odd powers of $R^{-1}$, the correct zero-order functions are infinite series of even powers of $R^{-1}$, and the second-order energy is an infinite series of even powers of $R^{-1}$. Relativistic corrections were also calculated as power series in $R^{-1}$. Explicit formulas and tables (extensions of those given by Knipp [23]) were given for the computation of the interaction energy of two atoms with one incomplete shell each. Dispersion energies were also computed approximately by means of London-type formulas [Eq. (49)], in order to compare the size of the first-order (electrostatic) and second-order (dispersion) contributions to the interaction energy. The dispersive energies were relatively unimportant for the range of $R$ considered, as were magnetic interactions.

### 4. Semiempirical Evaluation of van der Waals Constants

Now we consider second order, dispersion, or van der Waals forces. The term, "dispersion" is used because of the expression for $E^{(2)}$ [e.g., Eq. (44)] in terms of the oscillator strengths, which also enter the theory of dispersion. The expansion of $V$ in powers of $R^{-1}$ is almost always used, since evaluation of integrals involving $V$ itself is too difficult for any but the simplest systems. Furthermore, in the range of $R$ of interest (large enough so interatomic overlap can be neglected) a few terms in the energy series should suffice. Thus, most workers concern themselves with the evaluation of the coefficients in the expansion

$$E^{(2)} = \sum_i c_i R^{-i}$$

We have already noted that there is an inconsistency in taking more than a few terms of the $E^{(2)}$ series and neglecting $E^{(3)}$.

The approaches discussed in this section are concerned basically with manipulation of the formal sum-over-states expressions of perturbation theory to permit their estimation in terms of other experimentally obtainable quantities. Thus they must be classed as semiempirical. Nonempirical methods are discussed later.

For atoms in spherically symmetric ground states, the symmetry of the excited state in $\langle \Psi_0 \mid P_l^m \mid \Psi_{\mathrm{ex}} \rangle$ depends on $l$ and $m$, so that a given $\Psi_{\mathrm{ex}}$

is mixed in by only one $P_l^m$. Then only even powers of $R^{-1}$ appear in $E^{(2)}$. For neutral atoms, the leading term in $E^{(2)}$ is that in $R^{-6}$, sometimes referred to as *the* van der Waals term, and most work has dealt with evaluation of its coefficient. However, as emphasized by Fontana [26], even a small contribution of terms after the $R^{-6}$ term is extremely important for the interpretation of experimental data. When the interaction is approximated as $U = -CR^{-n}$, $n$ will be chosen as slightly greater than 6, since it includes part of the effect of higher terms for the range of $R$ interrogated by the measurement. Fontana showed that a change of 1% in $n$ can lead to a doubling of the calculated $C$ corresponding to it.

Margenau [27] gave one of the earliest and simplest treatments for the first three terms in the case of H–H, and later generalized the approximate formula to other systems. An average energy denominator [Volume I, Chapter III, Eq. (160)] was used to carry out the sums over excited states. Denoting it by $\beta$,

$$E^{(2)} = -(\beta R^6)^{-1}[\tfrac{2}{3}\langle r^2\rangle^2 + (2/R^2)\langle r^4\rangle\langle r^2\rangle + (14/5R^4)\langle r^4\rangle^2] \qquad (40)$$

where $\langle r^2\rangle$ and $\langle r^4\rangle$ refer to the individual atoms. An average over angles for each atom has been performed. For the hydrogen atom, $\langle r^2\rangle = 3a_0^2$ and $\langle r^4\rangle = 45a_0^4/2$. It must be noted that there is some inconsistency in including the quadrupole–quadrupole term in $V$ and not the dipole–octupole term, as both go as $R^{-10}$. The leading term from $E^{(3)}$ goes [3, Sect. 2.2.3(b)] as $R^{-11}$.

Now the problem is the estimation of the average energy or the choice of $\beta$. Naturally, an accurate determination of the correct average energy is equivalent to calculating $E^{(2)}$ exactly. Physically, we can argue for the validity of the use of a mean energy denominator if the states that give the bulk of the contribution to $E^{(2)}$ lie in a small range of energy [28], but, in general, the average energy depends on the perturbation. It is, however, possible to get an upper limit on the coefficient, since the average energy cannot be lower than the lowest excitation energy. For one H atom, this is $\tfrac{3}{8}$, making $\beta = \tfrac{3}{4}$. Eisenschitz and London [4] argued that the maximum value of $\beta$ was 1 because states with energies above the ionization potential (continuum states) could not make much of a contribution. It turned out later that the average energy exceeds the ionization potential for other atoms. For the H–H case, Margenau [27] took the average excitation energy for each atom as about the ionization potential, $\tfrac{1}{2}$. This means $\beta = 1$, and the leading term for the dispersion interaction between H atoms becomes $-6R^{-6}$, which is not far from the correct $-6.5R^{-6}$. Using this value of $\beta$

and the expectation values cited above, one has in atomic units,

$$E^{(2)} = -(6/R^6)(1 + 22.5/R^2 + \cdots)\tag{41}$$

It was argued [27, 29] that since $V$ is one-electron, the relevant excitations for other atoms are not too different from those for H. Then the quantity $-\beta$ should be replaced by $2E_0$, with $E_0$ the ground state energy. For $\langle r^2 \rangle$ one must put $N\langle r^2 \rangle$, with $N$ the number of electrons and $\langle r^2 \rangle$ referring to one electron. Similarly, $\langle r^4 \rangle$ becomes $N\langle r^4 \rangle$. Then for He–He Margenau [27] found $E^{(2)} = 1.62R^{-6}(1 + 7.9R^{-2} + \cdots)$. Another way of generalizing the H–H results to other atoms is to interpret their spectra in terms of a hydrogenic model. The electronic states involved for each atom are hydrogen atom functions with some effective charge or screening constant $\bar{Z}$. Then the computation of the interaction energy of atoms A and B may be carried out like that of two H atoms, using the mean energy denominator to give a series like Eq. (41):

$$E^{(2)} = -\frac{2}{\bar{Z}_A{}^2 + \bar{Z}_B{}^2}\left[\frac{6}{R^6\bar{Z}_A{}^2\bar{Z}_B{}^2} + \frac{135}{2R^8}\left(\frac{1}{\bar{Z}_A{}^2\bar{Z}_B{}^4} + \frac{1}{\bar{Z}_A{}^4\bar{Z}_B{}^2}\right)\right.$$
$$\left. + \frac{315}{2R^{10}}\left(\frac{9}{\bar{Z}_A{}^4\bar{Z}_B{}^4} + \frac{8}{\bar{Z}_A{}^2\bar{Z}_B{}^6} + \frac{8}{\bar{Z}_A{}^6\bar{Z}_B{}^2}\right)\right] + \cdots$$

For details, see Margenau and Kestner [3, Sect. 2.2.3(b)].

It was recognized early that the leading term of $E^{(2)}$ was closely related to the strengths of atomic electronic transitions and to atomic polarizabilities. We may write

$$E^{(2)} = -R^{-6}\sum_{k>0}\sum_{l>0}$$
$$\times \frac{\left|\begin{array}{c}\langle A0 \mid \mu_x \mid Ak\rangle\langle B0 \mid \mu_x \mid Bl\rangle + \langle A0 \mid \mu_y \mid Ak\rangle\langle B0 \mid \mu_y \mid Bl\rangle \\ -2\langle B0 \mid \mu_z \mid Ak\rangle\langle B0 \mid \mu_z \mid Bl\rangle\end{array}\right|^2}{E_{Ak} + E_{Bl} - E_{A0} - E_{B0}}\tag{42a}$$

Here we use $\mid Ak\rangle$ to mean the $k$th eigenstate of A, with eigenvalue $E_{Ak}$, and similarly for the states of B. The dipole moment operator in the $x$ direction is

$$\mu_x = \sum_i ex_i$$

the sum going over all the electrons in an atom. The polarizability can also be written as a sum of transition dipoles $\langle A0 \mid \mu_x \mid Ak\rangle$ divided by excitation energies. The sums over excited states include a sum over magnetic quantum numbers, i.e., over degenerate states. Later we will average (42a) over the magnetic quantum numbers of the ground state. The result is equivalent to an average over spatial directions, and causes the cross terms in the numerator of (42a) to vanish, while making the terms for $x$, $y$, and $z$ equal. Now, if we interpret the sums over $k$ and $l$ to mean sums over quantum numbers other than magnetic quantum numbers, we may write

$$E^{(2)} = -6R^{-6} \sum_{k>0} \sum_{l>0} \sum_{m_k} \sum_{m_l} \frac{|\langle A0 \mid \mu_x \mid Ak \rangle|^2 \, |\langle B0 \mid \mu_x \mid Bl \rangle|^2}{E_{Ak} + E_{Bl} - E_{A0} - E_{B0}} \qquad (42b)$$

In the following discussion, we will put

$$E^{(2)} = -C_{AB} R^{-6}$$

for the leading term in the dispersive interaction of atoms A and B. The explicit expression for $C_{AB}$ is given by (42b) wherein we now introduce the spherically averaged oscillator strengths for atomic transitions. Let $m_0$ refer to the magnetic quantum number of the initial state and $m_k$ the magnetic quantum number of the final state involved in a transition. We sum over degenerate final states and average over degenerate initial states. Let $L$ be the angular momentum of the initial state, so the degeneracy is $(2L + 1)$. Then the average oscillator strength for the transition in the presence of $x$-polarized radiation is

$$f_{0k}^{(x)} = (2L + 1)^{-1} \sum_{m_0} \sum_{m_k} \tfrac{2}{3} |\langle 0m_0 \mid \mu_x \mid km_k \rangle|^2 \, (E_k - E_0) \qquad (43)$$

for an atom and the effect of the sums is to make $f_{0k}^{(x)} = f_{0k}^{(y)} = f_{0k}^{(z)} = \tfrac{1}{3} f_{0k}$. Introducing the $f_{0k}$ into (42), we have [29]

$$C_{AB} = \frac{3}{2} \sum_{k>0} \sum_{l>0} \frac{f_{0k}^A f_{0l}^B}{(E_{Ak} + E_{Bl} - E_{A0} - E_{B0})(E_{Ak} - E_{A0})(E_{Bl} - E_{B0})}. \qquad (44)$$

Equation (44) may be formally written as an integral,

$$C_{AB} = \frac{3}{2} \int_{\omega_{A0}}^{\infty} d\omega_A \int_{\omega_{B0}}^{\infty} d\omega_B \, \frac{f^{(A)}(\omega_A) f^{(B)}(\omega_B)}{(\omega_A + \omega_B)\omega_A \omega_B}. \qquad (45)$$

Here, $\omega_{A0}$ and $\omega_{B0}$ are the lowest transition frequencies from the ground state for A and B, while $f^{(A)}(\omega)$ is the oscillator strength distribution function per unit frequency range for A, and $f^{(B)}(\omega)$ is the corresponding quantity for atom B. The contribution of a discrete state to such a distribution function is a $\delta$-function. Much of our discussion later in this section and in subsequent sections relates to attempts to derive $C_{AB}$ from what knowledge is available of the oscillator strength distribution functions of the atoms A and B. The calculation and measurement of these functions, particularly as they relate to atomic spectra, was the subject of an extensive recent review by Fano and Cooper [30].

The expression for the polarizability of an atom also involves the transition dipoles. It may likewise be written in terms of the oscillator strengths:

$$\alpha(\omega) = \sum_{k>0} \frac{f_{0k}}{(E_0 - E_k)^2 - \omega^2} \tag{46}$$

$\alpha(\omega)$ is the polarizability for light of frequency $\omega$. This expression is for unpolarized light, and includes an average over the magnetic quantum number (for $L \neq 0$) of the ground state. Expression (46) with $\omega = 0$ (static polarizability) bears much relation to (44), but the exact relation between the polarizability and van der Waals constants was not elucidated until relatively recently (Subsection 5).

If it turns out that $E_{Ak} - E_{A0}$ is large compared to $E_{Bl} - E_{B0}$ for the states making an important contribution to the sum in (44), we have

$$C = \frac{3}{2} \sum_{k>0} \frac{f_{0k}^{(A)}}{(E_{Ak} - E_{A0})^2} \sum_{l>0} \frac{f_{0k}^{(B)}}{(E_{Bl} - E_{B0})}$$

$$= \frac{3}{2} \alpha_A(0) \sum_{l>0} \frac{f_{0l}^{(B)}}{E_{Bl} - E_{B0}}$$

The sum may be evaluated in closed form by using the definition of the oscillator strengths and closure. Then

$$C = e^2 \alpha_A(0) \left\langle 0 \left| \sum_i (\mathbf{r}_i \cdot \mathbf{r}_i) \right| 0 \right\rangle_B$$

This works quite well for helium–alkali-metal interactions and fairly well for other noble-gas–alkali cases [31].

The dimensionless oscillator strengths give the intensities of the lines in the atomic spectrum. Thus knowledge of the intensities and frequencies of all the transitions, or at least of all the important ones, allows computation of $C_{AB}$. The continuous absorption must also be included; in this case the sums in (44) are integrals over energy. Thus Eisenschitz and London [4] were able to evaluate (44) from experimentally measured intensities to obtain

$$C_{HH} = 6.47 e^2 a_0^5$$

This corresponds to an average energy denominator $\beta$ of 0.925. Almost half the contribution came from transitions to the continuum. A similar calculation for the He–He case was given by Margenau [32, 33].

Ignorance of the oscillator strengths can be alleviated somewhat because one has the Thomas–Reiche–Kuhn sum rule

$$\sum_k f_{0k} = \text{number of electrons.} \tag{47}$$

The polarizability [Eq. (46)] is another sum rule. If it is known, or can be calculated, its value gives additional information about the oscillator strength distribution function. We discuss this use of the sum rules later.

The early workers, however, used knowledge of the static polarizabilities primarily to aid in estimation of the mean energy denominator. If $\Delta_\alpha$ is the mean excitation energy for the polarizability, we have from (46)

$$\alpha(0) = N/\Delta_\alpha{}^2. \tag{48a}$$

Here $N$ is the number of electrons and the sum rule (47) has been used. The corresponding expression for the van der Waals coefficient for a homo-nuclear pair, with $\Delta_c$ equal to the mean excitation energy per atom, is

$$C = \tfrac{3}{2}N^2/(2\Delta_c{}^3) \tag{48b}$$

from (44). Actually, $N$ might be interpreted as the number of valence electrons, assuming only their transitions contribute (see page 27), or just considered as a parameter. If $\Delta_\alpha$ and $\Delta_c$ are assumed equal, $C = \tfrac{3}{4}\alpha^2\Delta$. The reasoning was generalized by London [33] to the heteronuclear case A–B, in which case

$$C_{AB} = \tfrac{3}{2}[\Delta_A\Delta_B/(\Delta_A + \Delta_B)]\alpha_A\alpha_B \tag{49}$$

where $\Delta_A$ and $\Delta_B$ are the mean excitation energies for atoms A and B. (Similar expressions can be derived for higher van der Waals coefficients [27].) Equations of the form of (49) arise in a variety of contexts.

Somewhat different results are obtained when $C$ and $\alpha$ are written in terms of the dipole moment matrix elements instead of the oscillator strengths. This means Eq. (42) for $C$ and

$$\alpha(0) = 2 \sum_{k>0} \left[ \left| \left\langle 0 \left| \sum_i x_i \right| k \right\rangle \right|^2 \middle/ (E_k - E_0) \right]. \tag{50}$$

London [4, 33] derived bounds on $C_{AA}$ in terms of the static polarizability for atom A, using these formulas, and assuming that one could obtain bounds on the mean energy denominators. Now $\Sigma \mid \langle 0 \mid \Sigma x_i \mid k \rangle \mid^2$ is equal to, by closure, the expectation value of $x^2$ over the ground state, or (since we average over directions) to $\frac{1}{3}\langle \mathbf{r} \cdot \mathbf{r} \rangle$, where $\mathbf{r} = \sum_i \mathbf{r}_i$ with the sum running over all the electrons in the atom. With a mean excitation energy $\Delta'$

$$\alpha_A(0) = \frac{2}{3}(\langle \mathbf{r} \cdot \mathbf{r} \rangle_A / \Delta'_{A\alpha}). \tag{51a}$$

Similarly, $C_{AB}$ becomes

$$C_{AB} = \frac{2}{3}[\langle \mathbf{r} \cdot \mathbf{r} \rangle_A \langle \mathbf{r} \cdot \mathbf{r} \rangle_B / (\Delta'_{AC} + \Delta'_{BC})]. \tag{51b}$$

If (51a) is used to evaluate $\langle \mathbf{r} \cdot \mathbf{r} \rangle_A$ and $\langle \mathbf{r} \cdot \mathbf{r} \rangle_B$ for insertion in (51b), and if the average energies are assumed the same $(\Delta'_{A\alpha} = \Delta'_{AC} = \Delta_A')$,

$$C_{AB} = \frac{3}{2}[\Delta_A' \Delta_B' / (\Delta_A' + \Delta_B')]\alpha_A \alpha_B. \tag{51c}$$

Although this is of the form of Eq. (49), the average energies here are not identical to $\Delta_A$ and $\Delta_B$, because they are defined in terms of different sum rules: Compare Eqs. (48a) and (51a). The same objection could be made to equating $\Delta'_{A\alpha}$ and $\Delta'_{AC}$. Kramer [34] has systematically considered average energies from a wide variety of sum rules, and derived bounds on the van der Waals coefficients in terms of other properties. His work will be discussed in Section 6.

Neglecting "correlation" terms $(i \neq j)$,

$$\left\langle \sum_i \mathbf{r}_i \cdot \sum_j \mathbf{r}_j \right\rangle \cong \left\langle \sum_i \mathbf{r}_i \cdot \mathbf{r}_i \right\rangle = N\langle r^2 \rangle.$$

The diamagnetic susceptibility is:

$$\chi = -N_0 e^2 N \langle r^2 \rangle / 6mc^2$$

($N_0$ = Avogadro's number). Vinti [35] and Kirkwood [36] explored the connection between diamagnetic and dielectric susceptibilities (polarizabilities). If $\langle r^2 \rangle$ is obtained from $\chi$, knowledge of $\alpha(0)$ gives the mean excitation energy for use in (51). Assuming the same mean excitation energy may be used in $C$, Kirkwood [36] derived

$$\begin{aligned}
\frac{3}{2}C_{AB} &= \frac{N_A N_B \langle r^2 \rangle_A \langle r^2 \rangle_B}{N_A \langle r^2 \rangle_A / \alpha_A(0) + N_B \langle r^2 \rangle_B / \alpha_B(0)} \\
&= -\frac{6mc^2}{N_0} \frac{\chi_A \chi_B}{\chi_A / \alpha_A(0) + \chi_B / \alpha_B(0)} \\
&= -\frac{6mc^2}{N_0} \frac{\chi_A \chi_B}{\alpha_A(0)/\chi_A + \alpha_B(0)/\chi_B}. 
\end{aligned} \tag{52}$$

The energies obtained from this formula are generally too large in magnitude, partly from neglect of the correlation term [37] and partly because it overestimates the contribution of the inner electrons [2]. With respect to the first problem, we have

$$\left\langle\left(\sum_i \mathbf{r}_i\right)^2\right\rangle = -\frac{6mc^2}{N_0 e^2}\,\chi + \sum_{i \neq j}\langle\mathbf{r}_i \cdot \mathbf{r}_j\rangle$$

Salem [37] considered the calculation of the correlation term from an approximate wave function. It vanishes for a product of spherically symmetric one-electron orbitals, and its value is quite sensitive to the wave function used. Neglecting the correlation term gives an upper limit to the magnitude of the dispersion energy.

Use of a mean energy denominator is justified when the important transitions are associated with about the same energy of excitation. Then the oscillator strength distribution is dominated by a single term, with an effective oscillator strength $f$. Since the important transitions are due to the valence electrons, it is conceivable that their energies could all be about the same. The static polarizability would be just $f/\bar{\omega}^2$, where $\bar{\omega}$ is the effective or average transition frequency. If the same oscillator strength distribution is used in calculation of $C_{AB}$, we obtain

$$C_{AB} = \tfrac{3}{2}[\bar{\omega}_A\bar{\omega}_B/(\bar{\omega}_A + \bar{\omega}_B)]\alpha_A\alpha_B$$

which is just Eq. (49). To the extent that the oscillator strength distribution is representable by one term, there is no problem in substituting one average energy for another: They are all identical.

Slater and Kirkwood [38], in the course of variational calculations for the polarizability and van der Waals constants (see Subsection 7), noted that the dominant contributions arose from the outer electrons. They used this to derive another approximate relation between the polarizability and van der Waals constants, which was extended by Margenau [29] (who gave a simpler derivation) to the heteronuclear case:

$$C_{AB} = \frac{3}{2}\,\frac{\alpha_A\alpha_B}{(\alpha_A/\bar{n}_A)^{1/2} + (\alpha_B/\bar{n}_B)^{1/2}} \tag{53}$$

Here, $\bar{n}_A$ is the number of electrons in the outer shell of A and similarly for $\bar{n}_B$. Formula (53) generally gives values of $C_{AB}$ which are too small [37], but correct to within a factor of 2 or so [39]. By interpreting $\bar{n}_A$ and $\bar{n}_B$ as effective numbers of electrons, one can improve things somewhat [39]. Note that (53), like (52), is similar in form to (49). In addition, Pitzer [38a] derived an expression like (49) from variational calculations of $\alpha$ and $C_{AB}$. In this case, $\Delta$ involved matrix elements of $r_{12}^{-1}$ between atomic orbitals. It was suggested that $\Delta$ be more like twice the ionization potential than the ionization potential itself, which agreed with experience in using the London formula. Formulas of the form (49) or (53) have been discussed in general by Wilson and Crowell [39].

An approach which may be considered a generalization of the use of mean energies is to write $\alpha(\omega)$ in some approximate way, such as a small number of terms of the proper form:

$$\alpha(\omega) = \sum_{i=1}^{m} f_i/(\omega_i^2 - \omega^2) \tag{54}$$

Here, $f_i$ and $\omega_i$ $(i = 1 \cdots m)$ are parameters chosen to give agreement with experimental frequency-dependent polarizabilities, which are obtained from refractive-index measurements. This is known as a Sellmeier-type dispersion formula. Similarly, one may write $C$ in terms of the $f_i$ and $\omega_i$ for the two atoms and use the values obtained from the polarizabilities. The simplest case, $m = 1$ and $\alpha_A = f^A \omega_A^{-2}$, is the average energy approxima-tion, Eq. (53). Mavroyannis and Stephen [40] used $\alpha(\omega) = a/(b^2 - \omega^2)$, with $a$ and $b$ parameters, in conjunction with the correct exact relationship between $\alpha$ and $C$ (see Subsection 5).

Heller [41] considered the terms in $R^{-8}$, $R^{-10}$, and $R^{-12}$ as well as that in $R^{-6}$, using a one-term Sellmeier formula. Margenau [29] previously and Hornig and Hirschfelder [42] subsequently developed similar expressions. Heller gave formulas, analogous to Eq. (49), expressing the coefficients of $R^{-8}$, etc. in terms of polarizabilities and transition frequencies for the atoms. He computed the terms for the diatomic systems Hg–He, Hg–Ne, Hg–Ar, Hg–Kr, Hg–Xe, and Hg–Hg, using the one term formula for $c_2$, $c_3$, and $c_4$, and various Sellmeier formulas for $c_1$, in

$$U(R) = -c_1 R^{-6} - c_2 R^{-8} - c_3 R^{-10} - c_4 R^{-12} + Ae^{-R/\varrho}.$$

The last term is to represent the effect of exchange (overlap) repulsion. Heller [41] considered the energy of the lowest vibrational state and the equilibrium internuclear distance of the molecules formed for comparison with experiment. He noted that, whereas the effect of the $R^{-12}$ term is only a few percent, the $R^{-8}$ and $R^{-10}$ terms were appreciable near the minimum. But it is not clear that the expansion is valid there.

Fontana [43] used the expansion of $V$ in spherical harmonics, together with a harmonic oscillator model, to evaluate the first six terms in the interaction between two noble gases in terms of oscillator strengths. This corresponds to $R^{-6}$, $R^{-8}$, and $R^{-10}$ terms. The harmonic oscillator model was used in connection with this problem in London's earliest work (see Volume I, Chapter II, Section A). The atomic wave functions for the electrons are represented by harmonic oscillators, all of the same frequency. The selection rules mean there is only a single nonvanishing oscillator strength, so this corresponds to a one-term dispersion formula. Use of the properties of the harmonic oscillator wave functions allows explicit evaluation of the sums over states.

Dalgarno and co-workers [20, 31, 44, 45] have considered the possibility of augmenting the two sum rules (46) and (47) to improve further the semi-empirical calculation of $C$ [Eq. (44)]. We have already mentioned that

$$\sum_{k>0} f_{0k}/(E_k - E_0) = \tfrac{2}{3} \sum_{k>0} \left| \left\langle 0 \left| \sum_i \mathbf{r}_i \right| k \right\rangle \right|^2$$

$$= \tfrac{2}{3}\left[ \left\langle 0 \left| \left(\sum_i \mathbf{r}_i\right)^2 \right| 0 \right\rangle + \left\langle 0 \left| \sum_{i \neq j} \mathbf{r}_i \cdot \mathbf{r}_j \right| 0 \right\rangle \right] \quad (55)$$

since $\langle 0 | \sum_i \mathbf{r}_i | 0 \rangle = 0$. The first term in the square bracket is given by the diamagnetic susceptibility, but the other is not experimentally accessible. Vinti [46] showed that for an atom

$$\sum_{k>0} f_{0k}(E_k - E_0) = \tfrac{1}{3}\left\{ \left\langle 0 \left| \sum_{i \neq j} \mathbf{p}_i \cdot \mathbf{p}_j \right| 0 \right\rangle - 2E_0 \right\} \quad (56)$$

and

$$\sum_{k>0} f_{0k}(E_k - E_0)^2 = (4\pi Z/3)\left| \left\langle 0 \left| \sum_i \delta^{(3)}(\mathbf{r}_i) \right| 0 \right\rangle \right|^2. \quad (57)$$

For derivations of these sum rules, see Bethe [17, p. 147]. The expectation values must be calculated from an approximate atomic wave function. In addition, other such summations are experimentally accessible [44]. The index of refraction gives $\alpha(\omega)$, Eq. (46). If $\omega$ is less than the lowest transition frequency, we may expand

$$\alpha(\omega) = \sum_{k>0} f_{0k}(a_{k0}^{-1} + \omega^2 a_{k0}^{-2} + \omega^4 a_{k0}^{-3} + \cdots)$$

with $a_{k0} = (E_k - E_0)^2$. The first term gives just $\alpha(0)$, and successive terms give

$$S(j) = \sum_{k>0} f_{0k}(E_k - E_0)^j \quad (58)$$

for $j = -4, -6$, etc. The values of $S(j)$ for $j = 2, 1, 0, -1, -2$ are obtainable, respectively, from Eqs. (57), (56), (47), (55), and, with $\omega = 0$, (46). It can be shown that $S(j)$ is infinite for $j \geq 2.5$, for an $S$-state [45]. The Verdet constant [19, p. 223] depends on $d\alpha/d\omega$, so also can give $S(-4)$, $S(-6)$, etc. It was suggested [44] that $S(j)$ can be written as a function of $j$ for purposes of extrapolation to values of $j$ for which $S(j)$ is not known, using the values available. Thus one has a great deal of information on atomic oscillator strengths which can be used in the evaluation of the $R^{-6}$ dispersion interaction. Since considerably less reliable information about quadrupole and higher oscillator strengths is available, these procedures cannot now be extended to other terms in $E^{(2)}$.

Dalgarno and Kingston [31] started with the available experimental and theoretical values of oscillator strengths for the alkali metals and inert gases and made modifications where necessary to satisfy the sum rules. They then calculated the coefficients $C_{AB}$ from (44). Other calculations of this kind were done by Dalgarno and Lynn [45], for the helium atom, and by Dalgarno and Kingston [20], for interactions between H, He, Ne, and Ar. Later, Kingston [47] considered the rare gases, obtaining oscillator strengths from spectral and photoabsorption measurements, and adjusting them to fit $S(k)$ for $k = 0, -2, -4$, and $-6$, to improve and extend the computations of Dalgarno and Kingston. Barker and Leonard [48] performed similar calculations at the same time for the homonuclear rare-gas-atom pairs. They estimated their values for $C_{AB}$ to be good to better than 3%. Bell and Kingston [49] treated the He–He interaction after careful study of the oscillator strength distribution of He and found $C_{AB} = 1.4633$.

Bell [50] pointed out that a certain amount of intuition enters into the construction of the oscillator strength spectrum needed for these calculations. He proposed a more systematic procedure. Equation (44) was rewritten:

$$C_{AB} = \tfrac{3}{2}(e^4\hbar^4/m^2E^3) \sum_{k>0} \sum_{l>0} f_{0k}^A f_{0l}^B \phi(x_k, y_l) \tag{59}$$

where $E$ is the smallest energy difference for either atom,

$$x_k = E/(E_{Ak} - E_{A0}), \qquad y_l = E/(E_{Bl} - E_{B0})$$

and

$$\phi(x, y) = x^2y^2/(x + y).$$

Suppose $\phi(x, y)$ can be approximated, for $0 \leq x \leq 1$ and $0 \leq y \leq 1$, by a finite sum

$$\phi(x, y) \cong \sum_i \sum_j a_{ij}x^iy^j$$

Then $C_{AB}$ of (59) becomes

$$C_{AB} = \tfrac{3}{2}(e^4\hbar^4/m^2E^3)\sum_i\sum_j a_{ij}E^{i+j}S^A(-i)S^B(-j) \qquad (60)$$

The choices of $i$, $j$, and $a_{ij}$ were done by standard polynomial-fitting methods. The values $S(-3)$ and $S(-5)$ were obtained by interpolation of measured $S(-k)$ to obtain $S(k)$ for $0 = k = -6$. The results checked those of Barker and Leonard [48], whose values for $S(k)$ were used, to a few percent. The method was subsequently used to check variation-perturbation calculations (see Subsection 7) for the $R^{-6}$, $R^{-8}$, and $R^{-10}$ terms in the H–H long-range interaction. For the atoms H, He, Ne, Ar, Kr, and Xe, the set of $S(k)$ derived by Kingston [47] was used [51] with various choices of the $a_{ij}$ [Eq. (60)] to derive $C_{AB}$ for all pairs of these atoms.

The tables that follow give values of $C_{AB}$ derived by these methods for various pairs of atoms and ions in their ground states. The data in Table I are from Dalgarno and Kingston [31], those of Table II from Bell and Kingston [51].

Other semiempirical procedures, related to these we have discussed, will be discussed in later sections, in connection with the Casimir–Polder relation. The calculation of error bounds is of particular interest.

### Table I

*van der Waals Constants, Derived Semiempirically*

(a) *Interactions involving* H *or* He

|     | H   | H⁻  | He  | He⁺  | Li⁺  | Li  | Na  | K   | Rb  | Cs  |
|-----|-----|-----|-----|------|------|-----|-----|-----|-----|-----|
| H   | 6.5 | 89  | 2.9 | 0.66 | 0.52 | 67  | 74  | 99  | 100 | 120 |
| He  | 2.9 | 30  | 1.5 | 0.38 | 0.32 | 22  | 25  | 33  | 34  | 38  |

(b) *Rare gas and Alkali atoms*

|     | Li            | Na            | K             | Rb            | Cs            | He  | Ne  | Ar  | Kr  | Xe  |
|-----|---------------|---------------|---------------|---------------|---------------|-----|-----|-----|-----|-----|
| Li  | $1.4\times10^3$ |               |               |               |               | 22  | 48  | 200 | 300 | 480 |
| Na  |               | $1.6\times10^3$ |               |               |               | 25  | 52  | 210 | 330 | 530 |
| K   |               |               | $3.5\times10^3$ |               |               | 33  | 68  | 280 | 420 | 680 |
| Rb  |               |               |               | $3.8\times10^3$ |               | 34  | 69  | 280 | 430 | 700 |
| Cs  |               |               |               |               | $5.2\times10^3$ | 38  | 77  | 320 | 480 | 780 |

**Table II**

*van der Waals Constants, Derived Semiempirically*

|      | H     | He    | Ne    | Ar    | Kr    | Xe    |
|------|-------|-------|-------|-------|-------|-------|
| H    | 6.500 | 2.815 | 5.62  | 19.93 | 28.54 | 41.61 |
| He   |       | 1.456 | 3.012 | 9.62  | 13.44 | 18.67 |
| Ne   |       |       | 6.30  | 19.60 | 27.26 | 37.49 |
| Ar   |       |       |       | 65.3  | 92.1  | 130.3 |
| Kr   |       |       |       |       | 130.3 | 185.7 |
| Xe   |       |       |       |       |       | 218.2 |

## 5. The Casimir–Polder Relation

It turns out that there is an *exact* relation between the van der Waals constant $C$ and the polarizability, if one extends the definition of the frequency-dependent polarizability [Eq. (46)] to include imaginary frequencies. This relation arose in the work of Casimir and Polder [52] on the effect of retardation, due to the finite velocity of light, on the interaction. Discussions of this and related work, leading to the formula (62), are given by Margenau and Kestner [2, Chapter 6] and by Power [53]. Feinberg and Sucher [54] gave a more general treatment of the van der Waals interaction between two neutral spinless systems, using techniques developed for elementary particle scattering problems.

Casimir and Polder first considered the interaction with the zero-point radiation field of an atom enclosed in a large cubical box with perfectly conducting walls. They used quantum electrodynamics and calculated terms in the energy proportional to $e^2$. The atom was supposed always to be far from all the walls but one; its distance $R$ from that wall could vary. Let $\lambda_k$ represent a transition wavelength of the atom; $\lambda_k = \hbar c/(E_k - E_0)$. For $R$ large compared to all $\lambda_k$, the interaction energy was found to go as $R^{-4}$, and be proportional to the polarizability of the atom. Classically, one may say that the oscillating dipole of the atom induces a dipole in the wall, which interacts with the atomic dipole. This should lead to an energy decreasing as $R^{-3}$. The difference is due to the finite velocity of light. If the time it takes the field of the dipole to propagate to the wall and back is important, the oscillating dipole can get out of phase and the interaction energy will be decreased.

Clearly similar considerations obtain for the interaction between two atoms at a distance $R$. Casimir and Polder [52] went on to treat by quantum electrodynamics two atoms interacting with the zero-point radiation field. The perturbation included, in addition to the electrostatic interactions between the constituents of the two atoms, the interaction of both with the radiation field. Divergences which arose were removed by computing the difference in the energies for finite and infinite $R$. It was necessary to go to fourth-order perturbation theory, for $C$ involves the fourth power of the electronic charge and the interaction of a charged particle with the radiation field only the first power. The results of the lengthy calculation were that:

(1) When $R$ is small compared to all $\lambda_k$ for transitions on either atom, the interaction energy goes as $R^{-6}$; it is just the well-known London expression we have been discussing.

(2) For $R$ large compared to all $\lambda_k$, the energy varies as $R^{-7}$ due to retardation, and is proportional to the product of the static atomic polarizabilities:

$$E^{(2)}(R \to \infty) = (-23c/4\pi R^7)\alpha_A \alpha_B \tag{61}$$

after averaging over magnetic quantum numbers.

Both the London expression and (61) arose as limiting forms of the general formula

$$E^{(2)} = -\frac{c}{\pi} \int_0^\infty du\, u^4 e^{-2uR} \left[ \frac{1}{R^2} + \frac{2}{uR^3} + \frac{5}{u^2 R^4} + \frac{6}{u^3 R^5} + \frac{3}{u^4 R^6} \right]$$
$$\times \sum_{k>0} \frac{f_{0k}^A}{(E_{Ak} - E_{A0})^2 + u^2 c^2} \sum_{l>0} \frac{f_{0l}^B}{(E_{Bl} - E_{B0})^2 + u^2 c^2} \tag{62}$$

The integration over $u$ comes from the sum over the wave vectors for the waves of the radiation field. While the integral (62) can be evaluated [6] in terms of known functions, two limiting cases are of interest to us. For very large $R$, the effect of the exponential is to limit the range of $u$ to very small values. If $R \gg \lambda_k$, $u \ll \lambda_k^{-1}$, and the terms $c^2 u^2$, small compared to the energy difference, may be dropped. Then one can carry out the integration over $u$ to give $23/4R^7$. The sums over $k$ and $l$ give the polarizabilities [Eq. (46)]. For $R$ small compared to all $\lambda_k$, we can keep just the last term in the square bracket and approximate the exponential by unity. Then we have

$$E^{(2)} = -\frac{3}{\pi R^6} \sum_{k>0} f_{0k}^A \sum_{l>0} f_{0l}^B \int_0^\infty \frac{d\omega}{[(\omega_{0k}^A)^2 + \omega^2][(\omega_{0l}^B)^2 + \omega^2]} \tag{63}$$

where we have introduced the transition frequencies defined by

$$\omega_{0k}^A = E_k{}^A - E_0{}^A.$$

The value of the definite integral is $\frac{1}{2}\pi[\omega_{0k}^A\omega_{0l}^B(\omega_{0k}^A + \omega_{0l}^B)]^{-1}$, so that we recover Eq. (44).

To show that

$$\int_0^\infty \frac{d\omega}{(a^2 + \omega^2)(b^2 + \omega^2)} = (\pi/2)[ab(a + b)]^{-1} \tag{64}$$

we may extend the range of integration to $-\infty \le u \le \infty$ and close the contour by a semicircle at infinity in either the upper or lower half plane. The contribution of the semicircle to the integral vanishes, so we have only the residues of the poles at $u = ia, ib$ or $u = -ia, -ib$, depending on how the semicircle is drawn. Thus the integral is

$$\frac{1}{2}(2\pi i)\left(\frac{1}{(a^2 - b^2)(2ib)} + \frac{1}{(b^2 - a^2)(2ia)}\right) = \frac{\pi}{2}\frac{a - b}{ab(a^2 - b^2)}$$

as in (64).

Now the relation (64) could have been used to write $C_{AB}$, as in (44), in the form of Eq. (63), without the derivation of (63) by way of (62) and quantum electrodynamics. But this step was not taken until after Casimir and Polder's work, which is a pity, since Eq. (63) has produced a tremendous amount of fruitful work on the calculation of van der Waals constants. (The rate of production of such work, unfortunately for writers of monographs, shows no sign of abating!)

The integrand in Eq. (63) is the product of two terms which look very much like those entering the polarizability. Indeed, we can introduce the polarizability for complex frequency as a generalization of (46).

$$\alpha(i\omega) = \sum_{k>0} f_{0k}/[(\omega_{0k})^2 + \omega^2] \tag{65}$$

Then

$$C_{AB} = \frac{3}{\pi}\int_0^\infty d\omega\alpha_A(i\omega)\alpha_B(i\omega). \tag{66}$$

This is the exact relation between the van der Waals interaction energy and the polarizabilities, but the polarizabilities required are those for imaginary frequencies; i.e., they relate to an exponentially decaying electric field. Relations similar to Eqs. (65) and (66) may also be obtained [1, 2] for the coefficients of higher powers of $R^{-1}$ in $E^{(2)}$. The coefficients are expressed as integrals over frequency of the product of multipole polarizabilities for imaginary frequencies.

Tang [55] has pointed out that formula (66) allows one immediately to obtain an upper bound for $C_{AB}$ in terms of $C_{AA}$ and $C_{BB}$. Simply invoking the Schwarz inequality,

$$\left[\int_0^\infty d\omega\, \alpha_A(i\omega)\alpha_B(i\omega)\right]^2 \le \left[\int_0^\infty d\omega\, \alpha_A(i\omega)\alpha_A(i\omega)\right]$$

$$\times \left[\int_0^\infty d\omega\, \alpha_B(i\omega)\alpha_B(i\omega)\right]$$

and thus

$$C_{AB} \le (C_{AA}C_{BB})^{1/2}.$$

The approximation $C_{AB} \cong (C_{AA}C_{BB})^{1/2}$ has been commonly used,* suggested by the form of Eq. (45). If an average energy $\Delta_A$ can be used for the transitions of atom A in $C_{AA}$ and $C_{AB}$, and similarly 'or $\Delta_B$, we find from (45) that

$$(C_{AA}C_{BB})^{1/2}/C_{AB} = \tfrac{1}{2}(\Delta_A + \Delta_B)/(\Delta_A\Delta_B)^{1/2}$$

and the right-hand side is always greater than unity for positive $\Delta_A$ and $\Delta_B$. But $\Delta_A$ need not be the same in $C_{AA}$ and $C_{AB}$. Weinhold [56] generalized Tang's reasoning [55] as follows: Since

$$\sum_A \sum_B \chi_A\chi_B C_{AB} = (3/\pi)\int_0^\infty d\omega\left[\sum_A \chi_A\alpha_A(i\omega)\right]^2 \ge 0$$

the matrix $\mathbf{C}$, whose elements are $C_{AB}$, is positive semidefinite. Then the determinant of such a matrix is nonnegative. The $2\times2$ determinant is $C_{AA}C_{BB} - C_{AB}^2$, so that $C_{AB} \le (C_{AA}C_{BB})^{1/2}$, as stated earlier. Formulas for the $3\times3$ case were given by Weinhold, and numerical examples of the use of the $2\times2$ result were given. The matrices formed from higher van der Waals coefficients (dipole–quadrupole, quadrupole–quadrupole, etc.) were also shown to be positive semidefinite, by expressing the coefficients in terms of higher polarizabilities, as were the matrix whose $(i, j)$ element is $E_{i,j}^{(2)}$ (total second-order energy between atoms $i$ and $j$) and the matrix whose $(i, j)$ element is $C_{AA}^{i,j}$ ($2^i$-pole–$2^j$-pole dispersion interaction between atoms $A$).

---

* "Combination rules," which give formulas for one van der Waals constant in terms of others, will be discussed in Chapter III, Section B.

Mavroyannis and Stephen [40] indicated several methods for using the relations (65) and (66) to calculate van der Waals interaction energies. First, one can calculate the frequency-dependent polarizability directly. Consider the Hamiltonian

$$H = H_0 - \mathbf{\mu} \cdot \mathbf{E}[e^{i\omega t} + e^{-i\omega t}]$$

where $H_0$ is the unperturbed atomic Hamiltonian, $\mathbf{\mu}$ the dipole moment operator, and $\mathbf{E}$ the strength of the electric field. The time-dependent Schrödinger equation is

$$H\Psi(\mathbf{r}, t) = i\hbar \, \partial\Psi(\mathbf{r}, t)/\partial t.$$

A solution to first-order in the perturbation may be written:

$$\Psi = \exp(-iE_0 t/\hbar)[\Psi_0 + \mathbf{F}^+ \cdot \mathbf{E} \exp(i\omega t) + \mathbf{F}^- \cdot \mathbf{E} \exp(-i\omega t)] \quad (67)$$

where $H_0\Psi_0 = E_0\Psi_0$. Substitution into the time-dependent Schrödinger equation yields these first-order terms:

$$H_0(\mathbf{F}^+ \cdot \mathbf{E}e^{i\omega t} + \mathbf{F}^- \cdot \mathbf{E}e^{-i\omega t}) - \mathbf{\mu} \cdot \mathbf{E}(e^{i\omega t} + e^{-i\omega t})\Psi_0$$
$$= (E_0 - \hbar\omega)\mathbf{F}^+ \cdot \mathbf{E}e^{i\omega t} + (E_0 + \hbar\omega)\mathbf{F}^- \cdot \mathbf{E}e^{-i\omega t} \quad (68)$$

This leads to equations for $\mathbf{F}^+$ and $\mathbf{F}^-$, which may be solved by way of an equivalent variational principle [40]. From $\mathbf{F}^+$ and $\mathbf{F}^-$, the dipole moment induced by the oscillating field and hence the polarizability for frequency $\omega$ can be calculated. Then one puts $i\omega$ for $\omega$. We discuss such calculations in Subsection 6.

Dalgarno [1, Pt. II; 57] has formulated the computation somewhat differently, in terms of a function which is a combination of $\mathbf{F}^+$ and $\mathbf{F}^-$. He showed that if $\Psi$ satisfies

$$[(H_0 - E_0)^2 + \omega^2]\Psi + \mu_z\Psi_0 = 0 \quad (69a)$$

where the field is in the $z$-direction, then

$$\alpha(i\omega) = (8\pi/3)\Big\langle \Phi_0 \Big| \sum_i (\partial/\partial z_i) \Big| \Psi \Big\rangle. \quad (69b)$$

The sum over $i$ is over electrons. The validity of (69b) is demonstrated if $\Psi$ is expanded in the excited eigenfunctions $\Psi_k$ of $H_0$ with eigenvalues $E_k$. Alternatively, if $\Phi$ satisfies

$$[(H_0 - E_0)^2 + \omega^2]\Phi + \sum_i (\partial/\partial z_i)\Psi_0 = 0$$

one may express the polarizability as

$$\alpha(i\omega) = (8\pi/3)\langle \Psi_0 \mid \mu_z \mid \Phi \rangle.$$

These results were generalized [1, Pt. II] to higher-order polarizabilities and thus higher van der Waals constants. Approximate solutions to these equations are obtainable from a variational principle.

It is also possible to compute $\alpha(i\omega)$ directly from (65). Of course, the exact function for the atom is available only for the H atom. Suppose we start with a zero-order function such as the Hartree–Fock which is a determinant of one-electron spin-orbitals. The nature of the matrix element in $f_{0k}$ means that the excited states over which we sum are determinants differing from the zero-order function in a single spin-orbital. We now must sum over all occupied spin orbitals and over all virtual spin orbitals. The latter sum becomes an integral over continuum functions in the Hartree–Fock case, where there are no discrete virtual orbitals. The contribution to $\alpha(i\omega)$ of a continuum of states is more correctly written [58, 59]

$$\int_0^\infty d\varepsilon \, \frac{df/d\varepsilon}{[\omega(\varepsilon)]^2 + \omega^2}.$$

The corrections to the Hartree–Fock function due to correlation must be considered simultaneously. Such a calculation was carried through by Kelly [60] using diagrammatic many-body perturbation theory, for the coefficient of $R^{-6}$ for two oxygen atoms. Naturally, the appropriate modification was made for the fact that the ground state of O is $^3P$, so that there are several values for $C_{OO}$, depending on the values of $M_L$ for the two atoms. A review of the formalism, with references, was given [61] in Volume IIIS of the *International Journal of Quantum Chemistry*, in which results of several calculations of this kind are published. Among these is that of Dutta *et al.* [62] who obtained $\alpha(i\omega)$ for H and He and computed $C_{HH}$ $= 6.49903$, $C_{HHe} = 2.820$, $C_{He-He} = 1.500$, all in good agreement with the results from other methods. Extension of these calculations to heavier atoms seemed practical.

Mavroyannis and Stephen [40] also proposed the use of Sellmeier-type dispersion formulas [Eq. (54)] to approximate $\alpha(i\omega)$, and derived several approximate formulas. In the one-term formula,

$$\alpha(i\omega) \approx (e^2/m)[f/(\bar{\omega}^2 + \omega^2)] \tag{70}$$

$f$ and $\bar{\omega}$ may be determined so that the expression is correct in the limits

$\omega \to 0$ and $\omega \to \infty$. In the former case, $\alpha(i\omega)$ becomes the static polarizability, so

$$f/\bar{\omega}^2 = \alpha.$$

At very large frequency, $\bar{\omega}$ may be neglected in (70) and all $\omega_{0k}$ in (65). For the expressions to be equal, one requires

$$f = \sum_{k>0} f_{0k} = N$$

the number of electrons, invoking the Thomas–Reiche–Kuhn sum rule. Thus the formula (70) becomes

$$\alpha(i\omega) \approx N[\omega^2 + (N/\alpha)]^{-1}. \tag{71}$$

Using (71) for atoms A and B, we can perform the integral in (66):

$$E^{(2)} = \frac{-3N_A N_B}{\pi R^6} \int_0^\infty \frac{d\omega}{(\omega^2 + N_A/\alpha_A)(\omega^2 + N_B/\alpha_B)}$$

or, using Eq. (64),

$$C_{AB} = \frac{3}{2} \frac{\alpha_A \alpha_B}{(\alpha_A/N_A)^{1/2} + (\alpha_B/N_B)^{1/2}}. \tag{72}$$

This may be compared to the London formula (49), the Kirkwood formula (52), and the Slater–Kirkwood formula, Eq. (53). Calculated $C_{AB}$ were given by Mavroyannis and Stephen [40] for the homopolar rare gas cases.

The results were improved when they went to a two-term Sellmeier formula, in which the contribution of one excited state was put in explicitly:

$$\alpha(i\omega) = \frac{e^2}{m} \frac{f_{01}}{(\omega_{01})^2 + \omega^2} + \frac{e^2}{m} \frac{f'}{\omega'^2 + \omega^2} \tag{73}$$

After inserting experimental or theoretical values for $f_{01}$ and $\omega_{01}$, they used the sum rules to evaluate $f'$ and $\omega'$. The value $C_{AB}$ was calculated, using (73) in (66), for He–He and for H–H. If the information about $n$ excited states is available, one can go to a $(n+1)$-term formula.

One may also attempt to construct $\alpha(i\omega)$ from experimental data. This is closely related to work discussed earlier on construction of the distribution of oscillator strengths for direct calculation of $C_{AB}$. Dalgarno et al. [63] successfully treated the rare gas van der Waals interactions by fitting refractive index data to a two-term Sellmeier form, putting $\omega \to i\omega$, and using the Casimir–Polder formula. Dalgarno and Davidson [64] considered interactions between pairs (as well as triplets) of alkali and inert gas atoms. From the known oscillator strengths and transition frequencies for the valence electrons of an alkali atom, their contribution to $\alpha(i\omega)$ was constructed. It was found that it could be well represented by a two-term Sellmeier form. To this were added inner-shell contributions, estimated from van der Waals coefficients for rare gas atoms. These were scaled according to the ratios of static polarizabilities of the rare gas atoms and the alkali cores. It was noted that one may have to worry about valence-type forces (formation of singlet and triplet molecules) in applying these results to experiment.

A closely related approach was also suggested by Mavroyannis and Stephen [40]. Analogously to (66), they showed that the van der Waals energy could be written:

$$E^{(2)} = (-3/2R^6) \int_0^\infty du G_A(u)G_B(u) \tag{74}$$

where for each atom

$$G(u) = (\hbar e^2/m) \sum_{k>0} f_{0k}\omega_{k0}^{-1} \exp(-u\hbar\omega_{k0}). \tag{75}$$

The identity (74) follows immediately on substitution and integration, which yields (44). The functions $G_A(u)$ and $G_B(u)$ may be treated like $\alpha_A(i\omega)$ and $\alpha_B(i\omega)$. A Sellmeier one-term formula for $G(u)$ would be

$$G(u) \approx Be^{-bu}. \tag{76}$$

Here, $B$ and $b$ were determined by comparing (76) and (75) for $u \to 0$ (both go to zero for $u \to \infty$). The equality of (76) and (75) at $u = 0$ requires

$$B = \tfrac{2}{3} \sum_{k>0} |\langle 0 | \mu_x | k \rangle|^2 = \tfrac{2}{3}\langle 0 | \mu_x^2 | 0 \rangle \tag{77}$$

since $\langle 0 | \mu_x | 0 \rangle$ vanishes for an $S$ ground state. Equating the expressions for $dG/du$ at $u = 0$ gives

$$bB = \sum_{k>0} f_{0k} = N. \tag{78}$$

Using (77) and (78) in (76) and inserting the expressions for $G_A$ and $G_B$ in (74) yields:

$$C_{AB} = \frac{4}{9} \frac{\langle x^2 \rangle_A \langle x^2 \rangle_B}{N_A/\langle x^2 \rangle_A + N_B/\langle x^2 \rangle_B} \tag{79}$$

where $x = \sum_i x_i$. Equation (79) may be compared to (49) and to Kirk-wood's formula [first formula of Eq. (52)]. Mavroyannis and Stephen applied this to homonuclear rare gas pairs. Again here, one can [40] go to a two-term formula for $G$, in which knowledge of one excited state is used explicitly.

### 6. *Bounds on van der Waals Constants*

Recently, many authors have given theoretical developments leading to upper and lower bounds for the van der Waals coefficients, mostly based on the Casimir–Polder relation, equations (65)–(66). That of Tang [55], $C_{AB} \leq (C_{AA}C_{BB})^{1/2}$, has already been mentioned.

Langhoff and Karplus [58, 59] write the polarizability for imaginary frequencies as a series in $\omega$:

$$\alpha(i\omega) = \sum_{j=0}^{\infty} (-\omega^2)^j \alpha_j. \tag{80}$$

This has been used previously for $\alpha(\omega)$ [see the discussion preceding Eq. (58)]. Formally, each energy denominator is expanded:

$$(\omega_{0k}^2 + \omega^2)^{-1} = (\omega_{0k})^{-2} \sum_{j=0}^{\infty} (-1)^j (\omega/\omega_{0k})^{2j}$$

We note that $\alpha_j$ is just $S(-2j-2)$, Eq. (58). Thus the first few $\alpha_j$ are known from experiment. Padé approximants are used to give upper and lower bounds for (80), given the $\alpha_j$ values as constraints. Direct calculation is possible for the $\alpha_j$ in the case of $H_2$: the bounds obtained on $C_{HH}$ are quite good. For various pairs of noble gases, the upper and lower bounds obtained were compared [58, 59] with the experimental values and with the estimates of Bell [50] and Bell and Kingston [49, 51] which also use the $S(k)$ from experiment. In some cases, experimental $C_{AB}$ fell outside the bounds (which themselves are subject to errors in measurement, since empirical values are used). Subsequently, Tang [65] obtained upper and lower bounds on $\alpha(i\omega)$ and hence on van der Waals coefficients for H and rare gas atoms by using Padé approximants. The results are identical to what is obtained from Radau quadrature (see page 41) but the method has certain advantages, including the possibility of obtaining improved results by using additional experimental data on excitation energies. The bounds are comparable to those cited in Table III.

**Table III**

*van der Waals Constants for Rare Gases*

|    | He | Ne | Ar | Kr | Xe |
|----|----|----|----|----|----|
| He | $1.4618 \pm 0.0004$ | | | | |
| Ne | $3.059 \pm 0.088$ | $6.55 \pm 0.87$ | | | |
| Ar | $9.86 \pm 0.16$ | $20.47 \pm 0.78$ | $67.7 \pm 2.0$ | | |
| Kr | $13.65 \pm 0.06$ | $27.3 \pm 2.0$ | $94.6 \pm 1.7$ | $133 \pm 1$ | |
| Xe | $20.31 \pm 0.58$ | $41.3 \pm 4.3$ | $142 \pm 5$ | $199 \pm 5$ | $301 \pm 14$ |

Gordon [66] also considered the determination of bounds of $C_{AB}$, by way of bounds on the polarizabilities of the atoms A and B. The polarizability for frequency $i\omega$ was written as an integral [see Eq. (45)]:

$$\alpha(i\omega) = \int_{\omega_1}^{\infty} \frac{f(E)\,dE}{E^2 + \omega^2} \qquad (81)$$

Here, $f(E)$ is the oscillator strength per unit energy; $\omega_1$ is the lowest transition frequency, supposed to be known: $\omega_1 = E_1 - E_0$. A change of variable in the integral to $u = E^{-2}$ makes it

$$\alpha(i\omega) = \int_{0}^{u_1} u^2 (1 + \omega^2 u)^{-1} F(u)\,du \qquad (82)$$

where $u_1 = (\omega_1)^{-2}$ and $F(u) = \frac{1}{2} u^{-5/2} f(u^{-1/2})$. The sums $S(j)$ are moments of the distribution $F(u)$. Put

$$\mu_k = \int_{0}^{u_1} u^k F(u)\,du.$$

Then we have $\mu_0 = S(2)$, $\mu_1 = N$, $\mu_2 = \alpha(0)$, and $\mu_3$, $\mu_4$, etc. are derivatives of the frequency-dependent polarizability and accessible from refractive index and Verdet constant measurements. Gordon investigated the possible range of $\alpha(i\omega)$, given $u_1$ and $\mu_k$ ($k = 0 \cdots 4$). If the integral (82) is approximated by the Radau quadrature method [65] with one integration point fixed at $u_1$ it may be shown that

$$\alpha(i\omega) \geq W_1 g(u_1) + W_2 g(u_2) + W_3 g(u_3)$$

where $g(u) = u^2/(1 + \omega^2 u)$, and the weights $W_1$, $W_2$, and $W_3$ and points

$u_2$ and $u_3$ are chosen, according to the Radau formulas, in terms of the five $\mu_k$. Furthermore, if the Radau quadrature method is used with one integration point fixed at zero, it may be shown that an upper bound is obtained. It is interesting to note that the upper and lower bounds can be written as Sellmeier-type formulas for $\alpha(i\omega)$:

$$\sum_{i=1}^{3} W_i \frac{u_i^2}{1 + \omega^2 u_i} = \sum_{i=1}^{3} \frac{f_i}{\omega_i^2 + \omega^2}$$

with $f_i = W_i u_i$ and $\omega_i = (u_i)^{-1/2}$.

Upper and lower bounds to the polarizabilities at imaginary frequencies were computed for the rare gases as well as for II. The values for $S(j)$ derived from experiment by Barker and Leonard [48] were used. The upper and lower bounds for $C_{AB}$ generally differed by a few percent, and agreed well with experiment. McQuarrie *et al.* [67] used this method to calculate upper and lower bounds for $C_8$, the coefficient of $R^{-8}$ in the dispersion energy for H–H, He–H, and He–He interactions. The coefficient was written as an integral like (66) of the product of dipole and quadrupole polarizabilities. For He–He, they found $13.8 < C_8 < 14.2$.

Subsequently, Starkschall and Gordon [18] produced bounds (most of them tighter than Tang's [65]) on the polarizabilities for imaginary frequency and hence on the van der Waals constants for rare gas interactions, using moment theory and linear programming (see their paper for details). They considered the integral (81) assuming knowledge of the frequency of the lowest electronic transition, $\omega_1$, and also the integrals

$$\int_{\omega_1}^{\infty} d\bar{\omega} f(\bar{\omega}) = N, \qquad \int_{\omega_1}^{\infty} d\bar{\omega} f(\bar{\omega})\bar{\omega}^2 = (4\pi N/3)\left|\left\langle \sum_i \delta(\mathbf{r}_i) \right\rangle\right|^2,$$

and

$$\alpha(\omega) = \int_{\omega_1}^{\infty} d\bar{\omega} f(\bar{\omega})(\bar{\omega}^2 - \omega^2)^{-1}.$$

Linear programming gives bounds on $\alpha(i\omega)$. For He, additional information from theoretical calculations was available. Some of Starkschall and Gordon's results [18] for $C_{AB}$ are listed in Table III. Comparisons with other experimental and theoretical results were given. Not all the values for $C_{AB}$ given here are consistent with those of Tang [65].

Goscinski [68, 69] using operator inequalities, also derived upper and lower bounds to $\alpha(i\omega)$ in terms of the $S(j)$. He actually obtained bounds on the second-order energies for a variety of perturbation theory problems, but we shall not give any of the general treatment here, and limit ourselves to sketching the derivation for $\alpha(i\omega)$. Nor will we list additional references for this formalism, but refer the reader instead to Goscinski's article [68, 69]. One can formally rewrite $\alpha(i\omega)$ as an expectation value over the ground state [cf. Eqs. (43), (65)]:

$$\begin{aligned} \alpha(i\omega) &= 2\langle \psi_0 \mid \mu P(H_0 - E_0)[(H_0 - E_0)^2 + \omega^2]^{-1} P\mu \mid \psi_0 \rangle \\ &= 2\langle \psi_0 \mid \mu P\Omega P\mu \mid \psi_0 \rangle \end{aligned} \tag{83}$$

where $P$ is the projection operator*

$$P = \sum_{k>0} \mid \psi_k \rangle\langle \psi_k \mid$$

with the $\psi_k$ the eigenfunctions of the atomic Hamiltonian, and $\mu$ is $\mu_x$, $\mu_y$, or $\mu_z$. Since the Hermitian operator

$$\Omega(\omega) = (H_0 - E_0)[(H_0 - E_0)^2 + \omega^2]^{-1}$$

has no negative eigenvalues, one can define its inverse square root $\Omega^{-1/2}$ analogously to the way in which the inverse square root of the overlap matrix was constructed in Volume I, Chapter III, Section A, Eq. (44). From an arbitrary set of functions $\{\chi_i\}$, $i = 1 \cdots n$, spanning a linear subspace orthogonal to $\psi_0$, Goscinski formed [68, 69] the projection operator

$$Q = \sum_{k,l} \mid \Omega^{-1/2}\chi_k \rangle S_{kl}^{-1} \langle \Omega^{-1/2}\chi_l \mid. \tag{84}$$

Here, $S_{kl}^{-1}$ is the $(k, l)$ element of the inverse of the overlap matrix of the functions $\Omega^{-1/2}\chi_k$:

$$S_{mn} = \langle \Omega^{-1/2}\chi_m \mid \Omega^{-1/2}\chi_n \rangle = \langle \chi_m \mid \Omega^{-1} \mid \chi_n \rangle$$

Like all projection operators, $P$ and $Q$ are bounded by zero and unity. Thus

$$0 \leq \Omega^{1/2}Q\Omega^{1/2} \leq \Omega^{1/2}\Omega^{1/2} = \Omega$$

so that, taking the expectation value of $\Omega^{1/2}Q\Omega^{1/2}$ over an arbitrary function $g$,

$$0 \leq \sum_{k,l} \langle g \mid \chi_k \rangle S_{kl}^{-1} \langle \chi_l \mid g \rangle \leq \langle g \mid \Omega \mid g \rangle. \tag{85}$$

---

* Here, the operator $\langle \psi \mid$ means multiplication by the function $\psi^*$ and integration over coordinates. In $\mid \phi \rangle\langle \psi \mid$, the result of operation with $\langle \psi \mid$ is to be multiplied by the function $\phi$.

If we choose g as $P\mu \mid \psi_0\rangle$, we see that (85) reads, with the definition (83),

$$\alpha(i\omega) \geq 2 \sum_{k,l} \langle \psi_0 \mid \mu \mid \chi_k \rangle S_{kl}^{-1} \langle \chi_l \mid \mu \mid \psi_0 \rangle \qquad (86)$$

assuming that

$$P\chi_k = \chi_k \qquad (87)$$

i.e., that all $\chi_k$ are orthogonal to $\psi_0$. The matrix elements $S_{mn}$ may be written as

$$S_{mn} = \langle \chi_m \mid (H_0 - E_0) + \omega^2 (H_0 - E_0)^{-1} \mid \chi_n \rangle \qquad (88)$$

and are not infinite for a nondegenerate ground state because of (87).

A convenient special choice for the functions $\chi_k$ was [68, 69]

$$\chi_i^{(p)} = (H_0 - E_0)^{i-p} P\mu \mid \psi_0 \rangle, \qquad i = 1, 2, \ldots, n \qquad (89)$$

where $p$ is an integer. Note that these functions are automatically orthogonal to $\psi_0$. With this choice,

$$S_{kl} = \langle \psi_0 \mid \mu P[(H_0 - E_0)^{k+l-2p+1} + \omega^2 (H_0 - E_0)^{k+l-2p-1}] P\mu \mid \psi_0 \rangle$$
$$= \tfrac{1}{2} S(k+l-2p) + \tfrac{1}{2} \omega^2 S(k+l-2p-2) \qquad (90)$$

and

$$\langle \psi_0 \mid \mu \mid \chi_k \rangle = \tfrac{1}{2} S(k-p-1). \qquad (91)$$

We have introduced the sums over oscillator strengths defined by Eq. (58). Goscinski [68, 69] thus derived a whole family of lower bounds on $\alpha(i\omega)$ as special cases of (86). For instance, when $n = 1$ (one basis function in the set),

$$\alpha(i\omega) \geq \frac{2 \mid \langle \psi_0 \mid \mu (H_0 - E_0)^{i-p} P\mu \mid \psi_0 \rangle \mid^2}{S_{11}}$$
$$= \frac{[S(i-p-1)]^2}{S(2i-2p) + \omega^2 S(2i-2p-2)}. \qquad (92)$$

For $n \geq 2$, the inverse matrix must be constructed, which may be done numerically for fixed values of $\omega$.

In order to construct upper bounds to $\alpha(i\omega)$, Goscinski [68, 69] carried out manipulations like those above, but using another operator for $\Omega$. With the special choice of basis functions (88) and with $n = 1$, the bound is

$$\alpha(i\omega) \leq S(-2) - [S(i-p-1)]^2/[\omega^{-2}S(2i-2p+2) + S(2i-2p)]. \qquad (93)$$

When $i - p = -1$, we may use $S(0) = N$ and $S(-2) = \alpha$, the static polarizability. Then we recover one of the approximations to $\alpha(i\omega)$ constructed by Mavroyannis and Stephen [40], Eq. (71), which is shown to be an upper bound. With $i - p = -1$ in (92), the lower bound is

$$C_{AB} \geq \frac{3}{\pi} \int_0^\infty \frac{[S^A(-2)S^B(-2)]^2 \, d\omega}{[S^A(-2) + \omega^2 S^A(-4)][S^B(-2) + \omega^2 S^B(-4)]}$$

$$= \frac{3}{2} \frac{S^A(-2)S^B(-2)}{(S^A(-4)/S^A(-2))^{1/2} + (S^B(-4)/S^B(-2))^{1/2}}. \tag{94}$$

While (71) is only a rough approximation, (94) was claimed to be very close to $C_{AB}$. A number of other bounds, of various degrees of complexity, were given for $C_{AB}$. Goscinski also showed how one can get bounds for any $S(-k)$ in terms of the others. This should be useful in constructing a set of $S(k)$, such as is needed in the calculations of Dalgarno and coworkers.

Kramer [34] also derived bounds on van der Waals coefficients in terms of the $S(i)$, for the homonuclear case. He studied the behavior of various average energy approximations. Equation (49) is

$$C_{AA} = \tfrac{3}{4}\Delta\alpha_A{}^2. \tag{95}$$

[It may be generalized [27] to higher-pole interactions:

$$C_{AA}^{(l,l)} = \frac{(4l)!}{8[(2l!)]^2} \alpha_l{}^2 \Delta_l \tag{96}$$

where $\alpha_l$ is the $2^l$-pole static polarizability and $\Delta_l$ the appropriate average energy. Equation (95) corresponds to $l = 1$, the dipole–dipole interaction.] Many formulas for $C_{AB}$ given earlier are of the form (49), with different choices for the average energies. Kramer [34] pointed out that any quotient of sums, having the proper dimensions, can be used as an average energy in Eq. (95):

$$\Delta = \bar{\omega}(i, k) = [S(-i)/S(-k)]^{1/(k-i)}$$

Here, $i$ and $k$ need not be integers. The quantity $\bar{\omega}(i, i)$, defined by

$$\bar{\omega}(i, i) = \lim_{\varepsilon \to 0} \{\bar{\omega}(i - \tfrac{1}{2}\varepsilon, i + \tfrac{1}{2}\varepsilon)\}$$

is also a possible average energy.

Since all the $\bar{\omega}(i, j)$ are quotients of mean values of energies, using the same set of oscillator strengths as weights, certain inequalities must hold between them. For example, as shown by Kramer,

$$\bar{\omega}(i, k) < \bar{\omega}(i, k') \qquad \text{for} \quad k > k' \tag{97}$$

$$\bar{\omega}(i', k) < \bar{\omega}(i, k) \qquad \text{for} \quad i' > i \tag{98}$$

$$\bar{\omega}(k', k) < \bar{\omega}(i, k') \qquad \text{for} \quad k > i \tag{99}$$

$$\bar{\omega}(i, k) > \bar{\omega}(i', k') \qquad \text{for} \quad i > i', \quad k > k' \tag{100}$$

Inequality (98) shows that $\bar{\omega}(1, 2)$ is lower than $\bar{\omega}(0, 2)$, which implies that $\frac{3}{4}\alpha_A \langle \mathbf{r} \cdot \mathbf{r} \rangle$ [see Eq. (51)] gives a lower value for $C_{AA}$ than the Slater–Kirkwood formula, (53). The correct average energy for calculation of $C_{AA}$, defined by Eq. (95), is, like the $\bar{\omega}(i, j)$, a weighted mean energy with weighting factors derived from the same oscillator strength distribution. It also obeys inequalities; in particular, Kramer [34] cites

$$\Delta < \bar{\omega}(1.5, 2) = [S(-1.5)/S(-2)]^2$$

The corresponding bound on the van der Waals constant,

$$C_{AA} < \tfrac{3}{4}\bar{\omega}(1.5, 2)\alpha_A\alpha_A = \tfrac{3}{4}[S(-1.5)]^2, \tag{101}$$

was also derived (see page 47) by Pack [69a]. Since $\bar{\omega}(1.5, 2) < \bar{\omega}(1, 2)$, both $\frac{3}{4}\alpha_A\langle \mathbf{r} \cdot \mathbf{r} \rangle$ and Eq. (53) are actually upper bounds to $C_{AA}$. A better upper bound is provided by $\bar{\omega}(1.5, 2)$, but it requires $S(-1.5)$ and there is no sum rule to relate $S(-1.5)$ to experimental quantities. The possibility of obtaining it from other $S(j)$ exists (see Dalgarno [44] and Goscinski [68, 69], for example), and was investigated by Kramer [34].

For rare gas atoms and some molecules, Kramer fitted frequency-dependent dipole polarizabilities to functions of four or more parameters, and derived the sum $S(j)$ from $j = 0$ to $j = -4$ in half-integral steps. He used these $S(j)$ to compute a large number of $\bar{\omega}(i, k)$ and compared them with $\Delta$ of Eq. (95) for the 19 cases for which values of $C_{AA}$ were known. It turned out that $\bar{\omega}(2, 2)$ worked very well (average deviation from $\Delta$ of 0.25%). The problem is that $\bar{\omega}(2, 2)$ requires $dS(j)/dj$ for $j = -2$, which must be derived from the $S(j)$ by a numerical differentiation. Suggestions to cope with this problem were made.

Pack's derivation [69a] of (101) follows from the observation that

$$A_{mn}(A, B) = \frac{[(E_m^A - E_0^A)(E_n^B - E_0^B)]^{1/2}}{\tfrac{1}{2}[(E_m^A - E_0^A) + (E_n^B - E_0^B)]} \tag{102}$$

is not greater than unity. According to Eq. (44),

$$C_{AB} = \frac{3}{4} \sum_{k>0} \sum_{l>0} \frac{A_{kl}(A, B)f_{ok}^A f_{ol}^B}{[(E_m^A - E_0^A)(E_n^B - E_0^B)]^{3/2}} \tag{103}$$

which, using (102), becomes

$$C_{AB} \leq \tfrac{3}{4} S_A(-1.5) S_B(-1.5). \tag{104}$$

For rare gases, the right side actually approximates $C_{AB}$ well [69a], since $A_{mn}(A, B) \approx 1$. To obtain $S(-1.5)$, Pack suggested interpolation in the $S(j)$ for integral $j$, or the use of a one-term Sellmeier form with parameters fit to give $S(-1)$ and $S(-2)$ correctly.

Other bounds on the van der Waals coefficients are obtained from variation–perturbation theory (see Subsection 7). Suppose the second-order energy term $-C/R^6$ is being calculated by minimizing the functional

$$\mathscr{F}(\Phi) = \langle \Phi \mid H_0 - E_{A0} - E_{B0} \mid \Phi \rangle + 2\langle \Phi \mid v \mid \Psi_{A0} \Psi_{B0} \rangle \tag{105a}$$

with respect to variations in the trial function $\Phi$. This is Eq. (167) of Volume I, Chapter III, with the zero-order function the product of atomic eigenfunctions $\Psi_{A0} \Psi_{B0}$, the zero-order energy $E_{A0} + E_{B0}$, and the first-order energy (the expectation value of $v$ over $\Psi_{A0B0}$) $= 0$. Here, $v$ is the perturbation [the leading term in the expansion (32), in atomic units]:

$$v = R^{-3} \sum_i^{(A)} \sum_j^{(B)} (x_i x_j + y_i y_j - 2z_i z_j) \tag{105b}$$

When $\Phi$ is the correct first-order wave function, $\mathscr{F}(\Phi)$ equals the second-order energy, $-C_{AB}/R^6$; for any other $\Phi$, $\mathscr{F}(\Phi)$ is an upper bound to the second-order energy. Therefore a lower bound to $C_{AB}$ is obtained. As shown in Subsection 7, the individual terms in the expansion (32) can be treated separately for spherically symmetric atoms. Thus lower bounds can be found for any of the coefficients in the expansion of $E^{(2)}$ in powers of $R^{-1}$.

A complete review, with references, of methods for calculating bounds on van der Waals coefficients was given by Langhoff et al. [70]. Relations among different methods, such as those using Padé approximants and those using quadrature formulas, were brought out. It was concluded that, computationally, bounding methods using experimentally or semi-empirically obtained moments of oscillator strengths are superior to those requiring calculation of $\alpha(i\omega)$.

## 7. *Variation–Perturbation Calculations*

As explained in Section B of Chapter III, Volume I, one can obtain approximations to second- and higher-order energies of perturbation theory from variational principles, and such calculations are referred to as variation–perturbation calculations. Here we discuss variation–perturbation calculations for van der Waals coefficients. We have seen in the previous sections several ways in which perturbation theory may enter the calculation of dispersive interactions. First, the perturbation $V$ [Eq. (5)], acting on a product of atomic wave functions, may be considered. If $V$ is expanded as in (30) or (32), one may consider the effect of any term as a perturbation and compute the second order energy. Second, one may calculate polarizabilities of atoms for imaginary frequency, from which the van der Waals interactions between the atoms may be obtained via Eq. (66). As suggested by Mavroyannis and Stephen [Eqs. (67)–(68)], this may be obtained by considering the effect of a time-varying electric field on the atom. For either the first and the second case, one can attempt to obtain approximate solutions for the first-order wave function by variational methods. In the second case we have to deal with time-dependent perturbation theory. For the calculation of the effect of $V$, which is time-independent, perturbation theory as discussed in Volume I, Chapter III, Section B can be used. We will discuss this first.

Variation–perturbation calculations for this problem were early performed and provided the first accurate values of $C_{AB}$ for simple systems. Suppose that $V$ is expanded as in Eq. (30), and let the ground states of both atoms be spherically symmetric (extension of this argument to non-spherically symmetric ground states is not difficult). Let a variation–perturbation calculation be carried out using a set of product functions $\tilde{\Psi}_{Ap}\tilde{\Psi}_{Bq}$, where $\tilde{\Psi}_{Ap}$ is a function of the $n_A$ electronic coordinates of atom A and $\tilde{\Psi}_{Bq}$ a function of the $n_B$ electronic coordinates of atom B. The set includes $\Psi_{A0}\Psi_{B0}$, the product of ground state eigenfunctions. The functions are chosen with the following properties:

(1) The functions $\tilde{\Psi}_{Ap}\tilde{\Psi}_{Bq}$ are orthonormal.

(2) The unperturbed (atomic) Hamiltonian $H_A + H_B$ is diagonalized by the set.

(3) The $\tilde{\Psi}_{Ap}$ and $\tilde{\Psi}_{Bq}$ have symmetries of atomic states.

Property (2) is obtained when, for atom A, the set of $\tilde{\Psi}_{Ap}$ (including $\Psi_{A0}$) diagonalizes $H_A$, and correspondingly for atom B. From any set of functions, one can always form an orthonormal set which diagonalizes the matrix of $H_A + H_B$. The set of $\tilde{\Psi}_{Ap}$ may also be chosen so that its members have the symmetries of atomic wave functions. We may now use Eqs. (161) et seq. of Volume I, Chapter III, except that for every choice of $l_A$, $l_B$, and $m$ in (30) a different subset of the basis functions contributes. Our approximate second order energy [see Volume I, Chapter III, Eq. (165)] is a sum of terms:

$$E^{(2)} = \sum_{p,q} \frac{|\langle \Psi_{A0}\Psi_{B0} | (Q_{lA}^m)^* Q_{lB}^m | \tilde{\Psi}_{Ap}\tilde{\Psi}_{Bq}\rangle|^2}{E_{A0} + E_{B0} - \tilde{E}_{Ap} - \tilde{E}_{Bq}} R^{-2(l_A + l_B + 1)} \tag{106}$$

where $\tilde{E}_{Ap} = \langle \tilde{\Psi}_{Ap} | H_A | \tilde{\Psi}_{Ap}\rangle$, $\tilde{E}_{Bq} = \langle \tilde{\Psi}_{Bq} | H_B | \tilde{\Psi}_{Bq}\rangle$. The sum is only over the basis functions of the proper symmetry, and the subsets of different symmetries do not mix. Thus, the various interactions (dipole–dipole, dipole–quadrupole, etc.) may be treated separately. Also, if both induction and dispersion forces are present, we may carry out a variation-perturbation calculation for the latter only. Imagine the basis set to include all the products of the form $\Psi_{A0}\tilde{\Psi}_{Bm}$ and $\tilde{\Psi}_{An}\Psi_{B0}$, as well as functions $\tilde{\Psi}_{Ap}\tilde{\Psi}_{Bq}$. Property (1) means $\tilde{\Psi}_{Ap}$ is orthogonal to $\Psi_{A0}$ and $\tilde{\Psi}_{Bq}$ to $\Psi_{B0}$. Then $\tilde{\Psi}_{Ap}\tilde{\Psi}_{Bq}$ represents a situation where both atoms A and B are excited. Consider a particular term in $V$ which we write $W_A W_B$, the two factors referring to the two atoms. Then the approximate second-order energy (again referring to (165) of Volume I, Chapter III) is

$$E^{(2)} = \langle \Psi_{A0} | W_A | \Psi_{A0}\rangle \sum_m |\langle \Psi_{B0} | W_B | \tilde{\Psi}_{Bm}\rangle|^2 (E_{B0} - \tilde{E}_{Bm})^{-1}$$
$$+ \langle \Psi_{B0} | W_B | \Psi_{B0}\rangle \sum_n |\langle \Psi_{A0} | W_A | \tilde{\Psi}_{An}\rangle|^2 (E_{A0} - \tilde{E}_{An})^{-1}$$
$$- \sum_{p,q} |\langle \Psi_{A0} | W_A | \tilde{\Psi}_{Ap}\rangle\langle \Psi_{B0} | W_B | \tilde{\Psi}_{Bq}\rangle|^2$$
$$\times (\tilde{E}_{Ap} + \tilde{E}_{Bq} - E_{A0} - E_{B0})^{-1}.$$

The induction energy is the first two terms and the remainder, the dispersion energy, is of the form (106).

Because of the similarity of the perturbation $V$ to that involved in a calculation of a static polarizability, the early workers generally calculated atomic polarizabilities at the same time as van der Waals constants, using similar trial functions for the two cases. Then comparison of calculated polarizabilities with measured values gave an indication of the accuracy to which $C_{AB}$ (whose exact value was not generally known) was being calculated. Since the variational principle for the second-order energy is a minimum principle, the expression (106) is an upper bound for the second-order energy. For the perturbation term with $l_A = l_B = 1$, $E^{(2)} = -C_{AB}R^{-6}$, so this means a lower bound for $C_{AB}$. A derivation of the formulas, which makes only indirect reference to perturbation theory (and is in fact closer to the spirit of the early workers than the variation–perturbation theory as we have sketched it) is given by Margenau and Kestner [2, Sect. 2.4,; 71].

It must be noted, however, that the use of the variation–perturbation method implies the use of the exact unperturbed function. This is easy to accomplish for H–H, but for other atoms we require a good approximation to the exact function, which may be complex in form and lead to involved calculations. With an approximate function for $\Psi_0$, the upper bound is not guaranteed. This point was emphasized by Hassé [72, 73]. One possibility, if we have an approximate function, is to use $H_0$, an approximate Hamiltonian of which this wave function is an eigenfunction, and consider the effect of the perturbation along with the corrections to $H_0$ (double perturbation theory: see Volume I, Chapter III, Section A.5).

Slater and Kirkwood [38] were among the earliest workers to employ the variation–perturbation method for the $R^{-6}$ term. For the calculation of $C_{HH}$, the trial function for the first-order wave function was taken as a product of a function $\phi$ with the unperturbed product function for H–H. The form they used for $\phi$ was

$$\phi = \lambda r_1{}^\nu r_2{}^\nu (x_1 x_2 + y_1 y_2 - 2z_1 z_2) \tag{107}$$

with $\lambda$ and $\nu$ variational parameters and $r_1$ and $r_2$ the distances of the electrons from their nuclei. The last factor is the perturbation itself. The value for $C_{HH}$ was $6.49e^2a_0{}^5$, in excellent agreement with the London–Eisenschitz result, which used experimental data on oscillator strengths. A more elaborate calculation by Pauling and Beach [74] gave $C_{HH} = 6.49903$, and also the coefficients of the dipole–quadrupole ($R^{-8}$) and quadrupole–quadrupole ($R^{-10}$) terms. A still more extensive computation, using natural orbitals to simplify the trial wave function, was given relatively recently by Hirschfelder and Löwdin [75] for the $R^{-6}$ and $R^{-8}$ terms. The second-order energy was $-6.499026R^{-6} - 124.395R^{-8}$.

Slater and Kirkwood [38; 38a] extended their calculations to the He–He interaction, and other workers continued this work [39, 72, 76]. The results were reviewed by Buckingham [76] and Margenau [29]. Buckingham [76] was the first to use properly antisymmetrized zero-order wave functions, i.e., a determinant

$$\Psi_0 = \mathscr{A} \prod_i \psi_i(i) s_i(i)$$

for polarizability calculation and a product of determinants

$$\Psi_0 = \left[ \mathscr{A}_A \prod_{p=1}^{n_A} \psi_p{}^A(p)s_p(p) \right]\left[ \mathscr{A}_B \prod_{r=1}^{n_B} \psi_r{}^B(r)s_r(r) \right] \qquad (108)$$

for the van der Waals constant. The $s_i$, $s_p$, and $s_r$ are the $\alpha$ or $\beta$ spin functions. The antisymmetrizer $\mathscr{A}_A$ operates on the electrons of $A$ only. The wave function in the presence of the perturbation was obtained for polarizabilities by replacing the orbital $\psi_i(i)$ by $\psi_i(i)[1 + f(r_i)v_i]$, where the perturbation is written as $\sum_i v_i$. Buckingham considered $f = \lambda$, $f = \lambda_i$ (different for each orbital), and other choices. The first gave

$$\alpha = 4\langle r^2 \rangle^2/9n. \qquad (109)$$

Here, $\langle r^2 \rangle$ is a sum of expectation values over individual orbitals and $n$ is the number of electrons. For the calculation of the van der Waals constant, the wave function in the presence of the perturbation was obtained from (107) by replacing $\psi_p{}^A$ by $\psi_p{}^A(1+\lambda_p \sum_{r=1}^{n_B} v_{pr})$ and $\psi_r{}^B$ by $\psi_r{}^B(1+\lambda_r \sum_{p=1}^{n_A} v_{pr})$. Assuming $\lambda_p$ the same for all $p$ and $\lambda_r$ the same for all $r$ corresponds to $f = \lambda$ in the polarizability calculations and yielded

$$C_{AB} = \frac{4}{9} \frac{(\langle r^2 \rangle_A)^2(\langle r^2 \rangle_B)^2}{n_B\langle r^2 \rangle_A + n_A\langle r^2 \rangle_B} \qquad (110)$$

which is actually a lower bound, as was proved in the preceding section. Inserting the result (109),

$$C_{AB} = \frac{3}{2} \frac{\alpha_A\alpha_B}{(\alpha_A/n_A)^{1/2} + (\alpha_B/n_B)^{1/2}} \qquad (111)$$

which had previously been derived by Slater and Kirkwood [38], and is associated with their names. Correspondingly, the use of a different $\lambda_p$ for each orbital led to

$$C = \frac{4}{9} \sum_{p=1}^{n_A} \sum_{r=1}^{n_B} \frac{(\langle r^2 \rangle_p)^2(\langle r^2 \rangle_r)^2}{\langle r^2 \rangle_p + \langle r^2 \rangle_r}. \qquad (112)$$

Buckingham [76] also considered the dipole–quadrupole terms.

More recently, Davison [77] extended the Slater–Kirkwood calculation for some small systems. The purpose was to obtain higher accuracy to check the reliability of the results, which did not always agree with values for the interaction constants derived from experiment. The interaction was expressed in terms of irreducible tensors according to the formalism of

Rose [7]. With trial functions which were products of polynomials with ground state atomic wave functions, all integrals were expressed as expectation values over the unperturbed atomic wave functions. These were exactly calculable for hydrogenic systems, but for the two-electron atomic cases discussed very complex wave functions had to be used to give good values. The dipole and quadrupole polarizabilities (for the induction terms) were computed for H, He+, He, Li+, and H−, and the dipole–dipole and dipole–quadrupole dispersion terms for various pairs of these. The results for the latter agreed with those obtained using oscillator strengths. It was pointed out that trial functions which are sums of terms like $r_i r_j f(r_i, r_j)$ times the zero-order wavefunction (electron $i$ is on atom A, $j$ on B), make for calculations of a high degree of difficulty for larger systems. Table IV displays Davison's results. For the ionic cases, the coefficients of $R^{-4}$ are computed from the dipole polarizability [see Eq. (35)], while higher terms include both dispersion and induction contributions.

**Table IV**

*Coefficients in Expansion* $E^{(2)} = -\sum_{n=2} C_{AB}^{(2n)} (R/a_0)^{-2n}$

| A | B | $C_{AB}^{(4)}$ | $C_{AB}^{(6)}$ | $C_{AB}^{(8)}$ |
|---|---|---|---|---|
| H | H | | 6.49903 | 124.3991 |
| H | He+ | 2.250 | 8.15772 | 73.9628 |
| H | He | | 2.83 | 41.9 |
| H | Li+ | 2.250 | 7.99 | 71.9 |
| H | H− | 2.250 | 93 | $7 \times 10^3$ |
| He | Ne+ | 0.692 | 1.59 | 7.3 |
| He | He | | 1.47 | 14.2 |
| He | Li+ | 0.692 | 1.52 | 6.6 |
| He | H− | 0.692 | 31 | $2 \times 10^3$ |

A recent calculation by Murrell and Shaw [78] used formula (106). They found, for the $R^{-6}$, $R^{-8}$, and $R^{-10}$ terms in H–H and for the $R^{-6}$ term in He–He, that a very few excited atomic configurations of the proper symmetry were needed to get good results for the dispersion energy and van der Waals coefficient. The ground and excited states were approximated, for He–He, by determinants of Slater orbitals. As we discuss in Section B, the calculations were extended to consider the effect of overlap on the energy, which indeed was the goal of their work.

Variation perturbation calculations for dispersion energies, and also for polarizabilities and induction energies, can be simplified by the use of Gaussian basis functions (see Chapter II). Further simplification is obtained if the unperturbed atomic functions are also approximated in terms of Gaussians [79].

Now we discuss variation–perturbation calculations [80, 81] for $\alpha(i\omega)$, to be used to compute $C_{AB}$ from formula (66), the Casimir–Polder relation. This approach was investigated by Karplus and Kolker [82, 83]. We consider the function of Eq. (67). Equation (68) implies

$$(H_0 - E_0 \pm \hbar\omega)\mathbf{F}^{\pm} = \boldsymbol{\mu}\psi_0 \tag{113}$$

with $\mathbf{F}^{\pm}$ orthogonal to $\psi_0$. Where Eq. (113) is intractable to direct solution, one may consider the functionals

$$L_0{}^{\pm} = \langle \mathbf{F}^{\pm} \cdot \mathscr{E} \mid H_0 - E_0 \pm \hbar\omega \mid \mathbf{F}^{\pm} \cdot \mathscr{E} \rangle$$
$$+ 2 \, \mathrm{Re}\langle \mathbf{F}^{\pm} \cdot \mathscr{E} \mid -\boldsymbol{\mu} \cdot \mathscr{E} \mid \psi_0 \rangle \tag{114}$$

Karplus and Kolker showed that minimization of $L_0{}^{\pm}$ with respect to all variations in $\mathbf{F}^{\pm}$ is equivalent to solving Eq. (113). The polarizability is given by

$$\alpha(\omega) = \tfrac{1}{3} \, \mathrm{Re}[\langle \psi_0 \mid \boldsymbol{\mu} \cdot \mathbf{F}^{+} \rangle + \langle \psi_0 \mid \boldsymbol{\mu} \cdot \mathbf{F}^{-} \rangle]. \tag{115}$$

Upon inserting, for $\mathbf{F}^{+}$ and $\mathbf{F}^{-}$ in (113), expansions in the eigenfunctions of $H_0$, we readily recover $\alpha(\omega)$, Eq. (46).

If one uses a determinantal function for $\psi_0$, the one-electron character of $\boldsymbol{\mu}$ means that $\mathbf{F}^{+}$ involves only one-electron excitations. The situation is quite analogous to the time-independent Hartree–Fock perturbation theory. Karplus and Kolker [1, Pt. V; 82] used a Hartree–Fock function for $\psi_0$, and neglected self-consistency in computing the first-order wave function. Thus, theirs is an uncoupled theory (see Volume I, Chapter III, Section B.2) in which the perturbation of the oscillating electric field acts on a system which obeys the Hartree–Fock Hamiltonian in the absence of the perturbation. Writing $\mathbf{F}^{\pm}$ as a sum of determinants each differing from the Hartree–Fock determinant by replacement of the $i$th orbital by $\phi_i{}^{\pm}$ and substituting into $L_0{}^{\pm}$, they derived equations for the $\phi_i{}^{\pm}$. If we compare (113) with the situation discussed in Volume I, Chapter III, Eqs. 177 et seq.

we can see that instead of Eq. (180) of that chapter we will have

$$(h_0 - \varepsilon_l \pm \hbar\omega)\phi_l^{\pm} - \sum_k (\varepsilon_k - \varepsilon_l \pm \hbar\omega)\lambda_k\langle\lambda_k \mid \phi^{\pm}\rangle + \mu\lambda_l$$

$$= \sum_k^{(k \neq l)} \langle\lambda_k \mid \mu \mid \lambda_l\rangle\lambda_k \tag{116}$$

We refer to Volume I, Chapter III, Section B for definitions. If each $\phi_l^{\pm}$ is expanded in a convenient basis set,

$$\phi_l^{\pm} = \sum_i \beta_{li}^{\pm}\chi_i^{(l)}. \tag{117}$$

The coefficients will obey an equation of the form

$$(A^{(l)} \pm \hbar\omega B^{(l)})\beta_l^{\pm} = \gamma_l. \tag{118}$$

To extend [83] the result to imaginary frequencies, the dependence of the polarizability on frequency was first made explicit. Let $\mathbf{S}$ be the matrix of the eigenvectors of $\mathbf{B}$, so that $\mathbf{S}$ is unitary and $\mathbf{S}^{-1}\mathbf{B}\mathbf{S} = \mathbf{D}$ is diagonal. Karplus and Kolker constructed $\mathbf{D}^{1/2}$, the diagonal matrix whose elements are the square roots of those of $\mathbf{D}$, and multiplied (118) on the left by $\mathbf{D}^{-1/2}\mathbf{S}^{-1}$ to give

$$(\tilde{\mathbf{A}} \pm \hbar\omega\mathbf{I})\tilde{\beta}_l^{\pm} = \tilde{\gamma}_l \tag{119}$$

where $\tilde{\mathbf{A}} = \mathbf{D}^{-1/2}\mathbf{S}^{-1}\mathbf{A}\mathbf{S}\mathbf{D}^{-1/2}$ and is Hermitian, $\mathbf{I}$ is the unit matrix, $\tilde{\beta}_l^{\pm} = \mathbf{D}^{1/2}\mathbf{S}^{-1}\beta_l^{\pm}$, and $\tilde{\gamma}_l = \mathbf{D}^{1/2}\mathbf{S}^{-1}\gamma_l$. Let $\tilde{\beta}_l^{\pm}$ be expanded in the eigenvectors of $\tilde{\mathbf{A}}$, represented as $\alpha_k^{(l)}$, with corresponding eigenvalues $e_k^{(l)}$:

$$\tilde{\beta}_l^{\pm} = \sum_k b_{lk}^{\pm}\alpha_k^{(l)}$$

Substituting gives (the $\alpha_k^{(l)}$ are orthonormal):

$$b_{lk}^{\pm} = (\alpha_k^{(l)\dagger} \cdot \tilde{\gamma}_l)/(e_k^{(l)} \pm \hbar\omega) \tag{120}$$

or

$$\beta_l^{\pm} = \mathbf{S}\mathbf{D}^{-1/2} \sum_k \alpha_k^{(l)}(\alpha_k^{l\dagger} \cdot \tilde{\gamma}_l)/(e_k^{(l)} \pm \hbar\omega) \tag{121}$$

The frequency dependence has been made explicit. When (116), (117), and (121) are used in Eq. (115), the polarizability assumes a Sellmeier form:

$$\alpha(\omega) = \mathrm{Re}\left|\sum_l \sum_k \mid \tilde{\gamma}_l \cdot \alpha_k^{(l)} \mid^2 \frac{2e_k^{(l)}}{e_k^{(l)2} - \hbar^2\omega^2}\right| \equiv \sum_{l,k} \frac{f_l^{(k)}}{e_k^{(l)2} - \hbar^2\omega^2} \tag{122}$$

To use this for imaginary frequencies, Karplus and Kolker [83] simply replaced $\omega$ by $i\omega$. The $f_l^{(k)}$ and $e_k^{(l)}$ need be calculated only once. By these formulas, the $C_{AB}$ were computed for interactions between He, Ne, and Ar. Subsequently, the time-dependent *coupled* Hartree–Fock perturbation theory was developed and used [84, 85] for the frequency-dependent polarizability of He (real $\omega$). It was found that, compared with an accurate variational calculation [86, 87] the coupled theory gave results 4–8% too small, and the uncoupled theory 8% too large. The terms in $R^{-8}$ and $R^{-10}$ can also be treated using the coupled Hartree–Fock theory to compute the corresponding polarizabilities for imaginary frequencies. The first-order equations were again solved [88] variationally and, compared to variational calculations for the two-center problem, gave errors of 7% or so.

Recently, another formalism [89], based on the hydrodynamic analogy to quantum mechanics, has been given for the computation of such quantities as frequency-dependent polarizabilities. It is equivalent to the formalism of Karplus and Kolker [82, 83].

For He at least, one can go beyond the Hartree–Fock approximation. Chan and Dalgarno [86, 87] used an accurate zero-order function and a linear variation function for the first-order function, computed $\alpha(\omega)$, and fitted it to

$$\alpha(\omega) = \sum_{i=1}^{m} f_i/(\omega_i{}^2 - \omega^2) \tag{123}$$

Here $m$ equals the number of linear parameters in the trial function. Then $\alpha(i\omega)$ was obtained by replacing $\omega$ by $i\omega$, and $C$ was computed for H and He interactions.

In principle, the effects of neglecting correlation in calculating $\alpha(i\omega)$ can be computed by double perturbation theory, where one perturbation is the difference between Hartree–Fock and exact Hamiltonians. Hirschfelder *et al.* [90] have given the extension to time-dependent problems, so that it may be applied to discussion of the frequency-dependent polarizability. This was done for He by Musulin and Epstein [91]. Deal and Kestner [1, Pt. VI; 92] applied double perturbation theory to the calculation of $C_{AB}$ via equations (113). The van der Waals constants for $H^- - H^-$, He–He, and $Li^+ - Li^+$ were considered. The effect of correlation for He–He is to decrease $C$ from 1.65 to 1.38 (the accepted value is 1.46), and was essentially [1, Pt. VI] the effect of going from an uncoupled to a coupled theory.

Chan and Dalgarno [57] have avoided the use of $F^{\pm}$ by dealing with the function $\psi(\omega)$ which satisfies (in atomic units)

$$[(H_0 - E_0)^2 + \omega^2]\psi(\omega) + [\mu - \langle\psi_0 \mid \mu \mid \psi_0\rangle]\psi_0 = \omega^2\langle\psi(\omega) \mid \psi_0\rangle\psi_0 \tag{124}$$

$H_0$ is the isolated atom Hamiltonian, of which $\psi_0$ is the eigenfunction corresponding to eigenvalue $E_0$, and $\mu$ is the dipole moment operator. By expanding $\psi(\omega)$ in the eigenfunctions of $H_0$, one may show that

$$\langle \psi_0 \mid [H_0, \mu] \mid \psi(\omega) \rangle = \tfrac{1}{2}\alpha(i\omega). \tag{125}$$

Equation (124) is to be solved variationally, by minimizing the proper functional. The same formalism may be applied to the quadrupole and higher polarizabilities. (Other methods for calculating $\alpha(i\omega)$ have also been suggested by these authors [2, Chapter IV; 86].) By the procedure of equations (124)–(125), Chan and Dalgarno [57] computed the dipole, quadrupole, and octupole polarizabilities for imaginary frequency for the H atom, and thence the coefficients of $R^{-6}$, $R^{-8}$, and $R^{-10}$ for the H–H interaction. These equations have also been applied to helium and lithium [1, Pt. III]. Values for $C_{AB}$ for the six pairs of atoms, formed from H, He, and Li, were obtained. The computations of course become more involved as the complexity of the atoms increases.

Victor et al. [93] performed variational calculations to obtain $F^{\pm}$ for ground and metastable states of He. The linear variational form used led to a Sellmeier representation of $\alpha(\omega)$, as in the perturbed Hartree–Fock theory, which facilitated calculation of $\alpha(i\omega)$ and $C_{AB}$. It may be shown that, when $\alpha(\omega)$ is written as in Eq. (123), the oscillator strength sum rule is automatically fulfilled by the $\{f_i\}$. Computations by this method were carried through [94] for Li, except that, because of the importance of the $1s^2 2p\ ^2P$ excited state, an explicit representation of its wave function was included in the basis set. The van der Waals constants for H–Li, He–Li, and Li–Li were computed.

Dalgarno and Victor [95] have derived $\alpha(i\omega)$ for ground and excited states of Li⁺, and used the Sellmeier representation to compute various sums $S(k)$. These were compared with theoretical values, calculated as expectation values over the ground state wavefunction, as a check on accuracy. From the $\alpha(i\omega)$, they computed the van der Waals coefficients for interaction of Li⁺ with H, Li⁺, rare gases, alkali metals, and also some molecules for which theoretical or semiempirical representations of $\alpha(i\omega)$ were available. In Table V, some of the results derived by way of $\alpha(i\omega)$ are summarized.

Alexander [96] has recently outlined another related approach. The expression (45) for $C_{AB}$ may be written

$$C_{AB} = \tfrac{3}{2} \int_{\omega_{A0}}^{\infty} d\omega_A \omega_A^{-1} \alpha_B^{+}(\omega_A) f^{(A)}(\omega_A) \tag{126}$$

**Table V**

*Calculated van der Waals Constants Involving H, He, Li, and Li$^+$ (in Ground States), Atomic Units*[a]

| | H | He | Li | Li$^+$ | Ne | Ar | Kr | Xe | Na | K | Rb | Cs |
|---|---|---|---|---|---|---|---|---|---|---|---|---|
| H | 6.499 | 2.830 | 66.65 | 0.4931 | | | | | | | | |
| He | 2.830 | 1.471 | 22.49 | 0.302 | | | | | | | | |
| Li | 66.65 | 22.49 | 1391 | 3.32 | | | | | | | | |
| Li$^+$ | 0.4931 | 0.302 | 3.32 | 0.0782 | 0.660 | 1.87 | 2.58 | 3.42 | 3.60 | 4.86 | 5.21 | 6.09 |

[a] The polarizabilities $\alpha(i\omega)$ for atoms other than H, He, Li, and Li$^+$ are derived from empirical data.

where

$$\alpha_B{}^+(\omega_A) = \int_{\omega_{B0}}^{\infty} d\omega [f^{(B)}(\omega)/\omega(\omega_A + \omega)] \tag{127}$$

It may be noted that the dynamic polarizability for real frequency is $\alpha(\omega_A) = \frac{1}{2}[\alpha^+(\omega_A) + \alpha^+(-\omega_A)]$. Thus the procedures for computation of $\alpha(\omega)$ may be used to obtain approximate expressions for $\alpha^+(\omega)$. For example, corresponding to (121) we have

$$\alpha_B{}^+(\omega) \cong \sum_l \sum_k | \, \tilde{\gamma}_l \cdot \alpha_k^{(l)} \, |^2/(e_k^{(l)} + \omega) \equiv \sum_n a_n/(b_n + \omega). \tag{128}$$

If such a Sellmeier form for $\alpha^+(\omega_A)$ is used in (127),

$$C_{AB} = \tfrac{3}{2} \sum_n a_n \alpha_A{}^+(b_n). \tag{129}$$

Using previous calculations for $\alpha(\omega)$, this formula gave $C_{AB}$ for He–He as $1.4605 \pm 0.0025$. The error analysis leading to the bounds (which makes this the most precise prediction of the value for this quantity) was given. The distinguishable electron method (Volume I, Chapter III, Section B.3) has also been applied [97, 98] to the calculation of polarizabilities for imaginary frequency, as have many-body perturbation techniques [99].

With some simple expressions for trial functions in the variational principles, one obtains approximate formulas for $\alpha(i\omega)$ in terms of the oscillator strength sums $S(j)$, Eq. (58). Some of these expressions appeared in

Section 5. They lead to approximate formulas for the $C_{AB}$ in terms of the $S(j)$. The variational principle for $C_{AB}$ itself also yields such formulas when certain trial functions are used. Examples are Eqs. (114) *et seq.* Davison [100] has discussed and compared approximate formulas derived from these sources which give $C_{AB}$ in terms of $S(0)$, $S(-1)$, $S(-2)$, and $S(-3)$. Tests of their accuracies were made for H and the inert gas atoms. Among these, Davison found that

$$C_{AB} = \frac{12[S^A(-2)S^B(-2)]^2}{S^A(-2)S^B(-3) + S^A(-3)S^B(-2)} \tag{130}$$

worked well [though not as well as formulas involving all four $S(i)$].

It should also be mentioned here that one can obtain the van der Waals energy without explicitly considering the dispersion interaction itself. A good configuration interaction calculation for the total molecular energy will necessarily give the long range energy, $\Sigma c_i R^{-i}$, even though the individual terms in the series will not be computed separately. Such calculations are discussed by Das and Wahl [101]. It is only necessary to take into account double excitations involving single excitations on the two atoms (interatomic correlations). Double excitations on one atom or the other, while leading to changes in the energy several orders of magnitude larger (intraatomic correlation energy) generally have much less effect on the potential energy curve at large distances if a basis of *localized* orbitals is used. Das and Wahl show this analytically. They started with $U(R)$ derived from Hartree–Fock, and evaluated separately intra- and interatomic corrections to the long range forces. For HeH, inclusion of excitations of only the latter type gave energies in good agreement (to $2 \times 10^{-7}$ a.u. for $R \geq 7a_0$) with $-C^{(6)}R^{-6} - C^{(8)}R^{-8}$, using Davison's values for $C^{(6)}$ and $C^{(8)}$ (Table V). For LiHe, a fit of the energy for $R \geq 10a_0$ to this form yielded $C^{(6)} = 23.8$ and $C^{(8)} = 1261$.

As already mentioned, if we are interested in more than just the first few powers of $R^{-1}$ in the energy at large $R$, we should consider higher orders of perturbation theory than second. Arrighini *et al.* [102] calculated the coefficients of $R^{-6}$, $R^{-8}$, $R^{-10}$, $R^{-11}$, and $R^{-12}$ for interactions between He and Be atoms, the $R^{-11}$ term corresponding to the dipole–quadrupole–dipole interaction of third-order perturbation theory. As shown by Dalgarno [1], such a term may be written, analogously to equation (66), as a double integral over frequency of a product of three atomic quantities. The other interactions were calculated from atomic multipole polarizabilities for imaginary frequency, using integrals of the form of Eq. (66).

## 8. *The Hellmann–Feynman Theorem, Large R Corrections, etc.*

An interesting footnote to this discussion is the Hellmann–Feynman theorem and its relevance. An attracting force means a distortion of the electron density of each atom toward the other, since nonoverlapping charge distributions imply that the spherically symmetric electron cloud of an atom would appear to the other as a point charge. Yet the simple picture of correlated fluctuating dipoles seems to mean that there is no net dipole on either atom: The charge density remains spherically symmetrical. Consider the Slater–Kirkwood form for the first-order wave function in the H–H case [Eq. (108)], which gave a good value for $C$. The wave function is

$$\psi^{(0)} + \lambda\psi^{(1)} = \psi_A(1)\psi_B(2)(1 + \lambda r_1{}^\nu r_2{}^\nu v)$$

where $\psi_A$ is the 1s orbital centered on A, and $v = x_1 x_2 + y_1 y_2 - 2z_1 z_2$. The force on nucleus A is given by $2\lambda\langle\psi^{(0)} \mid \mathscr{F}_A \mid \psi^{(1)}\rangle + \lambda^2\langle\psi^{(1)} \mid \mathscr{F}_A \mid \psi^{(1)}\rangle$. It is not hard to see that the optimal value of $\lambda$ is proportional to $R^{-3}$ here. The force operator $\mathscr{F}_A$ is

$$\mathscr{F}_A = (z_1/r_1{}^3) + [(z_2 + R)/r_{A2}^3] - 1/R^2$$

where $r_{A2}^2 = x_2{}^2 + y_2{}^2 + (z_2 + R)^2$. Because $\mathscr{F}_A$ is one-electron and $\psi^{(1)}$ is orthogonal to $\psi^{(0)}$ in two electrons the term in $\lambda$ vanishes. Thus there is *no* force to first-order in perturbation theory. In $\langle\psi^{(1)} \mid \mathscr{F}_A \mid \psi^{(1)}\rangle$, the first term of $\mathscr{F}_A$ makes no contribution because of parity. On expansion of the second term, one finds the leading remaining contribution goes as $R^{-3}$. But so does $\lambda$, which means $\lambda^2\langle\psi^{(1)} \mid \mathscr{F}_A \mid \psi^{(1)}\rangle$ goes as $R^{-9}$. The force corresponding to $E^{(2)} = -CR^{-6}$ should go as $R^{-7}$. Thus the Hellmann–Feynman force is indeed *not* predicted by this wavefunction, even though the energy is given correctly. Indeed, it is easy to see that all terms in $\langle\psi_0 + \psi_1 \mid z_1 \mid \psi_0 + \psi_1\rangle$ vanish: The charge density remains symmetric about each atom.

The answer to the seeming paradox is that the Hellmann–Feynman theorem applies to the exact eigenfunction, which is $\psi^{(0)} + \psi^{(1)} + \psi^{(2)} + \cdots$ This is an example of the fact that it is often easier to get good energies than good forces. The situation has been discussed in some detail by Hirschfelder and Eliason [103].

They treated the H–H interaction, deriving the electron density from a wave function constructed by perturbation theory.

$$\varrho = 2 \int \Psi^* \Psi \, d\tau' \tag{131}$$

$$\Psi = \Psi^{(0)} + \sum_{n=3}^{\infty} \Psi_n R^{-n} \tag{132}$$

$\Psi^{(0)}$ is the zero-order wave function and $\Psi$ is normalized to unity. By putting (132) into (131) one obtains an expansion of $\varrho$ in powers of $R^{-1}$:

$$\varrho = \sum_{m,n=1}^{\infty} \varrho_{mn} R^{-(m+n)} + \varrho_{00}$$

The terms $\varrho_{0n}$ arising from the first-order wave function vanish because $\Psi^{(1)}$ is doubly orthogonal to $\Psi_0$. Of course $\varrho_{00}$ gives no force. The leading term in $\varrho$ which gives a force on a nucleus comes out correctly, going as $R^{-7}$. It arises from $\varrho_{34} = \varrho_{43}$ and $\varrho_{07} = \varrho_{70}$, but the latter comes from the second-order wave function plus the zero-order wave function. The former is a mixed first-order term, arising from the induced dipole and dipole-quadrupole terms of the first-order wave function, so that it does not appear if we use the leading term $(R^{-3})$ of the perturbation only. In fact, this term contributes 2.717 a.u./$R^7$ to the force, while $\varrho_{07}$ gives 36.284 a.u./$R^7$. To get anything like the correct force, we need the second-order wave function.

For very large values of $R$, we must consider the effect of retardation—the electric fields propagate through space with a finite velocity. The analysis of Casimir and Polder [3, Chapter 6; 52] was used to investigate this, and showed that the interaction energy went as $R^{-7}$ in the limit of very large $R$. Calculations of the numerical value of the interaction, including the effects of retardation, were performed by several authors [104–107]. For most atoms, the distance $R$ has to be of the order of a hundred Angstroms for the retardation effects to become important.

In addition to considering the coupling between the atoms and the radiation field, one must treat the electrons relativistically [3, Chapter 6; 53; 106; 108]. In addition to a correction in the coefficient of $R^{-6}$, a term in $R^{-4}$ arises, which may not always be negligible. With $\alpha$ the fine-structure constant, the term is

$$\tfrac{1}{2}\alpha^2 R^{-4} \sum_k \sum_l \frac{f_{0k}^A f_{0l}^B}{\omega_{0k}^A + \omega_{0l}^B}$$

where atomic units are used. Power [53] reviews the work which has been done in calculating these corrections and discusses the methods used. He also considers relativistic corrections to other (nondispersive) interatomic interactions. Chang and Karplus [108, 112] derived bounds, using Padé approximants, on the relativistic long-range interaction for H, rare gas atoms, and molecules.

Feinberg and Sucher [54] have recently discussed the van der Waals interaction in a general way. They derived an interaction potential which included relativistic corrections, higher multipoles, and retardation effects. From this, the potential for the important special cases, corresponding to the formulas discussed in this Section, were extracted. Interpolation formulas for the general potential were discussed. Linder [109] has discussed reaction field techniques, in which one computes the field produced at an atom by the medium, polarized by the atom in question. The atoms of the medium may be treated individually. The result of such a treatment is a generalized potential which includes retardation effects as well as dispersive and inductive forces. The frequency-dependent susceptibilities of the atoms are required.

## REFERENCES

1. A. Dalgarno, *Advan. Chem. Phys.* **12**, 143 (1967).
2. A. Dalgarno and W. D. Davison, *Advan. At. Mol. Phys.* **2**, 1 (1966).
3. H. Margenau and N. R. Kestner, "Theory of Intermolecular Forces," pp. 68, 71. Pergamon, Oxford, 1969.
4. R. Eisenschitz and F. London, *Z. Phys.* **60**, 491 (1930).
5. J. O. Hirschfelder, C. F. Curtiss, and R. B. Bird, "Molecular Theory of Gases and Liquids." Wiley, New York, 1964.
6. B. C. Carlson and G. S. Rushbrooke, *Proc. Cambridge Phil. Soc.* **46**, 626 (1950).
7. M. E. Rose, *J. Math. Phys. (Cambridge, Mass.)* **37**, 215 (1958).
8. P. R. Fontana, *Phys. Rev.* **123**, 1871 (1961).
9. F. C. Brooks, *Phys. Rev.* **86**, 92 (1952).
10. G. M. Roe, *Phys. Rev.* **88**, 859 (1952).
11. A. Dalgarno and J. T. Lewis, *Proc. Phys. Soc. Sect. A* **69**, 57 (1956).
12. K. Fukui and T. Yamabe, *Int. J. Quantum Chem.* **2**, 359 (1968).
13. H. C. Longuet-Higgins, *Proc. Roy. Soc. Ser. A* **235**, 537 (1956).
14. H. Kreek and W. J. Meath, *J. Chem. Phys.* **50**, 2289 (1969).
15. P. Bertoncini and A. C. Wahl, *Phys. Rev. Lett.* **25**, 991 (1970); G. Das and S. Ray, *Ibid.* **24**, 1391 (1970).
16. W. Kutzelnigg and M. Gelus, *Chem. Phys. Lett.* **7**, 296 (1970).
17. H. Bethe, "Intermediate Quantum Mechanics," p. 156. Benjamin, New York, 1964.
18. G. Starkschall and R. G. Gordon, *J. Chem. Phys.* **54**, 663 (1971).
19. D. W. Davies, "Theory of the Electric and Magnetic Properties of Molecules." Wiley, New York, 1967.
20. A. Dalgarno and A. E. Kingston, *Proc. Phys. Soc. London* **78**, 607 (1961).
21. S. Tani and M. Inokuti, *J. Chem. Phys.* **54**, 2265 (1971).
22. M. J. Feinberg, *Theor. Chim. Acta* **19**, 109 (1970).
23. J. K. Knipp, *Phys. Rev.* **53**, 734 (1938).

24. P. G. Burke, J. H. Tait, and A. Dalgarno, *Chem. Phys. Lett.* **1**, 345 (1967).
25. T. Y. Chang, *Rev. Mod. Phys.* **39**, 911 (1967).
26. P. R. Fontana, *in* "Atomic Collision Processes" (M. R. C. McDowell, ed.). North-Holland Publ., Amsterdam, 1964.
27. H. Margenau, *Phys. Rev.* **38**, 747 (1931); *J. Chem. Phys.* **6**, 896 (1938).
28. F. London, *Trans. Faraday Soc.* **33**, 8 (1937).
29. H. Margenau, *Rev. Mod. Phys.* **11**, 1 (1939).
30. U. Fano and J. W. Cooper, *Rev. Mod. Phys.* **40**, 441 (1968).
31. A. Dalgarno and A. E. Kingston, *Proc. Phys. Soc. London* **73**, 455 (1959).
32. H. Margenau, *Phys. Rev.* **56**, 1000 (1939).
33. F. London, *Z. Phys.* **63**, 245 (1930).
34. H. L. Kramer, *J. Chem. Phys.* **53**, 2783 (1970).
35. J. P. Vintl, *Phys. Rev.* **41**, 813 (1932).
36. J. G. Kirkwood, *Phys. Z.* **33**, 57 (1932).
37. L. Salem, *Mol. Phys.* **3**, 441 (1960).
38. J. C. Slater and J. G. Kirkwood, *Phys. Rev.* **37**, 127, 682 (1931).
38a. K. S. Pitzer, *Advan. Chem. Phys.* **2**, 59 (1959).
39. J. N. Wilson, *J. Chem. Phys.* **43**, 2564 (1965), **49**, 3325 (1968); A. D. Crowell, *Ibid.* **49**, 3324 (1968).
40. C. Mavroyannis and M. J. Stephen, *Mol. Phys.* **5**, 629 (1962).
41. R. Heller, *J. Chem. Phys.* **9**, 154 (1941).
42. J. F. Hornig and J. O. Hirschfelder, *J. Chem. Phys.* **20**, 1812 (1952).
43. P. R. Fontana, *Phys. Rev.* **123**, 1865, 1871 (1961).
44. A. Dalgarno and A. E. Kingston, *Proc. Roy. Soc. Ser. A* **259**, 424 (1960).
45. A. Dalgarno and N. Lynn, *Proc. Phys. Soc. London Sect. A* **70**, 802 (1957).
46. J. P. Vinti, *Phys. Rev.* **41**, 432 (1932).
47. A. E. Kingston, *Phys. Rev. A* **135**, 1018 (1964).
48. J. A. Barker and P. J. Leonard, *Phys. Lett.* **13**, 127 (1964).
49. R. J. Bell and A. E. Kingston, *Proc. Phys. Soc. London* **90**, 901 (1967).
50. R. J. Bell, *Proc. Phys. Soc. London* **86**, 17, 239 (1965).
51. R. J. Bell and A. E. Kingston, *Proc. Phys. Soc. London* **88**, 901 (1966).
52. H. B. G. Casimir and D. Polder, *Phys. Rev.* **73**, 360 (1948).
53. E. A. Power, *Advan. Chem. Phys.* **12**, 167 (1967).
54. G. Feinberg and J. Sucher, Tech. Rep. No. 70-116. Center for Theor. Phys., Univ. of Maryland, College Park, Maryland, 1970.
55. K. T. Tang, *J. Chem. Phys.* **49**, 4727 (1968).
56. F. Weinhold, *J. Phys. B* **2**, 517 (1969), *J. Chem. Phys.* **50**, 4136 (1969).
57. Y. M. Chan and A. Dalgarno, *Mol. Phys.* **9**, 349 (1965).
58. P. W. Langhoff and M. Karplus, *Phys. Rev. Lett.* **19**, 1461 (1967).
59. P. W. Langhoff and M. Karplus, *J. Opt. Soc. Amer.* **59**, 863 (1969).
60. H. P. Kelly, *Phys. Lett. A* **29**, 30 (1969).
61. H. P. Kelly, *Int. J. Quantum Chem.* **3S**, 349 (1970).
62. N. C. Dutta, T. Isihara, C. Matsubura, and T. P. Das, *Int. J. Quantum Chem. Symp.* **3**, 367 (1970).
63. A. Dalgarno, I. H. Morrison, and R. M. Pengelly, *Int. J. Quantum Chem.* **1**, 161 (1967).
64. A. Dalgarno and W. D. Davison, *Mol. Phys.* **13**, 479 (1967).
65. K. T. Tang, *Phys. Rev. A* **1**, 1033 (1970).

66. R. G. Gordon, *J. Chem. Phys.* **48**, 3929 (1968).
67. D. A. McQuarrie, J. N. Terebey, and S. J. Shire, *J. Chem. Phys.* **51**, 4683 (1969).
68. O. Goscinski, Uppsala Quantum Chem. Group Rep. No. 215 (1968).
69. O. Goscinski, *Int. J. Quantum Chem.* **2**, 761 (1968).
69a. R. T. Pack, *Chem. Phys. Lett.* **5**, 257 (1970).
70. P. W. Langhoff, R. G. Gordon, and M. Karplus, *J. Chem. Phys.* **55**, 2126 (1971).
71. J. I. Musher, "Electron Correlation in Atoms. To be published, 1972.
72. H. R. Hassé, *Proc. Cambridge Phil. Soc.* **26**, 542 (1930).
73. H. R. Hassé, *Proc. Cambridge Phil. Soc.* **27**, 66 (1930).
74. L. Pauling and Y. Beach, *Phys. Rev.* **47**, 686 (1935).
75. J. O. Hirschfelder and P.-O. Löwdin, *Mol. Phys.* **2**, 229 (1959). The $R^{-8}$ coefficient is incorrect (see Chan and Dalgarno [57]).
76. R. A. Buckingham, *Proc. Roy. Soc. Ser. A* **160**, 94, 113 (1937).
77. W. D. Davison, *Proc. Phys. Soc. London* **87**, 133 (1966).
78. J. N. Murrell and G. Shaw, *J. Chem. Phys.* **49**, 4731 (1968).
79. T. R. Singh and W. J. Meath, *J. Chem. Phys.* **54**, 1137 (1971).
80. H. Hameka, "Advanced Quantum Chemistry." Addison-Wesley, Reading, Massachusetts, 1965.
81. J. I. Musher, *Ann. Phys.* (*New York*) **32**, 416 (1965).
82. M. Karplus and H. J. Kolker, *J. Chem. Phys.* **39**, 1493, 2997 (1963).
83. M. Karplus and H. J. Kolker, *J. Chem. Phys.* **41**, 3955 (1964).
84. A. Dalgarno and G. A. Victor, *Proc. Roy. Soc. Ser. A* **291**, 291 (1966).
85. S. Kaneko, *J. Chem. Phys.* **54**, 819 (1971).
86. Y. M. Chan and A. Dalgarno, *Proc. Phys. Soc. London* **85**, 277 (1965).
87. Y. M. Chan and A. Dalgarno, *Proc. Phys. Soc. London* **86**, 777 (1966).
88. A. Dalgarno and A. L. Stewart, *Proc. Roy. Soc. London Ser. A* **238**, 269 (1956).
89. J. H. Weiner and A. Askar, *J. Chem. Phys.* **54**, 1108 (1971).
90. J. O. Hirschfelder, W. Byers-Brown, and S. T. Epstein, *Advan. Quantum Chem.* **1**, 256 (1964).
91. B. Musulin and S. T. Epstein, *Phys. Rev. A* **136**, 966 (1964).
92. W. J. Deal, Jr. and N. R. Kestner, *J. Chem. Phys.* **45**, 4014 (1966).
93. G. A. Victor, A. Dalgarno, and A. J. Taylor, *J. Phys. B* **1**, 13 (1968).
94. G. M. Stacey and A. Dalgarno, *J. Chem. Phys.* **48**, 2515 (1968).
95. A. Dalgarno and G. A. Victor, *J. Chem. Phys.* **49**, 1982 (1968).
96. M. H. Alexander, *J. Chem. Phys.* **52**, 3354 (1970).
97. T. J. Venanzi and B. Kirtman, to be published.
98. B. Kirtman and R. L. Mowery, *J. Chem. Phys.* **55**, 1447 (1971).
99. C. Matsubara, N. C. Dutta, T. Isihara, and T. P. Das, *Phys. Rev. A* **1**, 561 (1970).
100. W. D. Davison, *J. Phys. B* **1**, 597 (1968).
101. G. Das and A. C. Wahl, *Phys. Rev. A* **4**, 825 (1971).
102. G. P. Arrighini, F. Biondi, and C. Guidotti, *J. Chem. Phys.* **55**, 4090 (1971).
103. J. O. Hirschfelder and M. A. Eliason, *J. Chem. Phys.* **47**, 1164 (1967).
104. R. E. Johnson, S. T. Epstein, and W. J. Meath, *J. Chem. Phys.* **47**, 1271 (1967).
105. P. M. Getzin and M. Karplus, *J. Chem. Phys.* **53**, 2100 (1970).
106. W. J. Meath and J. O. Hirschfelder, *J. Chem. Phys.* **44**, 3197 (1966).
107. W. J. Meath and J. O. Hirschfelder, *J. Chem. Phys.* **44**, 3210 (1966).
108. T. Y. Chang and M. Karplus, *J. Chem. Phys.* **52**, 4698 (1970).
109. B. Linder, *Advan. Chem. Phys.* **12**, 225 (1967).

## B. Large $R$

In Section A of this chapter, we considered the interactions of atoms at distances so large that the atoms maintained their identities. Now we must eonsider what happens when $R$ becomes small enough so that exchange of clectrons between the atoms cannot be neglected. By antisymmetrizing a product of atomic wave functions [Eqs. (6), (7), *et seq.*] we obtain a wave function for the molecule at large $R$ (separated atom limit). In Subsection 1, we will characterize the molecular state from the viewpoint of symmetry properties and relate its symmetry properties to those of the atomic states. It is easy to show that the symmetry of a state does not change as $R$ is changed. As interatomic interaction becomes important, the wave function and energy of a state change, but it is possible to follow each state as a function of $R$ down to the united atom limit, $R = 0$, according to the non-crossing rule (Chapter I, Section A.4 of Volume I). Thus we can correlate a state of the diatomic molecule with a state of the united atom and states of the separated atoms. The united atom limit is discussed in Section C. The calculation of the wave function and energy of the molecular state, starting from the separated atom states, is discussed in Subsections 2–5 of this section. In Subsection 2, attempts to use perturbation theory for this purpose are reviewed. The overlap between the wave functions of the atoms plays an important role in determining the size of the interaction: Theories that try to make this role explicit are described in Subsection 3. The valence bond theory and its applications (Subsections 4 and 5), are generalizations of some of the results of the perturbation theories of Subsection 2. Although the coupling of the electronic motion with rotational motion of the nuclei is sometimes extremely important to determining the actual energy levels, we consider the nuclei fixed along the $z$ axis with the center of mass at the origin of coordinates.

### 1. *Symmetry of Molecular States from Separated Atoms*

The problem of deriving the symmetries of molecular states from the states of the constituent atoms was treated by Wigner and Witmer [1] (with whose names the resulting rules are often associated), Hund [2], and others. Discussions of the results are found in the articles of Mulliken [3] and the books of Sponer [4] and Herzberg [5]. Knipp's calculation of the quadru-pole–quadrupole interaction between two atoms, discussed in Section A, also involves the same ideas. We will first mention the possible symmetries of the molecular states, then consider which states arise from a particular set of states for the separated atoms.

We first neglect spin. When the nuclei are different, the symmetry group is $C_{\infty v}$, including all rotations about the $z$ axis and reflections in planes including the $z$ axis (vertical reflections). The possible irreducible representations are: two one-dimensional cylindrically symmetric representations (for molecular states, denoted by $\Sigma^+$ and $\Sigma^-$), and an infinite number of two-dimensional representations, labeled by the quantum number $\Lambda$. $\Lambda$ is a positive integer; the states with $\Lambda = 1, 2, 3, \ldots$ are denoted by $\Pi, \Delta,$ $\Phi, \ldots$. $\Sigma^+$ and $\Sigma^-$ differ in that wave functions for $\Sigma^+$ states are unchanged by a vertical reflection while those for $\Sigma^-$ states change sign. $\Lambda = 0$ may be assigned to these representations. The value of $\Lambda$ is associated with the magnitude of the electronic angular momentum about the $z$ axis. A pair of functions corresponding to some nonzero value of $\Lambda$ may be chosen so that ($a$) a rotation of coordinates by $\varphi$ about the $z$ axis multiplies one by $e^{i\Lambda\varphi}$ and the other by $e^{-i\Lambda\varphi}$, and ($b$) a reflection in the $yz$ plane changes one function into the other. Letting $R_\varphi$ represent such a rotation, one has

$$R_\varphi \Psi_{\Lambda+}(\mathbf{r}_1 \cdots \mathbf{r}_n) = e^{i\Lambda\varphi}\Psi_{\Lambda+}(\mathbf{r}_1 \cdots \mathbf{r}_n) \tag{133}$$

where $\mathbf{r}_1 \cdots \mathbf{r}_n$ are the coordinates of the electrons. Since the left member of (133) is $\Psi_{\Lambda+}$ as a function of the rotated coordinates, it may be expanded in a Taylor series

$$R_\varphi \Psi_{\Lambda+}(\mathbf{r}_1 \cdots \mathbf{r}_n) = \sum_{k=0} (\varphi^k/k!)(\partial^k/\partial\varphi^k)\Psi_{\Lambda+}(\mathbf{r}_1 \cdots \mathbf{r}_n) \tag{134}$$

Comparing (134) with (133), we may conclude

$$\partial\Psi_{\Lambda+}/\partial\varphi = i\Lambda\Psi_{\Lambda+}$$

so that $\Psi_{\Lambda+}$ is an eigenfunction of $-i\hbar\partial/\partial\varphi$, the angular momentum operator, with eigenvalue $\Lambda\hbar$.

The spin of the function must now be considered. If the coupling of the spin and orbital angular momenta is small, an additional label referring to the spin multiplicity is simply added, usually as a left superscript, to the capital Greek letter denoting orbital angular momentum. This is Hund's coupling case ($a$). The magnetic field due to the orbital angular momentum, if any, is of course along the internuclear axis, and the spin will be quantized along the axis. The assumption of small spin–orbit coupling will be made for the present discussion.

If the molecule is homonuclear, an additional symmetry operation, which is usually taken as an inversion through the molecular center, is present. The symmetry group becomes $D_{\infty h}$, and the number of representations doubles. Each representation of $C_{\infty v}$ corresponds to two representations in $D_{\infty h}$, which differ in their behavior under the inversion. The label g or u, denoting gerade or ungerade (even or odd) behavior under inversion, is affixed as a right subscript to the term symbols (e.g., $\Sigma_g$) to distinguish the representations which are unchanged (g) or change sign (u) on inversion. The same notation (right subscripts g and u) is used to indicate inversion behavior (parity) of atomic states. The states $\Sigma, \Pi, \Delta, \ldots$ with g or u subscripts, $+$ or $-$ superscripts for $\Sigma$ states, and multiplicity superscripts, give all possible symmetry species for the molecular electronic wave function. The one or two degenerate states (excluding spin) of a given symmetry species are referred to as a term.

When rotation energies are large, the dependence of the wavefunction on the two angles of orientation of the internuclear axis, as well as the positions of the electrons relative to this axis, must be considered. The wavefunction must be a representation of the three-dimensional rotation–reflection group. The quantum numbers of the wave function are the total angular momentum, its component along some space-fixed axis, and the component of electronic angular momentum along the molecular axis. We shall not discuss the various Hund's cases of electronic–rotational coupling. For a summary, see Sponer [4, Sect. 5b] or Herzberg [5, Chapter V.2, VI].

We now consider formation of molecular states from the ground states of the separated atoms. For a given set of atomic states, the molecular states which may be derived, on the basis of symmetry, are probably more numerous than those actually of interest in studying the molecule, since the energies of the states are not considered. We assume small spin–orbit coupling. The spins are unaffected by the spatial symmetry operations, and are temporarily dropped from consideration.

Let atom A ($n_A$ electrons) and atom B ($n_B$ electrons) have angular momentum quantum numbers $L_A$ and $L_B$, respectively. Then there are $2L_A + 1$ degenerate eigenfunctions for atom A, which may be labeled by $\mu_A$, the component of electronic angular momentum about the internuclear axis. We write the wave functions as $\Psi_{\mu_A}^{n_A}(\mathbf{r}_1 \cdots \mathbf{r}_{n_A})$, $\mu_A = -L_A, -L_A + 1, \ldots, L_A$, where $n_A$ stands for all the quantum numbers besides $\mu_A$. Similarly, there are $2L_B + 1$ degenerate eigenfunctions for atom B, $\Psi_{\mu_B}^{n_B}(\mathbf{r}_{n_A} + 1, \ldots, \mathbf{r}_{n_A+n_B})$. From these, $(2L_A + 1)(2L_B + 1)$ product functions may be formed. These must be made antisymmetric to all interchanges of electrons.

The atomic wave functions are already antisymmetric to permutations of the electrons on the same atom. The complete antisymmetrization of the molecular wave function is performed by the operator $\sum_{\mathscr{R}} \delta_{\mathscr{R}} \mathscr{R}$, where $\mathscr{R}$ is a permutation which mixes electrons between the atoms, and $\delta_{\mathscr{R}} = \pm 1$ according to the parity of $\mathscr{R}$. The result is

$$\Psi^{n_A, n_B}_{\mu_A, \mu_B}(\mathbf{r}_1, \ldots, \mathbf{r}_{n_A+n_B}; s_1, \ldots, s_{n_A+n_B})$$

$$\equiv \sum_{\mathscr{R}} \delta_R \mathscr{R} \Psi^{n_A}_{\mu_A}(\mathbf{r}_1, \ldots, \mathbf{r}_{n_A}) \Psi^{n_B}_{\mu_B}(\mathbf{r}_{n_A+1}, \ldots, \mathbf{r}_{n_A+n_B}) \tag{135}$$

The first $n_A$ arguments in $\Psi^{n_A, n_B}_{\mu_A, \mu_B}$ refer to electrons on A in the original assignment of the product function. Since all spatial symmetry operations commute with $\sum_{\mathscr{R}} \delta_{\mathscr{R}} \mathscr{R}$, their effect on $\Psi^{n_A, n_B}_{\mu_A, \mu_B}$ can be obtained by letting them operate on the product of $\Psi^{n_A}_{\mu_A}$ and $\Psi^{n_B}_{\mu_B}$.

The operator for total angular momentum about the figure axis is

$$\Lambda = -i\hbar \partial/\partial \varphi = \sum_{j=1}^{n_A+n_B} - i\hbar \partial/\partial \varphi_j = \Lambda_A + \Lambda_B \tag{136}$$

because the rotation by $\varphi$ involves rotation of all electronic coordinates by $\varphi$. Thus

$$\Lambda \Psi^{n_A, n_B}_{\mu_A, \mu_B} = \sum_{\mathscr{R}} \delta_{\mathscr{R}} \mathscr{R} (\Lambda_A + \Lambda_B) \Psi^{n_A}_{\mu_A} \Psi^{n_B}_{\mu_B} = (\mu_A + \mu_B) \Psi^{n_A, n_B}_{\mu_A, \mu_B}. \tag{137}$$

The possible eigenvalues of $\Lambda$ extend from $-L_A - L_B$ to $L_A + L_B$. Thus the index $\Lambda$ runs from 0 to $L_A + L_B$. When the same value of $\Lambda$ can be obtained in several different ways, there are several (nondegenerate) terms of the same symmetry. It is easy to verify that, for $L_A > L_B$, there are: one term for $\Lambda = L_A + L_B$, two terms for $\Lambda = L_A + L_B - 1, \ldots, 2L_B + 1$ terms for $\Lambda = L_A - L_B$, and $2L_B + 1$ terms for all $0 \le \Lambda \le L_A - L_B$. The terms for $\Lambda = 0$ consist of one state each; those for $\Lambda > 0$ consist of two degenerate states, differing in the sign of the orbital angular momentum.

Reflection in the $yz$ plane shows that $\Psi^{n_A, n_B}_{\mu_A, \mu_B}$ and $\Psi^{n_A, n_B}_{-\mu_A, -\mu_B}$ are degenerate partners in a two-dimensional representation, except where $\mu_A + \mu_B = 0$. In this case, we have nondegenerate $\Sigma$ terms.

$$\Psi_{\pm} = \Psi^{n_A, n_B}_{\mu_A, -\mu_A} \pm \Psi^{n_A, n_B}_{-\mu_A, \mu_A}$$

are also eigenfunctions of $\Lambda$ with eigenvalue zero, and thus unchanged by a rotation about the $z$ axis. Now a reflection of coordinates in *any* plane including the $z$ axis has the same effect as a rotation in the $yz$ plane (since

we may first rotate such a plane into the $yz$ plane without affecting $\Psi_{\pm}$).
A reflection in the $yz$ plane transforms $\Psi_{\mu_A,-\mu_A}^{n_A,n_B}$ into $\Psi_{-\mu_A,\mu_A}^{n_A,n_B}$. Thus $\Psi_+$
is of $\Sigma^+$ symmetry and $\Psi_-$ is of $\Sigma^-$ symmetry. Such a pair is obtained
for $\mu_A = 1, \ldots L_B$, so $L_B \Sigma^+$ and $L_B \Sigma^-$ terms have been accounted for.
Note that the preceding argument can only be applied to $\Sigma$ terms. If
$\mu_A + \mu_B \neq 0$, the rotation of the plane of reflection would change $\Psi_{\pm}$ into

$$\exp[i\varphi(\mu_A + \mu_B)]\Psi_{\mu_A,\mu_B}^{n_A,n_B} \pm \exp[-i\varphi(\mu_A + \mu_B)]\Psi_{-\mu_A,-\mu_B}^{n_A,n_B}$$

and the reflection and subsequent rotation back to the original coordinate
system would not yield a multiple of $\Psi_+$ or $\Psi_-$.

For $\mu_A = \mu_B = 0$, $\Psi_-$ vanishes: There is always an odd number of $\Sigma$
states. The plus–minus symmetry of this last $\Sigma$ state, $\Psi_+$ or $\Psi_{0,0}^{n_A,n_B}$, must
now be determined. The effect of a vertical reflection on $\Psi_{\mu_A}^{n_A}$ is the same
as that of an inversion through nucleus A (which multiplies $\Psi_{\mu_A}^{n_A}$ by its
parity $\omega_A$) followed by a rotation by $\pi$ about an axis perpendicular to the
reflection plane (which multiplies $\Psi_{\mu_A}^{n_A}$ by $(-1)^{L_A}$). Therefore the vertical
reflection multiplies $\Psi_{0,0}^{n_A,n_B}$ by $\omega_A\omega_B(-1)^{L_A+L_B}$ and the state is $\Sigma^+$ when
this quantity is positive and $\Sigma^-$ when it is negative. Herzberg [5] gives a
table of the possible molecular electronic states resulting from various
states of the separated atoms.

If A and B are identical atoms the g–u symmetry must be established.
First, consider the case where the atoms are in states which differ in prin-
cipal quantum number or angular momentum. Then $\Psi_{\mu_A,\mu_B}^{n_A,n_B}$ and $\Psi_{\mu_B,\mu_A}^{n_B,n_A}$,
which differ by an exchange of states between the atoms, are degenerate.
We consider the two linear combinations:

$$\Psi'_{\mu_A,\mu_B} = \Psi_{\mu_A,\mu_B}^{n_A,n_B} + \Psi_{\mu_B,\mu_A}^{n_B,n_A} \tag{138a}$$

and

$$\Psi''_{\mu_A,\mu_B} = \Psi_{\mu_A,\mu_B}^{n_A,n_B} - \Psi_{\mu_B,\mu_A}^{n_B,n_A} \tag{138b}$$

which have opposite symmetry with respect to inversion of the wave function through the midpoint. In considering the effect of the inversion on

$$\Psi_{\mu_A}^{n_A}(\mathbf{r}_1, \ldots, \mathbf{r}_{n_A})\Psi_{\mu_B}^{n_B}(\mathbf{r}_{n_A+1}, \ldots, \mathbf{r}_{2n_A}) \tag{139}$$

we take the vectors $\mathbf{r}_i$ relative to the molecular midpoint. Let nucleus A
be at $-\mathbf{z}$ and nucleus B at $+\mathbf{z}$. We write $\mathbf{r}_i = -\mathbf{z} + \mathbf{r}_i'$ for the first $n_A$
electrons and $\mathbf{r}_i = \mathbf{z} + \mathbf{r}_i'$ for the remaining $n_A$ electrons. The inversion of
coordinates converts $\Psi_{\mu_A}^{n_A}$ to the same wave function, but now centered
at $+\mathbf{z}$ and with the position vectors $\mathbf{r}_i'$, relative to the atomic center, of

opposite sign. The product function (139) thus becomes, under inversion,

$$\omega_A\omega_B\Psi^{n_B}_{\mu_B}(\mathbf{r}_1, \ldots, \mathbf{r}_{n_A})\Psi^{n_A}_{\mu_A}(\mathbf{r}_{n_A+1}, \ldots, \mathbf{r}_{2n_A}),$$

where $\omega_A$ and $\omega_B$ are the parities of the atomic states. The application of the antisymmetrizer $\sum_R \delta_R \mathscr{R}$ yields $(-1)^{n_A}\omega_A\omega_B$ times $\Psi^{n_B,n_A}_{\mu_B,\mu_A}$. The factor of $(-1)^{n_A}$ is due to the $n_A$ interchanges of electrons produced by the inversion. Therefore, $\Psi'_{\mu_A,\mu_B}$ is g if $(-1)^{n_A}\omega_A\omega_B$ is positive and u if it is negative. $\Psi''_{\mu_A,\mu_B}$ has the opposite behavior. If $\mu_A + \mu_B = 0$, the functions $\Psi'_{\mu_A,\mu_B}$ and $\Psi''_{\mu_A,\mu_B}$ are to be formed from $\Psi_+$ and $\Psi_-$; thus, we consider

$$(\Psi^{n_A,n_B}_{\mu_A,-\mu_A} \pm \Psi^{n_A,n_B}_{-\mu_A,\mu_A}) \pm (\Psi^{n_B,n_A}_{-\mu_A,\mu_A} \pm \Psi^{n_B,n_A}_{\mu_A,-\mu_A}). \tag{140}$$

One of the two states formed from $\Psi_+$ is g and the other u; the same is true for the two states formed from $\Psi_-$. When $\mu_A = \mu_B = 0$, there is again a g and a u state. Note that, for a given set $(n_A, n_B)$, there are twice as many states when the atoms are identical (but not in identical states) as when they are different, but that the same states result, for identical atoms, from $(n_A, n_B)$ as from $(n_B, n_A)$.

If we have identical atoms in identical states ($n_A = n_B$), and $\mu_A \neq \mu_B$, $\Psi'_{\mu_A,\mu_B}$ and $\Psi''_{\mu_A,\mu_B}$ may still be formed. Their symmetries are as in the preceding paragraph, but only $\mu_A > \mu_B$ need be considered. The states obtained for $\mu_A < \mu_B$ duplicate those for $\mu_A > \mu_B$. If $\mu_A = \mu_B$, $\Psi''_{\mu_A,\mu_B}$ vanishes identically. The inversion behavior of the state $\Psi'_{\mu_A,\mu_B}$, since $\omega_A = \omega_B$, depends only on $n_A$: The state is g if $n_A$ is even and u if $n_A$ is odd. Only two of the four functions of (140) are nonzero: they are just $\Psi_+$ and $\Psi_-$.

We must now consider spin. The coupling of the spin and orbital motions of the electrons having been assumed small, the possible sets of spin quantum numbers do not depend on $L_A$, $L_B$, $\mu_A$, or $\mu_B$, with the exception of cases where the inversion (which interchanges atoms) enters. Every other molecular state already derived will have the same possibilities for spin quantum numbers. If $S_A$ and $S_B$ are the total spins of atoms A and B, the possible values for the total spin of the system (EYRING, Chap. 9, PILAR, Chap. 12) run from $|S_A - S_B|$ to $S_A + S_B$ in integral steps, each value occurring once. Each value of $S$ can be associated with each of the spatial symmetries derived in preceding paragraphs, if the atoms are not identical. For identical atoms in different states, each of the $\Psi'$ and $\Psi''$ [Eq. (138)] can be associated with each spin value: The situation is like that for different atoms.

The case in which the states are identical is more complicated. We refer to Wigner and Witmer [1] for a general treatment, giving only part of the calculation here. All the pairs of components of orbital angular momentum $(\mu_A, \mu_B)$ with $\mu_A \geq \mu_B$, may be considered under four headings.

(a) $\mu_A > \mu_B$ but $\mu_A + \mu_B \neq 0$.
(b) $\mu_A > \mu_B$ and $\mu_A + \mu_B = 0$.
(c) $\mu_A = \mu_B \neq 0$.
(d) $\mu_A = \mu_B = 0$.

In a situation falling under (a), we have a g and a u state with $\Lambda > 0$. Each state is a member of a $\Pi, \Delta, \ldots$ pair, with wave functions as in Eqs. (138a,b). Each spatial symmetry can occur with each value of $S$. In case (b) we form $\Psi_+$ and $\Psi_-$ which are $\Sigma^+$ and $\Sigma^-$, respectively: Each function may be associated with each value of $S$. The effect of the inversion of coordinates is to interchange $n_A$ pairs of spatial electronic coordinates between atoms. If the same pairs of electrons in the spin function were also interchanged, the subsequent antisymmetrization would multiply $\Sigma^+$ and $\Sigma^-$ by $\pm(-1)^{n_A}$. It may be shown [1] that the interchanges of electrons in the spin function introduce a factor $q = (-1)^{n_A}(-1)^{-S}$ ($S$ is always an integer since a homonuclear molecule has an even number of electrons). Thus $\Sigma^\pm$ is multiplied by $\pm(-1)^S$ on inversion, and the parities for $\Sigma^+$ and $\Sigma^-$ are $(-1)^S$ and $(-1)^{S+1}$, respectively. (c) For $\mu_A = \mu_B \neq 0$, there is only $\Psi_{\mu_A}^{n_A}\Psi_{\mu_A}^{n_A}$, a member of the term with $\Lambda = 2 \mid \mu_A \mid$. The inversion produces $n_A$ interchanges of spatial electronic coordinates. If we carry out $n_A$ interchanges in the spin function, we obtain a factor of $q$, and the subsequent antisymmetrization gives a factor of $(-1)^{n_A}$. Thus the parity of the state with spin $S$ is $(-1)^S$. Case (d) is $\Psi_0\Psi_0$, a $\Sigma^+$ term. The number of $\Sigma^+$ states is $L_A + 1$ and that of $\Sigma^-$ states, $L_A$. The parity of the last $\Sigma^+$ term, by the arguments used for case (b), is $(-1)^S$.

To summarize: For $\Lambda$ odd [states falling under (a) only] we obtain equal numbers of g and u states for each value of $S$. For $\Lambda$ even and not zero [states of types (a) and (c)], there is one more g term than u term for even $S$ and the reverse for odd $S$. The states arising under (b) and (d) are $\Sigma$ states. For even $S$, the $\Sigma^+$ are g and the $\Sigma^-$ terms u, with one more $\Sigma_g^+$ than $\Sigma_u^-$; for odd $S$ we have $\Sigma_g^-$ and $\Sigma_u^+$ terms, with one more of the latter than the former. Herzberg [5] fortunately gives a table which lists the molecular states arising from identical atomic states in most common cases.

If, for the separated atoms, the coupling between $L$ and $S$ is strong enough so that only the total angular momentum $J$ and its component $M$ are defined, $\Lambda$ and $S$ will likewise be undefined for large $R$, and $\Omega$, the component of total electronic angular momentum about the figure axis, must be considered [6]. This is Hund's coupling case $(c)$. It is also appropriate when the coupling between $L$ and $S$ is large compared to the interatomic interaction energy. This always is the case for large enough $R$. In the case of $L$–$S$ coupling just discussed, $\Omega$ is the absolute value of the sum of $\pm\Lambda$ and $\Sigma$. (Here, $\Sigma$ is the component of $S$ along the figure axis, and takes on the values $-S, -S + 1, \ldots, S$.) For unlike atoms, the combination of $J_A$ and $J_B$ yields for $\Omega$ (analogous to the combination of $L_A$ and $L_B$ to give $\pm\Lambda$) $J_A + J_B$, $J_A + J_B - 1$ (twice), etc. The lowest value of $\Omega$ is $\frac{1}{2}$ if $J_A + J_B$ is half-integral and 0 if $J_A + J_B$ is integral. When $J_A + J_B$ is integral, the states for $\Omega = 0$, which arise when $M_A = -M_B$, must be considered. As for the analogous coupling of $L_A$ and $L_B$, there are $0^+$ and $0^-$ states, except where $M = 0$, in which case there is only one state. Its behavior on vertical reflection is even $(0^+)$ when $\omega_A\omega_B(-1)^{J_A+J_B}$ is $+1$, otherwise the state is $0^-$. For like atoms in unlike states, we obtain the same results as for unlike atoms, except that there is a g and a u state in each case and we need consider only $M_A = M_B$. The case of like atoms in like states follows the corresponding one discussed by Wigner and Witmer [1].

As $R$ decreases, the coupling is expected to become Hund's case $(a)$ or $(b)$, i.e., $\Lambda$ and $S$ become good quantum numbers. The states derived from $J$–$J$ coupling now must be correlated with those discussed by Wigner and Witmer [1]. Mulliken [6] gave some results for this difficult problem. Nikitin [7] and Umanskij and Nikitin [8] have also discussed the connections between the states generated for the two cases. They derived formulas for the matrix elements of the electronic energy between such states, which permit, in principle, calculation of the energies of states of the molecule for large $R$. Some special cases were worked out. Chang [9] also computed Coulombic and dispersion interactions between atoms by forming molecular states of the proper symmetry. The effects of spin–orbit coupling were considered, as well as other magnetic effects.

Having established the symmetries of the molecular states which can be derived from a given set of atomic states, we would like to predict $(a)$ the order of energies of the molecular states and $(b)$ the behavior of the energies of these states as the internuclear distance is changed. For this purpose it is sometimes useful to continue the adiabatic change of $R$ from infinity (separated atoms) through the molecular region, if any, to zero (united atom) [1, 2]. In this imaginary process, the Coulombic repulsion of the nuclei, which is a constant with respect to electronic coordinates, is not considered, so that there is no opposition to the coalescence of the nuclei.

A more detailed discussion of the united atom is given in Section C. The cylindrical symmetry group of the molecule is a subgroup of the spherical symmetry group of the united atom. Since $\Lambda$ is the component of orbital angular momentum along the internuclear axis, a united atom term with orbital angular momentum $L$ yields molecular terms with $\Lambda = L, L - 1,$ $\ldots, 0$. The plus–minus symmetry of the $\Sigma$ term follows from consideration of the effect of a vertical reflection on the united atom wave function. A vertical reflection is equivalent to an inversion followed by a rotation by $\pi$. Thus the $\Sigma$ term is $\Sigma^+$ when $L$ is even and the atomic term is of g parity, or when $L$ is odd and the atomic term is of u parity. It is $\Sigma^-$ otherwise. The spin quantum number $S$ is unchanged by separation of the nuclei. If the separation of the united atom nucleus is into two equivalent nuclei, the g or u behavior of a molecular state is the same as that of the united atom state from which it derives. In correlating molecular states with atomic states for $R \to \infty$ or $R \to 0$, one invokes the noncrossing rule (see Section A.4 of Chapter I, Volume I) for states of the same symmetry.

With knowledge of the energies of the separated atom and united atom states which correlate with a given molecular state, one can, in some simple cases, make qualitative predictions about the behavior of $U(R)$. Thus, for H + H the two $^2S_g$ atoms produce $^1\Sigma_g^+$ and $^3\Sigma_u^+$ states. The completely symmetrical $^1\Sigma_g^+$ correlates with the completely symmetrical ground state of He, the $(1s)^2\,{}^2S$. Since the electronic energy drops considerably in going from separated atoms to united atom, the ground state would be predicted to be binding. The lowest triplet state of He which has odd parity is the $(1s)(2p)\,{}^3P$, whose energy is 20 eV above the $(1s)^2\,{}^1S$. Thus the electronic energy rises sharply as $R$ goes from infinity to zero, so the $^3\Sigma_u^+$ state is indicated to be repulsive. As shown by Herzberg [10], this reasoning does not work for the ground state of $N_2$, for example, which would be predicted to be repulsive. Here, the behavior of the electronic energy as a function of $R$ is not monotonic due to an avoided crossing of states, where the state of lower energy on one side of the avoided crossing point $R^*$, takes on certain characteristics of the higher state on the other (see Volume I, Chapter I, Section A.4). It is possible, when $R$ changes sufficiently rapidly (breakdown of Born–Oppenheimer approximation), for the molecule in its ground state to dissociate to excited atoms, even when ground state atoms can yield a state of the same symmetry species as the molecular ground state. This occurs when the electron density of the ground state resembles that of the excited atoms, and the molecule does not have the time to rearrange its electron density as $R$ passes through $R^*$. Then, for a qualitative understanding of the shape of $U(R)$ on one side of $R^*$, potential curves obtained by allowing crossings may be more useful. According to the Heitler–London (valence bond) theory, a molecular state is stabilized when new electron pairs (bonds) are formed: The more the better. Then we expect, in general, that states of lowest spin multiplicity have lowest energy. But the theory predicts nothing about the relative stabilities of states of the same multiplicity. The Hund–Mulliken (molecular orbital) theory makes such predictions on the basis of knowledge of the one-electron orbitals of the molecule and their relative energies.

Mulliken early worked out a number of problems on this basis. These molecular orbitals are supposed to be eigenfunctions of some one-electron Hamiltonian, and it is a good approximation (cf. Chapter III of Volume I and Chapter II of this volume) to take this Hamiltonian as having the full symmetry of the molecular frame. Then the symmetry classification of the molecular orbitals is like that for molecular states, but lowercase Greek letters are used to denote the different species. In the nature of things, there can be no $\sigma^-$ behavior for a one-electron function, and the $+$ superscript is suppressed. Of course, the spin label is also unnecessary.

Correlation of the molecular orbital with that of the united atom is easy: The united atom orbital with angular momentum $l$ yields molecular orbitals with $\lambda = -l, \ldots, l$. In considering correlations with atomic orbitals of the separated atoms, it must be noted that the molecular orbital usually becomes a linear combination of atomic orbitals on the two atoms. Atomic orbitals can combine in this way only if they have the same component of angular momentum about the figure axis. Their mixing will be favored if their energies are comparable, and if they overlap strongly. These criteria do not necessarily provide a one-to-one matching of atomic orbitals on A and B. When several orbitals on an atom may enter the same molecular orbital, we may combine them into hybrid atomic orbitals. In the event that no orbital on B is qualified to combine with some orbital on A, the orbital on A remains essentially monocentric, and is referred to as nonbonding.

When two atomic orbitals do combine, two molecular orbitals must result. Of the two molecular orbitals, one will be bonding and have lower energy, because it tends to put electronic charge in the internuclear region. For a homonuclear diatomic, each pair of atomic orbitals yields a g and a u molecular orbital, with the g (u) being bonding for even (odd) orbital angular momenta. The other molecular orbital will be antibonding. This is because the inversion of coordinates is equivalent to a reflection in the midplane followed by a rotation by $\pi$ about the figure axis and the rotation gives a sign change for $\pi, \varphi, \ldots$ orbitals. The bonding or antibonding character depends on the symmetry or antisymmetry to reflection in the midplane. Mulliken [3] suggested that the bonding or antibonding character of molecular orbitals could be predicted by comparing principal quantum numbers in the united atom and separated atom limits, but Herzberg [10] showed that this reasoning did not always lead to the correct conclusions.

Herzberg [10] suggested that the correlation of separated atom states with molecular states be made by assuming (1) unchanged molecular quantum numbers, (2) conservation of the numbers of $\sigma$, $\pi$, ... electrons. To some extent, the energy ordering of the one-electron orbitals may be obtained without direct calculation, by correlating the molecular orbitals with atomic orbitals, as $R \to \infty$. It appeared [10] that correlations of the molecular eigenstate and the molecular orbitals with the united atom limits were not of much importance in prediction of molecular stability, because of curve crossings and avoided crossings for $R$ between $R_e$ and zero. On the other hand, the united atom is often a better guide to correlation energy than the separated atoms, since the pair structure is usually unchanged from $R = 0$ to $R = R_e$ but not from $R = R_e$ to $R \to \infty$. (See Volume I, Chapter III, Section D.3 and Chapter II, Section A.2 of this volume.) Herzberg was able to give theoretical justification for the correlations of the molecular with separated atom states (dissociation) which are found experimentally, distinguishing between attractive and repulsive states.

## 2. Perturbation Theories

To go beyond the qualitative treatment, we require a means of calculating corrections to nonoverlapping atoms. In Section A, we sketched a perturbation treatment, due to Eisenschitz and London [11], in which the nonoverlapping atoms constitute the zero-order wave function. The first-, second-, etc. order functions were expanded in the functions

$$\Psi_i = \mathscr{A}\{\Psi_{Ai'}\Psi_{Bi''}\}. \tag{141a}$$

$\Psi_{Ai'}$ is an atomic wave function for atom A, $\Psi_{Bi''}$ one for atom B, and

$$\mathscr{A} = \sum_R \delta_R \mathscr{R}. \tag{141b}$$

Since $H$ commutes with $\mathscr{A}$,

$$H\Psi_i = \sum_R \delta_R \mathscr{R}\{(H_A + H_B + V)\Psi_{Ai'}\Psi_{Bi''}\}$$

with $H_A$ and $H_B$ being the Hamiltonians for atoms A and B, and $V$ including all other terms in $H$. Since $H_A \Psi_{Ai'} = E_{Ai'} \Psi_{Ai'}$, and $H_B \Psi_{Bi''} = E_{Bi''} \Psi_{Bi''}$,

$$H\Psi_i = (E_{Ai'} + E_{Bi''})\Psi_i + \sum_R \delta_R \mathscr{R}\{V\Psi_{Ai'}\Psi_{Bi''}\}$$

The last term is small and was used to define the orders of perturbation theory. The molecular energy was written as a sum of terms of various orders,

$$E = E^{(0)} + E^{(1)} + E^{(2)} + \cdots \tag{142}$$

The energy $E^{(0)}$ was $E_{A0} + E_{B0}$, a sum of atomic energies. Where the initial atomic states were nondegenerate,

$$E^{(1)} = \langle \Psi_{A0}\Psi_{B0} | V | \Psi_0 \rangle / \langle \Psi_{A0}\Psi_{B0} | V | \Psi_0 \rangle \tag{143}$$

$$E^{(2)} = \sum_j^{(j \neq 0)} \frac{\langle \Psi_0 | V - E^{(1)} | \Psi_{Aj'}\Psi_{Bj''} \rangle \langle \Psi_j | V - E^{(1)} | \Psi_{A0}\Psi_{B0} \rangle}{(E_{A0} + E_{B0} - E_{Aj'} - E_{Bj''}) \langle \Psi_0 | \Psi_{A0}\Psi_{B0} \rangle} \tag{144}$$

Here, $\Psi_0$ was constructed from the atomic ground states via Eqs. (141a,b). At very large internuclear distance, only the terms corresponding to the identity permutation in $\mathscr{A}$ survive in $E^{(1)}$ and $E^{(2)}$. (Then, as we have discussed, $E^{(1)}$ gives the interactions between the permanent moments of the two atoms, while $E^{(2)}$ yields the dispersion and induction forces—$\Psi_j$ for $j \neq 0$ may include $\Psi_{A0}\Psi_{Bj''}$.) At smaller internuclear distances, terms corresponding to other permutations are important. The term

$$\langle \Psi_{A0}\Psi_{B0} | V | \mathscr{P}(\Psi_{A0}\Psi_{B0}) \rangle,$$

in $E^{(1)}$ is referred to as a single, double, etc. exchange integral according as $\mathscr{P}$ interchanges one, two, etc. pairs of electrons between the atoms. The term for $\mathscr{P} = I$ is called the Coulomb integral. The single exchange integral can be much larger than the Coulomb integral for a diatomic molecule when the internuclear distance is a few Bohr radii. This appears in the valence bond calculation of $H_2$, for example (see Volume I, Chapter II, Section B), where the exchange integrals determine the bonding or anti-bonding behavior. Integrals like $\langle \Psi_{A0}\Psi_{B0} | V - E^{(1)} | \mathscr{P}(\Psi_{Aj'}\Psi_{Bj''}) \rangle$ contribute to a term in the energy which may be referred to as second-order exchange energy. What we are discussing in this and the following subsections are perturbation theories for the exchange energy.

The Eisenschitz–London formalism has rarely been used for computations on many-electron molecules. The valence bond theory of Heitler and London, which grows out of this formalism, has also been more useful for qualitative predictions than for numerical calculation. One of the problems in using the perturbation approach involves the linear dependence which results when the molecular wave functions of a desired symmetry are constructed from the atomic wave functions by Eqs. (144a,b). Another is the difficulty in defining a perturbation Hamiltonian so that corrections to a desired accuracy may be derived in a systematic way. Whereas the valence bond theory (see Subsection 4) gives a scheme for construction of "bond" structures which may be used to describe a molecule, calculation of matrix elements of the Hamiltonian over such structures becomes complicated. The complications have, so far, prevented general use of the theory for computation, although progress has been made in adapting the formulas for numerical work [12].

Recently, there has been considerable interest in reformulating or modifying the Eisenschitz–London formalism, hopefully in such a way as to permit calculations. A number of interesting schemes have been developed, but none has yet demonstrated its ready applicability for widespread use. We will discuss several here. Hirschfelder has studied and compared such schemes in several articles [13, 14].

Amos and Musher [15, 16] have suggested that the linear dependence problem be avoided by writing the wave function as

$$\Psi = \Psi_0 + \sum_{m \neq 0} a_m u_m \qquad (145)$$

Here, $\Psi_0$, the antisymmetric projection of $\Psi_{A0}\Psi_{B0}$, serves as zero-order wave function, but corrections to $\Psi_0$ are expressed in terms of the products of atomic wave functions, $u_m = \Psi_{Am'}\Psi_{Bm''}$. The use of the $\{u_m\}$ rather than their antisymmetric projections leads to simplifications, since the $\{u_m\}$ form an orthonormal and not overcomplete set. Replacing $u_0$ by $\Psi_0$ still does not give overcompleteness. The Hamiltonian was written [15, 16]:

$$H = H_A + H_B + V \qquad (146)$$

where $V$ acts as the perturbation. Since $H$ always operates directly on the $\{u_m\}$, its division into unperturbed and perturbing parts is unique. The coefficients $a_m$ were written in a series as in perturbation theory:

$$a_m = a_m^{(1)} + a_m^{(2)} + \cdots$$

and the energy in a series like (142). The zero-order equation is

$$\mathscr{A}\{E_0 \ u_0\} = E^{(0)}\Psi_0$$

so $E^{(0)} = E_{A0} + E_{B0}$. The first-order equation is

$$\mathscr{A}\{Vu_0\} + \sum_{m \neq 0} a_m^{(1)}(E_{Am'} + E_{Bm''})u_m = E^{(0)} \sum_{m \neq 0} a_m^{(1)}u_m + E^{(1)}\Psi_0$$

Multiplication by $u_0{}^*$ and integration gives directly the expression (143) for $E^{(1)}$. There is no ambiguity in the expansion coefficients. Multiplication by $u_n{}^*$ and integration gives directly

$$a_n^{(1)} = \langle \Psi_n \mid E^{(1)} - V \mid u_0\rangle/(E_{An'} + E_{Bn''} - E_{A0} - E_{B0}) \qquad (147)$$

Using this in the second-order equation leads to

$$E^{(2)} = \sum_{m \neq 0} \frac{\langle u_0 \mid V \mid u_m\rangle\langle \Psi_m \mid E^{(1)} - V \mid u_0\rangle}{\langle u_0 \mid \Psi_0\rangle(E_{Am'} + E_{Bm''} - E_0)}$$

It was shown [15, 16] that, to second-order, this is identical to Eq. (144), i.e., we can make the replacement by

$$\sum_{m \neq 0} \langle u_0 \mid V - E^{(1)} \mid a_m^{(1)}u_m\rangle \rightarrow \sum_{m \neq 0} \langle u_0 \mid V - E^{(1)} \mid a_m^{(1)}\Psi_m\rangle$$

with the neglect only of quantities of higher than second order.

Amos and Musher [15, 16] compared their formalism with that of Eisenschitz and London [11] and that of van der Avoird [17, 18], which we discuss later. Since use of an overcomplete set, such as the functions of (141a,b), is avoided, the determination of coefficients by a secular equation is possible. Another convenience of the Amos–Musher method is the possibility of displaying the perturbation V explicitly. The Eisenschitz–London treatment and some of its modifications write $H\Psi_m$ as $\mathscr{A}\{H_0 u_m + Vu_m\}$ and consider $Vu_m$ as small; the antisymmetrization makes it impossible simply to write $H$ as $H_0 + V$, and the division of the Hamiltonian depends on the function on which it operates.

Jansen [19] attempted to show how $V$ could be formally defined independently of a particular assignment of electrons to atoms in the function on which $V$ operates. (Jansen also gave a review and discussion of previous work, including some references we have not mentioned.) First- and second-order energies from this formalism were computed for H–H and compared with results of other treatments. Jansen's formalism is closely related to that of Corinaldesi (see page 78).

Claverie [20] has recently discussed the problems connected with the use of the Rayleigh–Schrödinger perturbation theory for interatomic forces, starting with a zero-order function not totally antisymmetric. Here, the unperturbed problem would be a sum of atomic Hamiltonians, as in Eq.

(146). While one cannot prove the convergence of the perturbation series, Claverie noted that there was no reason to believe that it did not converge or that the convergence problem could not be circumvented. Then by perturbation theory one might obtain the correct energy and a wave function from which the exact wave function could be derived. A detailed discussion of how the perturbed function (approximating an eigenfunction of $H$) relates to the unperturbed function (an eigenfunction of $H_0$) was given for the two-electron problem. One sees from the discussion that the connection is far from trivial. Starting with the ground state of $H_0$, one can be led to a physically inadmissible state of $H$. While there are states of $H_0$ from which perturbation theory can lead to desired eigenstates of $H$, we would not know how to choose them in any given situation. Furthermore, high orders of perturbation theory would be needed to get accurate $U(R)$, except possibly at large $R$.

Corinaldesi [21] avoided the problem connected with antisymmetrization (i.e., that the antisymmetrized product of atomic eigenfunctions is not an eigenfunction of the zero-order Hamiltonian corresponding to any particular assignment of electrons to the two atoms), by introducing a "modified Schrödinger equation." The eigenfunctions of the modified equation are column vectors of $(n_A + n_B)!$ elements, each element being a function of the space and spin coordinates of the $N = n_A + n_B$ electrons, corresponding to a particular permutation of the electronic coordinates, but not antisymmetric. The Hamiltonian of the modified equation, $H_M$, is correspondingly a $N! \times N!$ matrix operator. It was shown [21] that to each eigenfunction of the modified equation there corresponded a properly antisymmetric solution to the ordinary Schrödinger equation with the same eigenvalue, and that $H_M$ could be partitioned into $H^{(0)} + H^{(1)}$ such that the eigenfunction of $H^{(0)}$ corresponded to a product of atomic eigenfunctions.

Let the column vector $\Psi$ have, as its $\alpha$th element,

$$\Psi_\alpha = (N!)^{-1/2} P_\alpha \varphi(1, \ldots, N)$$

where $P_\alpha$ is a permutation of the $N$ electron labels and $\varphi$ is a function of the $N$ sets of space and spin coordinates. The modified Schrödinger equation is

$$H_M \Psi = (H^{(0)} + H^{(1)}) \Psi = E \Psi \qquad (148)$$

where $H^{(0)}$ and $H^{(1)}$ are matrix operators representing the unperturbed and perturbing Hamiltonians.

$$H^{(0)}_{\alpha\beta} = -\tfrac{1}{2} \sum_i \nabla_i^2 \, \delta_{\alpha\beta} + (N!)^{-1} P_\alpha U_0 P_\beta^{-1}$$

with

$$U_0 = -Z_A \sum_{i=1}^{n_A} r_{Ai}^{-1} - Z_B \sum_{i=n_A+1}^{n_A+n_B} r_{Bi}^{-1} + \sum_{i=1}^{n_A} \sum_{j=1}^{n_A} {}^{(i \neq j)} r_{ij}^{-1} + \sum_{i=n_A+1}^{n_A+n_B} \sum_{j=n_A+1}^{n_A+n_B} {}^{(i \neq j)} r_{ij}^{-1}$$

(potential energy terms in $H_A$ and $H_B$). It is easy to show that, if $\Psi_{Ai'}$ is an eigenfunction of $H_A$ with eigenvalue $E_{Ai'}$ and $\Psi_{Bi''}$ is an eigenfunction of $H_B$ with eigenvalue $E_{Bi''}$, the column vector

$$\Psi_\alpha^{(0)} = (N!)^{-1/2} P_\alpha \{\Psi_{Ai'} \Psi_{Bi''}\}$$

is an eigenfunction of $H^{(0)}$ with eigenvalue $E_{Ai'} + E_{Bi''}$, with $\langle \Psi^{(0)} \mid \Psi^{(0)} \rangle = 1$. (Note that $\langle \Psi^{(0)} \mid \Psi^{(0)} \rangle$ includes a sum over elements as well as an integration over coordinates.) Corinaldesi defined $H^{(1)}$ by

$$H_{\alpha\beta}^{(1)} = (N!)^{-1} \sum_\gamma \Gamma_{\alpha\gamma} P_\gamma V P_\beta^{-1}$$

where, as previously in this chapter,

$$V = -Z_B \sum_{i=1}^{n_A} r_{Bi}^{-1} - Z_A \sum_{i=n_A+1}^{n_A+n_B} r_{Ai}^{-1} + \sum_{i=1}^{n_A} \sum_{j=n_A+1}^{n_A+n_B} r_{ij}^{-1} + Z_A Z_B R^{-1},$$

and $\boldsymbol{\Gamma}$ is a square matrix of dimension $N!$ which satisfies

$$\sum_\beta (-1)^\beta (\Gamma_{\beta\gamma} - \delta_{\beta\gamma}) = 0.$$

Suppose that we have $\Psi$ which is an eigenfunction of $H_M$ with eigenvalue $E$,

$$H_M \Psi = E\Psi$$

The function

$$\tilde{\Psi} = \sum_\alpha (-1)^\alpha \Psi_\alpha = \mathscr{A}\varphi(1 \cdots N).$$

clearly is antisymmetric, and it is now shown that it is an eigenfunction of the ordinary Schrödinger equation with eigenvalue $E$ and Hamiltonian $H = H_A + H_B + V$. Multiplying (148) by the row vector $\Phi$ with elements $\Phi_\alpha = (-1)^\alpha$ we obtain

$$\sum_{\alpha,\beta} (-1)^\alpha (H_{\alpha\beta}^{(0)} + H_{\alpha\beta}^{(1)}) \Psi_\beta = E \sum_\alpha (-1)^\alpha \Psi_\alpha$$

so that it remains to be shown that the left side is $H \sum_\alpha (-1)^\alpha \Psi_\alpha$. The left side is, in fact,

$$\sum_{\alpha,\beta} (-1)^{\alpha}\left(-\tfrac{1}{2}\sum_{i=1}^{N} \nabla_i^2 \delta_{\alpha\beta} + (N!)^{-1}P_{\alpha}U_0 P_{\beta}^{-1}\right.$$

$$\left. + (N!)^{-1}\sum_{\gamma} \Gamma_{\alpha\gamma}P_{\gamma}VP_{\beta}^{-1}\right)(N!)^{-1/2}P_{\beta}\varphi$$

$$= \sum_{\alpha} (-1)^{\alpha}\left(-\tfrac{1}{2}\sum \nabla_i^2\right)\Psi_{\alpha} + \sum_{\alpha} (-1)^{\alpha}P_{\alpha}U_0\Psi_1 + \sum_{\alpha,\gamma} (-1)^{\alpha}\Gamma_{\alpha\gamma}P_{\alpha}V\Psi_1$$

where $\Psi_1$, the first element of $\Psi$, corresponds to the identity permutation. By virtue of the properties of $\Gamma$,

$$\sum_{\alpha} (-1)^{\alpha}\Gamma_{\alpha\gamma} = \sum_{\alpha} (-1)^{\alpha}\,\delta_{\alpha\gamma}\,.$$

Then, remembering that $U_0 + V$ commutes with all permutations, we have

$$\sum_{\alpha,\beta} (-1)^{\alpha}(H_{\alpha\beta}^{(0)} + H_{\alpha\beta}^{(1)})\Psi_{\beta} = \sum_{\alpha} (-1)^{\alpha}H\Psi_{\alpha}$$

as was to be shown.

Symmetry in addition to that demanded by the Pauli principle, such as the g–u symmetry for a homonuclear molecule, can be handled in a similar manner. The dimension of the modified Schrödinger equation is increased correspondingly (it becomes $(2N)!$ for the homonuclear case), but can be reduced in some cases by the choice of zero-order function.

Some interesting properties of this formalism are involved with the arbitrariness allowed in $\Gamma$, since the perturbation Hamiltonian $H^{(1)}$ contains parameters whose values may be chosen arbitrarily. Their choice does not affect the eigenvalue $E$, but does affect the individual energies $E^{(i)}$ for $i > 1$ [21, 22]. Indeed, for $H_2$, where $\Gamma$ (a $2\times2$ matrix) contains one arbitrary parameter $\alpha$, the first-order energy is given by

$$E^{(1)} = J + \alpha(K - J)$$

where $K$ and $J$ are the Coulomb and exchange integrals of the valence bond theory. It is clear that any desired result can be obtained by the choice of $\alpha$. With $\alpha = 1.07$, $E^{(0)} + E^{(1)}$, as a function of $R$, has a minimum at $R = 1.406$ $a_0$ (experimental $R_e = 1.402\ a_0$), and at the same time predicts $D_e = 4.74$ eV, the experimental value, and gives a good value for the quadrupole moment [23]. The Heitler–London calculation gives $R_e = 1.66\ a_0$ and $D_e = 3.17$ eV. Higher-order energies were not calculated. We clearly need a method for choosing $\alpha$, and several were suggested by Corinaldesi [27].

A problem is that $H^{(1)}$ is not Hermitian except when $\alpha = 0$, which may make computation of higher-order energies complicated. When $\alpha = 0$, $E^{(0)} + E^{(1)}$ is in very poor accord with experiment. The only other molecule that has been treated by this formalism is $He_2^+$, which has three electrons [24]. Here, $E^{(1)}$ involved a single arbitrary parameter,

$$E^{(1)} = J + 2\alpha(J - K)$$

The correct value of $D_e$ was obtained with $\alpha = 0.962$, but the predicted $R_e$ was 0.2 $a_0$ higher than the correct value. However, the He atom functions used were only rough approximations, and it was hoped that use of better approximations would improve the results.

A. van der Avoird [17] introduced projection operators into the Eisenschitz–London treatment. The expansion of an antisymmetric function (see Section A.1)

$$\Psi = \sum_i g^{-2}\Psi_i\langle\Psi_i \,|\, \Psi\rangle$$

means that $g^{-2}\sum_i \Psi_i\langle\Psi_i|$ acts as the identity operator (is a resolution of the identity) for antisymmetric functions, despite the fact that the $\{\Psi_i\}$ are overcomplete. In our notation, $\langle\Psi_i|$ is the operator which multiplies by $\Psi_i^*$ and integrates over the coordinates of $\Psi_i$. In Section A, $\Psi_i = \mathscr{A}u_i$, $\mathscr{A}$ being the antisymmetrizer. It is easy to generalize to a projection operator which gives other symmetry properties (behavior on rotation, reflection, or inversion of the coordinate system), besides antisymmetry. For example, $\frac{1}{2}(I + R)\phi$, where $R$ is a reflection and $I$ the identity, is a function which is even on reflection (or zero) regardless of the nature of $\phi$. If $\phi$ is an antisymmetrized product, $R\phi$ is an antisymmetrized product of the one-electron functions generated from those of $\phi$ by the reflection. Thus,

$$\mathscr{B}_R = \frac{1}{2}(I + R)$$

is a projection operator which generates the even representation of the group containing $I$ and $R$. The factor of $\frac{1}{2}$ makes it idempotent: $\mathscr{B}_R^2 = \mathscr{B}_R$. The operator $\mathscr{B}_R$ can also be seen to be Hermitian: $\langle\phi \,|\, \mathscr{B}_R\psi\rangle = \langle\mathscr{B}_R\psi \,|\, \phi\rangle$. Construction of such projection operators is done by standard formulas of group theory [18]. Functions of a desired spin may also be constructed using such projection operators, since the spin is intimately related to the behavior of the wave function with respect to permutation of spatial coordinates only (see Chapter II). Eisenschitz and London's formalism [11], which treats representations of the permutation group other than the totally antisymmetric one, is designed to treat states of different spin.

Starting from a complete orthonormal set of functions $u_i$ (products of atomic wave functions), van der Avoird [17] used an idempotent and Hermitian projection operator $\mathscr{B}$ to project out functions corresponding to some desired representation of the molecular symmetry, spin, etc., groups. If $G$ is any function of the symmetry under consideration, $\mathscr{B}G = G$. Using the completeness of the $\{u_i\}$ and the properties of $\mathscr{B}$ we may write the expansion of $G$ in the $\{u_i\}$

$$G = \sum_i u_i \langle u_i \mid G \rangle = \sum_i u_i \langle u_i \mid \mathscr{B}G \rangle = \sum_i u_i \langle \mathscr{B}u_i \mid G \rangle.$$

Operating on both sides with $\mathscr{B}$,

$$G = \sum_i \mathscr{B}u_i \langle \mathscr{B}u_i \mid G \rangle \tag{149}$$

Thus $\sum_i (\mathscr{B}u_i)\langle \mathscr{B}u_i \mid$ is a resolution of the identity in the space of functions of the desired symmetry. The possible overcompleteness of the $\{\mathscr{B}u_i\}$ is no problem if we use expansion (149). This prescription for expansion of $G$ is that given by Eisenschitz and London except for a difference in normalization. In (149), $\mathscr{B}$ is idempotent so that $\langle \mathscr{B}u_i \mid G \rangle = \langle u_i \mid G \rangle$, while a factor of $f/g^2$ entered in Section A, where $f$ is the dimension of the desired representation and $g$ is the number of operations in the full group.

The formalism of van der Avoird now follows that of Eisenschitz and London. The operator $\mathscr{B}$ commutes with the Hamiltonian $H = H_A + H_B + V$, but not with its parts. The exact eigenfunction $\Psi$ is expanded in the $\{\mathscr{B}u_i\}$. Then $H\Psi = E\Psi$ is written, expanding expansion coefficients and energy in orders of perturbation theory:

$$\sum_i (c_i^{(0)} + c_i^{(1)} + \cdots)\mathscr{B}(H_A + H_B + V)u_i$$
$$= (E^{(0)} + E^{(1)} + \cdots) \sum_i (c_i^{(0)} + c_i^{(1)} + \cdots)\mathscr{B}u_i$$

Again, $(H_A + H_B)u_i = (E_{Ai'} + E_{Bi''})u_i = E_i u_i$, while $Vu_i$ is a first-order term. Separating terms of each order, we obtain

$$\sum_i c_i^{(0)} E_i \mathscr{B}u_i = E^{(0)} \sum_i c_i^{(0)} \mathscr{B}u_i \tag{150a}$$

$$\sum_i c_i^{(0)}(\mathscr{B}Vu_i - E^{(1)}\mathscr{B}u_i) = \sum_i c_i^{(1)}(E^{(0)} - E_i)\mathscr{B}u_i \tag{150b}$$

$$\sum_i c_i^{(0)}(-E^{(2)})\mathscr{B}u_i + \sum c_i^{(1)}(\mathscr{B}Vu_i - E^{(1)}\mathscr{B}u_i) = \sum c_i^{(2)}(E^{(0)} - E_i)\mathscr{B}u_i \tag{150c}$$

In the absence of degeneracy, we satisfy (150a) by taking $E^{(0)} = E_0$ and $c_i^{(0)} = \delta_{i0}$ (0 referring to the ground state). These equations are used in equations (150b) etc. Then in (150b) the left side is expanded in the $\{\mathscr{B}u_i\}$ according to (149), and coefficients of each $\mathscr{B}u_i$ on each side are set equal. This yields $E^{(1)}$ (from $\mathscr{B}u_0$) and $c_i^{(1)}$ (from $\mathscr{B}u_i$ with $i \neq 0$). The left side of (150c) is likewise expanded in the $\{\mathscr{B}u_i\}$, and equating coefficients of each $\mathscr{B}u_i$ yields $E^{(2)}$ and $c_i^{(2)}$.

The transparency of the projection operator formalism makes extensions of the theory more tractable, and van der Avoird [18] developed a partitioning or wave operator formalism (see Volume I, Chapter III, Section B) for this problem. The wave operator $W$ was defined by

$$W\Psi_0 = \Psi \tag{151}$$

where $\Psi_0 = \mathscr{B}\{\Psi_{A0}\Psi_{B0}\} = \mathscr{B}u_0$ and $\Psi = \mathscr{B}\Psi$ is the desired eigenfunction. A perturbation-like expression was obtained for $W$, and this implied expansions of the energy and wave function. The operator $W$ commutes with $\mathscr{B}$, and the normalization condition

$$\langle \Psi_0 \mid \Psi \rangle = \langle \Psi_0 \mid W\Psi \rangle = \langle \Psi_0 \mid \Psi \rangle \tag{152}$$

was imposed. The projection operator $\mathscr{P}$ was defined [18] by

$$\mathscr{P} = \mathscr{B} - \langle \Psi_0 \mid \Psi_0 \rangle^{-1}\Psi_0\langle \Psi_0 \mid. \tag{153}$$

$\mathscr{P}$ is idempotent:

$$\mathscr{P}^2 = \mathscr{B} - \frac{2\Psi_0\langle \Psi_0 \mid}{\langle \Psi_0 \mid \Psi_0 \rangle} + \frac{\Psi_0\langle \Psi_0 \mid \Psi_0 \rangle\langle \Psi_0 \mid}{\langle \Psi_0 \mid \Psi_0 \rangle^2} = \mathscr{P}$$

and Hermitian. It commutes with $\mathscr{B}$ ($\mathscr{B}\mathscr{P} = \mathscr{P}\mathscr{B} = \mathscr{P}$) while its effect on $\Psi_0$ or $u_0$ is to annihilate it. If $\mathscr{Q} = \mathscr{B} - \mathscr{P}$, $\mathscr{Q}\Psi = \Psi_0$ because of the condition (152). The operators $\mathscr{P}$ and $\mathscr{Q}$ correspond to the partitioning of the subspace of $\mathscr{B}$ into unperturbed function and corrections.

The eigenvalue equation $(E - H)\Psi = 0$ may be written, since $\Psi = \mathscr{B}\Psi$ and $\mathscr{B} = \mathscr{P} + \mathscr{Q}$;

$$(E - H)\mathscr{P}\Psi = (H - E)\mathscr{Q}\Psi = (H - E)\Psi_0 \tag{154}$$

A formal solution to this equation is [18]

$$\mathscr{P}\Psi = TH\Psi_0$$

where

$$T = \mathscr{P}[\alpha(1 - \mathscr{B}) + \beta\mathscr{Q} + \mathscr{P}(E - H)\mathscr{P}]^{-1}\mathscr{P} \qquad (155)$$

as we will show. Here, $\alpha$ and $\beta$ are nonzero scalars whose presence is necessary if $T$ is to be well-defined for all functions $\psi$, including functions annihilated by $\mathscr{P}$. Differentiating $T$ with respect to $\alpha$ gives a factor $\mathscr{P}(1 - \mathscr{B})$ which vanishes, whereas differentiating with respect to $\beta$ gives $\mathscr{P}\mathscr{Q}$, which also vanishes. Thus $T$ is formally independent of $\alpha$ and $\beta$. To verify the solution to (154), start from the obvious equation

$$\mathscr{P}[\alpha(1 - \mathscr{B}) + \beta\mathscr{Q} + \mathscr{P}(E - H)\mathscr{P}]$$
$$\times [\alpha(1 - \mathscr{B}) + \beta\mathscr{Q} + \mathscr{P}(E - H)\mathscr{P}]^{-1}\mathscr{P}H\Psi_0 = \mathscr{P}H\Psi_0$$

Since $\mathscr{P}(1 - \mathscr{B})$, $\mathscr{P}\mathscr{Q}$, and $\mathscr{P}\Psi_0$ vanish, the equation is:

$$\mathscr{P}(E - H)TH\Psi_0 = \mathscr{P}(H - E)\Psi_0 \qquad (156)$$

Equation (154) will hold, with $\mathscr{P}\Psi = TH\Psi_0$, provided that

$$\mathscr{Q}(E - H)TH\Psi_0 = \mathscr{Q}(H - E)\Psi_0.$$

This becomes a condition on $E$. Since $\mathscr{Q} = (|\Psi_0\rangle\langle\Psi_0|)\langle\Psi_0|\Psi_0\rangle^{-1}$, rearranging gives

$$E = \frac{\langle\Psi_0| H + HTH |\Psi_0\rangle}{\langle\Psi_0|\Psi_0\rangle}. \qquad (157)$$

The total wave function is

$$\Psi = \Psi_0 + TH\Psi_0. \qquad (158)$$

Thus

$$W = \mathscr{B} + TH. \qquad (159)$$

To develop $\Psi$ and $E$ in a perturbation series, van der Avoird noted that $H$ always operates directly on $\Psi_0 = \mathscr{B}u_0$, in (157) and (158). Commuting $H$ with $\mathscr{B}$, he put, as was done in other theories,

$$H\Psi_0 = \mathscr{B}Hu_0 = \mathscr{B}(H_0 + V)\Psi_0 = E_0\Psi_0 + \mathscr{B}Vu_0 \qquad (160)$$

$\mathscr{B}V_0$ was considered as small. Using the vanishing of $T\Psi_0$, Eq. (158) became

$$\Psi = \Psi_0 + TVu_0 \qquad (161)$$

and Eq. (156)

$$E = E_0 + \frac{\langle \mathscr{B}u_0 \mid \mathscr{B}Vu_0 \rangle}{\langle \mathscr{B}u_0 \mid \mathscr{B}u_0 \rangle} + \frac{\langle \mathscr{B}Vu_0 \mid T \mid \mathscr{B}Vu_0 \rangle}{\langle \mathscr{B}u_0 \mid \mathscr{B}u_0 \rangle}. \tag{162}$$

The last term is at least second order in $Vu_0$, so that the first two terms give the zeroth and first order energies, the latter being the generalization of (143). The higher-order energies were obtained from an expansion of $T$. A. van der Avoird defined $V'$ by

$$E - H = E_0 - H_0 - V' \tag{163}$$

and showed that

$$(E_0 - H_0)T = \mathscr{P} + UT \tag{164}$$

where

$$U = V' - \langle \Psi_0 \mid \Psi_0 \rangle^{-1} \mid \Psi_0 \rangle \langle u_0 \mid V'\mathscr{B} \tag{165}$$

The operator

$$R_0 = \sum_{k>0} u_k \langle u_k \mid (E_0 - E_k)^{-1}$$

where $H_0 u_k = E_k u_k$ and $\Psi_k = \mathscr{B}u_k$, has the property that $\mathscr{P}R_0(H_0 - E_0)\mathscr{P} = \mathscr{P}$. Multiplying (164) on the left by $\mathscr{P}R_0$ yields

$$T = \mathscr{P}R_0\mathscr{P} + \mathscr{P}R_0UT. \tag{166}$$

Thus the zeroth approximation to $T$ is $\mathscr{P}R_0\mathscr{P}$ and higher approximations are obtained by iterating (166) to obtain

$$T = \sum_{n=0}^{\infty} (\mathscr{P}R_0U)^n\mathscr{P}R_0\mathscr{P}.$$

Use of this formula gives wave functions and energies to any order. For instance, the second order energy is, from Eq. (162),

$$E^{(2)} = \frac{\langle \mathscr{B}Vu_0 \mid \mathscr{P}R_0\mathscr{P} \mid \mathscr{B}Vu_0 \rangle}{\langle \Psi_0 \mid \Psi_0 \rangle} = \sum_{k>0} \frac{\mid \langle \mathscr{B}Vu_0 \mid \mathscr{B}u_k \rangle \mid^2}{\langle \Psi_0 \mid \Psi_0 \rangle(E_0 - E_k)} \tag{167}$$

A variational principle equivalent to this expression can be derived [18].
van der Avoird [18] was able to show that all energy expressions are real and that, for the ground state, all even-order energies are negative. It is evident from the form of $T$ that the wave functions to all orders have the proper symmetry (which is not so in a closely related formalism [25]).

The theory was applied [18] to the hydrogen molecular ion, with $u_0$ a 1s
function centered on either proton and $\mathscr{B} = \frac{1}{2}(I \pm R)$, where $R$ is a reflection through the midplane. The upper and lower signs are for the $\sigma_g$
and $\sigma_u$ states. Except for small internuclear distances, the energy through
$E^{(2)}$ was very satisfactory for both states.

The extension to the case where $E_0$ is degenerate, even after a projection
operator is used to select states of a given symmetry type, was given by
Micha [26]. The $g$ different degenerate states were denoted by $u_{0i}$, $i = 1$,
..., g, and are eigenfunctions of $(H_A + H_B)$:

$$(H_A + H_B)u_{0i} = H_0 u_{0i} = (E_{A0} + E_{B0})u_{0i}, \qquad i = 1, \ldots, g \qquad (168)$$

After projection with $\mathscr{B}$, the $g$ states may no longer be linearly independent,
but since the set of functions is finite it is not difficult to select a linearly
independent set. A method for doing this was given by Micha [26]. Let
$\{\Psi_{0p}\}$ be a set of $h$ functions ($h \leq g$) formed from the $\{\mathscr{B}u_{0i}\}$ by such a
method. The $\Psi_{0p}$ are symmetry-adapted and linearly independent (in fact,
orthonormal). Micha used a projection operator $\mathscr{Q}$ formed from the $\Psi_{0p}$.
Instead of (153),

$$\mathscr{P} = \mathscr{B} - \mathscr{Q} = \mathscr{B} - \sum_{j=1}^{h} \frac{\Psi_{0j}\langle\Psi_{0j}|}{\langle\Psi_{0j}|\Psi_{0j}\rangle}. \qquad (169)$$

The zero-order functions could be expected to be a linear combination of
the $\Psi_{0i}$ ($i = 1 \cdots h$) (any $\mathscr{B}u_{0j}$ can be written as a linear combination of
the $\Psi_{0i}$), so that $\mathscr{P}\Psi_0 = 0$:

$$\Psi_0 = \sum_{p=1}^{h} c_p \Psi_{0p} \qquad (170)$$

The wave operator $W$ which satisfies (151) could be written [26] as in
the nondegenerate case, using the $\mathscr{P}$ and $\mathscr{Q}$ of Eq. (169). Then the wave
function and energy are given by (157) and (158). Substituting (170) into
the eigenvalue equation, multiplying by $\Psi_{0q}^*$, and integrating,

$$\sum_{p} \langle\Psi_{0q}|(H-E)W|\Psi_{0p}\rangle c_p = 0 \qquad (171)$$

This is a kind of secular equation, giving different eigenvalues, say $\varepsilon_m$,
and corresponding sets of eigenvectors $c_p{}^m$. The corresponding eigenfunctions of $H$ would be

$$\sum_{p} c_p{}^m (\mathscr{Q} + TH)\Psi_{0p}.$$

Micha [26] showed it was possible to extend van der Avoird's formalism to obtain perturbative expansions of $T$, the energy $E$, and the coefficients. Variational principles were given which are equivalent to the differential equations corresponding to the various orders of perturbation theory. We will not give the details here.

Perhaps it should be noted that operator formalisms of this kind are deceptively simple. Manipulations may be carried through with ease and may, in fact, lead to new results. But implementing the formulas by numerical calculation may be difficult. From the formal definitions of operators, it is not always easy to produce them explicitly (say as integrodifferential operators) or in a form suitable for computation.

Hirschfelder and Silbey [27] developed a perturbation formalism related to van der Avoird's. Starting from an unsymmetrized function $F^{(0)}$ (which could be a product of atomic wavefunctions or simply a product of one-electron functions) they produced zero-order functions of various symmetries, using the group theoretical projection operators (EYRING, Chap. 10, PILAR, Sect. 14.7)

$$\mathscr{B}_{kj}^{(\alpha)} = (h_\alpha/g) \sum_R D^{(\alpha)}(R)_{jk}^* R. \tag{172}$$

Here, $D^{(\alpha)}(R)_{kj}$ is the $kj$th matrix element of the matrix corresponding to the operator $R$ in the $\alpha$th irreducible representation, $h_\alpha$ is the dimension of the representation, and $g$ the number of operations in the group. Operation of $\mathscr{B}_{kj}^{(\alpha)}$ on any function produces a function which transforms as the $j$th partner in a basis for the $\alpha$th representation. Hirschfelder and Silbey supposed that the exact eigenfunction of $H$ of symmetry $\alpha$ could be written as the projection of some $F_1$. Taking $k = j$, they demanded that $\psi_{kk}^{(\alpha)} = \mathscr{B}_{kk}^{(\alpha)} F_1$ satisfy

$$H\psi_{kk}(\alpha) = E_k(\alpha)\psi_{kk}(\alpha). \tag{173}$$

To derive the equation satisfied by $F_1$, they used

$$F_1 = \sum_{\alpha,k} \psi_{kk}(\alpha) \tag{174}$$

and the fact that any operator $S$ in the group commutes with $H'$ so that

$$HSF_1 = SHF_1 = \sum_{\alpha,k} SE_k(\alpha)\psi_{kk}(\alpha) \tag{175}$$

Since

$$S\psi_{kk}(\alpha) = (h_\alpha/g) \sum_R D^{(\alpha)}(R)_{kk}^* SRF_1$$

$$= (h_\alpha/g) \sum_T D^{(\alpha)}(S^{-1}T)TF_1 \tag{176}$$

$H(SF_1)$ is a linear combination of the functions $TF_1$. Writing the linear combination coefficients in terms of the $E_k(\alpha)$, Hirschfelder and Silbey [27] derived a set of coupled equations for the $TF_1$.

The connection with van der Avoird's formalism comes in because the equations are treated by a perturbation scheme. Each $TF_1$ is expanded as $F_{1T}^{(0)} + F_{1T}^{(1)} + \cdots$ (zero-order, first-order, etc., functions) and $E_k(\alpha) = \varepsilon_{k\alpha}^{(0)}$ $+\varepsilon_{k\alpha}^{(1)} + \cdots$ . Hirschfelder and Silbey defined a Hamiltonian, different for each $T$, by $H^{(T)}F_{1T} = \varepsilon_k^{(0)}(\alpha)F_{1T}^{(0)}$, and derived coupled equations for the higher-order corrections. For $H_2{}^+$, the lowest $\sigma_g$ and $\sigma_u$ states were considered simultaneously as projections of the hydrogen atom functions. The gerade energy through third order was relatively poor.

Hirschfelder [13] gave a new derivation of the Hirschfelder–Silbey formalism [27] and a modification of van der Avoird's treatment which related it to that of Hirschfelder and Silbey. Hirschfelder showed that the energy expressions agreed with those of van der Avoird only through second order, which suggests that the grouping by orders is not an invariant process in these formalisms. Indeed, there seems to be no unique perturbation theory for exchange forces. Hirschfelder [13] also considered Brillouin–Wigner perturbation theories for this problem and found slightly different second-order energy expressions. The nonuniqueness is connected with the fact (see Section A) that the set of functions of equation 9a is overcomplete. A generalized treatment which points this up has been given [28].

Since the Brillouin–Wigner theories work with matrix elements only, separation of $H$ into $H_0$ and $V$ is avoided. Orders of perturbation theory are defined by the smallness of certain matrix elements. Such formalisms are used in the work of Murrell *et al.* [29], Musher and Salem [30], and Dalgarno and Lynn [31] (which we will discuss later), as well as that of other authors (see Hirschfelder [13] and Certain *et al.* [14] for bibliography). The energy to second order in the Brillouin–Wigner theory is given by [Volume I, Chapter III, Eq. (172)]:

$$E = \frac{H_{00}}{S_{00}} - \sum_{i \neq 0} \frac{|H_{i0} - ES_{i0}|^2}{S_{00}(H_{ii} - ES_{ii})}. \tag{177}$$

Here, the eigenfunction is written as

$$\Psi = \Psi_0 + \sum_{i \neq 0} c_i \Psi_i \tag{178}$$

where the $\{\Psi_i\}$ are a set of functions which need not be eigenfunctions of any unperturbed operator, and $\Psi_0$ is an approximation to $\Psi$. In (177), $S_{ij} = \langle \Psi_i \mid \Psi_j \rangle$ and $H_{ij} = \langle \Psi_i \mid H \mid \Psi_j \rangle$.

In the present problem, $\Psi_0$ would be the wave function in the absence of interaction; $\Psi_0$ and the $\{\Psi_i\}$ are of the proper symmetry: $\Psi_0 = \mathscr{B}u_0$ and $\Psi_i = \mathscr{B}u_i$. The first term in (177) is just the sum of the zeroth-order and first-order energies of the Schrödinger theories.

$$\frac{H_{00}}{S_{00}} = \frac{\langle \Psi_0 \mid \mathscr{B}(H_A + H_B + V) \mid u_0 \rangle}{\langle \Psi_0 \mid u_0 \rangle}$$

$$= E^{(0)} + \frac{\langle \Psi_0 \mid \mathscr{B}V \mid u_0 \rangle}{\langle \Psi_0 \mid u_0 \rangle} = E^{(0)} + E^{(1)} \tag{179}$$

As shown in Chapter III, Section B of Volume I, the second term in (177) reduces to the Schrödinger formula when the $\{\Psi_i\}$ are eigenfunctions of $H_0$. The $\{\Psi_i\}$ will not be eigenfunctions in the exchange perturbation theories.

Musher and Salem [30] used the Brillouin–Wigner perturbation theory to show how the Eisenschitz–London result could be obtained without invoking the separation of the Hamiltonian into $H_0 + V$. The matrix elements which appear in the secular equation may be written as

$$H_{jk} - ES_{jk} = \langle \Psi_j \mid H - E \mid \Psi_k \rangle$$

$$= (E_k - E)\langle \Psi_j \mid \Psi_k \rangle + \langle \Psi_j \mid \mathscr{B}Vu_k \rangle \tag{180}$$

or as

$$\langle \Psi_j \mid H - E \mid \Psi_k \rangle = (E_j - E)\langle \Psi_j \Psi_k \rangle + \langle \mathscr{B}Vu_j \mid \Psi_k \rangle. \tag{181}$$

While the matrix of $H - E$ is Hermitian, the two parts clearly are not. This points up the difficulty in defining $V$. Using Eq. (181) in (177), and neglecting certain terms, one can derive Eq. (144) from the summation. Higher order, and hopefully more accurate, approximations to the energy, were discussed [30], and the formulas compared to those derived by other workers. Use of the formulas where Hartree–Fock functions are employed to describe the separate systems was briefly discussed. It was concluded that the analysis of the energy into overlap, Coulombic, and exchange contributions could not be maintained in an accurate calculation.

While the problems of partitioning the Hamiltonian into $H_0 + V$ are avoided, the linear dependence of the $\{\Psi_i\}$ still leads to problems [15, 16]. In obtaining the first approximation to the expansion coefficients [see Volume I, Chapter III, Eq. (171)],

$$c_i^{(1)} = -\langle \Psi_i \mid H - E \mid \Psi_0 \rangle / \langle \Psi_i \mid H - E \mid \Psi_i \rangle \tag{182}$$

we neglect

$$\sum_{m \neq 0, i} c_m \langle \Psi_i \mid H - E \mid \Psi_m \rangle$$

as being of the second order of smallness. The matrix element here is $(E_i - E_0)\langle \Psi_i \mid \Psi_m \rangle$ plus first-order terms. Now $\sum_{m \neq 0} \langle \Psi_i \mid \Psi_m \rangle$ is not at all small when the $\{\Psi_j\}$ are linearly dependent, since one $\Psi_i$ may be expressed in terms of the others. Amos and Musher [15, 16] suggested that one use a basis set consisting of $\Psi_0$ and the product functions $u_i$ ($i \neq 0$) [see Eq. (145)]. They were able to rederive the results [Eqs. (147), et seq.] by the Brillouin–Wigner formalism. At the same time, application of the theory to the H–H problem showed that the first-order wave function was actually a mixture of the singlet and triplet wave functions, and the second-order wave function was necessary to get meaningful results.

A new "partitioning perturbation theory" for the electron exchange problem was presented by Certain and Hirschfelder [32]. It builds on previous work by Kirtman [33] on the use of the van Vleck perturbation theory for degenerate states. The van Vleck theory differs from the Rayleigh–Schrödinger in that the latter diagonalizes $H$ over the degenerate zero-order states before computing energies of second and higher order [see Volume I, Chapter III, Eqs. (153)–(154)], while the van Vleck theory treats the coupling between different energy levels before diagonalizing. This maintains the identity of the zero-order states and therefore allows one to use unsymmetrized zero-order functions and zero-order Hamiltonian, and antisymmetrize after computing corrections. There is a clear relation to the formalisms of Corinaldesi [21] and Hirschfelder and Silbey [27]. In the application of the theory to $H_2^+$ and $H_2$, the computations necessary and the results resembled those from the other methods [14].

An interesting semiempirical application of the formula (177) for interatomic interaction was made by Dalgarno and Lynn [31], who used an average energy denominator $\varepsilon$ in the summation of (177) [see Volume I, Chapter III, Eq. (160)] to give

$$E = H_{00} + \varepsilon^{-1}[\langle \Psi_0 \mid H^2 \mid \Psi_0 \rangle - 2E\langle \Psi_0 \mid H \mid \Psi_0 \rangle + E^2] \qquad (183)$$

Here $E$ may be approximated as $H_{00}$ on the right side. For a calculation of the interaction of two He atoms, the He ground state was approximated by a simple one-parameter function, with the parameter chosen by variation on the atomic problem. The wave function $\Psi_0$ was produced by antisymmetrization of a product of such functions. The value of $\varepsilon$ was chosen [31] so the asymptotic form of the energy (the $R^{-6}$ van der Waals term) agreed with that from accurate calculations. Then (183) gave an estimate of the interaction energy, including the effect of exchange. Dalgarno and Lynn also discussed other empirical ways of choosing $\varepsilon$.

Hirschfelder [13], in comparing the different formalisms for the treatment of exchange by perturbation theory, concluded that there was no real way to choose between them on formal grounds alone. Rather, numerical results would decide the superiority of one over another. Such tests were carried out by Certain *et al.* [14] for the case of $H_2$. The energies for the lowest $^1\Sigma_g^+$ and $^3\Sigma_u^+$ states of $H_2$ (the states which correlate with ground state H atoms) were computed through second order by four formalisms. These were: Hirschfelder's modification of van der Avoird's theory, the Hirschfelder–Silbey theory, the Musher–Amos theory (and a closely related one due to Murrell and Shaw [34]), and a Rayleigh–Schrödinger perturbation treatment. In the Rayleigh–Schrödinger treatment, a zero-order Hamiltonian was constructed such that the symmetrized product of hydrogen atom functions was an eigenfunction. The first three formalisms differ only in the expressions for the second-order energy. Calculations were made on $H_2$ for $R = 4, 6$, and $8\ a_0$.

It was found that the van der Avoird formalism did not give a good ground state energy for $R = 8\ a_0$, yielding only about half the correct dispersion energy. The other schemes were of comparable accuracy. For $R = 4\ a_0$ (and presumably smaller $R$) the Rayleigh–Schrödinger scheme worked best. None of the methods tested led to more efficient computations than a straightforward variational calculation.

In all cases a quantity corresponding to the first-order wave function was calculated by variation–perturbation theory. The bulk of the computational effort went into calculating matrix elements. Having these, one could vary the entire wave function to minimize $\langle H \rangle$ rather than to approximate the first-order wave function. The former procedure would necessarily have given better results.

3. *Expansions in the Overlap*

If indeed the perturbation theories discussed in the previous subsection are not particularly advantageous for the computation of the total energy of molecular systems, their main use may be as sources for ideas which may be incorporated in other calculations. An example is the idea of the valence bond, discussed in Subsection 4. The energy expressions of perturbation theory may be analyzed into contributions of different kinds (although several authors [13, 30] have warned that labels such as "Coulomb energy" or "exchange energy" may have no meaning for some of the higher-energy expressions). The idea of the overlap between atoms determining the strength of a bond is made precise in the formalisms of this section.

Murrell *et al.* [29] have given a formalism in which the role of the interatomic overlap is made explicit. In addition to classifying contributions to the energy by orders in $V$, they considered orders in overlap, $S$. A matrix element like $\langle \Psi_{Ai'} \Psi_{Bi''} \mid \Psi_{Ai'} \Psi_{Bi''} \rangle$, which equals 1, is zero order. (Note that $\langle \Psi_{Ai'} \Psi_{Bi''} \mid \Psi_{Aj'} \Psi_{Bj''} \rangle$ vanishes by our choice of basis set.) If $\mathscr{I}$ is an interchange of two electrons between atoms, $\langle \Psi_{Ai'} \Psi_{Bi''} \mid \mathscr{I} \Psi_{Ai'} \Psi_{Bi''} \rangle$ was taken to be of order $S^2$. This is because, if atomic wave functions were determinants of one-electron orbitals, this matrix element would reduce to a product of two one-electron overlap integrals. Each of these would be equal to the overlap between the atomic orbitals (one on each atom) between which the electrons are exchanged. The order in $S$ of $\langle \Psi_{Ai'} \Psi_{Bi''} \mid \mathscr{R} \Psi_{Ai'} \Psi_{Bi''} \rangle$ when $\mathscr{R}$ is more than a single exchange can be similarly considered, but Murrell *et al.* [29] considered terms only through second order in $S$. The permutations which merely rearrange electrons on each atom give rise to zero-order terms, since the atomic functions are separately antisymmetric in their electrons.

The calculation outlined by Murrell *et al.* [29] differs in several respects from the Eisenschitz–London calculation. First, only a finite number of states $\Psi_i$ are used, so there is no linear dependence problem. Second, the atomic states are augmented by a number of ionic or "charge transfer" states

$$\Psi_i^+ = \sum_R \delta_R \mathscr{R} \{ \Psi_{A+i'} \Psi_{B-i''} \} \tag{184a}$$

$$\Psi_i^- = \sum_R \delta_R \mathscr{R} \{ \Psi_{A-i'} \Psi_{B+i''} \} \tag{184b}$$

$\Psi_{A+i'}$ is a state of the ion $A^+$, and a function of $n_A - 1$ electrons; $\Psi_{B-i''}$ is a function of $n_B + 1$ electrons and corresponds to the ion $B^-$; and so on. The overlap $\langle \Psi_{Ai'} \Psi_{Bi''} \mid \Psi_{A+j'} \Psi_{B-j''} \rangle$ is at least of order $S$, since the atomic and ionic functions differ at least by the transfer of one electron from an orbital on A to one on B. If all other electrons were unchanged (an idealized situation) this integral would reduce to the overlap between the orbitals involved in the transfer. The exact function was written

$$\Psi = \Psi_0 + \sum_i c_i \Psi_i \tag{185}$$

where the $\Psi_i$ include the $\Psi_i^+$ and $\Psi_i^-$. The terms may be classified as Heitler-London ($\Psi_0$), induction (one-atom excitations), charge transfer, and van der Waals [29]. There is a relation to the Brillouin–Wigner theories, and also to the atoms-in-molecule methods (see Chapter III). Murrell et al. [29] determined the coefficients $c_i$ by the variation method. The energy was written as

$$E = \sum_{i,j} E_{ij} \tag{186}$$

where the first subscript refers to order in $V$ and the second to order in $S$. They considered terms for $i$ and $j$ equal to 0, 1, and 2. The $E_{ij}$ were derived by considering the matrix elements entering the secular equation.

Rather than repeat their manipulations, we will attempt to show the origins of the terms in $E^{(1)}$ (the $E_{1j}$) and $E^{(2)}$ (the $E_{2j}$). For $E^{(1)}$ we use the Brillouin–Wigner expression [Eq. (179)]:

$$E^{(1)} = \frac{\langle \mathscr{B}\{\Psi_{A0}\Psi_{B0}\} \mid V \mid \Psi_{A0}\Psi_{B0} \rangle}{\langle \mathscr{B}\{\Psi_{A0}\Psi_{B0}\} \mid \Psi_{A0}\Psi_{B0} \rangle} \tag{187}$$

and for $E^{(2)}$ we put $E \sim E_{A0} + E_{B0}$ in the summation of (177) to get:

$$E^{(2)} = \sum_{i \neq 0} \frac{\mid \langle \Psi_{A0}\Psi_{B0} \mid V \mid \mathscr{B}\{\Psi_{Ai'}\Psi_{Bi''}\} \rangle \mid^2}{S_{00} S_{ii} (E_{A0} + E_{B0} - E_{Ai'} - E_{Bi''})} \tag{188}$$

Here, $\mathscr{B}$ includes interchanges of electrons between atoms A and B. We will assume the atomic or ionic wave functions are normalized, so the leading term in $S_{ii}$ is 1. There are no $E_{0j}$ for $j > 0$. The terms zero order in the overlaps arise from the identity permutations in $E^{(1)}$ and $E^{(2)}$.

$$E_{10} = \langle \Psi_{A0}\Psi_{B0} \mid V \mid \Psi_{A0}\Psi_{B0} \rangle / \langle \Psi_{A0}\Psi_{B0} \mid \Psi_{A0}\Psi_{B0} \rangle$$

$$E_{20} = \sum_{j}^{(j \neq 0)} \frac{\mid \langle \Psi_{A0}\Psi_{B0} \mid V \mid \Psi_{Aj'}\Psi_{Bj''} \rangle \mid^2}{E_{A0} + E_{B0} - E_{Aj'} - E_{Aj''}}$$

Charge transfer states, which would give $E_{22}$ terms, do not contribute to the second expression. Thus $E_{10} + E_{20}$ are just the electrostatic, inductive, and dispersive forces we calculate in the no-overlap limit. Evidently, $E^{(1)}$ contributes no terms first order in the overlap. Neither does $E^{(2)}$, because of the squares of matrix elements in the numerator. All remaining terms are then second order in overlap.

The $E_{12}$ terms are obtained from the single exchange terms in numerator and denominator of (187). Here, $E_{12}$ is what we normally call the exchange energy, as in valence bond theory. The terms in $V^2 S^2$ ($E_{22}$) can come from charge transfer states in the second-order energy sum as well as from $\Psi_i$ corresponding to neutral atoms. The contribution of the neutral atom states to $E_{22}$ arises because the normalization integrals $S_{ii}$ have terms zero and second order in $S$, as does each matrix element in the numerator.

If $\Psi_{A0}$ and $\Psi_{B0}$ are single determinants, Murrell et al. [29] wrote $E_{12}$ as a sum of integrals involving electron densities from the occupied orbitals and transition densities $\phi_A\phi_B$, where $\phi_A$ is an orbital on A and $\phi_B$ one on B. The possibility of approximating or leaving out some of the integrals was investigated with inconclusive results. It was argued that the contribution of neutral states to $E_{22}$ would be small compared to $E_{10}$ and $E_{12}$, so such terms could be neglected in calculations of interatomic forces. The charge transfer terms in $E_{22}$ were reduced to contributions of one-electron functions, assuming the atomic and ionic wave functions constructed from determinants. It is expected that these terms will be important when one atom has a low ionization potential and the other a high electron affinity, so the energy denominator can be small.

Salem [35] gave related analysis of the terms in $E^{(1)}$ and $E^{(2)}$ for the case where higher-order exchanges can be neglected in all integrals, and where the exact wave function is expanded in antisymmetrized products of atomic wave functions (no charge transfer states). The energy $E^{(1)}$ consists of the electrostatic interaction $E_{10}$ and the exchange energy $E_{12}$. The energy $E^{(2)}$ consists of the induction and dispersion terms, $E_{20}$, plus terms involving exchange of electrons between atoms. Each of the matrix elements appearing in (188) was written as a no-exchange part (zero order in $S$) and single exchange terms (of order $S^2$) so that there are terms in $S^2$ and $S^4$ in addition to $E_{20}$. Terms in $S^2$, arising from products of matrix elements, one with and one without exchange, were referred to as the second order Coulomb-exchange energy. Terms in $S^4$ were called pure second-order exchange energy. Salem argued that the Coulomb-exchange terms give a positive (repulsive) contribution. A more extensive analysis was given using the assumption that the $\{\Psi_i\}$ are orthogonal.

It was subsequently pointed out [25] that the classification of terms according to order in overlap is not rigorous, since no perturbation expansion in powers of $S$ is possible. A new formalism, closely related to that of van der Avoird [18], was introduced by Murrell and Shaw [25] to derive the same energy expressions. They also analyzed the first- and second-order energies into the contributions of Coulomb, exchange, dispersion, etc. terms, when multiple exchange terms were neglected. The first-order energy was written as the Coulombic energy $E_{10}$ plus the first-order exchange energy $E_{12}$, as in (187). Similarly, the second-order energy was divided into induction and dispersion energies, second-order Coulomb-exchange energies (as defined by Salem), and second-order exchange energies. Explicit expressions were given. This formalism was one of those tested by Certain *et al.* [14] for $H_2$. Tests on $H_2$ and $He_2$ by Murrell and Shaw [25] showed that the effect of overlap and exchange on the dispersion energy was small for $R$ near the van der Waals minimum, so that it would not be necessary to antisymmetrize a wave function to exchanges of electrons between atoms. (The error in some terms in the series in $R^{-1}$ may be substantial, but the energy is dominated by the dipole–dipole term.)

Margenau [36], in some early work, had investigated the validity of simply adding the van der Waals attraction, calculated by second order perturbation theory on a wave function not antisymmetric to electron interchange between atoms, to the exchange repulsion, calculated from the effect of antisymmetrizing the function. This is also essentially what was done in recent calculations by Umanskij and Nikitin [8] (following paragraph). While simply superposing the two calculated effects was quite incorrect for H–H near the van der Waals minimum, it seemed that the correction (second-order exchange effect) was small for He–He. Murrell and Shaw [25] derived a simple variational wave function which gave good results for the dispersion interaction (van der Waals coefficient); it involved one doubly excited configuration built from Slater orbitals of variable exponent. Representing it as $\mathscr{B}\{\tilde{\Psi}_{A1}\tilde{\Psi}_{B1}\}$, they computed

$$| \langle (V - E^{(1)})u_0 \mid \tilde{\Psi}_1 \rangle |^2 \langle \Psi_0 \mid \Psi_0 \rangle^{-1}(E_0 - \tilde{E}_1)^{-1}$$

where $\tilde{E}_1 = \langle \tilde{\Psi}_{A1} \mid H_A \mid \tilde{\Psi}_{A1} \rangle + \langle \tilde{\Psi}_{B1} \mid H_B \mid \tilde{\Psi}_{B1} \rangle$. It differed from $| \langle u_0 \mid V \mid \tilde{\Psi}_{A1}\tilde{\Psi}_{B1} \rangle |^2/(E_0 - E_k)$ by less than 3% for distances greater than the van der Waals minimum, the difference representing the second-order exchange and Coulomb-exchange contributions.

Relatively recently, Umanskij and Nikitin [8] gave formulas for matrix elements of the electronic energy between molecular states constructed from atomic states by the procedures of Subsection 1. They neglected multiple exchange terms, and overlaps in normalizing factors. It was then possible to define an effective Hamiltonian which consisted, aside from Coulombic and spin–orbit terms involving the individual atoms, of operators for Coulombic, dispersion and exchange interactions. The expectation value of this operator was taken over a wave function which is not antisymmetric to interatomic exchanges. The Coulombic term corresponded to interactions between permanent moments on the two atoms, the dispersion operator gave the interactions of the van der Waals type, and the exchange operator was written:

$$H_{ex} = -n_A n_B H_{AB}(\tfrac{1}{2} + 2\mathbf{S}_{1A} \cdot \mathbf{S}_{1B})P^x_{1A,1B} \tag{189}$$

Here, atoms A and B have $n_A$ and $n_B$ electrons, $\mathbf{S}_{1A}$ is the spin operator (a vector with $x$, $y$, and $z$ components) for electron 1 on A, and $P^x_{1A,1B}$ interchanges the spatial coordinates of two electrons, one on each atom. $H_{AB}$ is the interaction operator V of our previous discussions. The factor $(\tfrac{1}{2} + 2\mathbf{S}_{1A} \cdot \mathbf{S}_{1B})$ gives 1 or 0 according to whether or not the electrons have parallel spins. Except for the dispersion operator, the Hamiltonian corresponds to first-order perturbation theory, so that the effect of distortions of the wave function on the exchange energy was not considered.

### 4. Valence Bond Theory

As we have mentioned, the valence bond theory, whose ideas (if not method of calculation) have proved useful in many computations, derives from the treatment of exchange forces given by Eisenschitz and London [11], incorporating the notion of bond formation by electron pairing associated with the names of Heitler and London.

In our discussion of the bond in $H_2$ (Volume I, Chapter II, Section B), we saw that a wave function describing the bond required two electrons of opposite spin, one on each atom, to start with. The molecular wave function has no net spin. We can say that bond is described by the symmetric spatial wave function

$$\Psi_r = \phi_A(1)\phi_B(2) + \phi_B(1)\phi_A(2) \tag{190}$$

(where $\phi_A$ and $\phi_B$ are the 1s atomic orbitals on A and B) multiplied by the spin singlet function

$$\Psi_s = 2^{-1/2}[\alpha(1)\beta(2) - \beta(1)\alpha(2)] \tag{191}$$

which gives overall antisymmetry. The exact form of the atomic orbitals may be varied to minimize the energy of the molecule. It is clear that the wave function (190) describes a buildup of charge between the atoms, and at the same time reduces repulsion between the bond electrons by keeping one of them on A when the other is on B and vice versa. Equation (191) describes the pairing of unpaired spins on bond formation. For a polyelectronic molecule, the two electrons in a bond would be described by something like $\Psi_r\Psi_s$, and electrons not involved in a bond would be described by atomic orbitals. The entire wave function must, of course, be antisymmetrized.

In terms of determinants, we can write the two-electron wave function $\Psi_r \Psi_s$ as

$$\mathscr{A}\{\phi_A(1)\alpha(1)\phi_B(2)\beta(2) - \phi_A(1)\beta(1)\phi_B(2)\alpha(2)\}. \tag{192}$$

The two terms in (192) differ by an exchange of spins between $\phi_A$ and $\phi_B$ and are degenerate in the absence of interatomic interaction even if A and B are not identical atoms. The combination of the two degenerate wave functions describes the bond. If there were several electrons on each atom, we would expect to be able to write a molecular wave function corresponding to a bond between an atomic orbital on A and one on B (supposing that one was associated with spin $\alpha$ and the other with spin $\beta$) as a linear combination of antisymmetrized products of atomic wave functions, differing by the interchange of spins between atoms. Only exchanges of opposite spins need be considered since an exchange of identical spins produces no new function. If a spatial orbital is doubly occupied, exchange of (opposite) spins between one of the two spin orbitals and some spin orbital on the other atom produces a determinant with two electrons in the same spatial orbital and with the same spin, which would automatically vanish. Thus, only exchanges between singly occupied orbitals need be considered.

In general, spin and symmetry properties of the atomic states make it necessary to use several determinants to describe the separated atoms, complicating the molecular problem. We could use multideterminant atomic wave functions to build up the states instead of single determinants. On the other hand, we could form the functions of different molecular symmetry species first (as in Subsection 1) and then consider linear combinations of functions of the same symmetry. Secular equations of much lower order result; sometimes (as in $H_2$) spatial and spin symmetry suffice to solve the problem completely.

We will first consider the formation of molecular states from atomic configurations without derivation of atomic terms. If we calculate the energy for a linear combination of determinants, there will be contributions from Coulomb and exchange integrals such as those which were responsible for binding in $H_2$, as well as from exchange integrals between orbitals on the same atom. The latter are responsible for the splitting of atomic configurations into terms. The former give rise to the $R$-dependent interaction energy between A and B. For certain linear combinations, the interatomic exchange integrals will enter with the proper sign to give an energy lowering. Generalizing from our experience with the two-electron problem, we expect that the lowest energy of the molecule corresponds to forming the maximum number of bonds. By considering only functions corresponding to the maximum possible number of bonds, we can reduce the dimension of our problem. How important the energy lowering will be depends basically on the extent to which the orbitals on the two atoms overlap. In particular, two orbitals which have different symmetries about the internuclear axis will have exchange and overlap integrals identically zero, and thus cannot be paired in a bond.

We want to evaluate matrix elements of the Hamiltonian between the determinantal functions, and set up the resulting secular equation. This has been worked out by Eyring *et al.* [37]. We write down all the determinants corresponding to particular configurations for the atoms A and B. Since it will not be necessary to consider explicitly the doubly filled orbitals, which will be the same in all the determinants, we denote such a determinant by

$$\Psi_k = \mathscr{A}\{a(1)S_1(1)b(2)S_2(2) \cdots g(n_A'+n_B')S_{n_A'+n_B'}(n_A'+n_B')C\} \qquad (193)$$

Here, $C$ represents the core functions, and $a, b, c, \ldots$ the other orbitals. There are $n_A'$ valence orbitals on A and $n_B'$ on B, thus $n_A' + n_B'$ valence electrons. By construction, each $\Psi_k$ is an eigenfunction of $S_z$ with eigenvalue equal to $\frac{1}{2}\hbar$ times the difference between the number of $\alpha$'s and the number of $\beta$'s among the $\{S_i\}$. The secular determinant is block-diagonal, corresponding to the different values of $S_z$: $\langle \Psi_k \mid \Psi_l \rangle$ and $\langle \Psi_k \mid H \mid \Psi_l \rangle$ vanish when $\Psi_k$ and $\Psi_l$ have different values for $S_z$. Since a bond pairs off unpaired spins, one on each atom, and since we are assuming that more bonds mean more energy lowering, the eigenvalue of total spin of interest is $\frac{1}{2}\mid n_A' - n_B' \mid$. This corresponds to making all spins not involved in the bonds parallel, which should give lowest energy according to Hund's rule for atoms.

We consider in particular the case where $n_A' = n_B'$ (all electrons are involved in bonds) or where $\mid n_A' - n_B' \mid = 1$. In the latter case, we can introduce another orbital on the atom which needs it to make $n_A' = n_B'$, go through the calculations, and eventually set all integrals involving the extra orbital equal to zero. Suppose $n_A' = n_B' = n$, so that there are $n/2\ \alpha$ spins and $n/2\ \beta$ spins. The number of $\Psi_k$ equals the number of ways of distributing $n/2\ \alpha$'s over $n$ orbitals,

$$n_D = \frac{n!}{(n/2)!(n/2)!}. \qquad (194)$$

The bond function will be of the form

$$\Psi_b = \sum_k c_k \Psi_k. \qquad (195)$$

If the bonds are to be between orbital $i$ (on A) and $i'$ (on B), between $j$ (on A) and $j'$ (on B), etc., then the only $\Psi_k$ which appear in (195) with nonzero coefficients are those for which $i$ and $i'$ have opposite spins, $j$ and $j'$ have opposite spins, etc. For $n$ bonds, there are $2^n$ functions. Furthermore, the coefficients of determinants, differing by an exchange of opposite spins between functions in a bond, must differ by a sign change. We may write

$$c_k = \prod p_{ij}(k) \tag{196}$$

where there is one factor $p_{ij}$ for each pair of orbitals forming a bond; and $p_{ij}(k) = 0$ if orbitals $i$ and $j$ have the same spin in $\Psi_k$, $p_{ij}(k) = +1$ if $i$ has spin $\alpha$ and $j$ spin $\beta$, $p_{ij}(k) = -1$ if the reverse is true. Equation (195) may now be written, using (193) and (196), as

$$\Psi_b = \mathscr{A}\left\{a(1)b(2)\cdots g(n_A' + n_B')\left[\sum_k \prod p_{ij}(k)T_k\right]\right\} = \mathscr{A}\{PQ\}$$

where $T_k$ is the product of spin functions corresponding to $\Psi_k$. The $T_k$ are all obtained from a single function, say $T_0$, by interchanging the spins of the two electrons in one or more bonds.

The spin function $Q$ may be written as

$$\prod_i (1 - \mathscr{I}^s_{i\,n_A+i})\{\alpha(1)\alpha(2)\cdots\alpha(n_A)\beta(n_A+1)\cdots\beta(n_A+n_B)\} \tag{197}$$

where $\mathscr{I}^s$ interchanges spins only, or as

$$Q = [\alpha(1)\beta(n_A+1) - \beta(1)\alpha(n_A+1)]$$
$$\times [\alpha(2)\beta(n_A+2) - \beta(2)\alpha(n_A+2)]\cdots \tag{198}$$

Then it may be shown that the bond function corresponds to zero eigenvalue of $S^2$. To show this, we write $S^2$ as (EYRING, Sect. 32, PILAR, Sect. 6-4):

$$S^2 = S_z^2 + S_z + \sum_j [S_x(j) + iS_y(j)]\sum_k [S_x(k) - iS_y(k)] \tag{199}$$

This is to be operated on $\Psi_b$, Eq. (195). Each determinant in (195) is an eigenfunction of $S_z$ with eigenvalue zero. The double sum in $S^2$ commutes with permutations, so we can operate it on the function $PQ$ directly. The spatial function $P$ need not be considered, and we now show that the double sum annihilates the spin function $Q$. Note that $S_x(k) - iS_y(k)$ annihilates $\beta(k)$ and changes $\alpha(k)$ to $\beta(k)$. Then

$$\left(S_x(k) - iS_y(k)\right)\left(\alpha(k)\beta(l) - \beta(k)\alpha(l)\right) = \beta(k)\beta(l)$$

But $S_x(l) - iS_y(l)$, operating on the same spin function, gives $-\beta(k)\beta(l)$, so that the sum of the two operators annihilates the spin function. It may then be seen that

$$\sum_k \left(S_x(k) - iS_y(k)\right)Q = 0 \qquad (200)$$

The bonding function $\Psi_b$ (Eq. (195)] corresponds to one bonding structure, i.e., one way of forming bonds, each between an atomic orbital on A and an atomic orbital on B. Where several bonding structures seem equally reasonable and of comparable energy, the wave function will be a linear combination of these wavefunctions, and one must consider a secular equation.

We require first overlap and Hamiltonian integrals between the determinantal functions $\Psi_k$ and $\Psi_l$. Atomic orbitals on the same atom are orthogonal to each other. If overlaps between two orbitals on different atoms are small enough compared to unity to be neglected, we may use the formulas of Volume I, Chapter III, Section A.7. Since $\Psi_k$ and $\Psi_l$ differ by at least one spin exchange between orbitals, the overlap $\langle \Psi_k | \Psi_l \rangle$ vanishes, and $\Psi_k$ is normalized when $\mathscr{A}$ is $(N!)^{-1/2} \sum_p \delta_p P$. The matrix elements of $H$ may be written:

$$\sum_p \langle \{a(1)S_1(1) \cdots g(n_A{}' + n_B{}')S_{n_A+n_B}(n_A{}' + n_B{}')\} \mid H \mid \delta_P P\{a(1)S_1{}'(1) \cdots$$

$$g(n_A{}' + n_B{}')S'_{n_A{}'+n_B{}'}(n_A{}' + n_B{}')\}\rangle \qquad (201)$$

For the diagonal element of $H$, $\langle \Psi_k | H | \Psi_k \rangle$, only a permutation $P$ which interchanges parallel spins gives a nonzero term. Furthermore, if $P$ permutes more than two electrons the contribution to (201) involves overlap factors which we have agreed to neglect and vanishes. Limiting $P$ to the identity and single exchanges of electrons between orbitals of parallel spins, we find for $\langle \Psi_k | H | \Psi_k \rangle$ a Coulomb integral,

$$Q = \int \{a(1) \cdots g(n_A{}' + n_B{}')\}^* H\{a(1) \cdots g(n_A{}' + n_B{}')\}\, d\tau_r$$

minus a sum of exchange integrals such as

$$\int \{a(1) \cdots c(k)d(l) \cdots g(n_A{}'+n_B{}')\}^* Ha(1) \cdots c(l)d(k) \cdots g(n_A{}'+n_B{}')\, d\tau_r$$

$$\qquad (202)$$

where $c$ and $d$ were associated with parallel spins in the original assignment. Since we are neglecting overlap integrals, the off-diagonal matrix element

$\langle \Psi_k \mid H \mid \Psi_l \rangle$ is zero unless $\Psi_k$ and $\Psi_l$ differ by a single spin interchange. Let the orbitals involved in the spin exchange be $c$ and $d$ (associated with opposite spins). Then the only permutation in (201) which gives a non-vanishing contribution is that which interchanges electrons between $c$ and $d$. The expression (201) reduces to the negative of the exchange integral (202).

The formulas for matrix elements between bond structures,

$$\langle \Psi_b \mid H \mid \Psi_c \rangle = \sum_k \sum_l c_k^{(b)} c_l^{(c)} \langle \Psi_k \mid H \mid \Psi_l \rangle \tag{203a}$$

$$\langle \Psi_b \mid \Psi_c \rangle = \sum_k \sum_l c_k^{(b)} c_l^{(c)} \langle \Psi_k \mid \Psi_l \rangle = \sum_l c_k^{(b)} c_l^{(c)} \tag{203b}$$

may now be derived using the properties of the coefficients, Eq. (196). One way of doing this is discussed by Eyring, *et al.* [37] (see p. 243 for the final formulas). Another derivation of these formulas is discussed by Mc-Weeny and Sutcliffe [38], using results derived by Cooper and McWeeny [39]. A recent alternative derivation, which shows that the orbitals need not be orthogonal, was given by Shull [12]. It was suggested that the formulas may now make the method useful for automatic computation. Another set of formulas suited for computer calculation is derived by Davidson [40].

Now we consider the use of wave functions for the atomic terms (which may even have a form more complex than a linear combination of determinants). The determination of the proper combinations of atomic wave functions to form molecular states, using the properties of the permutation group of electrons, was investigated by Heitler [41]. The results were used by him to discuss bonding in homopolar diatomics [42]. Born [43] subsequently gave a simpler treatment, deriving an important formula for the energy of the diatomic system in terms of exchange integrals.

Born [43] started from two sets of atomic functions, $\Phi_1^A, \Phi_2^A, \ldots$ and $\Phi_1^B, \Phi_2^B, \ldots$ and formed all products. He operated with the operator $\mathscr{A}'$, $\mathscr{A}' = (n_A! n_B!/N!)^{1/2} \sum_Q \delta_Q Q$, to get a set of fully antisymmetric $N$-electron functions for the diatomic system. The permutations $Q$ are those which permute electrons between the atoms and $\delta_Q$ is the parity of $Q$. The number of such permutations is $N!/(n_A! n_B!)$. The solution to the Schrödinger equation for the diatomic system was written as a linear combination of the $\Psi_K^0$, where

$$\Psi_K^0 = (n_A! n_B!/N!)^{1/2} \sum_Q (-1)^Q Q(\Phi_j^A \Phi_k^B). \tag{204}$$

We first calculate matrix elements between the $\Psi_K{}^0$. For the overlap, we have

$$S_{KL} \equiv \int \Psi_K^{0*}\Psi_L^0 \, d\tau = (n_A! n_B!/N!) \sum_Q \sum_{Q'} (-1)^{Q+Q'}$$

$$\times \int Q(\Phi_j{}^A\Phi_k{}^B)^* Q'(\Phi_l{}^A\Phi_m{}^B) \, d\tau$$

By arguments already used in connection with determinantal functions [see Volume I, Chapter III, Eqs. (50) et seq.], we simplify this expression to

$$S_{KL} = \sum_Q (-1)^Q \int \Phi_j{}^A(1, \ldots, n_A)^* \Phi_k{}^B(n_A + 1, \ldots, N)^*$$

$$\times Q(\Phi_l{}^A(1, \ldots, n_A)\Phi_m{}^B(n_A + 1, \ldots, N)) \, d\tau.$$

The contribution of the identity permutation is a product of overlaps between atomic states:

$$\int \Phi_j^{A*}\Phi_l^A \, d\tau_{1\cdots n_A} \int \Phi_k^{B*}\Phi_m^B \, d\tau_{n_A+1,\ldots,N} = \delta_{KL} \tag{205}$$

The permutations making one interchange between atoms contribute terms proportional to squares of overlaps between an atomic orbital on A and one on B. The permutations making two interchanges give terms proportional to the fourth power of overlaps, and so on. These overlaps decrease exponentially with $R$ (see Murrell et al. [29]). For the Hamiltonian operator, we write

$$H = H_A + H_B + V \tag{206}$$

as previously, where $V$ contains only potential energy terms, corresponding to interactions between pairs of particles, one on A and one on B. The sum, $H_A + H_B$, is like a zero-order Hamiltonian of which $\Phi_j{}^A\Phi_k{}^B$ is an eigenfunction with eigenvalue $W_j{}^A + W_k{}^B$, while $Q(\Phi_j{}^A\Phi_k{}^B)$ is an eigenfunction of $Q(H_A + H_B)$. Except for the change in notation, we may use the formulas of Section 2. Defining $W_K{}^0$ as $W_j{}^A + W_k{}^B$,

$$H_{KL} = W_K{}^0 \sum_Q (-1)^Q \int (\Phi_j{}^A\Phi_k{}^B)^* Q(\Phi_l{}^A\Phi_m{}^B) \, d\tau$$

$$+ \sum_Q (-1)^Q \int (V\Phi_j{}^A\Phi_k{}^B)^* (Q\Phi_l{}^A\Phi_m{}^B) \, d\tau \tag{207}$$

The first group of terms is $W_K{}^0 S_{KL}$. In the second group of terms, write $V_{KL}^0$ for the contribution where $Q$ is the identity; $V_{KL}^0$ corresponds to the

Coulomb term. Following this are single exchange, double exchange, etc., terms, which contain one, two, or more many exchange integrals. The values of the exchange integrals decrease exponentially with $R$. Using a superscript to denote number of exchanges,

$$S_{KL} = \delta_{KL} + S_{KL}^{(1)} + \cdots \tag{208}$$

$$H_{KL} = W_K^{\,0}\delta_{KL} + V_{KL}^{(0)} + (W_K^{\,0}S_{KL}^{(1)} + V_{KL}^{(1)}) + \cdots \tag{209}$$

The superscripts can be interpreted as "order of smallness" in the spirit of perturbation theory, except that we anticipate—from numerical results in the H–H case—that $V_{KL}^{(1)}$ and $V_{KL}^{(0)}$ will be of the same size (first order). Substitution of the expansion

$$\Psi = \sum_K C_K \Psi_K^{\,0} \tag{210}$$

into the Schrödinger equations leads to

$$\sum_L C_L(H_{KL} - ES_{KL}) = 0.$$

Born [43] expanded the energy and expansion coefficients in terms of different orders, and collected zero-order and then first-order terms (see Volume I, Chapter III, Section B.1).

$$\sum_L (W_K^{\,0}\delta_{KL} - E^{(0)}\delta_{KL})C_L^{(0)} = 0 \tag{211}$$

$$\sum_L [(W_K^{\,0}\delta_{KL} - E^{(0)}\delta_{KL})C_L^{(1)}$$
$$+ (V_{KL}^{(0)} + W_K^{\,0}S_{KL}^{(1)} + V_{KL}^{(1)} - E^{(0)}S_{KL}^{(1)} - E^{(1)}\delta_{KL})C_L^{(0)}] = 0 \tag{212}$$

Equation (211) is satisfied if $E^{(0)}$ equals some zero-order energy, say $W_J^{\,0}$, and $C_K^{(0)} = 0$ for all $\Psi_K^{\,0}$ whose zero-order energy is not equal to $W_J^{\,0}$. The $\Psi_K^{\,0}$ whose zero-order energy equals $W_J^{\,0}$ will now be denoted by $\Psi_{J\lambda}^{0}$. This change of indexing means that $\sum_K$ must be written $\sum_K \sum_\lambda^{(K)}$. In this notation, $C_K^{(0)} = 0$, $K \neq J$, and Eq. (212) is written

$$\sum_\mu^{(J)} [V_{K\lambda,J\mu}^{(0)} + (W_K^{\,0} - W_J^{\,0})S_{K\lambda,J\mu}^{(1)} + V_{K\lambda,J\mu}^{(1)}]C_{J\mu}^{(0)}$$

$$= E^{(1)}C_{K\lambda}^{(0)} - (W_K^{\,0} - W_J^{\,0})C_{K\lambda}^{(1)}. \tag{213}$$

For $K = J$, Eq. (213) determines the coefficients $C_{J\lambda}^{(0)}$ and the energy $E^{(1)}$.

The secular determinant is

$$| V^{(0)}_{J\lambda,J\mu} + V^{(1)}_{J\lambda,J\mu} - E^{(1)} \delta_{J\lambda,J\mu} | = 0 \qquad (214)$$

The spin and orbital symmetry may be used to block-diagonalize the matrix. Neglecting spin–orbit coupling, the atomic states $\Phi_j{}^A$ and $\Phi_k{}^B$ may be characterized by eigenvalues of $S^2$, $S_z$, $L^2$ and $L_z$ (total spin, component of total spin, total orbital angular momentum, component of total orbital angular momentum). With the internuclear axis as the $z$ axis, the product functions and hence the $\Psi^0_{K\lambda}$ remain eigenfunctions of $S_z$ and $L_z$, where these operators include sums over all electrons in the molecule. The eigenvalue for $S_z$, $\sigma$, is the sum of those for the atomic wave functions, and similarly for the eigenvalue of $M_z$, $m$. Overlap and Hamiltonian matrix elements vanish between $\Psi^0_{K\lambda}$ and $\Psi^0_{L\mu}$ if they belong to different values of $m$ and/or $\sigma$. Instead of (214), we need only consider separately the smaller secular equations for each possible set of $m$ and $\sigma$. The symmetry requires that the energies depend only on $|m|$ and not on $m$, while the spin-independence of $H$ means that the energies for a given $|m|$ are independent of $\sigma$, so only one value of $\sigma$ need be considered for each $|m|$. The largest dimension of such a secular equation is equal to the number of ways of forming $\sigma$ by adding the $S_z$-quantum numbers for the two atoms, multiplied by the number of ways of forming $m$ by adding the $L_z$-quantum numbers for the two atoms. The states for the molecule that result from these secular equations are those discussed at the beginning of this section.

It should be noted that this scheme assumes that the interaction energy between atoms is large compared to the spin–orbit coupling. This cannot be the case for very large $R$, and we must then consider coupling of the total angular momenta of the atoms ($J$–$J$ coupling). The effect of the spin–orbit coupling on the interaction energy in such a case is discussed by Nikitin [7].

If $L_A$ or $L_B$ is 0 ($S$-state), Born showed [43] that no secular equation of more than one dimension need be solved. Other roots could be obtained by subtraction of traces of submatrices (EYRING, p. 250). Born used the symmetry of the spatial parts of the atomic functions [44] to reduce the roots to Coulomb and exchange integrals. He wrote the Coulomb integral as $J_E^{(0,m)}$ and distinguished three contributions to the exchange term:

(a) those due to exchange of an unpaired electron on B with an unpaired electron on A;

(b) those due to exchange of an unpaired electron on A with a paired electron on B or vice versa;

(c) those due to exchange of a paired electron on A with a paired electron on B.

The contributions (all negative) were denoted, respectively, by $J_x^{(0,m)}$, $J_y^{(0,m)}$, and $J_z^{(0,m)}$. The energy corresponding to total spin $S$, where $S$ runs from $|S_A - S_B|$ to $S_A + S_B$, is

$$E^{(S)} = J_E^{(0,m)} - kJ_x^{(0,m)} - \sum_y J_y^{(0,m)} - 2\sum_z J_z^{(0,m)} \tag{215}$$

where

$$k = S + S^2 - S_A - S_B - (S_A - S_B)^2 \tag{216}$$

(see also McWeeny and Sutcliffe [38]).

Usually, the unpaired orbitals belong to the valence shell, and have larger differential overlaps with the orbitals on the other atoms than do the closed-shell orbitals. Thus only $J_E$ and $J_x$ integrals need be considered—$J_y$ and $J_z$ being small—and (215) becomes:

$$E^{(S)} = J_E^{(0,m)} - kJ_x^{(0,m)}$$

$J_x$ and $J_E$ are negative, with $J_x$ of larger magnitude than $J_E$.

The most stable electronic state will correspond to the lowest value of $k$. Equation (216) tells us that, for given $S_A$ and $S_B$, this corresponds to smallest total spin, which means maximum pairing of electrons on molecule formation. When $S$ takes on its smallest value of $|S_A - S_B|$, $k$ is equal to $(-2)$ times $S_L$, the smaller of $S_A$ and $S_B$. Since $k$ is negative, the exchanges leading to $J_x^{(0,m)}$ (between unpaired electrons on the two atoms) lower the energy. Exchange of unpaired electrons with closed-shell electrons leads to an increase of energy, as does exchange of closed-shell electrons on one atom with closed-shell electrons on the other. This is seen in Eq. (215), where $J_y^{(0,m)}$ and $J_z^{(0,m)}$ enter with negative signs. This "repulsion of closed shells" eventually becomes important as $R$ becomes small. The value of $S_L$ gives the number of pairings, i.e., bonds, which have been formed. In general, greater stability is associated with higher $S_A$ and $S_B$, or maximum unpairing of electrons in the atoms A and B. This maximum unpairing usually corresponds to the ground state of an atom according to Hund's rules. If there are no unpaired electrons, $S_A = S_B = S = 0$, and there are no exchange integrals contributing to stability, but only the much smaller Coulomb integral.

An approximate equation for the bond energy where a single bond structure dominates is

$$\Delta E = \sum J_x^{(0,m)} \tag{217}$$

each term in the sum being an exchange integral between orbitals involved in a bond. This expression and the approximations involved in its justification, are discussed by Davidson [40].

The preceding discussions dealt with formation of covalent bonds. As has been mentioned previously, structures corresponding to "ionic bonding" may be important in some cases. The molecular wave function will generally be written as a linear combination of covalent and ionic wave functions. The latter are formed from the ions exactly as the covalent wave functions are formed from the neutral atoms, except that the Coulomb integral will include an important energy-lowering term due to the attraction of the net charges on the two centers. Inclusion of wave functions corresponding to several different bonding structures, covalent or ionic, is often referred to as "resonance," and the necessarily positive difference between the energy computed from one bonding structure and that from the more complex wave function is referred to as "resonance energy."

Where there are excited atomic configurations whose energies are not much above the energy of the ground-state configuration, bonding structures involving these excited configurations may also be included in the wave function. If one starts with atomic terms, the same statement may be made for terms with low energies relative to the ground state term. Inclusion of excited terms and configurations is especially important when they allow the possibility of formation of more bonds than can be formed from the atomic ground states. The additional exchange integrals may give a substantial negative contribution to the molecular energy. The atoms which are bonded together to form the molecule, when excited states have been included, are said to have been "prepared" for bonding. An alternate description is that their wave functions are built up from one-electron functions which are combinations of orbitals appropriate to states of the free atom. Such functions are called "hybrid orbitals." From this viewpoint, the increased stability of the molecule is explained in terms of the stronger bonds which may be formed from hybrid orbitals that overlap strongly with orbitals on the other atom. At what point the valence bond function "improved" by consideration of excited states for the two atoms, should simply be referred to as a configuration interaction, is a question of taste.

## 5. Use of Valence Bond Theory

In the 1920s and 1930s, there was competition between the valence bond theories, developed by Heitler, London, and others, and the molecular orbital theories of Hund, Mulliken, and others. Articles by Lennard-

Jones [45] and Herzberg [10] in 1929 give an idea of what were felt, at that time, to be the strong and weak points of each. The competition was largely related to the abilities of the two theories to explain molecular band spectra, i.e., to predict the symmetries and stabilities of the ground and excited molecular states. There was little possibility for quantitative work on atomic interaction energies. For the simple case of the $H_2$ ground state, the simplest valence bond function gave a better energy than the simplest molecular orbital function. However, subsequent calculations on multielectronic molecules resulted in much more widespread use of molecular orbital (MO) methods, because of the greater simplicity of calculations. Because the valence bond (VB) function is a sum of determinants of nonorthogonal atomic orbitals, the energy expectation value for this function becomes a complicated expression. The molecular orbitals are mutually orthogonal, and the energy expressions in the MO theory are simpler. The calculations by Karo and Olson [46, 47] for LiH, using the two methods, provide illustrations of the comparison. The fact that, in the valence bond theory, concepts such as chemical bonds play a fundamental role, and the possibility of giving physical meaning to terms in the energy expression, are insufficient to compensate for the computational complications.

Slater [48] gave a discussion of the problems of the valence bond method in 1953, in which he concluded that it was unsuitable for systems of "more than a very few electrons." In addition to emphasizing the computational difficulties due to nonorthogonality, he pointed out that approximate treatments which neglected the resulting troublesome terms gave quite incorrect results. Furthermore, an attempt to work [49] with a basis of orthogonal orbitals formed from the atomic orbitals did not lead to bonding even for $H_2$.

McWeeny [50] gave some further analysis of this problem as part of a general outline and comparison of valence bond and molecular orbital methods. He compared the charge density of the valence bond function

$$\Phi_{VB} = [2 + 2S^2]^{-1/2}(\phi_A(1)\phi_B(2) + \phi_B(1)\phi_A(2)] \tag{218}$$

to a sum of atomic charge densities, $\frac{1}{2}(\phi_A(1)^2 + \phi_B(1)^2)$. The buildup of additional charge between the nuclei leads to molecular stability according to the Hellmann–Feynman theorem (see Volume I, Chapter III, Section D). The process of mutually orthogonalizing $\phi_A$ and $\phi_B$, however, removes electron density from each orbital just in the crucial overlap region. If we include "ionic" structures in the wave function (still using orthogonalized atomic orbitals), the cross terms in the energy expression between the "ionic" and the "covalent" structures lead to an energy decrease and to bonding [50]. However, the interpretation of the bonding in terms of "exchange energy" disappears.

McWeeny emphasized that it was inconsistent to neglect orthogonality of atomic orbitals on different atoms in deriving energy formulas while retaining exchange integrals to explain bonding, as was done by many approximate valence bond theories. He developed a consistent theory in terms of orthogonalized atomic orbitals. Bond structures like those mentioned earlier were set up and matrix elements computed, but a reinterpretation and reclassification of the structures was necessary. It was suggested [50] that one now had a consistent basis for semiempirical valence bond schemes.

In some cases, the simplest MO function contains a serious error, since it describes the molecule at large values of $R$ incorrectly. For example, for $H_2$, it leads to a wave function representing a superposition of H + H, $H^+ + H^-$, and $H^- + H^+$. The valence bond function in this case correctly gives H atoms at large $R$. (However, it also yields increasingly poor values for the separation between the singlet and triplet $H_2$ states which correlate with ground state H atoms [51–53].) The $U(R)$ curve derived from single configuration molecular orbital energies will be distorted due to the error at large $R$; in particular, the vibrational frequency will be poorly predicted. It is necessary to go to a configuration interaction function to correct this (although, as will be discussed in Chapter II, Section C, the problem is well enough understood that ad hoc corrections to the energy and other properties may be introduced). However, it is *always* necessary to go beyond single-configuration calculations or simple valence bond calculations to get accurate dissociation energies. On the basis of a large number of calculations on diatomic molecules formed from first-row atoms, Ransil [54] stated, in 1960, that a single determinant wave function, in which molecular orbitals were constructed from valence and inner-shell atomic orbitals, gave insufficiently accurate results for dissociation energies. The molecular and atomic energies were each inaccurate to a percentage point or so, which was generally larger than their difference. For the case of $F_2$, a negative dissociation energy was computed because the error in the molecular energy was much larger than that in the atomic energy (computed from atomic Roothaan-type wave functions).

Comparisons of the valence bond and molecular orbital methods were given by Kotani *et al.* [55] for homonuclear diatomic molecules. While the valence bond method usually yielded better energies for large values of $R$, it was not possible to state, in general, which approach would be superior for a given case. Often it was possible to distinguish a critical value of $R$, beyond which the valence bond was superior to the molecular orbital method. These authors also investigated the use of "core approximations," in which the effect of the core electrons was expressed as an electrostatic field acting on the valence electrons. The effect of such approximations was about the same on the molecular orbital and valence bond energies. For more discussion and comparison of the two theories, we refer to the books of Coulson, Slater, and Pilar [56].

We will now discuss briefly some of the earlier valence bond calculations for systems with more than two electrons. These are meant to be illustrative, and no attempt is made to cover all such work. In 1960, Allen and Karo [57] summarized the *ab initio* valence bond calculations (as well as *ab initio* calculations by other methods) on molecules with more than two electrons. By *ab initio* they meant treating all electrons explicitly, using the correct nonrelativistic fixed nucleus Hamiltonian, and attempting to evaluate all integrals rigorously. Semiempirical schemes for evaluation of integrals, it was stated, were not sufficiently good and consistent to give useful results.

Pauling and Weinbaum [58] made some of the earliest valence bond calculations for a multielectronic system, $He_2^+$. The wave function here corresponds to a three-electron bond, and was written as a linear combination of two determinants:

$$\Psi = \mathscr{A}\{\phi_A(1)\alpha(1)\phi_A(2)\beta(2)\phi_B(3)\alpha(3)\}$$
$$\pm \mathscr{A}\{\phi_A(1)\alpha(1)\phi_A(2)\beta(2)\phi_A(3)\alpha(3)\} \qquad (219)$$

The atomic orbitals were hydrogen-like 1s orbitals. Some improvement in the energy was obtained by allowing the exponential parameter in the orbitals to vary. The symmetric combination of the determinants [plus sign in (219)] gives rise to a stable molecule. The most complete calculations [58] predicted: equilibrium internuclear distance = 1.097 Å, dissociation energy = 2.22 eV, and vibrational frequency = 1950 cm$^{-1}$, in good agreement with experiment. The antisymmetric function, with a minus sign in (219), led to a purely repulsive $U(R)$ curve. The situation is quite analogous to the one-electron bond in $H_2^+$, since it corresponds to "resonance" between the structures He–He$^+$ and He$^+$–He. Already in these calculations one notes that the valence bond theory was not used to give the proper linear combination of determinants. Symmetry requires, in this case, that the coefficients of the determinants be of equal magnitude, and one could predict that the positive sign in (219) would yield the lower energy because it leads to a buildup of charge between the nuclei while the negative sign has a node at the molecular midpoint. Even where symmetry is absent, most of the computational labor arises in evaluation of the matrix elements of the Hamiltonian between determinantal functions (which involves two-electron integrals over the orbitals), and not in determining the linear combination coefficients, so little additional work is required to find these coefficients by linear variation.

The wave function for the homonuclear four-electron case is very simple. For He–He, all the four atomic orbitals, $\phi_A\alpha$, $\phi_A\beta$, $\phi_B\alpha$, and $\phi_B\beta$, are filled. The only wave function one can write is

$$\Psi_{\text{HeHe}} = \mathscr{A}\{[\phi_A(1)\phi_A(2)]\mathscr{S}(1, 2)[\phi_B(3)\phi_B(4)]\mathscr{S}(3, 4)\} \qquad (220)$$

Here, $\mathscr{S}$ is the singlet spin function:

$$\mathscr{S}(i, j) = \tfrac{1}{2}[\alpha(i)\beta(j) - \beta(i)\alpha(j)] \qquad (221)$$

and $\phi_A$ and $\phi_B$ are 1s orbitals appropriate to the He atom. For the product $\phi_A(1)\phi_A(2)$, a more accurate spatial wave function for the ground state of He, say $u_A(1, 2)$, could be used. No exchange of spins between the atom is possible and no bond may be formed.

A computation using a wave function like (220), although with no reference to the valence bond formalism, was done by J. C. Slater [59] in 1928. In the calculation of $\int \Psi^* H\Psi \, d\tau$ and $\int \Psi^*\Psi \, d\tau$, (220) was expanded into products and a large number of terms appeared, but it sufficed to evaluate integrals over the spatial wave function

$$u = u_A(1, 2)u_B(3, 4) - u_A(3, 2)u_B(1, 4) - u_A(1, 4)u_B(3, 2) + u_A(3, 4)u_B(1, 2)$$

(The symmetry of the functions $u_A$ and $u_B$ to interchange of electrons and the properties of $\mathscr{S}$ were used.) The integrals arising in the calculation involved spatial functions such as $|u_A(1, 2)u_B(3, 4)|^2$ (Coulomb-type), $u_A^*(1, 2)u_B^*(3, 4)u_A(3, 2)u_B(1, 4)$ (single exchange), and $u_A^*(1, 2)u_B^*(3, 4)$ $\times u_A(3, 4)u_A(1, 2)$ (double exchange). The double exchange terms, expected to be small, were dropped from Slater's calculation [59]. With some further approximations, it was possible, using a simple form for the pair functions $u_A$ and $u_B$, to give all the integrals in the energy calculation as explicit functions. It turned out, however, that the neglected exchange integrals were too important to neglect. Much later, Rosen [60] repeated the calculation, using a simpler function for He, a symmetric sum of products of 1s orbitals of different effective charges. All the exchange integrals could now be computed and the neglected ones were important. The potential curve $U(R)$ was generated and compared to Slater's results and to experiment.

The lithium hydride molecule has attracted much interest because it is a simple (four electron) heteronuclear diatomic of chemical interest, with a number of experimentally measured properties. Accurate calculation of its wave function became one of the important challenges of modern quantum chemistry. We will only mention a few of the published calculations. For more complete bibliography, we refer to the authors cited in what follows and to Chapter II. The atomic orbitals that are considered in the simplest valence bond calculation of LiH are the 1s orbitals of hydrogen ($\phi_H$), and the 1s and 2s orbitals of lithium ($\phi_{1s}$, $\phi_{2s}$). The simple covalent wave function is

$$\psi_{cov} = \mathscr{A}\{\phi_{1s}(1)\alpha(1)\phi_{1s}(2)\beta(2) \\ \times [\phi_{2s}(3)\alpha(3)\phi_H(4)\beta(4) - \phi_{2s}(3)\beta(3)\phi_H(4)\alpha(4)]\} \tag{222}$$

a difference of two determinants. The ionic wave function for $Li^+H^-$,

$$\psi_{ion} = \mathscr{A}\{\phi_{1s}(1)\alpha(1)\phi_{1s}(2)\beta(2)\phi_H(3)\alpha(3)\phi_H(4)\beta(4)\}$$

is also expected to be important, but that for $Li^-H^+$ less so. The 2p orbitals, which are not occupied in the Li ground state, would not be employed in the simplest valence bond functions, but, since their energy is relatively low, bond structures involving them really should be considered as well. Alternatively, s–p hybrids could be used instead of the atomic s orbitals.

Fischer [61] carried out calculations on the simplest covalent wave function, and also on a wave function which included the covalent and both ionic forms. Only the valence electrons were considered, the inner shells being treated as point charges at the Li nucleus. Consideration was also given to the use of an s–p hybrid on Li. With the simplest VB function and a pure s orbital, a binding energy of 0.065 a.u. was obtained, the experimental value being 0.097 a.u. Inclusion of s–p hybridization gave 0.102 a.u. The binding energies here were obtained by subtraction of the calculated energy for the molecule from that of the atoms. From calculations at several internuclear distances, the equilibrium internuclear distance and vibrational frequency were derived.

Karo and Olsen [46] performed all-electron calculations for LiH, by configuration interaction. The functions they treated included valence bond functions of different degrees of complexity. The simplest covalent function, corresponding to a bond between the H 1s and Li 2s orbitals,

gave a well whose depth was only about a third of the experimental dissociation energy, and the position of the minimum was about 0.5 $a_0$ larger
than the experimental equilibrium internuclear distance, 3.0 $a_0$. The curve
$U(R)$ for the ionic (Li$^+$H$^-$) wave function showed a shallow minimum at
about 5 $a_0$, but this curve lay above that for covalent LiH at all $R$ considered. Introduction of hybridization into the covalent function improved
the dissociation energy considerably, as did inclusion of a contribution
from the ionic structure Li$^+$H$^-$. For the hybridization, the orbital $c_s\phi_{2s}$
$+c_p\phi_{2p}$ was substituted for $\phi_{2s}$ in the LiH bond function and $c_s$ and $c_p$
were determined variationally. The Li$^-$H$^+$ configurations (1s$^2$2s$^2$, 1s$^2$2s2p,
and 1s$^2$2p$^2$) gave little contribution. The depth of the potential well was
doubled by the joint effects of hybridization and ionic–covalent resonance.
Of course, since LiH dissociates to spherically symmetric atomic systems,
the amount of admixture of $p$ decreases to zero at large $R$. Since the wavefunctions used gave an energy too high by about 0.1 a.u. for large $R$ (this
is about 0.6% of the total energy), the dissociation energies were necessarily
evaluated throughout using calculated, rather than experimental values,
for the energy as $R \to \infty$. The calculated properties are summarized in
Table VI.

### Table VI

*Results of VB Calculation for LiH*

| | Dissoc. energy $D_e$ (eV) | $R_e(a_0)$ | Vib. freq. $\omega_e$ (cm$^{-1}$) |
|---|---|---|---|
| Covalent s-bond | 0.871 | 3.50 | |
| With hybridization | 1.365 | 3.40 | |
| s-bond with ionic structure | 1.324 | 3.238 | 1165 |
| Six configurations | 1.669 | 3.245 | 1212 |
| Experiment | 2.515 | 3.014 | 1406 |

Valence bond calculations for H$_2$ and LiH, employing a minimum basis
set of Slater orbitals, but with exponential parameters optimized at each
$R$, have recently been presented [62]. Calculations were performed for (*a*)
the simplest bond function; (*b*) function *a* plus the Li$^+$H$^-$ function; (*c*)
function *a* plus the 2p-H bond function; (*d*) function *c* plus Li$^+$H$^-$; (*e*) all
the six configurations obtainable using these orbitals and keeping the
1s$_{Li}$ orbital doubly filled. The exponent of the 1s$_H$ orbital showed the
greatest change from the free-atom value. Optimizing the exponents for

each $R$ was important in decreasing the calculated $R_e$ closer to the true value (3.127 calculated versus 3.015 experimental for best calculation) and improving $D_e$ (1.8118 eV versus 2.516 eV). With respect to $\omega_e$, the situation is not as clear. For small molecules, it was stated [62] that the VB approach gives better results than the single configuration MO when a few basis functions are used.

The HF molecule has 10 electrons and a $^1\Sigma^+$ ground state. The F atom in its ground state may be described by a single determinant corresponding to the configuration $(1s)^2(2s)^2(2p)^5$. There are three degenerate states, corresponding to an empty $2p_x$, $2p_y$, or $2p_z$ orbital, but symmetry allows only the p-orbital along the internuclear axis to form a bond with the H 1s orbital. It may also be anticipated that a structure corresponding to $H^+F^-$ will be important. The $F^-$ ion has the closed shell configuration $(1s)^2(2s)^2(2p)^6$. A valence bond calculation on HF was given in 1953 by Kastler [63]. He found that, in addition to structures involving the ground state configurations of F and $F^-$, it was necessary to consider the configuration $(1s)^2(2s)(2p)^6$, which is also constructed from orbitals occupied in the ground state of the fluorine atom and hence of low energy. Mixing between this and the ground configuration is equivalent to the introduction of hybrid orbitals. Kastler also included functions corresponding to $H^-F^+$, in which the hydrogen orbital is doubly filled and either both $2p_z$ orbitals, both 2s orbitals, or one $2p_z$ and one 2s orbital of F were empty, $z$ being the bond direction. There are thus two covalent states, corresponding to formation of a bond between the 1s orbital on H and either the $2p_z$ or the 2s orbital on F, each written as a combination of two determinants, and four ionic states: $H^+F^-$, and $H^-F^+$ formed three ways.

The atomic orbitals employed for the HF calculation [63] were orthogonalized Slater orbitals with several different sets of exponential parameters. The results obtained using only functions for HF with p-bonding and $H^+F^-$ were quite close to what was obtained when all six states were included. In all cases, however, the calculated bond energy was poor unless a semiempirical correction (which is suggestive of atoms-in-molecules theories, $q. v.$) was made. It was noted that the energy calculated for $H^+F^-$ at infinite nuclear separation differed by almost half an atomic unit from the value derived from experimental measurements. Thus the diagonal element of the Hamiltonian matrix corresponding to this state was corrected by this amount for all values of $R$, other matrix elements being left unchanged. Then a bond energy of 0.2049 or 0.2121 a.u. (for different basis sets) was obtained from the full six-term wave function. When only two terms were considered, the results were 0.1847 or 0.1856 a.u. The correct value is about 0.24 a.u.

The interaction of two atoms or ions with rare gas (closed shell) configurations was treated, using the Heitler–London method, by Bleick and Mayer [64]. Only one molecular state can be formed from $^1S$ atoms. With the zero-order wave function $\Psi_0$ represented by an antisymmetrized product of $n_A + n_B$ atomic orbitals, the first-order energy involved a Coulomb integral and a repulsive contribution from exchanges of paired electrons. Bleick and Mayer evaluated the energy, neglecting multiple exchanges. In the language of Eq. (215),

$$E^{(0)} = J_E^{(0,0)} - 2 \sum J_z^{(0,0)}.$$

The Coulomb and exchange integrals were written out explicitly and evaluated using Hartree wave functions for neon. The Ne–Ne interaction potential was computed for several values of $R$, and could be fitted to $2\%$ by an exponential form, $U = Be^{-R/\varrho}$. Rough agreement with experiment was obtained. Subsequently, the Ar–Ar interaction potential was computed by this method [65]. The results were compared with the repulsive parts of various experimentally determined potentials. Agreement was good, especially with the exponential-six potential.

For the interaction of two identical rare gas atoms, the valence bond (covalent) wave function is identical to the simplest molecular orbital wave function, if the molecular orbitals are written as linear combinations of a minimal basis set of atomic orbitals. We can imagine a transformation from the molecular orbitals to a basis of atomic-like orbitals, but these new orbitals would be orthogonal. They would resemble the atomic orbitals to the extent that nonorthogonality between orbitals on different atoms is negligible. For a more general wave function, the use of additional basis functions would give some distortion of the atomic orbitals due to the interaction, but this is small, especially for large $R$. Then $U(R)$ from the valence bond and molecular orbital calculations should be the same for the region of small overlap; this is borne out in actual results [66].

The valence bond theory at present is rarely, if ever, used as such to calculate the energy of a molecule. In a configuration interaction calculation, it may be possible to simplify computations by forming bond functions and computing matrix elements of the Hamiltonian between them according to the formulas of valence bond theory [38]. Sometimes, formulas like (215) may be used as a basis for qualitative discussion. Generally, it is the ideas of the theory which are employed in conjunction with other formalisms, such as configuration interaction, or semiempirically (see Chapter III).

For example, Palke and Goddard [67] treated LiH by generalizing the VB function and considering the resulting function in terms of Goddard's

G1 theory (see Chapter II). Here, the trial wave function is written as a projection of a product of space and spin functions in such a way that one obtains an antisymmetric function which is an eigenfunction of spin. In the present case, the spin function is $\alpha(1)\beta(2)\alpha(3)\beta(4)$ and the space function $\phi_{1a}(1)\phi_{1b}(2)\phi_{2a}(3)\phi_{2b}(4)$. On projection, this gives the valence bond function

$$\mathscr{A}\{[\phi_{1a}(1)\phi_{1b}(2) + \phi_{1b}(1)\phi_{1a}(2)][\phi_{2a}(3)\phi_{2b}(4)$$
$$+\phi_{2b}(3)\phi_{2a}(4)]\alpha(1)\beta(2)\alpha(3)\beta(4)\}$$

wherein bonds or pairings are present between $\phi_{1a}$ and $\phi_{1b}$ and between $\phi_{2a}$ and $\phi_{2b}$. In the simple valence bond picture, $\phi_{1a}$ and $\phi_{1b}$ would both be Li 1s orbitals, $\phi_{2a}$ would be the Li orbital involved in the bond (some sort of s–p hybrid) and $\phi_{2b}$ the H orbital involved in the bond (essentially a 1s orbital). Relaxing this restriction, Palke and Goddard [67] found the optimum spatial orbitals, which could in general be delocalized. (With $\phi_{1a} = \phi_{1b}$ and $\phi_{2a} = \phi_{2b}$, one would obtain the Hartree–Fock wave function.) It turns out that $\phi_{2a}$ is a hybrid of Li s, p, and d orbitals, with some H 1s orbital, while $\phi_{2b}$ is primarily H 1s. As $R$ increases $\phi_{2a}$ approaches Li 2s. The inner-shell orbitals $\phi_{1a}$ and $\phi_{1b}$ resemble the 1s orbitals for Li, treated as an open shell to allow for correlation splitting.

In Chapter II, we will discuss the kinds of wave functions most commonly used at present for calculation of $U(R)$: single determinant (restricted Hartree–Fock) and configuration interaction wave functions. It is quite rare to come across a simple valence bond calculation for a diatomic system in the modern literature. Valence bond calculations involving several bond structures are most conveniently thought of as particular examples of configuration interaction calculations. We shall have occasion to refer to the VB theory when its ideas are used in connection with other work.

## REFERENCES

1. E. Wigner and E. E. Witmer, *Z. Phys.* **51**, 859 (1928) (the contributions of earler workers are reviewed here).
2. F. Hund, *Z. Phys.* **63**, 719 (1930).
3. R. S. Mulliken, *Rev. Mod. Phys.* **4**, 1 (1932).
4. H. Sponer, "Molekulspektrum," Pt. II, 6. Springer-Verlag, Berlin and New York, 1936.
5. G. Herzberg, "Molecular Spectra and Molecular Structure," Vol. I. "Spectra of Diatomic Molecules," Chapter VI. Van Nostrand-Reinhold, Princeton, New Jersey, 1950.
6. R. S. Mulliken, *Phys. Rev.* **36**, 1440 (1930).

7. E. E. Nikıtin, *Opt. Spectrosc.* (*USSR*) **10**, 227 (1961); *Opt. Spektrosk.* **10**, 443 (1961).
8. S. J. Umanskij and E. E. Nikitin, *Theor. Chim. Acta* **13**, 91 (1969).
9. T. Y. Chang, *Rev. Mod. Phys.* **39**, 911 (1967).
10. G. Herzberg, *Z. Phys.* **57**, 601 (1929).
11. R. Eisenschitz and F. London, *Z. Phys.* **60**, 491 (1930).
12. H. Shull, *Int. J. Quantum Chem.* **3**, 523 (1969).
13. J. O. Hirschfelder, *Chem. Phys. Lett.* **1**, 325, 363 (1968).
14. P. R. Certain, J. O. Hirschfelder, W. Kołos, and L. Wolneiwicz, *J. Chem. Phys.* **49**, 24 (1968); *Int. J. Quantum Chem. Symp.* **2**, 125 (1968).
15. A. T. Amos and J. I. Musher, *Chem. Phys. Lett.* **1**, 149 (1967).
16. J. I. Musher and A. T. Amos, *Phys. Rev.* **164**, 31 (1967).
17. A van der Avoird, *Chem. Phys. Lett.* **1**, 24 (1967).
18. A. van der Avoird, *J. Chem. Phys.* **47**, 3649 (1967); *Chem. Phys. Lett.* **1**, 411, 429 (1968).
19. L. Jansen, *Phys. Rev.* **162**, 63 (1967).
20. P. Claverie, *Int. J. Quantum Chem.* **5**, 273 (1971).
21. E. Corinaldesi, *Nuovo Cimento* **30**, 105 (1963).
22. E. Corinaldesi, *Nuovo Cimento* **25**, 1190 (1962); E. Corinaldesi and H. E. Lin, *Ibid.* **28**, 654 (1963); E. Corinaldesi, *Ibid.* **31**, 1258 (1964).
23. E. Corinaldesi and G. J. Hurford, *Phys. Lett.* **9**, 135 (1964).
24. E. Corinaldesi, E. Remiddi, and F. Strocchi, *Nuovo Cimento* **30**, 1563 (1963).
25. J. N. Murrell and G. Shaw, *J. Chem. Phys.* **46**, 1768 (1967); **49**, 4731 (1968).
26. D. A. Micha, *J. Chem. Phys.* **48**, 3639 (1968).
27. J. O. Hirschfelder and R. Silbey, *J. Chem. Phys.* **45**, 2188 (1966); R. B. Hake, R. Silbey, and J. O. Hirschfelder, Rep. WIS-TCI-180. Theor. Chem. Inst., Univ. of Wisconsin, Madison, Wisconsin, 1966.
28. H. N. W. Lekkerkerker and W. G. Laidlaw, *J. Chem. Phys.* **52**, 2953 (1970).
29. J. N. Murrell, M. Randić, and D. R. Williams, *Proc. Roy. Soc. Ser. A* **284**, 566 (1965); H. Margenau and N. R. Kestner, "Theory of Intermolecular Forces," Sect. 4.3. Pergamon, Oxford, 1969.
30. J. I. Musher and L. Salem, *J. Chem. Phys.* **44**, 2943 (1966).
31. A. Dalgarno and N. Lynn, *Proc. Phys. Soc. London Sect. A* **69**, 821 (1956). N. Lynn, *ibid.* **72**, 201 (1958).
32. P. R. Certain and J. O. Hirschfelder, *J. Chem. Phys.* **52**, 5977, 5992 (1970).
33. B. Kirtman, *J. Chem. Phys.* **49**, 3890, 3895 (1968).
34. J. N. Murrell and G. Shaw, *J. Chem. Phys.* **46**, 1768 (1967).
35. L. Salem, *Discuss. Faraday Soc.* **40**, 150 (1965).
36. H. Margenau, *Phys. Rev.* **56**, 1000 (1939).
37. H. Eyring, J. Walter, and G. E. Kimball, "Quantum Chemistry," Chapter XIII, Wiley, New York, 1944.
38. R. McWeeny and B. T. Sutcliffe, "Methods of Molecular Quantum Mechanics," Sects. 6.3, 6.4. Academic Press, New York, 1969.
39. I. L. Cooper and R. McWeeny, *J. Chem. Phys.* **45**, 226, 3484 (1966).
40. E. R. Davidson, *in* "Physical Chemistry: An Advanced Treatise (D. Henderson, ed.), Vol. III, pp. 166–168. Academic Press, New York, 1969.
41. W. Heitler, *Z. Phys.* **46**, 47 (1927).
42. W. Heitler, *Z. Phys.* **47**, 835 (1928).
43. M. Born, *Z. Phys.* **64**, 729 (1930).

44. J. I. Musher, *J. Phys. (Paris)* **31**, C4-51 (1970).
45. J. E. Lennard-Jones, *Trans. Faraday Soc.* **25**, 668 (1929).
46. A. M. Karo and A. R. Olson, *J. Chem. Phys.* **30**, 1232 (1959).
47. A. M. Karo, *J. Chem. Phys.* **30**, 1241 (1959).
48. J. C. Slater, *Rev. Mod. Phys.* **25**, 199 (1953).
49. J. C. Slater, *J. Chem. Phys.* **19**, 220 (1951).
50. R. McWeeny, *Proc. Roy. Soc. Ser. A* **223**, 63, 306 (1954).
51. M. H. Alexander and L. Salem, *J. Chem. Phys.* **46**, 430 (1967).
52. C. Herring, *Rev. Mod. Phys.* **34**, 631 (1962).
53. A. C. Wahl, *J. Chem. Phys.* **41**, 2600 (1964); P. E. Cade, *Ibid.* **47**, 2390 (1967); E. Clementi, *Ibid.* **38**, 2780 (1963).
54. B. J. Ransil, *Rev. Mod. Phys.* **32**, 245 (1960).
55. M. Kotani, Y. Mizuno, and K. Kayama, *Rev. Mod. Phys.* **32**, 266 (1960).
56. C. A. Coulson," Valence." Oxford Univ. Press, London and New York, 1952; J. C. Slater, "Quantum Theory of Molecules and Solids," Vol. 1. McGraw-Hill, New York, 1963; F. Pilar, "Elementary Quantum Chemistry." McGraw-Hill, New York, 1968.
57. L. C. Allen and A. M. Karo, *Rev. Mod. Phys.* **32**, 275 (1960).
58. L. Pauling, *J. Chem. Phys.* **1**, 56 (1933). S. Weinbaum, *J. Chem. Phys.* **3**, 547 (1935).
59. J. C. Slater, *Phys. Rev.* **32**, 349 (1928).
60. P. Rosen, *J. Chem. Phys.* **18**, 1182 (1950).
61. I. Fischer, *Ark. Fys.* **5**, 349 (1952).
62. J. N. Murrell and C. L. Silk, *Symp. Faraday Soc.* **2**, 84 (1968).
63. D. Kastler, *J. Chim. Phys.* **50**, 556 (1953).
64. W. E. Bleick and J. E. Mayer, *J. Chem. Phys.* **2**, 252 (1934).
65. M. Kunimune, *Progr. Theor. Phys.* **5**, 412 (1950).
66. D. Butler and N. R. Kestner, *J. Chem. Phys.* **53**, 1704 (1970); and references therein.
67. W. A. Palke and W. A. Goddard, III, *J. Chem. Phys.* **50**, 4524 (1969).

## C. United Atom Methods

As $R$ approaches zero, the electronic wave function for a diatomic molecule must approach a wave function of the atom whose nuclear charge equals the sum of the nuclear charges in the molecule ("united atom"), while the energy, exclusive of the additive term $Z_A Z_B e^2 / R^2$, must approach the energy of the united atom. Like the separated atom limit, the united atom limit has been investigated from several viewpoints. First, there is the correlation of the electronic states of the united atom with those of the molecule, already discussed in part in Section B. Similarly, one wants to consider the correlation between atomic and molecular orbitals. The results are important for the use of atomic orbitals or other one-center functions as basis functions for Hartree–Fock or more complicated calculations. The united atom notation for molecular orbitals was introduced by Hund [1] as being more relevant than the separated atom notation for the outer electrons. Finally, one can treat the effect of the separation of the united atom nucleus into two parts by perturbation theory. In principle, this could lead to a scheme for calculation of a molecular wave function in terms of the more easily obtainable atomic wave functions.

### 1. United Atom–Molecule Correlations

In going from the spherical symmetry group of an atom to the $C_{\infty v}$ (or $D_{\infty h}$) group for heteronuclear (or homonuclear) diatomic molecules, the only rotation operations which remain are those about the internuclear axis. Thus the component of angular momentum along the figure axis, but not total angular momentum, is a good quantum number. There is a double degeneracy corresponding to the sign of this component when its value is nonzero.

Mulliken [2] discussed the correlation between united atom and molecular states in some detail. When $L$ and $S$ couple to give total angular momentum quantum number $J$, and the field of the separated nuclei is weak (small $R$), each $J$ state splits into $J + 1$ states. These have $\Omega = J$, $J - 1, \ldots, 0$ or $\frac{1}{2}$, $\Omega$ being the component of $J$ along the figure axis. They are doubly degenerate except for $\Omega = 0$. The behavior on vertical reflection for the state with $\Omega = 0$ is even or odd depending on whether $J$ plus the sum of the angular momentum quantum numbers for the atomic electrons is even or odd. The effect of the nuclear field on the electronic energy for small $R$ is proportional to $R^2$, and was supposed to be given by

$$k_J(3\Omega^2 - J(J + 1))R^2.$$

When $L$ and $S$ are good quantum numbers, the atomic state with angular momentum $L$ gives rise to molecular states with $\Lambda = L, L - 1, \ldots, 0$, where $\Lambda$ is the magnitude of the component of $L$ along the figure axis. This was referred to [2] as the "strong field case." The $\Lambda = 0$ state is $\Sigma^+$ if $L + \sum_i l_i$ is even and $\Sigma^-$ if $L + \sum_i l_i$ is odd, where $l_i$ is the angular momentum of the $i$th electron. The effect of the field on the energy was expected to be given by

$$k_L(3\Lambda^2 - L(L + 1))R^2 + A\Lambda\Sigma,$$

the second term representing the spin–orbit coupling. Mulliken also conjectured that $k_J/k_L$ should be positive for $J = L + S$ and $|L - S|$ and negative for some valves in between. Bingel [3] derived the expressions in $k_J$ and $k_L$ rigorously and showed that Mulliken's conjecture was correct. In the "very strong field" case, the coupling between the angular momenta of individual electrons is broken down, so the $m$-value for each electron is what is important. Their sum gives $\Lambda$. This was expected to be the situation for most molecules when $R$ is near the equilibrium value. Mulliken gave examples of how one can obtain the molecular state from the united atom state.

The correlation of atomic orbitals of the united atom with their molecular counterparts was given by Hund [1]. An atomic orbital with angular momentum quantum number $l$ gives $l + 1$ molecular orbitals, for $m = 0$, $\pm 1, \pm 2, \ldots, \pm l$. Those for $m > 0$ are doubly degenerate. For a homonuclear diatomic, parity is preserved. Thus, for example, the 2p united atom orbital gives a $\sigma_u$ and a $\pi_u$ molecular orbital.

The quantum numbers of the many-electron state may be derived from the specification of which orbitals are occupied, including the component of angular momentum along the internuclear axis for each. We may either derive the atomic terms and correlate them with molecular states, or correlate the atomic orbitals with molecular orbitals first. The states derived from a given atomic configuration (no $m$-specification) are independent of whether the first (strong field) or second (very strong field) procedure is used. For example, the configuration $(ms)(np)$ gives $^1P$ and $^3P$ atomic states. According to the strong field rules, the former yields $^1\Pi$ and $^1\Sigma^+$ states, and the latter $^3\Pi$ and $^3\Sigma^+$ states. The configurations are $(ms)(np\pi)$ and $(ms)(np\sigma)$, which become $(\sigma)(\pi)$ and $(\sigma)(\sigma')$ in the molecule (the prime distinguishes two different $\sigma$ orbitals). The $(\sigma)(\pi)$ configuration gives a $\Pi$ state, which may be a triplet or a singlet; the $(\sigma)(\sigma')$ gives $^{1,3}\Sigma^+$. Other examples were discussed by Hund [1]. The leading term in the energy splitting, which goes as $R^2$, is a sum of orbital contributions and is given later.

The one-electron diatomic ion is a special case because of the separability of the Schrödinger equation. The noncrossing rule (see Volume I, Chapter I, Section A.4) is not the usual one. Baber and Hassé [4] discussed this case in detail, and were unable to give the united atom expansion of the energy in powers of $R$, although they did obtain the term in $R^2$ (the term in $R$ vanishes). By comparison with a result derived by Bethe [5] for the homonuclear case, they were able to give the energy through terms in $R^2$:

$$E = - \frac{(Z_A + Z_B)^2}{2n^2} - \frac{2Z_A Z_B (Z_A + Z_B)^2 R^2 [l(l + 1) - 3m^2]}{n^3 (l + 1)(2l - 1)(2l + 1)(2l + 3)} \tag{223}$$

(atomic units are used here and $n$, $l$, and $m$ are the quantum numbers of the united atom state). Note that the $R^2$ term does not vanish for $l = m = 0$, but becomes $2Z_A Z_B(Z_A + Z_B)^2 R^2/3n^3$. Other results for this case will be discussed later. We turn to the quantitative treatment of the molecule starting from the united atom.

## 2. United Atom Perturbation Theory

Much of the development of the perturbation formalism for the energy and wave function is due to Bingel [6–9]. Initially [6], this was an attempt to derive theoretical results to remove some large errors in the atoms-in-molecules theories, which used united atom properties (see Chapter III, Section A.3). However, a full-scale perturbation formalism was then developed to compute the energy difference between the case of electrons in the field of two nuclei of charges $Z_A$ and $Z_B$ a distance $R$ apart and the case of the same number of electrons in the field of a single nucleus of charge $Z_A + Z_B$. This difference would hopefully be expressed as a power series in $R$. The series should not depend on the location of the united atom nucleus relative to the others. If the perturbation series is to converge rapidly, the diatomic Hamiltonian must differ by a small quantity from that of the united atom. This requires that the nuclei be placed near each other.

Imagine the nucleus of charge $Z_A + Z_B$ to split into fragments of charges $Z_A$ and $Z_B$ and these to be separated to a distance $R$ apart. The perturbation is just the difference of the nuclear-electronic attraction terms for the molecule and the atom. In atomic units, it is

$$H_1 = \sum_{i=1} (-Z_A/r_{Ai} - Z_B/r_{Bi} + (Z_A + Z_B)/r_i) \qquad (224)$$

Here, $r_{Ai}$, $r_{Bi}$, and $r_i$ are the distances of electron $i$ from nucleus A, nucleus B, and the united atom nucleus. To be more definite, we must specify how the separation is made. The three nuclear positions are naturally chosen collinear. To complete the specification, we must give the ratios of $R_A$ and $R_B$, the distances of nuclei A and B from the united atom nucleus, to the A–B distance $R$. Writing

$$R_A = x_A R, \qquad R_B = x_B R, \qquad x_A + x_B = 1 \qquad (225)$$

we may choose $x_A$ and $x_B$ for convenience. Virtually all the authors in this field choose $x_A$ and $x_B$ so the center of charge remains fixed as the nuclei separate. This means

$$x_A = Z_B/(Z_A + Z_B), \qquad x_B = Z_A/(Z_A + Z_B). \tag{226}$$

Bingel [8, 9] has discussed the invariance of the energy expression to the position of the united atom nucleus. He was able to show explicitly that for linear molecules the energy expression through terms in $R^2$ was unchanged by a change in $x_A$ and $x_B$. With a choice other than that of (226) off-diagonal matrix elements enter, whose contributions were treated by sum rules. Another reasonable choice of coordinates would be to move one nucleus only, keeping the origin of coordinates fixed on the other. This corresponds to the perturbation of a negative ion by a point charge at a distance $R$. For the one-electron case, an analytical treatment of this problem is possible [10–12].

Using a spherical coordinate system, centered at the united atom and with the polar axis pointing toward B, the terms $r_{Ai}^{-1}$ and $r_{Bi}^{-1}$ were expanded [8, 9] in Legendre polynomials:

$$Z_A/r_{Ai}+Z_B/r_{Bi} = \sum_{k=0}^{\infty} P_k(\cos\theta)[Z_A(-1)^k f_k(r_i, R_A)+Z_B f_k(r_i, R_B)] \tag{227a}$$

Here, $r$, $\theta$, and $\phi$ are the polar coordinates with $\phi$ the azimuthal angle, and

$$f_k(s, t) = \begin{cases} s^k/t^{k+1}, & s \le t \\ t^k/s^{k+1}, & s \ge t \end{cases} \tag{227b}$$

Including the last term in (224),

$$H_1 = \sum_i \sum_k P_k(\cos\theta_i)$$
$$\times [(Z_A+Z_B)r_i^{-1}\delta_{k,0} - Z_A(-1)^k f_k(r_i, R_A) - Z_B f_k(r_i, R_B)] \tag{228}$$

All the terms in $H_1$ are proportional to powers of $r_i$ and $R$. It was shown [8, 9] that all the matrix elements of $H_1$ between atomic states are proportional to $R^n$ with $n \ge 2$. The distance $R$ appears to be a natural perturbation parameter.

Let us consider the one-electron case first. The united atom is described by the hydrogen-like atomic orbital

$$\psi_j = \psi(r)P_l^m(\cos\theta)e^{im\phi} \tag{229}$$

where $P_l^m$ is an associated Legendre function. The first-order energy is the expectation value of $H_1$ over (229). It is straightforward to show, then, that the leading term in the change from the united atom energy goes as $R^2$. By using the properties of the spherical harmonics, the $R^2$ term in the energy may be expressed in terms of integrals over the radial function $\psi$. The result [1] is, for $l > 0$,

$$E^{(1)} = -R^2 K[l(l + 1) - 3m^2] \tag{230}$$

where $K$ involves these integrals. This corresponds to the leading term in the energy change due to a linear electric field proportional to $R$. For $l = 0$, there is an additional contribution proportional to

$$\tfrac{1}{6}a(Z_B R_B{}^2 + Z_A R_A{}^2) = \tfrac{1}{6}uR^2 Z_A Z_B/(Z_A + Z_B) \tag{231}$$

where $| \psi |^2$ is written as $a + br + cr^2 + \cdots$. Of course $a$ will also depend on $(Z_A + Z_B)$. The final result is that of Eq. (223). For a many-electron problem, one could approximate the energy at small $R$ as the energy of the united atom plus a sum of terms like that above. This corresponds to Mulliken's "very strong field" case.

Let the electronic energy of the atom be written as a power series in $R$ at $R = 0$,

$$E(R) = \sum_i W_i R^i \tag{232}$$

so that $W_0$ is the united atom energy. It was recognized by early workers that $W_1 = 0$, in general. This may be seen from consideration of the electronic force, $-dE/dR$. According to the Hellmann–Feynman theorem (Volume I, Chapter III, Section D) this may be calculated as the expectation value of the force operator. For $R = 0$, the force is $-W_1$, and the expectation value is over the united atom wave function. Since the electronic density for the atom is symmetric to an inversion through the origin, the expectation value vanishes and

$$W_1 = 0. \tag{233}$$

Bingel [7] derived expressions for higher terms in (232) without assuming any particular form for the wave functions. In the spirit of the Brillouin–Wigner perturbation theory, he considered the secular equations

$$(\mathbf{H_U} + \mathbf{H_1})\mathbf{c} = E\mathbf{c}$$

A basis set of united atom eigenfunctions was used. Here $\mathbf{H_U}$ is the matrix of the united atom Hamiltonian in this basis, $\mathbf{H_1}$ the matrix of the perturbation, and $c$ is the vector of expansion coefficients of the exact wave function in these basis functions. The matrix $H_U$ is diagonal with eigenvalues $E_n$. The matrix elements of $H_1$ are to be calculated as power series in $R$. Because $H_1$ is one-electron, the matrix element between states with wave functions $\psi_m$ and $\psi_n$ may be calculated from the spinless first-order transition density

$$\varrho_{mn}(r, \theta, \phi) = N \int \psi_m^*(x_1 \cdots x_N)\psi_n(x_1 \cdots x_N) \, ds_1 \, dx_1 \, dx_2 \cdots dx_N . \quad (234)$$

This quantity is conveniently expanded in spherical harmonics centered at the united atom nucleus

$$\varrho_{mn} = \sum_{L=0}^{\infty} \sum_{M=-L}^{L} \frac{(L - |M|)!}{(L + |M|)!} \frac{2L + 1}{4\pi} P_L^{|M|}(\cos \theta)e^{-iM\phi} \, r^L\varrho_{LM}^{mn}(r). \quad (235a)$$

Here,

$$r^L\varrho_{LM}^{mn}(r) = \int \varrho_{mn} P_L^{|M|}(\cos \theta)e^{iM\phi} \, d(\cos \theta) \, d\phi \quad (235b)$$

In principle, the atomic wave functions are the exact wavefunctions and are mutually orthogonal. Where it is necessary to use approximations to the $\psi_m$, these should also be chosen orthogonal. In either case, the functions may be chosen to be bases for representations of the molecular symmetry group. The perturbation is symmetric to rotations and vertical reflections so $\varrho_{mn}$ may be taken as zero unless it has a symmetric component. If determinantal wave functions are used, $\varrho_{mn}(R)$ vanishes between two determinantal functions which differ by more than one electron. Bingel [8, 9] showed that there is no complication if the united atom state is degenerate, since spin commutes with the perturbation so spin is conserved from atom to molecule, while the $L$-degeneracy is split. The $\Sigma$ terms are nondegenerate, and the twofold degeneracy of $\Pi$, $\Delta$, etc. terms leads to no problems because there are no matrix elements of $V$ between the degenerate $\Pi^+$ and $\Pi^-$ functions.

Using the expansions (235a,b) and (228) Bingel wrote the matrix elements as

$$(H_1)_{mn} = \sum_{L=0}^{\infty} \int_0^{\infty} r^L\varrho_{L0}^{mn}(r)[(Z_A + Z_B)r^{-1}\delta_{L,0} - (-1)^L Z_A$$

$$\times f_L(r, R_A) - Z_B f_L(r, R_B)]r^2 \, dr \quad (236)$$

The term for $L = 0$ is as follows:

$$\int_0^{R_A} \varrho_{00}^{mn} Z_A (r^{-1} - R_A^{-1}) r^2 \, dr + \int_0^{R_B} \varrho_{00}^{mn} Z_B (r^{-1} - R_B^{-1}) r^2 \, dr$$

$$= \int_0^1 \varrho_{00}^{mn}(tR_A) Z_A R_A^{-1}(t^{-1} - 1) t^2 R_A^{3} \, dt$$

$$+ \int_0^1 \varrho_{00}^{mn}(tR_B) Z_B R_B^{-1}(t^{-1} - 1) t^2 R^3 \, dt$$

$$= [Z_A Z_B/(Z_A + Z_B)^2] R^2 \int_0^1 (t - t^2)[Z_B \varrho_{00}^{mn}(tR_A) + Z_A \varrho_{00}^{mn}(tR_B)] \, dt \quad (237)$$

A term for $L > 0$ may be written:

$$\int_0^\infty r^L \varrho_{L0}^{mn}(r)[-Z_A(-1)^L R_A^L - Z_B R_B^L] r^{-L+1} \, dr + Z_A(-1)^L$$

$$\times \int_0^{R_A} r^2 \varrho_{L0}^{mn}(r)\left[\frac{R_A^L}{r^{L+1}} - \frac{r^L}{R_A^{L+1}}\right]$$

$$+ Z_B \int_0^{R_B} r^L \varrho_{L0}^{mn}(r)\left[\frac{R_B^L}{r^{L+1}} - \frac{r^L}{R_B^{L+1}}\right] r^2 \, dr$$

$$= -\left[\frac{Z_A Z_B^L(-1)^L + Z_B Z_A^L}{(Z_A + Z_B)^L}\right] R^L \int_0^\infty \varrho_{L0}^{mn}(r) r \, dr + Z_A(-1)^L$$

$$\times \int_0^1 t^2 R_A^L \varrho_{L0}^{mn}(tR_A)\left[\frac{1}{t^{L+1} R_A} - \frac{t^L}{R_A}\right] R_A^3 t^2 \, dt$$

$$+ Z_B \int_0^1 t^2 R_B^L \varrho_{L0}^{mn}(tR_B)\left[\frac{1}{t^{L+1} R_B} - \frac{t^2}{R_B}\right] R_B^3 t^2 \, dt \quad (238)$$

For $L = 1$, the first term is proportional to $Z_A Z_B - Z_B Z_A$ and vanishes. The other two integrals are proportional to $R_A^{L+2}$ and $R_B^{L+2}$, respectively, so give terms at least third power in $R$. It may thus be stated that $(H_1)_{mn}$ is at least second power in $R$.

Thus the two leading terms in the energy are the united atom energy and a term in $R^2$ arising from a diagonal matrix element of $H_1$. Returning to the secular equation, we note that the contributions of off-diagonal elements depend on the squares of matrix elements, which go at least as $R^4$. Thus it appears that the terms in $R^3$ in the energy expansion arise only from the diagonal elements. But this turns out not to be true, as was shown by Levine [13] for the one-electron homonuclear diatomic. The $R^2$ term in an off-diagonal matrix element between the ground state and an excited $S$ state may be calculated from (237), since $\varrho^{mn}$ is spherically symmetric.

Putting $Z_A = Z_B = \frac{1}{2}Z$ and expanding $\varrho_{00}^{0n}(r)$ as $\varrho(0) + r\varrho_r(0) + \cdots$, this gives

$$\frac{1}{4}ZR^2 \int^1 (t - t^2)\varrho_{00}^{0n}(\tfrac{1}{2}Rt)\, dt = \frac{1}{4}ZR^2\varrho(0)\frac{1}{6} + \cdots$$

where only the $R^2$ term is kept. The wave functions are those for the hydrogenic atom of charge $Z$. Now the state $n$ may be in the continuum. For the excited state of energy $E$, normalized on the energy scale, $\varrho(0)$ becomes [13] $4Z^2[1 - \exp(-2^{1/2}\pi Z E^{-1/2})]^{-1/2}$. The contribution of all the continuum states is

$$\int_0^\infty \frac{dE}{-\frac{1}{2}Z^2 - E}\,|\,(H_1)_{0n}\,|^2 = -\int_0^\infty \frac{\frac{1}{6}Z^3R^2\,dE}{[\frac{1}{2}Z^2 + E][1 - \exp(-2^{1/2}\pi Z E^{-1/2})]^{1/2}}$$

The integral diverges as $E \to \infty$. Levine [13] showed that the continuum states actually contribute to the $R^3$ term in the energy in this case, entering in second-order perturbation theory.

Byers-Brown and Steiner [11, 12] have discussed the electronic energy for the heteronuclear one-electron case more completely. They treated the Schrödinger equation in ellipsoidal coordinates. Among their conclusions was that the terms in $R^2$ and $R^3$, but not higher terms, could be obtained from the united atom perturbation treatment if one went to second order, using Levine's formula for the $R^3$ term. They also showed

$$W_3 = -(Z_A + Z_B)W_2 = -\tfrac{2}{3}Z_AZ_B(Z_A + Z_B)^3 \qquad (239)$$

The simple relation between $W_2$ and $W_3$ can be proved [12] by means of the Hellmann–Feynman theorem (see page 128); Bingel [8, 9] had demonstrated it for a united atom in an $S$ state. The expression for $W_2$ is identical to that given by Eq. (223) with $n = 1$. Perhaps their most surprising conclusion was that a term in $R^5 \log R$ is present in the expansion. Thus the energy is not an analytic function of $R$ at $R = 0$. This means problems are likely to be encountered in any united atom perturbation treatment, and that $R$ is *not* a good perturbation parameter.

Byers-Brown [13a] discussed the problems with the usual united-atom perturbation theories. These include the difficulties of representation of the off-center cusps in the electronic density (Volume I, Chapter III, Eqs. (397)–(398)), the lack of correspondence between the orders of perturbation theory and powers of $R$, and the fact that such theories lead to energies and wavefunctions which are not symmetric to simultaneous exchange of the nuclei and inversion of the electrons. This symmetry, present in the electronic

Hamiltonian, was considered by Wigner and Witmer and their formulation of the correlation rules (Section B.1). He noted the relation between these problems, and proposed a new perturbation theory, starting with a two-center united-atom function. The zero-order function was a linear combination of united-atom functions centered on the two nuclei, the linear combination coefficients being chosen to satisfy the cusp conditions and the symmetry to simultaneous exchange of nuclei and inversion of electrons. This required the use of projection operators and a non-Hermitian perturbation operator [see the discussion of Eqs. (148) *et seq.*]. The theory, and calculations using it, were described by Byers-Brown and Power [13a]. They showed that each term in the expansions of the energy and wavefunction had the proper exchange-inversion symmetry, that convergence was better than for other theories, and that, to first order, results superior to those obtained by other theories could be derived. Thus, the interaction energies for HeH, $He_2$, and HeLi$^+$ were accurate to a few per cent even when $R$ was $1a_0$ or more. However, for neutral or positively charged united atoms in $S$ states, the first-order theory could give no binding.

We conclude this subsection by discussing the leading terms in the usual united-atom energy expansion. As shown by Bingel [8, 9], the term in $R^2$ is obtained from the diagonal element $(H_1)_{00}$ (0 refers to the united atom ground state). A contribution proportional to $R^2$ may arise from the $L = 0$ and $L = 2$ terms, as appears from Eqs. (237) and (238). If $\varrho_{00}^{00}$ is expanded in a power series in $r$, only the leading term gives an $R^2$ energy contribution from (237). Combining this with the first term of (238), we obtain the terms proportional to $R^2$ in $(H_1)_{00}$:

$$W_2 = [Z_A Z_B/(Z_A + Z_B)]\varrho_{00}^{00}(0) \int_0^1 (t - t^2)\, dt$$

$$- [Z_A Z_B/(Z_A + Z_B)] \int_0^\infty \varrho_{20}^{00}(r) r\, dr.$$

In terms of the total atomic charge density $\varrho$, $\varrho_{00}^{00}(0) = 4\pi\varrho(0)$. For a spherically symmetric atom, $\varrho_{20}$ vanishes, and the $R^2$ term in the energy is just

$$W_2 = [Z_A Z_B/(Z_A + Z_B)](2\pi/3)\varrho(0) \tag{240}$$

For the one-electron molecule, this gives Eq. (223) when the proper expression for $\varrho(0)$ is used. For the other $R^2$ term,

$$\int_0^\infty r\varrho_{20}^{00}(r)\, dr = \int P_2(\cos\theta) r^{-3}\varrho(\mathbf{r})\, dt = \tfrac{1}{2} V_{zz}$$

The quantity $V_{zz}$ is the second derivative with respect to $z$, at the position of the nucleus, of the electrostatic potential due to the electrons. It is sometimes referred to as the field gradient in the $z$-direction. Thus, the energy of the molecule through terms in $R^2$ may be predicted from united atom properties. One requires the united atom energy, the united atom electron density at the nucleus, and the field gradient at the nucleus. When the atomic state "0" has $L > 0$ and spin–orbit coupling is small, the $M$-degeneracy is split by the separation of the nuclei, and several molecular states result, differing by the value of $\varLambda = |M_L|$. Bingel [3] derived the dependence of $W_2$ on $\varLambda$ by using the properties of the vector-coupling coefficients. He showed that $\varrho_{0,0}^{M_L,M_L}$, was independent of $M_L$, while

$$\varrho_{2,0}^{M_L,M_L} = \varrho_{2,0}^{L,L} \frac{3M_L{}^2 - L(L+1)}{3L^2 - L(L+1)}$$

The superscript 0 has been replaced, in these formulas, by the value of $M_L$. It followed that

$$W_2(\varLambda) = \frac{Z_A Z_B}{Z_A + Z_B} \frac{1}{6} \varrho_{00}^{L,L}(0) - \frac{3\varLambda^2 - L(L+1)}{3L^2 - L(L+1)} \int_0^\infty r\varrho_{2,0}^{L,L}\,dr$$

This is not quite of the form conjectured by Mulliken [2] (see Subsection 1), which omitted the term in $\varrho_{00}^{L,L}$. For the examples of BH, for which Bingel calculated $W_2(\varLambda)$ from approximate wavefunctions for C, the term in $\varrho_{00}^{L,L}$ was by far more important than the other. While the ordering of the molecular states predicted by the preceding formula (e.g., for the $^3P_g$ ground state, $^3\varPi$ should lie lower than $^3\varSigma^-$) was in accord with experiment, this may be coincidental. Bingel [3] also considered the case where spin–orbit coupling makes $J$, instead of $M_L$, a good quantum number. Here, $\varrho_{0,0}^{M_J,M_J}$ was independent of $M_J$ (component of $J$ along the internuclear axis) and

$$\varrho_{2,0}^{M_J,M_J} = \varrho_{2,0}^{J,J} \frac{3M_J{}^2 - J(J+1)}{3J^2 - J(J+1)}$$

where the states are labeled by $M_J$. Then

$$W_2(\varLambda) = \frac{Z_A Z_B}{Z_A + Z_B} \frac{1}{6} \varrho_{0,0}^{J,J} - \frac{3\varLambda^2 - J(J+1)}{3J^2 - J(J+1)} \int_0^\infty r\varrho_{2,0}^{J,J}\,dr$$

which again differs from Mulliken's conjecture [2] by the first term. From the properties of the vector coupling coefficients, Bingel [3] calculated the ratio

$$\int_0^\infty r\varrho_{2,0}^{J,J}\,dr \bigg/ \int_0^\infty r\varrho_{2,0}^{L,L}\,dr$$

to verify Mulliken's conjecture about its sign as a function of $J$.

The term in $R^3$ in the diagonal matrix element may come from the $L = 0$ term in (236). From (238), we obtain terms in $R^L$ and $R^{L+2}$ ($L$ even), so there is no $R^3$ contribution, while (237) gives a contribution from the second term in the expansion of $\varrho_{00}^{00}$ about the origin. Denoting $d\varrho_{00}^{00}/dr$ evaluated at $r = 0$ by $\varrho_r(0)$, the term becomes $\frac{1}{12}(Z_A R_A^3 + Z_B R_B^3)\varrho_r(0)$. Note that $\varrho_r(0)$ is $4\pi$ times the derivative of the electron density with $r$ at $r = 0$. This is not the $R^3$ term in the energy expression, since it does not take into account the contribution of the off-diagonal matrix elements.

However, the term in $R^3$ may be derived, for the general diatomic molecule, from the term in $R^2$, by use of the Hellmann–Feynman theorem, as shown by Byers–Brown [12]. The sum over matrix elements need not be evaluated. The electronic force operator may be taken as

$$\mathscr{F} = -(\partial/\partial R)(Z_A/r_{Ai} + Z_B/r_{Bi}) \tag{241}$$

with the understanding that an increase of $R$ by $dR$ is produced by increasing $R_A$ by $Z_B\,dR/(Z_A + Z_B)$ and $R_B$ by $Z_A\,dR/(Z_A + Z_B)$ so that the center of nuclear charge remains fixed. Let $r$ and $\theta$ be the polar coordinates, centered on the united atom nucleus, which define an electron's position: These are unchanged as $R$ changes. Then, according to Volume I, Chapter III, Section D [Eqs. (337) and (339), and Subsection 4],

$$\mathscr{F} = [Z_A Z_B/(Z_A + Z_B)](\cos\theta_A/r_A^2 + \cos\theta_B/r_B^2). \tag{242}$$

Byers-Brown expressed the electron density $\varrho$ and the force operator in confocal elliptical coordinates. The force then becomes

$$dU/dR = [Z_A Z_B/(Z_A + Z_B)]2R\pi \int_1^\infty d\xi \int_{-1}^1 d\eta\, f(\xi, \eta)\varrho(\xi, \eta) \tag{243}$$

where

$$f(\xi, \eta) = \frac{1}{2}\left[\frac{(1 + \xi\eta)(\xi - \eta)}{(\xi + \eta)^2} + \frac{(1 - \xi\eta)(\xi + \eta)}{(\xi - \eta)^2}\right].$$

He expanded $f$ and $\varrho$ in Legendre polynomials in $\eta$. Then the integral in (243) became a sum of terms for $l = 0, 2, \ldots$. Only even $l$ appears because $f$ is symmetrical. It was shown that terms with $l > 2$ give contributions to

$dU/dR$ proportional to $R^n$ with $n \geq 3$. On integration, they lead to terms in $U(R)$ proportional to the fourth or higher power of $R$, so they need not be considered here. The $l = 0$ and $l = 2$ contributions were evaluated to give the terms in $dU/dR$ going as $R$ and $R^2$, which give the $R^2$ and $R^3$ terms in $U$ near $R = 0$ on integration. The $R^2$ term in $U$ agreed with that derived by Bingel but the other did not, since excited-state matrix elements contribute. Results for the one-electron case confirm Byers-Brown's result. The $R^2$ term comes from $l = 0$. Its contribution to (243) is

$$T = [Z_A Z_B/(Z_A + Z_B)] 4\pi R \int_1^\infty f_0(\xi) \varrho_0(\xi) \, d\xi$$

where $f_0(\xi) = 2Q_2(\xi)$. (Here, $Q$ is a Legendre function of the second kind.) The condition that the local energy be finite at the nuclei for the exact wave function requires a cancellation between singularities from kinetic and potential energy contributions, as discussed in Volume I, Chapter III, Section E. Equation (398) of that chapter may be written [14] as a condition on the electron density $\varrho_{00}$, defined by Eq. (11):

$$[\partial \varrho/\partial r]_{r=0} = -2(Z_A + Z_B)[\varrho]_{r=0} \tag{244}$$

Byers-Brown expanded $\varrho_0$ about the origin:

$$\varrho_0 = \varrho_0(0) + \tfrac{1}{2} R\xi (\partial \varrho_0/\partial r)_{r=0} + \cdots$$

Using (244) for the second term and substituting into $T$, we have

$$[Z_A Z_B/(Z_A + Z_B)] 8\pi R \int_1^\infty Q_2(\xi)[\varrho_0(0) + R\xi(Z_A + Z_B)\varrho_0(0)] \, d\xi$$

$$= [Z_A Z_B/(Z_A + Z_B)] 4\pi R\varrho_0(0)[\tfrac{1}{3} - \tfrac{1}{2}R(Z_A + Z_B)]$$

The term in $R$ gives Eq. (240) (the term in $V_{zz}$ comes from $l = 2$ in the expansion); the other term is

$$W_3 = -(2\pi/3)Z_A Z_B \varrho_0(0) R^3$$

This is just $-(Z_A + Z_B)$ times Eq. (240), the spherical contribution to the $R^2$ coefficient. Bingel's result [8, 9] thus is general, even when $W_2$ includes a term in $V_{zz}$ in addition to the $L = 0$ term.

Buckingham [15] used the united atom expression for the interaction potential,

$$U = (Z_A Z_B/R) + E_0 + E_2 R^2 \tag{245}$$

to partly determine some of the constants in a proposed potential

$$U = (Z_A Z_B/R)p(R)e^{-aR} \tag{246}$$

where $p(R)$ is a polynomial in $R$ with constant term unity. Equation (245) must hold for small $R$, but (246) is supposed to represent the potential for all $R$. If $E_0$ and $E_2$ are known, as from calculations on the united atom, the terms in $R$, $R^2$, and $R^3$ in $p(R)$ can be expressed in terms of $a$. The form (246) proved to be a useful expression for the potential for several purely repulsive situations. Buckingham and Duparc [16] later used (246) with an additional term $E_3 R^3$ to investigate short-range interactions of hydrogen and helium atoms. Here, the values for $W_2$ and $W_3$ were extracted, for several electronic states, from energies computed from approximate wave functions. The relationship between $W_3$ and $W_2$ was checked. Note that the expression for $W_3$ cited by Buckingham and Duparc [16] for the one-electron atom is our Eq. (241), with

$$\varrho_r(0) = 4\pi(d/dr)[(Z^3/\pi)e^{-2Zr}].$$

## 3. *Applications of United Atom Ideas*

For systems in which the spherical symmetry of the electron cloud is not much disturbed by the separation of the nuclei we might expect that the united atom charge density could give information about molecular properties with very little additional computation. Diatomic hydrides are such systems, and Platt [17] showed that the equilibrium interatomic distances and force constants could be predicted quite well for a large number of such systems from rather crude application of perturbation theory to the united atom wave function. For the hydride MH, he imagined the separation of the proton from the united atom nucleus to take place without disturbing the atomic charge density. This corresponds to computation of energy by first-order perturbation theory. Let $r_{10}$ be the radius of the sphere centered on M which contains $M$ of the $M + 1$ electrons. Simple electrostatics shows that the proton will be repelled from the other nucleus so long as its distance is less than $r_{10}$, whereas it will be attracted by the excess negative charge when its distance exceeds $r_{10}$. Platt then identified the equilibrium internuclear distance, $R_e$, with $r_{10}$. To consider the force constant, he let the proton be moved outward by $dr$, again without changing the electronic density. The attractive force is

$$dF = -(4\pi r^2 \varrho_{10} \, dr)/r^2 \tag{247}$$

since $4\pi r^2 \varrho_{10} dr$ is the excess negative charge which attracts the proton at $r + dr$, where $\varrho_{10}$ is the electron density of the united atom at a distance $r_{10}$. Then the force constant is

$$k = -dF/dr = 4\pi e^2 \varrho_{10} \tag{248}$$

He suggested also that Eq. (248) might perhaps be more fairly tested by using, instead of $\varrho_{10}$, the united atom density at a distance equal to the true value of $R_e$. This density will be denoted by $\varrho_{11}$.

Rather good agreement was observed for a wide range of molecules with respect to both $R_e$ and force constant (using $\varrho_{11}$ for the latter). It was even possible for Platt [17] to analyze properties in terms of contributions of the different united atom orbitals. Platt's model was subsequently extended [18] to polyatomic hydrides.

Bratož *et al.* [18] were able to give a formal derivation of the Platt formulas in terms of perturbation theory. They correspond to neglecting terms higher than first order, so that the energy is simply the expectation value of the molecular Hamiltonian over the atomic wave function centered on the nucleus M. The force is the negative of the derivative of the energy with respect to the distance of the proton from M, and, by the Hellmann–Feynman theorem, is calculable from classical electrostatics. A second differentiation with respect to $R$ recovers Platt's formula for the force constant. Bratož *et al.* discussed the contributions of different orbitals to the force constant. In this connection, they showed that the valence orbitals often contribute little: For many two-electron bonds, they may be neglected, the force constant then being the second derivative of $1/R$, evaluated at the equilibrium value of $R$. They also suggested that some other spherically symmetrical wave function, possibly including parameters depending on $R$, be used instead of the united atom wave function. If the parameters were chosen variationally for each $R$, the Hellmann–Feynman theorem (Volume I, Chapter III, Section D) would be obeyed.

From a theoretical viewpoint, there are some unsatisfactory features to Platt's theory. Platt [17] noted that the Born–Oppenheimer picture (Volume I, Chapter I) has the electrons adjusting instantaneously to nuclear displacements, so it is the nuclei which should be virtually standing still while the electrons move rather than the reverse. Subsequently, Clinton [19] showed that the results of Platt's theory were inconsistent with the virial theorem. It follows from the virial theorem [Volume I, Chapter III, Eqs. (308)–(309)] that a nonvanishing force constant can occur *only* by virtue of the change in electron density with $R$. An explanation for the accuracy of Platt's results in terms of a cancellation of terms was given.

Bader and Ginsburg [20] have recently derived Platt's result from an exact equation for the force constant, Eq. (364) of Volume I, Chapter III, and given some detailed analysis. They stated that the good results are due to cancellation of two errors: (1) the charge density should not be rigidly fixed to the heavy atom and (2) the united atom density is not a good approximation to the density of the hydride.

Hall and Rees [21] pointed out that the inconsistency with the virial theorem may be avoided by allowing the united atom density to adjust to the change in $R$ by a scaling variation (see Volume I, Chapter III, Section C.1). The scaling factor is chosen to minimize the energy of the *molecular* system calculated with the scaled *atomic* wave function. We may use Eq. (297) of Volume I, Chapter III, where $V$ now is the potential energy operator for the united atom and $\lambda W$ is $H_1$ of Eq. (224). Now we must put $Z_A = 1$, $Z_B = Z$, and $R_A = R$. Using Eq. (227) and assuming that $\varrho_{UA}$ is spherically symmetrical,

$$s = 1 + 4\pi\left\{\int_0^\infty \varrho_{UA}(r)[-r_A^{-1} + r^{-1}]r^2\,dr + ZR^{-1}\right\}\Big/(2E_{UA})$$

$$= 1 + 4\pi\left\{\int_0^R \varrho_{UA}(r)[r - R^{-1}r^2]\,dr + ZR^{-1}\right\}\Big/(2E_{UA}).$$

The internuclear repulsion, $ZR^{-1}$, must be included in the potential energy. The new internuclear distance is $R/s$, and the equilibrium value will be determined by

$$Z = 4\pi \int_0^{R/s} s^3\varrho_{UA}(sr)r^2\,dr = 4\pi \int_0^R \varrho_{UA}(r)r^2\,dr$$

(the factor of $s^3$ is from normalization), so that the scale factor is given simply by

$$s = 1 + 4\pi \int_0^R \varrho_{UA}(r)r\,dr/(2E_{UA}).$$

Hall and Rees [21] also derived the equations for the force constant in the scaled theory. They showed that the scaling gave noticeable improvement in predicted equilibrium distances and force constants for $H_2$, LiH, and HF, compared to the unscaled results of Platt. More extensive calculations with the scaled theory by McDugle and Brown [22] on the series of hydrides from $H_2$ to HBr also showed excellent agreement with experiment for $R_e$ and $k_e$. Hall and Rees also suggested a "separated ion" theory, wherein the wave function of the negative ion formed when the proton is removed to infinity is used in the place of that of the united atom. Ion wavefunctions are harder to come by, but it was argued [21] that the separated ion should give better results. Tests showed [21, 22] that this was not at all the case.

Recently, a united atom perturbation theory was used for a calculation of the energy of the HF molecule [23]. The powerful linked-cluster perturbation theory was used. Here, the zero-order wave function was taken as a single determinant and the wavefunctions corresponding to higher orders were linear combinations of determinants formed from a predetermined set of one-electron functions. For HF, the Hartree–Fock wave function for the united atom, neon, was the zero-order function, and the real and virtual Hartree–Fock orbitals were used to construct the determinants. Using the methods of many-body theory, including diagrammatic representation of the contributions to the energy, an expression for the energy as sums of contributions of orbitals was derived. The perturbation in this calculation consists of the change in the nuclear potential plus the difference between the true interelectronic interaction term and the Hartree–Fock approximation to it. The calculated energy of $-100.4186$ a.u. obtained for HF is to be compared with the Hartree–Fock value of $-100.0703$ a.u., the experimental value of $-100.4485$, and a value obtained from a configuration calculation, $-100.3564$ a.u. Furthermore, the computed Hellmann–Feynman force on the $F$ nucleus due to the electrons differed from that due to the proton by only 5% (they should be equal at the equilibrium internuclear distance, which was used in the calculations).

Another perturbation treatment, starting from a spherically symmetric system, is the "molecular puff" treatment of Hauk et al. [24]. This was expected to be useful for $AH_n$-type molecules. The unperturbed wave function, instead of being for the atomic system of nuclear charge $Z_A + n$, is for $Z_A + n$ electrons in the field of a nucleus of charge $Z_A$ surrounded by a spherical shell of positive charge. This is as if the nuclear charges of the protons had been smeared out over a sphere. Spherical symmetry is maintained, and the molecular puff eigenfunction resembles the united atom function far from the center. The corrections were calculated from perturbation theory. Successful calculations were carried out for $H_2^+$, where the central nucleus is absent. Some problems were discussed for the application of this method to homonuclear diatomic molecules at large $R$ [25].

Richardson and Pack [26] suggested the use of united atom wave functions in an approach to molecular energies related to the integral Hellmann–Feynman theorem (Volume I, Chapter III, Section D.4). They showed that the difference in energy between the molecule and the united atom is exactly equal to $\langle \Psi_U \mid H_1 \mid \Psi_M \rangle$, where $H_1$ is the difference between the Hamiltonians of the molecule and the atom, $\Psi_U$ is an exact wavefunction for the united atom, and $\Psi_M$ is an exact wavefunction for the molecule. The hope was that, with approximations to the exact wavefunctions, reliable values for this quantity could be obtained. The computation is relatively simple, since $H_1$ is one-electron.

There are a large number of variational calculations which employ united atom orbitals, to take advantage of the simplifications associated with a one-center system. Integral evaluation is much simplified. These calculations would be most useful for diatomic systems for small $R$, but, if enough basis functions are used, any value of $R$ is possible.

Miller and Present [27] carried out a configuration interaction calculation with united atom basis functions for the $He_2$ ground state, for which $U(R)$ is repulsive. Their results were very good for $R$ less than about one Bohr radius. A previous single configuration calculation by Huzinaga [28] had given unsatisfactory results. The configuration considered by Huzinaga was $(1s)^2(2p\sigma)^2$, with which the ground state molecular orbital configuration, $(1\sigma_g)^2(1\sigma_u)^2$, correlates as $R$ approaches zero (each molecular orbital is considered separately). Here, $1\sigma_g$ means the first (lowest energy) orbital of $\sigma_g$ symmetry. The poorness of Huzinaga's results was due to the fact that the configuration $(1s)^2(2p\sigma)^2$ lies considerably above the ground state configuration, $(1s)^2(2s)^2$, for the united atom, Be. This configuration $(1s)^2(2s)^2$, like the other configuration, gives rise to a $^1\Sigma_g^+$ term. However, the 2s atomic orbital correlates with the $2\sigma_g$ molecular orbital so that the ground state configuration correlates with $(1\sigma_g)^2(2\sigma_g)^2$, which in turn dissociates to He atoms in the $1s2s\,^3S$ state. Thus there is a crossing of the approximate potential curves generated by the two united atom *configurations*, and their interaction must be included in the computation to get good results (see Volume 1, Chapter I, Section B.4). For very small $R$, the molecular wavefunction is dominated by the configuration $(1s)^2(2s)^2$, but, for larger $R$, the configuration $(1s)^2(2p\sigma)^2$ dominates.

Hirschfelder [29] has suggested that avoided crossings of potential energy curves for small $R$ is probably a widespread occurrence, characteristic of many molecules. In the case of $H_2$, there is a crossing between the $(1\sigma_u)^2$ and $(2\sigma_g)^2$ potential curves, associated with a change in the nature of the dominant contribution to the electronic correlation: from left–right to in–out. Where avoided crossings occur for small $R$, the united atom would not be a very appropriate guide to molecular wave functions, except at very small $R$. However, for the first-row diatomic hydrides, Cade and Huo [30] found that correlation energy, which seems to depend more on the number of electron pairs than on their nature, changed less in going

from united atom to molecule than in going from molecule to separated
atoms. Thus the $R$-independence of correlation energy (see Volume I,
Chapter III, Eq. (356)] is more valid for small $R$.

Miller and Present [27] calculated energies for $He_2$ from three united
atom configurations: $(1s)^2(2s)^2$, $(1s)^2(2p\sigma)^2$, and $(1s)^2(3d\sigma)^2$, choosing exponential parameters in the atomic orbitals to minimize the energy in each
case. The crossing between the potential curves corresponding to the first
two configurations occurred at $R \approx 0.56\ a_0$. At $R = 0.5\ a_0$, the single-
configuration energies (including internuclear repulsion) were $-2.14$ a.u.
and $-2.08$ a.u., while the configuration interaction between the two
lowered the energy to $-2.18$ a.u. For $R = 0.75\ a_0$, the single-configuration
energies and the CI energy were, respectively, $-3.18$ a.u., $-3.36$ a.u., and
$-3.36$ a.u., i.e. the CI gave negligible improvement. Similarly, addition
of the $(1s)^2(3d\sigma)^2$ configuration gave very little energy lowering. However,
a wavefunction including just the configurations $(1s)^2(2s)^2$ and $(1s)(3d\sigma)(2s)^2$
gave a much better result: at 0.5 and 0.75 $a_0$, the energies were $-2.365$
a.u. and $-3.489$ a.u. Indeed, below about 0.7 $a_0$, this wave function is
superior to fairly elaborate configuration interaction wave functions using
two-center molecular orbitals. The addition of the $(1s)(3d\sigma)(2s)^2$ configuration provides a distortion of the electron density of the inner orbital from
spherical symmetry. The 3d orbital used here is not at all the Be 3d orbital;
the value of the exponential parameter as determined by variation corresponds to a charge density tightly bound about the nucleus.

Whether one refers to a molecular orbital or other calculation, which
uses atomic-like orbitals centered at one point between the nuclei, as a
united atom calculation or a one-center calculation, is a matter of taste.
For our part, we use the former name when the orbitals actually are those
which would describe the united atom, or when the knowledge of the united
atom wavefunction serves as a guide in setting up the molecular wave
function. Otherwise, we refer to these calculations as single-center calculations (see Chapter II).

## REFERENCES

1. F. Hund, *Z. Phys.* **63**, 719 (1930).
2. R. S. Mulliken, *Rev. Mod. Phys.* **2**, 60, 506 (1930); **4**, 1 (1932).
3. W. A. Bingel, *J. Chem. Phys.* **49**, 1931 (1968).
4. W. G. Baber and H. R. Hassé, *Proc. Cambridge Phil. Soc.* **31**, 564 (1935).
5. H. Bethe, *in* "Handbuch der Physik" (S. Flügge, ed.), 24, 1, p. 527. Springer-Verlag,
   Berlin and New York, 1933.

6. W. A. Bingel, *Z. Naturforsch.* **12a**, 59 (1957).
7. W. A. Bingel, *J. Chem. Phys.* **30**, 1250, 1254 (1959).
8. W. A. Bingel, *Z. Naturforsch.* **16a**, 668 (1961).
9. W. A. Bingel, *J. Chem. Phys.* **38**, 274 (1963).
10. A. Dalgarno and N. Lynn, *Proc. Phys. Soc. London Sect. A* **70**, 223 (1957); P. D. Robinson, *Ibid.* **71**, 828 (1958).
11. W. Byers-Brown and E. Steiner, *J. Chem. Phys.* **44**, 3934 (1966).
12. W. Byers-Brown, *Discuss. Faraday Soc.* **40**, 140 (1966).
13. I. N. Levine, *J. Chem. Phys.* **40**, 3444 (1964); **41**, 2044 (1964).
13a. W. Byers-Brown, *Chem. Phys. Lett.* **1**, 655 (1968). W. Byers-Brown and J. D. Power, *Proc. Roy. Soc. Ser. A* **317**, 545 (1970).
14. E. Steiner, *J. Chem. Phys.* **39**, 2365 (1963).
15. R. A. Buckingham, *Trans. Faraday Soc.* **54**, 453 (1958).
16. R. A. Buckingham and D. M. Duparc, *J. Chem. Phys.* **38**, 275 (1963).
17. J. R. Platt, *J. Chem. Phys.* **18**, 932 (1950).
18. S. Bratož, R. Daudel, M. Roux, and M. Allavena, *Rev. Mod. Phys.* **32**, 412 (1960).
19. W. L. Clinton, *J. Chem. Phys.* **33**, 1603 (1960).
20. R. F. W. Bader and J. L. Ginsburg, *Can. J. Chem.* **47**, 3061 (1970).
21. G. G. Hall and D. Rees, *Mol. Phys.* **5**, 279 (1962).
22. W. G. McDugle and T. L. Brown, *J. Amer. Chem. Soc.* **89**, 3114 (1967).
23. T. Lee, N. C. Dutta, and T. P. Das, *Phys. Rev. Lett.* **25**, 204 (1970).
24. P. Hauk, R. G. Parr, and H. F. Hameka, *J. Chem. Phys.* **39**, 2085 (1963); P. Hauk and R. G. Parr, *Ibid.* **43**, 548 (1965).
25. Z. Dvořáček and Z. Horák, *J. Chem. Phys.* **43**, 874 (1965).
26. J. W. Richardson and A. K. Pack, *J. Chem. Phys.* **41**, 897 (1964).
27. R. V. Miller and R. D. Present, *J. Chem. Phys.* **38**, 1179 (1963).
28. S. Huzinaga, *Progr. Theor. Phys.* **17**, 169 (1957).
29. J. O. Hirschfelder, *J. Chem. Phys.* **43**, S199 (1965).
30. P. E. Cade and W. M. Huo, *J. Chem. Phys.* **47**, 614 (1968).

## SUPPLEMENTARY BIBLIOGRAPHY

Below are listed articles which came to our attention after completion of the manuscript. We have given journal references, titles, and, in some cases, additional information about content. At the end of each reference, the Section of this Chapter, to which the reference is relevant, is indicated in parentheses.

E. A. Andreev, On Asymptotic Calculation of the Exchange Interaction. *Theor. Chim. Acta* **28**, 235 (1973). (B2, B3)

M. V. Basilevsky and M. M. Berenfeld, Intermolecular Interactions in the Region of Small Overlap (dispersion interaction by perturbation theory). *Int. J. Quantum Chem.* **6**, 23 (1972). (A)

R. Boehm and R. Yaris, Van der Waals Forces Including Exchange in the Small Overlap Region (using many-body Green's Function techniques). *J. Chem. Phys.* **55**, 2620 (1971). (B)

R. A. Bonham, ...Optical Sums from Experimental Generalized Oscillator Strengths and ... the X-ray Incoherent Scattering Factor. *Chem. Phys. Lett.* **18**, 454 (1973). (A5, A6)

E. Brändas, A Remark on Perturbation Theory for Intermolecular Forces. *Int. J. Quantum Chem.* **4**, 285 (1971). (B2)

E. Brändas and O. Goscinski, Is Second-Order Perturbation Theory Sufficient to Treat Second-Order Properties (such as dispersion forces)? *Int. J. Quantum Chem.* **3S**, 383 (1970). (A1, B2)

P. A. Braun and T. K. Rebane, Variational Bounds for Imaginary Frequency Polarizability and Dispersion Interaction Constant. *Int. J. Quantum Chem.* **6**, 639 (1972). (A6, A7)

C. R. A. Catlow and M. R. Hayns, A Computational Study of the $F^--F^+$ Interionic Potential. *J. Phys. Ser. C,* **5**, 1237 (1972). (A3)

Y.-N. Chiu, Reformulation and Extension of Wigner-Witmer and Mulliken Correlation Rules for ... Diatomic Molecules Upon Dissociation. *J. Chem. Phys.* **58**, 722 (1973). (B1)

A. Conway and J. N. Murrell, The Dependence of Exchange and Coulomb Energies on Wave Functions of the Interacting Systems (for $He_2$). *Mol. Phys.* **23**, 1143 (1972). (B)

N. C. Dutta, C. M. Dutta, and T. P. Das, Many-Body Approach to the Properties of Interacting Atoms II. *Int. J. Quantum Chem.* **4**, 299 (1971). (A, B)

J. O. Hirschfelder, Primitive Exchange Perturbation Theory IV. *Int. J. Quantum Chem.* **4**, 257 (1971). (B2)

J. O. Hirschfelder and D. M. Chipman, On Optimizing the Treatment of Exchange Perturbations. *Chem. Phys. Lett.* **14**, 293 (1972). (B2)

S. Kaneko, Van der Waals Constants... for He and Ne by the Coupled Hartree-Fock Method, *J. Chem. Phys.* **56**, 3417 (1972). (A6)

D. J. Klein, Symmetry-Adapted Perturbation Theory for Interatomic and Intermolecular Exchange Interactions. *Int. J. Quantum Chem.* **4**, 271 (1971). (B2)

U. Landman and R. Pauncz, Remarks on Projection Operators Used in the Theory of Intermolecular Interactions, *Chem. Phys. Lett.* **9**, 489 (1971). (B2)

P. W. Langhoff, Sum-Rule Moment Bounds on (van der Waals Coefficients). *Chem. Phys. Lett.* **12**, 217 (1971). (A6)

P. W. Langhoff, Moment Theory Bounds on Long-Range Casimir-Polder Interactions. *Chem. Phys. Lett.* **12**, 223 (1971). (A6)

W. C. Mackrodt, Approximate Single-Center Perturbation Calculations. I. Ground States (of two-electron systems). *J. Chem. Phys.* **54**, 2952 (1971). (C3).

D. R. McLaughlin and H. F. Schaefer III, Interatomic Correlation Energy and the van der Waals Attraction Between Two Helium Atoms (by configuration interaction). *Chem. Phys. Lett.* **12**, 244 (1971). (A)

M. A. J. Michels and L. G. Suttorp, Multipole Expansion of the Retarded Interatomic Dispersion Energy. *Physica* **59**, 609 (1972). (A)

J. H. Miller and H. P. Kelly, Dipole Polarizability of the Neutral Carbon Atom and the Dipole-Dipole Interaction Between Carbon Atoms (by many-body perturbation theory). *Phys. Rev.* **A5**, 516 (1972). (A5)

J. N. Murrell and G. Shaw, The Helium-Helium Potential in the Region of the van der Waals Minimum. *Mol. Phys.* **15**, 325 (1968). (A, B)

R. T. Pack, Upper and Lower Bounds to van der Waals Force Constants from Half-Integer Oscillator Strength Sums. *Chem. Phys. Lett.* **13**, 205 (1972). (A6)

P. Pechukas and R. N. Zare, Elementary Derivation of Some of the Wigner-Witmer Rules. *Am. J. Phys.* **40**, 1687 (1972). (B1)

T. R. Singh, J. F. Buhta and W. J. Meath, Long Range Interaction Energies Using Gaussian Basis Sets and a One Center Method. *Int. J. Quantum Chem.* **6**, 201 (1972). (A)

G. Starkschall, Improved Rigorous Bounds to H Atom-Rare Gas Dispersion Forces. *J. Chem. Phys.* **56**, 5728 (1972). (A6)

G. Starkschall and R. G. Gordon, Error Bounds for $R^{-8}$ Dispersion Forces Between Atoms. *J. Chem. Phys.* **56**, 2102 (1972). (A)

# Chapter II  Intermediate R

In this chapter, we discuss the methods generally used to derive $U(R)$ for values of $R$ where expansions in inverse powers of $R$ do not converge rapidly. These methods involve variational calculations in which the trial function is either a single determinant (some sort of Hartree–Fock calculation) or a linear combination of determinants. The basic ideas involved were already discussed in Volume I, Chapter III, Section A, and we shall refer to that material. Here, we shall be going into more detail, emphasizing problems of particular relevance to diatomic molecules in their ground electronic states.

Section A is concerned with single-determinant functions. We begin by discussing the properties of solutions to the Hartree–Fock equations and methods used for their determination (Subsections 1 and 2). We then discuss the spin and symmetry properties of a single determinant, and the restricted Hartree–Fock function.

In Section B we turn to wave functions that involve several determinants. We refer to a multideterminant function as a configuration interaction (CI) function when not all the determinants belong to the same orbital configuration (orbitals differing only in projection of spin or orbital angular momentum being considered as identical). Multideterminant wave functions belonging to a single configuration are necessary, in some cases, to give proper spin and symmetry behavior. Methods for generating such functions are discussed in Subsections 1 and 2 of Section B. By configuration interaction, we can, in principle, approach the exact wave function. As discussed in Subsection 3, the choice of configurations to be included is made on the basis of experience or theoretical considerations.

The usual way of obtaining approximations to the orbitals in single-determinant and multideterminant functions is by expanding the orbitals in a set of basis functions. The choice of basis functions is of great importance, since it determines how accurately the desired functions can be approximated with a given amount of computer time. These problems are discussed in Section A, Subsections 2 and 3.

Section C considers the results of calculations by the methods of Sections A and B. These calculations do not give an explicit $R$-dependence of the energy, but a sequence of energies for different values of $R$. In Section C, Subsection 1, we discuss problems related to the derivation of $U(R)$ and spectroscopic properties from such results. The errors in restricted Hartree–Fock calculations, and to a lesser extent in other calculations, are well enough understood so that we know how to make modifications which lead to accurate predictions of properties of $U(R)$, as discussed in Section C.2. In Sections C.3 and C.4 we survey the accomplishments of the Hartree–Fock and configuration interaction calculations for ground state diatomic interaction potentials.

## A. Single Determinant Wave Functions

### 1. Unrestricted Hartree–Fock Functions

In Volume I, Chapter III, Section A, we minimized the expectation value of the electronic Hamiltonian over a single determinant with respect to variations in the spin orbitals from which the determinant was formed. The results was a set of eigenvalue equations satisfied by the spin orbitals $\{\lambda_k\}$, Eqs. (68) and (69). These equations are:

$$h^{\mathrm{HF}}(i)\lambda_k(i) = \big(h(i) + F(i)\big)\lambda_k(i) = \varepsilon_k\lambda_k(i) \tag{1}$$

$$F(i) = \sum_{l=1}^{N} \big(\langle\lambda_l(j) \mid g(i,j) \mid \lambda_l(j)\rangle_j - \langle\lambda_l(j) \mid g(i,j)\mathscr{P}_{ij} \mid \lambda_l(j)\rangle_j\big) \tag{2}$$

We will refer to Chapter III, Section A.5 for explanation of the notation. A determinant of lowest energy formed from spin orbitals satisfying Eqs. (1) and (2) is an unrestricted Hartree–Fock function. It is the best function of determinantal form, since no restrictions were imposed on the spin orbitals. Since the single-determinant function is to be an approximation to the exact eigenfunction, we would like the former to share some properties the latter is known to possess. Among these are the fact that it is an eigenfunction of total electronic spin (spin–orbit coupling is ignored) and belongs to some symmetry species of the diatomic molecule group. We shall consider such solutions in later sections. Here, we consider determinants built up from solutions to (1) and (2) which may not have these properties.

The most general form for a spin orbital is

$$\lambda_i(1) = \varphi_i^+(\mathbf{r}_1)\alpha(1) + \varphi_i^-(\mathbf{r}_2)\beta(1) \tag{3}$$

where $\varphi_i^+$ and $\varphi_i^-$ are independent spatial functions satisfying

$$\int |\varphi_i^+|^2 \, d\tau + \int |\varphi_i^-|^2 \, d\tau = 1.$$

However, it is virtually universally assumed that each spin orbital in the determinant is of the more restricted form

$$\lambda_i(1) = \phi_i(\mathbf{r}_1)\alpha(1) \qquad \text{or} \qquad \phi_i(\mathbf{r}_1)\beta(1) \tag{4}$$

It is true that such spin orbitals form a complete set for the discussion of any one-electron problem, but assuming that each spin orbital in a determinant is a product of a spatial function and a spin function constitutes a restriction beyond determinantal form. Although some authors have expressed reservations, solutions to the Hartree–Fock equations with the condition (4) are generally referred to as unrestricted Hartree–Fock functions, and we shall follow this usage. The term "restricted Hartree–Fock" is used in a more specific sense, to be indicated later.

The restriction (4) does not prevent formulation of the canonical Hartree–Fock equations (1) and (2). In their derivation, we note that the off-diagonal Lagrange multiplier $\varepsilon_{lk}$ [Volume I, Chapter III, Eq. (66)] vanishes when $\lambda_l$ and $\lambda_k$ have different spins. The matrix of the $\varepsilon_{lk}$ is block-diagonal and can be diagonalized by a transformation that mixes only spin-orbitals of the same spin. Then the solutions to the canonical Hartree–Fock equations will also be of the form (4).

The spin functions $\alpha$ and $\beta$ have the properties (EYRING, Sect. 9a; PILAR, Sect. 11-3; KAUZMANN, Sect. 9Db)

$$S_z\alpha = \tfrac{1}{2}\alpha, \qquad\qquad\qquad S_z\beta = -\tfrac{1}{2}\beta \tag{5}$$

$$(S_x + iS_y)\alpha = 0, \qquad\qquad (S_x + iS_y)\beta = \alpha$$

$$(S_x - iS_y)\alpha = \beta, \qquad\qquad (S_x - iS_y)\beta = 0 \tag{6}$$

$$S^2\alpha = (S_x^2 + S_y^2 + S_z^2)\alpha = \tfrac{3}{4}\alpha, \qquad\qquad S^2\beta = \tfrac{3}{4}\beta \tag{7}$$

where $S_x$, $S_y$ and $S_z$ are the operators for the components of spin (in units of $\hbar$) along three spatial directions, the $z$-direction being the axis of quantization. Where there are several electrons, we will write

$$S_z(i)\alpha(i) = \tfrac{1}{2}\alpha(i)$$

and so on. With the assumption (4), the determinant is automatically an eigenfunction of $S_z$ (total). Another simplification is that when $F$ [Eq. (2)] operates on a spin orbital $\lambda_k$, the last operator (exchange operator) gives a vanishing contribution when $\lambda_k$ and $\lambda_l$ are associated with different spins, since the integration over $j$ includes a formal integration over the spin variable. There is also a consequence for the orthonormality of the spatial functions.

Since they are eigenfunctions of the operator $h^{HF}$, spin-orbitals associated with different orbital energies are automatically orthogonal. Now if $\lambda_i$ and $\lambda_j$ have the same spin, this implies that

$$\int \phi_i^* \phi_j \, d\tau = 0$$

where $\phi_i$ and $\phi_j$ are the spatial parts of $\lambda_i$ and $\lambda_j$. Orthogonality of the spatial parts is not assured when $\lambda_i$ and $\lambda_j$ have opposite spins. It is sometimes convenient to go over to so-called corresponding orbitals for which the spatial parts of most spin orbitals are orthogonal. This can be accomplished by a transformation of the $\{\lambda_i\}$ of spin $\alpha$ among themselves and similarly for those of spin $\beta$ [1, 1a]. The corresponding orbitals are closely related [1, 2] to the natural orbitals, which diagonalize the first-order density matrix, and to the orbitals from a restricted Hartree–Fock calculation, in which the spatial parts of some orbitals for $\alpha$ and $\beta$ spin are forced to be identical.

Considering the widespread use of the Hartree–Fock method, it is surprising that some basic questions about the properties of the solutions are only now being treated. The problem is that the equations determining the orbitals are nonlinear because of the occurrence of the orbitals in the Fock operator. It may be expected that there are more solutions to these equations than to ordinary eigenvalue equations. Different Hartree–Fock functions for the same system are eigenfunctions of different operators. The Hartree–Fock equations may be referred to as pseudoeigenvalue equations because, for a fixed choice of occupied orbitals, they become eigenvalue equations. In addition, the determinant constructed from solutions to the Hartree–Fock eigenvalue equations is guaranteed only to make the energy stationary with respect to small variations in the orbitals, not necessarily a true minimum. This is referred to as the stability problem, and is discussed below.

It seems that usually one can obtain a good approximate ground state wave function by filling the orbitals that are likely to be of lowest energy (on the basis of correlations with atomic orbitals, nodal properties, etc.) and considering variations of the orbitals maintaining the configuration. But this does not mean another configuration cannot be found which leads to a lower energy. The implicit assumption is that one gets the determinant of lowest energy by filling the spin orbitals of lowest orbital energies (for instance, at each stage of an iterative process). Examination of the energy expressions shows this need not be the case, and one can exhibit situations where it is not, as mentioned later.

First, we want to indicate the variety of possible solutions to Eqs. (1) and (2) even for the simplest systems. For example, a solution to the equations for $H_2$ at infinite $R$ is

$$\psi^\infty = \mathscr{A}\{1s_A(1)\alpha(1)1s_B(2)\beta(2)\}$$

where $1s_A$ and $1s_B$ are hydrogen atom 1s orbitals on the two centers. The "molecular orbitals" are not symmetry orbitals ($1s_A$ and $1s_B$ form a two-dimensional reducible representation) and $\psi^\infty$ is not an eigenfunction of spin or of spatial symmetry operators. Perhaps the fact that $\psi^\infty$ gives the correct energy as $R$ becomes infinite, unlike the conventional "Hartree–Fock function," which approaches

$$\tfrac{1}{2}\mathscr{A}\{[1s_A(1) + 1s_B(1)]\alpha(1)[1s_A(2) + 1s_B(2)]\beta(2)\}$$

for $R \to \infty$, compensates for its unaesthetic qualities.

Functions like $\psi^\infty$ were introduced by Coulson and Fischer [3] in a study of the molecular orbital method. They first constructed approximate molecular orbitals for $H_2$ as the symmetric and antisymmetric combinations of screened 1s orbitals on the two centers. The screening constant $c$ was sometimes taken as 1 (hydrogen atom orbitals) and sometimes chosen variationally:

$$\phi_A(1) = (c^3/\pi)^{1/2} \exp(-cr_{A1}) \tag{8}$$

Doubly occupying the lower-energy molecular orbital gives

$$\psi_0 = [2(1 + S)]^{-1}\mathscr{A}\{(\phi_A(1) + \phi_B(1))\alpha(1)(\phi_A(2) + \phi_B(2))\beta(2)\}$$

which is a good approximation to the conventional (restricted) Hartree–Fock wave function. Here, $S$ is the overlap between $\phi_A$ and $\phi_B$. The function

$\psi_0$ is totally symmetric under the spatial symmetry operations (as is the occupied orbital itself), and is an eigenfunction of total spin with eigenvalue zero. (EYRING, Sect. 12a, LEVINE, Chap. 10). Factoring out the spin (as can always be done in the two-electron case) we have

$$\psi_0 = [2(1 + S)]^{-1}[\phi_A(1) + \phi_B(1)][\phi_A(2) + \phi_B(2)]\mathscr{S}(1, 2). \tag{9}$$

Here, $\mathscr{S}(1, 2)$ is the singlet function, $2^{-1/2}[\alpha(1)\beta(2) - \beta(1)\alpha(2)]$. For large R, (9) represents a mixture of ionic states [such as $\phi_A(1)\phi_A(2)$] and covalent states [such as $\phi_A(1)\phi_B(2)$], which is incorrect, and leads to too high an energy. For $R$ near $R_e$, it is a fairly good wave function. Next Coulson and Fischer considered the spatial wave function

$$\eta[\phi_A(1) + d\phi_B(1)][\phi_B(2) + d\phi_A(2)] \tag{10}$$

where $\eta$ is a normalizing factor. When $d = 1$, this corresponds to (9), while $d = 0$ gives the proper large-$R$ (covalent) function of the valence bond theory. When $d < 1$, multiplying (10) by $\mathscr{S}(1, 2)$ and antisymmetrizing does not give a single determinant. A single determinant is obtained by multiplying by $\alpha(1)\beta(2)$ and antisymmetrizing; for $d = 0$, this corresponds to $\psi^\infty$.

Coulson and Fischer [3] determined $d$ variationally for different values of $R$, with $c$ [Eq. (8)] equal to previously calculated values for the molecular orbital function. They found $d$ close to unity up to $R = 2.27\,a_0$, with $d$ suddenly decreasing and dropping to zero rapidly thereafter. This emphasizes that one may stay within the molecular orbital method and still get proper behavior for large $R$, if we abandon certain restrictions on the orbitals. The "incorrect dissociation" ascribed to the Hartree–Fock function for $H_2$ actually applies to the restricted Hartree–Fock function which resembles $\psi_0$, Eq. (9). This function also satisfies (1) and (2), and has proper symmetry behavior of the one-electron and determinantal functions (see Section A4).

The slope of the curve of $d$ versus $R$ can, in fact, be discontinuous. This is because the determinantal wave function is an eigenfunction of $H^{HF}$ [see Volume I, Chapter III, Eq. (70)], which involves the spin orbitals. Thus different wavefunctions obeying the Hartree–Fock equation are actually eigenfunctions of *different* Hamiltonians. Therefore they need not be orthogonal to each other, and there is no noncrossing rule. Then the following could happen: Suppose one obtains several determinantal functions, say $\psi_0$ and

$$\psi_1 = \det[\phi_A(1)\alpha(2)\phi_B(2)\beta(2)]$$

(corresponding to $\psi^\infty$), for several values of $R$, and tabulates their energies. This will generate two potential curves $U_0(R)$ and $U_1(R)$, the former lying lower for small $R$ and the latter for large $R$. Suppose they cross at $R_c$. A calculation which finds the lowest energy determinant for each value of $R$ will give $\psi_0$ for $R < R_c$ and $\psi_1$ for $R > R_c$. Thus $d$ in (10) can change discontinuously at $R_c$, where one goes to another Hartree–Fock function. On the other hand, if we used $\psi_0$ at each $R$ to assure proper spin and spatial symmetry (see Section A4), and varied parameters in the atomic orbitals to minimize the energy, we would never find $\psi_1$.

King and Stanton [4] showed examples of related behavior in other systems. They treated closed-shell molecular systems at infinite nuclear separations so that atomic orbitals on different centers have zero differential overlap. Then, expanding the molecular orbitals in these atomic orbitals, they showed that the molecular Hartree–Fock equations reduce to Hartree–Fock equations for the atomic fragments, coupled by the requirement that the orbital energy be a constant from one fragment to another. The number of electrons in a fragment, which enters as a weight on the Fock operator, need not be integral. For the heteronuclear two-electron diatomic hydride HZ, there were solutions to be equations corresponding to $H^+Z^{z-2}$, $H^-Z^z$, and $H^{1-2q}Z^{z-2p}$ (two states), where the charge on nucleus Z is $z$ and $q + p = 1$. As Z changed, the lowest energy state went from one to another of these, and discontinuously. It is also interesting that the Hartree–Fock energy of these solutions for large $R$ was of the form $E(\infty) - Q/R$, even for the $H_2$ case, for which the fragments are nominally neutral.

Stanton [5] was able for the first time to derive an expression for the number of solutions to the Hartree–Fock equations for the special case of a closed-shell, two-electron system, in which the occupied orbital is expanded in a set of $N$ basis functions (these are the Roothaan equations of Volume I, Chapter III, Eqs. (74)–(76)). For a linear variation problem, we would find $N$ solutions. For the Hartree–Fock problem, Stanton showed there may be as many as $\frac{1}{2}(3^N - 1)$, not all corresponding to energy minima. For $N = 2$, there may be from two to four solutions, depending on the basis functions and the Hamiltonian, and cases were presented where two minimum energy solutions existed. Worse, in some of these the occupied orbital had higher energy than the virtual one, so that the ground state was formed by doubly occupying the orbital of higher energy. Other cases of failure of the simple idea that filling the orbitals of lowest energy minimizes the total energy have been discussed by Adams [6] and Watson [7]. In the closed-shell two-electron case, the condition for stability, i.e., that small variations in the orbitals necessarily raise the energy, could be written down explicitly and this solution was shown to correspond to a true energy minimum [5].

Work on the stability problem is more often than not concerned with atomic, rather than molecular, problems [6–8]. One goal of such work would be to prove, if possible, that certain solutions to the Hartree–Fock equations actually correspond to minima, rather than maxima or saddle points, on the energy surface. It is necessary to consider changes in the energy through second-order terms.

Adams [6] considered the problem in a density matrix formulation. He began with a Hartree–Fock function having the density matrix $\rho$ [see Eq. (77) of Volume I, Chapter III], and supposed it to change to a function having the density matrix $\rho'$. The difference of $\rho'$ from $\rho$ corresponds to arbitrary first- and second-order variations in the orbitals, but it is assumed that $\rho'$ is idempotent through second-order terms, so that it still represents a normalized single determinant to this order. Adams [6] wrote

$$\rho' = \rho + \lambda\rho^{(1)} + \lambda^2\rho^{(2)} \tag{11}$$

where $\rho^{(1)} = \mathbf{v} + \mathbf{v}^\dagger$ and $\rho^{(2)} = \mathbf{v}\mathbf{v}^\dagger - \mathbf{v}^\dagger\mathbf{v} + \mathbf{v}' + \mathbf{v}'^\dagger$; here,

$$\mathbf{v} = (1 - \rho)\Delta\rho, \qquad \mathbf{v}' = (1 - \rho)\Delta'\rho \tag{12}$$

with $\Delta$ and $\Delta'$ being arbitrary matrices. Idempotency through terms in $\lambda^2$ is assured by these conditions. This is easily seen by writing out $\rho'\rho' - \rho'$; using the idempotency of $\rho$, the terms in $\lambda$ and $\lambda^2$ may be shown to vanish. Normalization (tr $\rho' = 1$) is also assured. $E(\rho')$ was computed [6] through terms in $\lambda^2$ (second order). Terms linear in $\lambda$ vanish because $\rho$ represents a Hartree–Fock function, for which the energy is stationary. The terms in $\lambda^2$ were:

$$E^{(2)} = \text{tr}[\mathbf{h}(\mathbf{v}\mathbf{v}^\dagger - \mathbf{v}^\dagger\mathbf{v}) + \tfrac{1}{2}\mathbf{G}^{(1)}(\mathbf{v} + \mathbf{v}^\dagger)]$$

where $\mathbf{G}^{(1)}$ is $\mathbf{G}$ [the Fock matrix, Eq. (89) of Volume I, Chapter III], computed with $\rho^{(1)}$ instead of $\rho$. The condition that $E^{(2)}$ be positive is that the eigenvalues of the matrix in square brackets be positive.

Adams obtained a simpler expression by writing the matrices in the representation of the real and virtual Hartree–Fock orbitals. Then he was able to write $E^{(2)}$ as a sum of two parts, one of which is positive whenever the orbital energies of the occupied orbitals are below those of the virtual orbitals. The second part can lead to instability in this case if it is of larger size than the first. It involves Coulomb and exchange integrals, and Adams was able to argue that it would not be expected to give trouble except where differences in orbital energies are small.

Čížek and Paldus [8] also considered the effect on the energy of a Slater determinant, $\Phi_0$, of small variations of the occupied spin orbitals. They started with a complete orthonormal set of spin orbitals $\lambda_i$, the first $N$ of which are occupied in $\Phi_0$.

$$\Phi_0 = \mathscr{A} \prod_{i=1}^{N} \lambda_i(i) \tag{13}$$

The small variation of $\lambda_i$ can be taken as addition of an arbitrary linear combination of the spin-orbitals not occupied in $\Phi_0$, i.e.,

$$\lambda_i \rightarrow \lambda_i + \sum_{m=N+1}^{\infty} c_i^m \lambda_m$$

where the $c_i^m$ are small (first order). Then $\phi_0$ becomes $\Phi_0 + \Phi_1 + \Phi_2 + \cdots$ where $\Phi_1$ consists of a sum of determinants differing from $\Phi_0$ by one spin orbital, $\Phi_2$ a sum of determinants differing from $\Phi_0$ by two spin orbitals, and so on. The sum $\Phi_1$ is of first order of smallness, $\Phi_2$ of second order. The determinantal functions of $\Phi_1$ and $\Phi_2$ are orthogonal to $\Phi_0$, $\langle \Phi_0 | \Phi_0 \rangle = 1$, and, if $\Phi_0$ is a Hartree–Fock determinant, $\langle \Phi_0 | H | \Phi_1 \rangle$ vanishes by the Brillouin theorem. Using these facts, the change in energy to second order is

$$\Delta E = \langle H \rangle_\phi - \langle H \rangle_{\Phi_0} = \langle \Phi_1 | \tilde{H} | \Phi_1 \rangle + \langle \Phi_0 | \tilde{H} | \Phi_2 \rangle + \langle \Phi_2 | \tilde{H} | \Phi_0 \rangle \tag{14}$$

where $\tilde{H} = H - \langle H \rangle_{\Phi_0}$. The last two terms on the right are linear combinations of matrix elements of $H$ between $\Phi_0$ and determinants differing from it in two spin orbitals, while $\langle \Phi_1 | H | \Phi_1 \rangle$ consists of matrix elements between determinants differing from $\Phi_0$ in one spin orbital. The coefficient of the determinant $\Phi^{i \rightarrow m}$ [see Volume I, Chapter III, Eqs. (100) et seq., for notation] in $\Phi_1$ is $c_i^m$ and the coefficient of $\Phi^{i \rightarrow m, j \rightarrow n}$ is $c_i^m c_j^n$. Thus

$$\langle \Phi_1 | \tilde{H} | \Phi_1 \rangle = \sum_{i,m} \sum_{j,n} (c_i^m)^* c_j^n \langle \Phi^{i \rightarrow m} | H | \Phi^{j \rightarrow n} \rangle$$

and

$$\langle \Phi_0 | \tilde{H} | \Phi_2 \rangle + \langle \Phi_2 | \tilde{H} | \Phi_0 \rangle = \sum_{i,m} \sum_{j,n} (c_i^m c_j^n) \langle \Phi_0 | \tilde{H} | \Phi^{i \rightarrow m, j \rightarrow n} \rangle$$

$$+ \sum_{i,m} \sum_{j,n} (c_i^m c_j^n)^* \langle \Phi^{i \rightarrow m, j \rightarrow n} | \tilde{H} | \Phi_0 \rangle$$

The matrix elements on the right may be calculated by formulas such as Eqs. (100)–(102) of Volume I, Chapter III.

$\Delta E$ is a quadratic form in the $\{c_i{}^m\}$. If we write the $c_i{}^m$ as a column vector (or matrix) $\mathbf{D}$,

$$\Delta E = \tfrac{1}{2}(\mathbf{D}^*\mathbf{D})\begin{pmatrix} \mathbf{A} & \mathbf{B} \\ \mathbf{M} & \mathbf{N} \end{pmatrix}\begin{pmatrix} \mathbf{D} \\ \mathbf{D}^* \end{pmatrix} \tag{15a}$$

where inspection of the formulas for the matrix elements shows that $\mathbf{M} = \mathbf{B}^*$ and $\mathbf{N} = \mathbf{A}^*$. If $\Phi_0$ is to be stable, $\Delta E$ must be positive. Since the coefficients in $\mathbf{D}$ are arbitrary, this requires that

$$\mathbf{E} = \begin{pmatrix} \mathbf{A} & \mathbf{B} \\ \mathbf{B}^* & \mathbf{A}^* \end{pmatrix} \tag{15b}$$

be positive definite, which means that it must have only positive eigenvalues. Suppose $\mathbf{E}$ has a negative eigenvalue $f_n$ with eigenvector $\mathbf{F}_n$, normalized so that $\mathbf{F}_n{}^\dagger\mathbf{F}_n = 1$. Choosing $(\mathbf{D}\mathbf{D}^*)^\dagger$ as $K\mathbf{F}_n$, the energy difference becomes

$$\Delta E = \tfrac{1}{2}K^2\mathbf{F}_n{}^\dagger\mathbf{E}\mathbf{F}_n = \tfrac{1}{2}K^2 f_n$$

For $K = 0$, $\Delta E$ and $d(\Delta E)/dK$ naturally vanish, but $d^2(\Delta E)/dK^2$ is negative, so $\Delta E$ will be negative for small $K$. For very large $K$, higher terms than second order will become important, and the energy will start to increase, so that the energy will actually go through a minimum at some value below $\langle H\rangle_{\Phi_0}$. These derivations were carried out somewhat more concisely by Čížek and Paldus in the second quantization formalism [8].

Related to the stability problem is the question of the effect of addition of an electron on the Hartree–Fock energy of an $N$-electron system. The electron, before being added, may be thought of as lying in a Rydberg-type orbital of very large principal quantum number and effectively zero energy, so having no effect on the other electrons. One has to compare the energy of this situation with that of a true determinantal function for the $(N + 1)$-electron system. If the latter energy is lower, one has a kind of instability for the $N$-electron system. Kaplan and Kleiner [9] showed that for a one-center system the energy of the determinant for which the last electron is distant (the $N$-electron system plus an unbound electron) is never lower than the energy of the determinant for which all $N + 1$ electrons interact. In the single-determinant approximation, electron affinities are thus always nonnegative. The relation of this result to the stabilities of certain Hartree–Fock determinants was discussed [9].

## 2. Roothaan Equations and Basis Functions

In this section we consider procedures for finding solutions to Eqs. (1) and (2). For atoms, the spherical symmetry allows a great simplification in Hartree–Fock calculations, since the orbitals may be written as spherical harmonics multiplied by radial functions, with only the radial part to be determined variationally. Because of the properties of the spherical harmonics, the Hartree–Fock equations can be reduced to coupled integro-differential equations in the radial variable. Numerical solution of these equations is tractable [10]. For molecules, the loss of spherical symmetry makes life more difficult. (It has been suggested [11] that direct numerical solution of the equations deserves investigation, since such a solution, though difficult and time-consuming, could replace several hundred calculations needed to optimize exponents when the expansion methods described later are used.)

It is still possible, however, to choose a center in the molecule and write each molecular orbital as an expansion in spherical harmonics about this center. Formally, there will be an infinite number of terms for each orbital; the number that are actually important will depend on how much the field, for a given molecular orbital, differs from spherical symmetry. For each molecular orbital, there will be a number of "radial" functions, one for each term in the expansion in spherical harmonics. Coupled integrodifferential equations for these functions may be derived. This one-center expansion has been investigated by several authors. In particular, Keefer et al. [12] explored its applicability to Hartree–Fock calculations of $U(R)$ for LiH and $HeH^+$. We will discuss this and other one-center calculations later.

Most all molecular calculations, however, treat the Hartree–Fock equations by expansion of the molecular orbitals in a set of functions (basis set), often atomic orbitals centered on the nuclei of the molecule. The unknown quantities are the expansion coefficients, and one can obtain matrix equations for them. We refer to such equations as Roothaan [13] equations. The procedure is sometimes referred to as "matrix Hartree–Fock" or MO–LCAO (molecular orbital–linear combination of atomic orbitals). In Volume I, Chapter III [Eqs. (74)–(76)] we obtained the Roothaan equations by substituting the expansions of the molecular orbitals into the Hartree–Fock integrodifferential equations.

This is a valid derivation only if the expansions are exact, i.e., if the orbitals used form a complete set. Of course, in practice we must use a basis set of limited size, and the equations for the coefficients must be derived by minimizing the energy with respect to variations in the coefficients.

As in Volume I, Chapter III, we will write $C_{lk}$ for the coefficient of the $l$th basis function in the $k$th molecular spin orbital. The $k$th spin orbital is:

$$\lambda_k = \sum_{l=1}^{p} C_{lk}\chi_l \tag{16}$$

where there are $p$ functions in the basis set. The energy expression [Eq. (62) of Volume I, Chapter III] becomes

$$
\begin{aligned}
\langle \Phi \mid H \mid \Phi \rangle = & \sum_{i=1}^{N} \sum_{k=1}^{p} \sum_{l=1}^{p} C_{ki}^{*} C_{li} h_{kl} \\
& + \sum_{l<j=1}^{N} \sum_{k=1}^{p} \sum_{l=1}^{p} \sum_{m=1}^{p} \sum_{n=1}^{p} C_{ki}^{*} C_{lj}^{*} C_{mi} C_{nj} \\
& \times (\langle \chi_k \chi_l \mid g \mid \chi_m \chi_n \rangle - \langle \chi_k \chi_l \mid g \mid \chi_n \chi_m \rangle)
\end{aligned}
\tag{17}
$$

and the orthonormality condition becomes

$$\sum_{k=1}^{} \sum_{l=1}^{} C_{ki}^{*} C_{lj} S_{kl} = \delta_{ij}, \qquad i, j = 1, \ldots, N \tag{18}$$

The notation of Volume I, Chapter III, Eqs. (74)–(76), is adopted. The procedure now follows precisely that of Volume I, Chapter III, Eqs. (62)–(68), with $\delta C_{ki}$ and $\delta C_{ki}^{*}$ being taken as independent variations. The result is

$$(\mathbf{h} + \mathbf{G})\mathbf{C}_k = \varepsilon_k \mathbf{S}\mathbf{C}_{k'} \tag{19}$$

as in Volume I, Chapter III, except that all sums over basis functions run from 1 to $p$ instead of over an infinite range. For further discussion of the derivation, see Roothaan's basic articles [13, 14].

The determinant $\Phi^R$ formed from the spin orbitals (16) has an energy that is stationary to any variation of the spin orbitals permitted by the basis set. Corresponding to Brillouin's theorem one has

$$\langle \Phi^R \mid H \mid \delta\Phi \rangle = 0.$$

Here, $\delta\Phi$ is a determinant differing from $\Phi^R$ by the substitution, for one of the $\{\lambda_i\}$, of another spin orbital which can be constructed from the same basis set. If $\delta\Phi$ is a singly excited determinant involving a contribution from outside the basis set, $\langle \Phi^R \mid H \mid \delta\Phi \rangle$ does not necessarily vanish. The consequences for statements about the accuracy of expectation values of one-electron operators, calculated with $\Phi^R$, are obvious.

If it is demanded that $\lambda_k$ be of the form: Spatial function multiplied by $\alpha$ or $\beta$, all the basis functions appearing in (16) are of this form. Usually, the same set of *spatial* basis functions is used to expand all the molecular orbitals. Sometimes, there may be an advantage in using disjoint basis sets to expand the spin orbitals of $\alpha$ and $\beta$ spin. Similarly, if $\lambda_k$ is a molecular orbital of a particular symmetry, only basis functions of that same symmetry would be used in its expansion.

It is evident that we may approach the exact Hartree–Fock functions as closely as desired by using larger and larger basis sets. But larger sets require computation of more of the two-electron integrals $\langle \chi_j \chi_k \mid g \mid \chi_m \chi_n \rangle$. If there are $p$ spatial orbitals in the basis set, there are $p^4$ such integrals, but not all are distinct. The result is that the amount of work increases roughly as $p^3$. Thus, a great deal of effort has gone into doing one or more of the following:

(*a*) Evaluating the two-electron integrals more rapidly (this includes gaining access to a faster computer);

(*b*) Choosing "good" basis functions, so that a smaller number are needed to describe the molecule;

(*c*) Finding basis functions for which the two-electron integrals are easy to evaluate.

The remainder of this subsection is devoted to progress in these areas. No one approach has been able to make the other two unimportant.

With respect to (*a*), we may say that many of the leaps forward of recent years are associated with the availability of new hardware (computing machines) and software (techniques and programs for integral evaluation) for this purpose. Huzinaga [15] has given an interesting review of the principal developments in integral evaluation, and we shall give further references in what follows. The review of Allen and Karo [16] summarizes many of the problems connected with the choice of basis functions.

Since the SCF Hamiltonian for the molecule contains the kinetic energy and nuclear attraction operators, exactly as do the Hamiltonians of the atoms of which the molecule is composed, the molecular orbitals should resemble the atomic orbitals to some extent, especially near the nuclei. Then it is reasonable to take as basis functions the atomic SCF functions for the separate atoms.

Roughly, overlaps and SCF Hamiltonian matrix elements between inner-shell orbitals on the same atom should be the same for the molecule as for the atom, while matrix elements between inner-shell orbitals and other orbitals should be relatively small. Then solution of the molecular problem will yield some SCF orbitals closely resembling those for the atoms, i.e., the inner shells are not much affected by molecule formation. For a homonuclear diatomic, there will be degeneracy between inner-shell orbitals on the two atoms, and any interaction, however small, will split it. The energies will be changed but slightly, but the orbitals that diagonalize the interaction will be essentially the symmetric and antisymmetric combinations of the atomic orbitals. Therefore the orbitals needed to describe the atomic cores should be included essentially unchanged in the basis set for the molecular calculation.

In general, there is an advantage in using basis functions having the molecular symmetry. If the atomic functions do not have the full molecular symmetry, we can generate from them basis functions belonging to the various irreducible representations of the molecular symmetry group. Overlaps and matrix elements of the Hartree–Fock Hamiltonian may vanish (if the Hartree–Fock Hamiltonian is totally symmetric—see page 166) between basis functions of different symmetry. Then the secular equations factor, and we have several smaller problems, for orbitals of each symmetry. As a trivial example, if $H_2$ is treated with a basis set consisting of the 1s orbitals on the two centers, the symmetry orbitals are the sum and difference of atomic orbitals. They are of $\sigma_g$ and $\sigma_u$ symmetry, respectively, and do not interact; the secular equations are one dimensional.

The atomic SCF functions are almost always calculated either in numerical form or as expansions in some basis set. The latter is more convenient. If the atomic orbitals are themselves linear combinations of basis functions, integrals over the atomic orbitals require integrals over the basis functions. Having performed these, one could consider using the set of basis orbitals themselves as the basis set for the molecular problem. This will obviously give better results because of the greater flexibility. However, to get really good results, approaching the true Hartree–Fock solutions, additional basis functions are also needed, i.e., it is more difficult to approach, via an expansion, the Hartree–Fock functions for molecules than for atoms.

Basis orbitals which have been successful for atomic problems, and are much used for molecules, are the Slater orbitals, of the form

$$\phi_{nlm}(\mathbf{r}) = \mathcal{N} r^{n-1} e^{-ar} Y_{lm}(\theta, \varphi). \tag{20}$$

Here, $\mathcal{N}$ is a normalizer, $Y_{lm}$ is a spherical harmonic, and the spherical coordinates on the right side of (20) are usually with reference to the nucleus of an atom. These orbitals have the proper behavior close to and far away from the nucleus. While $\phi_{nlm}$ has no radial nodes, they are put in by orthogonalizing it to orbitals of lower principal quantum number. Such orthogonalization may be regarded as having been performed when $\phi_{nlm}$ appears with the other orbitals in a determinant, since the determinant is unchanged when $\phi_{nlm}$ is replaced by a linear combination of the orbitals. The orthonormalization can be produced by such a linear combination. Atomic SCF functions, to a first approximation, can be constructed by representing each SCF orbital by a single Slater orbital.

The choice of exponential parameters in Slater and other basis functions is of great importance if accurate results are to be obtained without the basis set's becoming too large (see page 154). The earlier workers used "Slater screening constants," values for the exponential parameters chosen according to a formula derived semiempirically for atoms [17]. At present, values for exponential parameters are chosen on the basis of previous experience and by variation of the energy. Because these parameters enter the energy in a nonlinear fashion, this requires a succession of calculations with different values for the parameters. Systematic procedures have been suggested, such as deriving the dependence of the energy on the parameters as a power series. It has also been proposed [18] that the SCF perturbation theory (Volume I, Chapter III, Sections B.2 and B.4) be used to compute this dependence. Note that $R$ is a nonlinear parameter entering the energy, like the exponential parameters. Semiempirical schemes for determining exponential parameters, by interpolating between values for the united atom and separated atom orbitals, have been investigated [19]. The principal quantum number $n$ may also be considered a parameter to be varied, but usually it is fixed because integrals involving the $\phi_{nlm}$ increase in complexity when $n$ is not an integer.

For the ground state of a molecule, a reasonable initial choice of basis functions would be the set of approximations to the atomic orbitals occupied in states belonging to the lowest configurations of each constituent atom. In particular, one Slater orbital might be used for each such atomic orbital. This is referred to as a minimal basis set. For example, diatomics formed from atoms from B to Ne would require 1s, 2s, and 2p-like orbitals in the basis set. Minimal basis set calculations are important because they are quite practicable for fairly large molecules and because they yield approximations to the SCF orbitals of simple form. For many molecules, it is possible to optimize the minimal basis set function with respect to all the exponential parameters.

Where both atoms are closed shells (with respect to spin and orbital angular momentum), so that the number of occupied atomic orbitals is exactly half the number of electrons, it is not necessary to determine the molecular orbitals in a minimal basis set calculation. Let the orbitals of atom A be $\phi_1, \ldots, \phi_{n_A}$, those of B $\phi_{n_A+1}, \ldots, \phi_{n_A+n_B}$. Then the only single-determinant wave function we can construct corresponds to filling each of the $n_A + n_B$ orbitals with two electrons.

$$\Phi = \mathscr{A}\{\phi_1(1)\alpha(1)\phi_1(2)\beta(2) \cdots \phi_{n_A}(2n_A)\beta(2n_A)$$
$$\times \phi_{n_A+1}(2n_A + 1)\alpha(2n_A + 1) \cdots \phi_{n_A+n_B}(N)\beta(N)\}.$$

The orbitals are not all mutually orthogonal, but we can imagine making them so by a transformation among themselves which leaves the determinant invariant.

In going beyond the minimal basis set, we can distinguish roughly between two kinds of basis functions that need to be added. There are those which improve the description of the atoms and those which are more important for the molecule. We must use a "balanced set" of orbitals, particularly if we want to avoid errors that distort the shape of $U(R)$. This was emphasized by Mulliken [20] in a discussion of criteria for the choice of basis sets. Basis sets replacing each orbital in a minimal basis set by two Slater orbitals differing in exponential parameter are referred to as "double-zeta" basis sets [zeta $(\zeta)$ is often used to represent the exponential parameter], and are often employed in atomic and molecular Hartree–Fock calculations. The completeness of such a set (ability to represent atomic wave functions) has recently been discussed [21]. Basis functions particularly important for description of a molecule and calculation of $U(R)$ ("polarization functions") are those with $l$ values one greater than atomic orbitals occupied in the atoms.

It must be noted that the atomic orbitals on either center form a complete set: Any orbital on one center may be expanded in terms of those on the other. Thus a basis composed of many atomic orbitals from each center approaches a formally redundant or overcomplete set, with attendant problems of linear dependence when the set is used in expansions. The complete set of atomic orbitals for an atom includes orbitals of the continuum, which are difficult to manipulate in matrix calculations, but complete sets which consist wholly of discrete functions can be constructed from Slater orbitals [22]. However, the overcompleteness problem exists whenever we approach a complete set on each center. The linear dependence problem actually arises, in most cases, because of the accuracy of evaluation of integrals rather than from the actual overcompleteness. If the lowest eigenvalue of the overlap matrix is not larger than the error in an overlap integral, the basis set is effectively linearly dependent, since the transformation to orthogonal basis functions, which is implicit in most calculations, will be highly inaccurate [23].

An additional problem with large basis sets is that the calculation time consumed in matrix manipulations, as well as in the evaluation of integrals, becomes extremely large. The choice of optimum values for the exponential parameters serves to keep down the number of basis functions needed.

### 3. Basis Functions for Molecular Calculations

In this subsection, we discuss the basis functions which are used for calculations by the Hartree–Fock and configuration interaction methods. We will return specifically to the single-determinant calculations in the following subsections.

The fact that calculations are currently appearing using Slater orbitals, Gaussians, and other basis functions indicates that no one scheme has shown overall superiority in dealing with the integral problem. Slater orbitals are perhaps the most commonly used, and have the longest history.

Many authors have given methods for the evaluation of integrals over Slater orbitals. The bibliography given by Kotani et al. [24] is very complete up to 1962, as is that of Preuss [25].

The one-electron integrals involving Slater orbitals on one or two centers can, in general, be evaluated in closed form or in terms of auxiliary functions for which tables and straightforward formulas are available [24–27]. These formulas are derived most readily by writing the integrals in ellipsodial coordinates. With the Neumann expansion [24] of $r_{12}^{-1}$, many of the two-electron integrals also become tractable. They may be reduced to auxiliary functions which may be tabulated [24, 25], or to a small number of numerical quadratures plus recursion relations [28, 29]. The Barnett–Coulson expansion [30, 31] for a Slater orbital in one center in terms of spherical coordinates on the other was made the basis of schemes for integral evaluation. The various methods were compared by Huzinaga [15], Harris and Michels [30a], and other authors presenting new schemes for calculation. Harris and Michels [30a] gave methods which attempted to minimize the amount of computer time needed for evaluation of all electron repulsion integrals over a large basis set. The integrals were analyzed into interactions of charge distributions, and all integrals,

$$\int \chi_1(1)^*\chi_2(2)^*g(1, 2)\chi_3(1)\chi_4(2)\,d\tau$$

were treated in the same way. Wahl *et al.* [29] discussed methods for all two-center integrals (involving one or two electrons) on the basis of experience with their use for molecular calculations. They presented an analysis thought to be well-adapted to automatic computation by digital computers. Auxiliary functions as well as numerical integrations, are involved. Wahl and Land [32] discussed a "polished brute force" scheme and gave a discussion of programming it for computers. In addition to comparisons with other methods, they gave assessments of accuracy and of computer time required. Fourier transforms have been employed by several authors [33] to derive formulas for electron repulsion integrals. O-Hata and Ruedenberg [34] gave analytical formulas for two-center Coulomb integrals ($\chi_1 = \chi_3$, $\chi_2 = \chi_4$ in the preceding expression) over Slater orbitals, and recently, Christoffersen and Ruedenberg [35] gave numerical techniques especially suited for evaluation of the hybrid integrals over Slater orbitals. These integrals, in which three of the four atomic orbitals appearing belong to one center and the fourth to the other, are the most common two-electron integrals in diatomic molecule calculations.

Ormond and Matsen [36] and Ebbing [23] (as well as other authors) reintroduced basis functions in ellipsoidal coordinates (two-center basis functions) for calculations on diatomic molecules. Such functions had been used in the earlier calculations of James and Coolidge [37] and Knipp [37a]. These functions are very flexible and should be well-suited for the diatomic problem [38]. Browne and Matsen [36] used a mixed basis of ellipsoidal and Slater functions in their work on LiH, and Harris and Taylor [38] have shown their utility in a series of calculations. An ellipsoidal basis function may be written [23]

$$\psi_{ij}^{\alpha\beta} = (2\pi a^3)^{-1/2} e^{-\alpha\mu} e^{-\beta\eta} \mu^i \eta^j [(\mu^2 - 1)(1 - \eta^2)]^{(1/2)|m|} e^{im\varphi}$$

where $a = \frac{1}{2}R$. Here,

$$\mu = R^{-1}(r_A + r_B), \qquad \eta = R^{-1}(r_A - r_B)$$

are the coordinates of an electron. The factor in square brackets ensures a finite kinetic energy. When $\beta = \pm\alpha$, $\psi_{ij}^{\alpha\beta}$ is an atomic (one-center) Slater orbital; when $\beta = 0$, $\psi_{ij}^{\alpha\beta}$ is symmetric or antisymmetric (depending on $j$ and $m$) to inversion, and suitable for a homonuclear problem. Overlap integrals for these functions are independent of $R$ because of the $(2\pi a^3)^{-1/2}$ factor, while matrix elements of the Hamiltonian may be written

$$H_{ij,kl} = a^{-2}T_{ij,kl} + a^{-1}V_{ij,kl}$$

when these orbitals are used to set up wave functions, with $T_{ij,kl}$ and $V_{ij,kl}$ $R$-independent. This allows all integrals to be calculated once only, for a series of calculations at different $R$ [23]. Integrals over ellipsoidal basis functions are related to integrals over Slater orbitals; many of the same methods for evaluation are used.

The two-electron integrals over Slater orbitals get difficult. If the orbitals are of the Gaussian type [39], i.e., with $\exp(-\alpha r^2)$ replacing $e^{-\alpha r}$ in (20), the integrals are considerably simplified. Because $r^2 = x^2 + y^2 + z^2$, $\exp(-\alpha r^2)$ is a product of functions for the three coordinates and most integrals can be evaluated in closed form. This led to the suggestion by Boys [40] that Gaussian orbitals centered on the nuclei be used instead of Slater orbitals. Browne and Poshusta [41] gave formulas for all energy integrals over the Gaussian functions

$$x^i y^j z^k \exp(-\alpha_x x^2 - \alpha_y y^2 - \alpha_z z^2),$$

where $x$, $y$, and $z$ are distances of the electron from a nucleus along the three coordinate axes. This is a generalization of the Gaussian mentioned earlier, since it includes three exponential parameters. Functions of this form are obtained by differentiating the s-type Gaussian ($i = j = k = 0$) with respect to $x$, $y$, or $z$, and this is used in evaluating the integrals. The explicit formulas of Browne and Poshusta [41] require at most the numerical evaluation of a one-dimensional integral, which, it was stated, can be done to eight figures by a 10-point quadrature.

Indeed, the simplicity with which the integrals may be calculated is the basis of the Gaussian transform method [39, 42] for evaluating integrals over Slater orbitals. The Slater orbitals are expanded in an infinite complete set of Gaussians, over which the integrations are carried out. The Gaussians in the set differ in the values of parameters which vary continuously, so the expansion of the Slater orbital is actually an integral transform.

Gaussian orbitals have neither the cusp at the nucleus nor the exponential fall-off far from the nuclei expected for the exact wave function. Therefore, a minimal basis calculation with Gaussian orbitals is not likely to produce very good results. The integrals are so much simpler, however, that one can afford to use much larger basis sets. The choice of exponential parameters is still important in keeping the size of the basis sets down. Results of atomic calculations provide a good guide to guessing these parameters, and suggested values have been tabulated by Huzinaga [43], Veillard [44], and recently for third-row atoms by Roos et al. [45]. The last authors discuss modifications needed when using these functions for molecular calculations.

Dunning recently [46] presented sets good enough to approach Hartree–Fock for molecules.

Allen [47] discussed the advantages and disadvantages of Gaussian orbitals: the simplicity of multicenter integrals and the ease of programming the resulting analytical expressions; the discouraging results for Gaussians compared to an equal number of Slater orbitals; the necessity of using many Gaussians to describe the orbital cusps and the exponential tail. It was estimated [47] that the Gaussian set should be 1.5 times as large as the Slater set for comparably accurate results. Later authors have tended to increase this ratio somewhat. However, it has been argued [48] that the ratio should be less for molecules than for atoms, because there is relatively more density in nonexponential fall-off regions. Bishop [49] has proposed addition to a Gaussian basis set of certain functions designed to put in the cusps at the nuclei without unduly complicating the integrals.

Huzinaga [43] studied the expansion of a Slater orbital in Gaussians as a guide toward the choice of Gaussian basis functions for a calculation. If a Slater orbital can be represented by a small number of Gaussians, integrals over Slater orbitals may be evaluated as sums of integrals over Gaussians [42, 50]. This may be considered as a simplified and approximate version of the Gaussian transform method. More recently, other authors have given least-squares representations of Slater orbitals in terms of Gaussians and tested their utility for molecular calculations [51–54]. With the Slater functions replaced by three-term Gaussian approximations, one can carry out calculations on molecular systems with the integrals much simplified, and generally reproduce in results of the Slater orbital calculations [55]. A two-center integral over Gaussian functions can be evaluated over 100 times as fast as one over Slater functions [53] and multiplication of the number of basis functions by 3 increases the number of integrals by a factor between $3^3$ and $3^4$. Polarization functions (see Section C) in Gaussian expansions have been given [48].

It is also possible to construct atomic and molecular functions from a basis set consisting solely of s-type Gaussians, centered on and off nuclei. Integrals involving s-type Gaussians are extremely simple [56]. The exponential parameters and the locations of the Gaussians are varied to minimize the energy. Exploratory calculations have been made [57] for two-electron molecules, while atomic SCF functions have been fitted [58] to expansions in these functions.

The Gaussian lobe technique [59] uses expansions in s-type Gaussians to represent Slater orbitals in molecular calculations. Each basis function is a set of Gaussians centered at different points in space. The relative positions of the centers, as well as all parameters needed to define the linear combination of Gaussians, are kept fixed during a molecular calculation.

New kinds of basis functions may be generated [60] by an integral transform, starting from a trial function appropriate for the system. Gaussian and Slater orbitals are still used almost exclusively.

The hydrogen atom wave functions are, of course, linear combinations of Slater orbitals, but they have some special characteristics. Using properties of their Fourier transforms, Alper [61] derived a new scheme for calculation of multicenter molecular integrals, giving formulas for two-center overlap, two-center Coulombic, and three- and four-center Coulombic integrals over hydrogenic orbitals of integral principal quantum number.

It has also been suggested [62] that numerical integration by crossed Gauss quadrature methods be substituted for the part-analytic, part-numerical schemes generally used (except for Gaussian functions). The use of purely numerical techniques would permit calculations for any basis functions desired. For instance, each basis function could be expressed as a sum of atomic orbitals without the integrals becoming complicated sums of integrals, or basis functions given as numerical tables could be employed.

Another attempt to get around the difficult integrals involves the use of one-center expansions. In these expansions, all the functions in the basis are centered at the same point in the molecule, generally the position of a heavy nucleus, although other positions may be used. For example, Hagstrom and Shull [63] chose the midpoint of the molecule for some one-center calculations on $H_2$ (see Section B). No multicenter two-electron integrals arise. The only two-center integrals come from the nuclear potential energy operators, and are easy to evaluate. Because each basis function may be taken as a product of a radial function with a spherical harmonic, the integrals may be considerably simplified by using the properties of the representations of the three-dimensional rotation group (spherical symmetry). Only integrations over one or two radial coordinates need be carried out.

One-center calculations go back to the earliest years of quantum chemical calculations [64]. A review article by Bishop [65] in 1967 discussed one-center calculations in a systematic way. Another, by Hayes and Parr [66], included a complete annotated bibliography. It may be expected that one-center expansions work best when the actual molecular charge density is close to spherical, as for the hydrides among the diatomic molecules.

The basis functions for one-center calculations are generally Slater functions. Since they resemble the orbitals of the united atom in many cases these calculations are sometimes called "united atom calculations." However, parameters in the basis functions are generally chosen to minimize the energy of the molecular calculation. There may be some advantage to using hydrogen-like functions or other combinations of Slater orbitals as basis functions. Bishop [67] compared Fock–Petrashen, hydrogenic and Schmidt-orthogonalized Slater orbitals as basis functions for one-center calculations. All are linear combinations of Slater orbitals; the first two form complete orthonormal sets. He found that the orthogonalized Slater orbitals worked best for single-determinant calculations.

Even if the proper fall-off of the wave function at large distance is obtained by Slater functions, the difficulty of simulating the cusps in the wave function at nuclei other than the expansion center remains. This means a large basis set is needed. The choice of exponential parameters is important, and may be done [68] using results of atomic calculations and multicenter molecular calculations. As for the case of Gaussians, one can introduce into the basis set functions specifically designed to put in the cusps at the off-center nuclei, without losing the simplicity of integral evaluation characteristic of one-center calculations [49].

Moccia [68] has discussed the problem of inadequate representation of cusps at off-center nuclei, as well as the choice of orbitals, in connection with his calculations on HF and other molecules. Other authors [69] have studied the expansion of a Slater orbital on one center in terms of orbitals on another center in order to learn about problems associated with the off-center cusps. Among these expansions is that of Barnett and Coulson [30], which has been made the basis of a method to evaluate multicenter integrals. If a Slater orbital can be well-represented as a finite expansion on another center, one can use a multicenter basis set, but reduce all integrals to very tractable one-center integrals [70].

The basis functions in a one-center calculation belong to representations of the spherical symmetry group, so they are characterized by quantum numbers for the total angular momentum ($l$) and the component of angular momentum ($m$) along an axis, which may be taken as the internuclear axis. The molecular orbitals belong to representations of the cylindrical symmetry group, a subgroup of the spherical symmetry group (see Chapter I, Section C). This means that basis functions corresponding to different $l$ may be combined in the same molecular orbital, as long as they have the same value of $m$.

Hake and Banyard [71] compared their previous calculations [72], which used a single atomic orbital per atom, with the more general one-center calculations of Moccia [73]. The molecules considered were those whose united atoms were Ne or Ar; the heavy atom was used as expansion center. As shown in the results for HF and HCl, some of which appear in Table I, writing each molecular orbital as a sum of atomic orbitals gives noticeably better results. This is due partly [71] to the improved description of the off-center cusps. However, the great simplicity of the calculations using a single basis function for each molecular orbital is a feature in their favor.

**Table I**

*Results of One-center Calculations*

| | Energy at $R_e$ (a.u.) | | $R_e(a_0)$ | | $k$ (10^5 dynes/cm) | |
|---|---|---|---|---|---|---|
| | HF | HCl | HF | HCl | HF | HCl |
| Banyard–Hake | −99.0045 | −457.8012 | 1.563 | 2.426 | 13.4 | 5.68 |
| Moccia | −100.0053 | −458.8378 | 1.728 | 2.404 | — | — |
| Experimental | −100.48 | −462.81 | 1.733 | 2.408 | 9.66 | 5.16 |

With respect to computations of the potential curve $U(R)$, one-center calculations possess two convenient features. First, unless the expansion center is rechosen for each $R$ (an unlikely procedure), all parameters which depend on R would normally be chosen variationally. Then the Hellmann–Feynman theorem would be satisfied (Volume I, Chapter III, Section D). The "hidden parameters" of multicenter expansion calculations are not present. A second advantage is that force constants may be easily obtained from calculations at a single value of $R$ because differentiation of the energy expression with respect to $R$ is much simplified. We will discuss this point further in Section C.

### 4. Spin and Symmetry Properties

Returning to the single determinant, we consider what restrictions must be placed on spin orbitals, such as solutions to Eqs. (1) and (2), to ensure that the determinant formed from them has the desired symmetry properties. In some cases, it will be possible to find a solution (not necessarily that of lowest energy) to the unrestricted Hartree–Fock Eqs. (1) and (2) within these restrictions. Where this is not so, some modification will be necessary in deriving equations for the spin orbitals.

Spin is considered first. Since a spin-independent Hamiltonian is used, the exact eigenfunctions will be simultaneously eigenfunctions of total spin and total $z$-component of spin. Our choice of spin orbitals, $\lambda_i(i) = \phi_i(i)s_i(i)$ ($s_i = \alpha$ or $\beta$ and $\phi$ a function of space coordinates only) ensures that the determinant is an eigenfunction of $S_z$:

$$S_z \mathscr{A}\left[\prod_{i=1}^{N} \lambda_i(i)\right] = \mathscr{A}\left[\sum_{l=1}^{N} S_z(l)\prod_{i=1}^{N} \phi_i(i)s_i(i)\right] = \mathscr{A}\left[\sigma\prod_{i=1}^{N}\lambda_i(i)\right] \quad (21)$$

The eigenvalue $\sigma$ is equal to $\tfrac{1}{2}(n_\gamma - n_\beta)$, where $n_\alpha$ is the number of $\alpha$ spins and $n_\beta$ the number of $\beta$ spins. Under what conditions is the determinant on eigenfunction of $S^2$?

Since $S^2$ is symmetric in the electron coordinates, it commutes with $\mathscr{A}$ and we consider $S^2\prod_{i=1}^{N}\lambda_i(i)$, bearing in mind that $\mathscr{A}$ operates on this afterward. For this purpose, we write

$$S^2 = \left(\sum_j S_z(j)\right)^2 + \sum_j [S_x(j) + iS_y(j)]$$

$$\times \sum_k [S_x(k) - iS_y(k)] - i\left[\sum_j S_y(j), \sum_k S_x(k)\right]. \quad (22)$$

The commutator which constitutes the last square bracket may be simplified since $S_y(j)$ and $S_x(k)$ commute unless $j = k$. It reduces to

$$-i\sum_j [S_y(j), \quad S_x(j)] = -\sum_j S_z(j) = -S_z$$

Thus, the first and last terms in (22) equal $S_z^2 - S_z$. The operator $S_x(k)$ $-iS_y(k)$ annihilates $\lambda_k(k)$ if electron $k$ is in a spin orbital with $\beta$ spin, and changes its spin to $\beta$ if it is $\alpha$, while $S_x(k) + iS_y(k)$ annihilates $\lambda_k(k)$ if its spin is $\alpha$ and changes it to $\alpha$ if its spin is $\beta$. Thus operation of $\sum_k(S_x(k)$ $-iS_y(k))$ on the product $\prod \lambda_i(i)$ produces $n_\alpha$ products, in each of which there are $(n_\beta + 1)$ $\beta$-spins. Subsequent operation with $\sum_j(S_x(j) + iS_y(j))$ produces $n_\beta + 1$ products from each of the $n_\alpha$ products, each with $n_\beta$ $\beta$-spins. A number $n_\alpha$ of the products will be identical to the original $\prod \lambda_i(i)$; the others differ from it by having the spins on two $\lambda_i$ interchanged.

Let us suppose that $n_\alpha \geq n_\beta$. The Pauli principle, enforced by the determinantal form of the wave function, allows no more than two occupied orbitals to have the same spatial part, and these must have opposite spins.

Let there be $n_p$ doubly occupied spin orbitals, $n_u$ singly occupied ones with spin $\alpha$, $n_d$ singly occupied with spin $\beta$.

$$n_\alpha = n_p + n_u; \qquad n_\beta = n_p + n_d; \qquad n_\alpha + n_\beta = N \qquad (23)$$

We will show that, for an eigenfunction of spin, $n_d$ must vanish. $S_z^2 - S_z$ operating on the product of spin orbitals multiplies it by $\sigma(\sigma - 1)$, with $\sigma = \frac{1}{2}(n_\alpha - n_\beta)$. The remaining operators in (22) give $n_\alpha(n_\beta + 1)$ products, in $n_\alpha$ of which the same $\alpha$ spin was demoted to $\beta$ and then promoted back to $\alpha$. The others involve exchanges of opposite spins. There are: (a) $n_p$ in which $\alpha$ and $\beta$ spins within a pair are interchanged; (b) $n_u n_p$ in which one singly occupied orbital trades an $\alpha$ spin for a $\beta$ from a pair, (c) $n_d n_p$ in which a singly occupied orbital trades a $\beta$ spin for an $\alpha$ from a pair; (d) $n_u n_d$ in which singly occupied orbitals exchange opposite spins; and (e) $n_p(n_p - 1)$ in which opposite spins are exchanged between pairs. The products of (b), (c), and (e) have two electrons with the same space and spin function, so they are annihilated by $\mathscr{A}$. The exchanges of case (a) each give a determinant differing in sign from the original. Spin exchanges of type (d) give spin–orbital products which are not annihilated by $\mathscr{A}$ and lead to determinants not multiples of the original. Thus, if we want an eigenfunction of $S^2$ we must have (in the case $n_\alpha \geq n_\beta$) $n_d$ equal to zero. Then we obtain, putting $\Phi$ for $\mathscr{A} \prod_i \lambda_i(i)$,

$$S^2\Phi = \sigma(\sigma - 1)\Phi + (n_\alpha - n_p)\Phi. \qquad (24)$$

This means that for $n_\alpha \geq n_\beta$ a single determinant will not be an eigenfunction of spin unless all singly occupied orbitals are associated with the same spin. In such a case, the eigenvalue of $S^2$ is, according to (23) and (24), $\sigma(\sigma - 1) + n_u = \sigma(\sigma + 1)$, where $\sigma$ is the eigenvalue of $S_z$. The spin-paired electrons do not contribute to the eigenvalue of $S^2$. For proofs of this using density matrices, Löwdin's article [74] on the correlation problem may be consulted. Having chosen $n_u$ and $n_p$, we may be able to determine spatial orbitals so that the unrestricted Hartree–Fock equations (1) are satisfied, but not all solutions to (1) will produce determinants that are eigenfunctions of $S^2$.

We now consider spatial or orbital symmetry. The many-electron functions must transform according to the representations of the $C_{\infty v}$ group for a heteronuclear diatomic or, for a homonuclear diatomic, the group $D_{\infty h}$. This may be accomplished by choosing the spin orbitals to belong to such representations but this is not necessary. Sometimes, one may be in-

terested in molecular orbitals which do not have the full molecular symmetry
(see page 143), but for the purposes of calculation, there are advantages to
working with spatial orbitals that belong to the irreducible representations
of the molecular symmetry group. Because of cylindrical symmetry, such
orbitals may be written as (see Chapter I, Section B.1)

$$f(r_A, r_B)e^{im\varphi} \tag{25}$$

where $\varphi$ is the angle of rotation about the internuclear axis and $r_A$ and $r_B$
are the distances to the nuclei A and B. The orbitals are called $\sigma, \pi, \delta, \ldots$
according as $m = 0, \pm 1, \pm 2, \ldots$ There is a twofold degeneracy corresponding to the sign of $m$, so that the representations are either one dimensional ($m = 0$) or two dimensional. Rotation by an angle $\Phi$ about the
internuclear axis multiplies the orbital by $e^{\pm im\Phi}$. On reflection in a plane
containing the internuclear axis, $\sigma$ orbitals are unchanged and an orbital
belonging to any other representation is transformed into a multiple of its
partner in the representation. If this plane is at an angle $\Phi$ to the $x$–$z$ plane
(internuclear axis along $z$), the reflection changes $\varphi$ to $2\Phi - \varphi$, so that
$e^{im\varphi}$ becomes $e^{2im\Phi}$ times $e^{-im\varphi}$. In the homonuclear case, the function $f$
must be odd or even (g or u) to inversion in the molecular midpoint, giving
rise to the species $\sigma_g, \sigma_u, \pi_g, \pi_u$, etc.

To obtain the symmetry of the many-electron determinant it suffices to
consider the operation of the symmetry operator $R$ on the product of spin
orbitals $\prod_i \lambda_i(i)$, as in (21). The operator $R$ being a transformation of
spatial coordinates, its operation on a product of one-electron functions
means operation on each factor simultaneously. Then it is easy to see that
the product, and hence the determinant, is automatically an eigenfunction
of rotation about the figure axis, with eigenvalue $e^{iM\Phi}$. Here, $M$ is the sum
of the $m$-values of the occupied orbitals, and the determinant is an eigenfunction of angular momentum about the figure axis with eigenvalue $M\hbar$.
Similarly, the determinant will be an eigenfunction of the inversion operator
for the homonuclear case, with eigenvalue (parity) equal to the product of
eigenvalues for the spin orbitals present. The capital letters $\Sigma, \Pi, \Delta, \ldots$
indicate the value of $M$, and the inversion behavior is appended as a right
subscript (g or u). For $\Sigma$ states, the wave function is nondegenerate (except
for spin), so it must be an eigenfunction of reflection in a plane containing
the figure axis. This will be the case if the reflection converts $\prod \lambda_i(i)$ into a
function differing from $\prod \lambda_i(i)$ by at most a permutation of electrons.
According to whether the permutation is even or odd, the antisymmetrization will yield $+1$ or $-1$ times the original determinant ($\Sigma^+$ or $\Sigma^-$ state).

Since function (25) is transformed to

$$e^{2im\Phi}f(r_A, r_B)e^{-im\varphi}$$

by the reflection, the reflection will permute electrons in the product of occupied spin orbitals when the spin orbitals for $m \neq 0$ appear in pairs: If (25) occurs in the determinant with some value of $m$ and some spin, the spin orbital differing from it only in the sign of $m$ must also occur. The factors of $e^{2im\Phi}$ and $e^{-2im\Phi}$ cancel. The state is positive if an even number of such pairs occur, negative if an odd number.

When the orbital (25) and its partner (for $m \neq 0$) appear only once in $\prod_i \lambda_i(i)$ (both associated with the same spin) we speak of a half-filled shell; when both contain two electrons, we refer to it as a filled or closed shell. According to the preceding discussion, all half-filled shells must have the same spin if $\mathscr{A} \prod \lambda_i(i)$ is to be an eigenfunction of $S^2$. If an orbital (25) appears without its partner ($m \neq 0$) we speak of an open shell (sometimes half-filled shells are referred to as open shells as well); all open-shell orbitals must be associated with the same spin for an eigenfunction of $S^2$. Since the ground states of diatomic molecules generally have low spin, single determinant wave functions for such systems usually involve one open or half-filled shell at most. It may be seen that the problems of orbital symmetry are much simpler for diatomic molecules than for atoms.

Let us assume that the spin orbitals $\lambda_k$, from which our determinants are constructed, transform according to the irreducible representations of the molecular group. Then the effect of the symmetry operator $R$ is as follows:

$$R\lambda_k = \sum_{j=1}^{d_k} [D^{(\mu_k)}(R)]_{jk}\lambda_j^{(\mu_k)} \tag{26}$$

Here, $D^{(\mu_k)}(R)$ is the matrix corresponding to $R$ in the irreducible representation $\mu_k$, of dimension $d_k$. The question is, under what circumstances can such determinants be solutions to the unrestricted Hartree–Fock equations, so that they make the energy stationary? The answer involves further restrictions on the spin orbitals (in addition to those involving the spin, derived earlier). It is perhaps remarkable that determinants constructed to obey these restrictions, usually referred to as restricted Hartree–Fock (RHF) functions, constitute good approximations to the electronic ground states of many diatomic molecules. The RHF energies often appear to be close to those of the unrestricted Hartree–Fock (UHF) functions. The question of whether the UHF energy is lower than that of the RHF will be discussed below and in Subsection 5.

The spin orbitals can transform according to (26) if $h^{\mathrm{HF}}$ has the full molecular symmetry. To establish this, we evaluate the commutator $[R, h^{\mathrm{HF}}]$ for a symmetry operator $R$; if the commutator vanishes, $h^{\mathrm{HF}}$ has the full molecular symmetry and the assumption that the spin orbitals $\lambda_k$ appearing in $F$ obey (26) is a self-consistent assumption. Since $h$ is totally symmetric, only the Fock operator $F$ can give a nonvanishing commutator with $R$. We first consider $[R, F]f = R(Ff) - F(Rf)$ where $f$ is an arbitrary function.*

$$R(Ff(i)) = \sum_{l=1}^{N} \sum_{m_l=1}^{d_l} \sum_{k_l=1}^{d_l} [D^{(\mu_l)}(R)_{m_l l}]^* D^{(\mu_l)}(R)_{k_l l}$$

$$\times [\langle \lambda_{m_l}(j) \mid g(i, j) \mid \lambda_{k_l}(j) \rangle Rf(i)$$

$$- \langle \lambda_{m_l}(j) \mid g(i, j) \mid g(i, j) \mid Rf(j) \rangle \lambda_{k_l}(l)] \tag{27}$$

since $Rg(i, j) = g(i, j)$. Suppose the occupied orbitals include all the $d_\nu$ orbitals spanning the $\nu$ representation. Their contribution to the sum over $l$ will be

$$\sum_{l}^{(d_\nu)} [D^{(\nu)}(R)_{m_l l}]^* D^{(\nu)}(R)_{k_l l} = \delta_{m_l k_l}$$

since the representation is unitary. If all the occupied orbitals may be divided into such sets, the sum over $l$ yields a sum of terms $\delta_{m_l k_l}$, one for each set, and (27) is

$$RFf(i) = \sum_{(\mu_l)} \sum_{m_l=1}^{d_l} (\langle \lambda_{m_l}(j) \mid g(i, j) \mid \lambda_{m_l}(j) \rangle Rf(i)$$

$$- \langle \lambda_{m_l}(j) \mid g(i, j) \mid Rf(j) \rangle \lambda_{m_l}(j))$$

$$= FRf(i)$$

The sum $\sum_{(\mu_l)}$ means a sum over the representations which are represented in $F$ so $(\sum_{(\mu_l)} \sum_{m_l=1}^{d_l})$ means a sum over all orbitals in $F$.

Thus we have shown: If the occupied orbitals may be divided into sets, each of which spans a representation of the symmetry group, the Fock operator commutes with the symmetry operators, and it is permissible (self-consistent) to choose solutions to the one-electron Hartree–Fock equations which transform as symmetry orbitals. The occurrence of such sets corresponds to filled or half-filled shells. If we now consider self-consistency with respect to spin, however, we can see that only in the case

* This proof is essentially that of Slater [75].

of filled shells (double occupation) can we have a solution to the UHF equations. If there are unequal numbers of electrons with $\alpha$ and $\beta$ spins, the Fock Hamiltonians [Eq. (2)] will be different for spin orbitals associated with $\alpha$ spin and those associated with $\beta$ spin. The solutions to the Hartree–Fock equations will not be the same, so it is inconsistent to assume the same spatial orbitals for $\alpha$ and $\beta$ spins. Thus, aside from closed-shell situations, the solutions to Eq. (1) cannot give a determinant of desired spin and symmetry properties.

Since it is almost always advantageous to use symmetry orbitals (matrix elements of most operators vanish between orbitals of different symmetry, for instance), one attempts to arrange things, even where there are open shells, so that symmetry orbitals are obtained. Here, one assumes the desired symmetry species for the occupied orbitals, and allows in the energy minimization only variations which maintain these restrictions. The result will be a set of equations to determine the orbitals, (restricted Hartree–Fock equations) but the Hartree–Fock Hamiltonian will not be totally symmetric. In order to produce such a Hamiltonian, whose eigenfunctions are symmetry orbitals, one must average over occupation of the orbitals in the incomplete shells [76], to produce a totally symmetric operator.

The eigenfunctions of this operator are chosen with the desired symmetry properties, but the determinant formed from them does not really make the expectation value of the Hamiltonian stationary. (Nesbet [77] has proposed an alternative formalism in which one can see explicitly what matrix elements of the Hamiltonian are being neglected.) This means that the energy of the single determinant may be lowered by using orbitals which do not transform according to the representations of the symmetry group (see the discussion of stability on page 176). The resulting determinant will not have the symmetry behavior of a true eigenfunction (although we çan then project out the proper symmetry component—see page 172). We usually prefer to have proper symmetry behavior, even if we do not have the best single determinant. The restriction on the variation means that Brillouin's theorem is not satisfied. This has been investigated [78] by considering the difference between the correct Hartree–Fock Hamiltonian and the Hamiltonian obtained from it by averaging over orbitals of the open shell.

A procedure essentially equivalent to averaging the Hartree–Fock Hamiltonian over occupation of orbitals in open shells is to minimize, instead of the energy of the particular state desired, the average energy of all the states formed from the same configuration. [The configuration is specified by giving the numbers of electrons in each shell, e.g., $(1\sigma)^2(1\pi)$

specifies a configuration for which four determinantal functions may be written.] Then one is not obtaining the best energy for the particular state of interest. Furthermore, the different states from the same configuration will have the same kinetic energy because the kinetic energy operator is one-electron, spin-independent, and totally symmetric [see Volume I, Chapter III, Eq. (60), for the expectation value over a determinant]. Since the total energies of the different states will differ because of the inter-electronic repulsion [Volume I, Chapter III, Eq. (61)], the virial theorem will be violated by this procedure [79]. It is still often used for open shells (see page 193). Open-shell Hartree–Fock procedures have been discussed and compared by several authors [80].

For a review, with many references, of problems related to the symmetry properties of Hartree–Fock functions, see articles by Löwdin [81] and Nesbet [82]. Nesbet has noted that the canonical UHF orbitals are symmetry orbitals with respect to the symmetry group of the Hartree–Fock Hamiltonian, which is a subgroup of the molecular symmetry group.

Before closing this subsection, we return to the use of molecular orbitals not having the molecular symmetry. For homonuclear diatomics, we might use atomic-like orbitals to describe the inner shells, rather than their symmetric and antisymmetric (g and u) combinations. As another example, the HF molecule could be described in terms of an inner-shell F orbital, a bond orbital, and a lone pair F orbital, all of $\sigma$ symmetry and each containing two electrons, and a doubly filled pair of $\pi$ orbitals, expected to be concentrated on the F atom. Alternatively, one could combine the last three mentioned orbitals into a set of equivalent orbitals arranged in a pyramid with apex on the F atom and base away from the FH direction [83]. This would conform to the tetrahedral description of the $F^-$ ion (isoelectronic with neon), and put HF comfortably into the series: Ne, HF, $H_2O$, $NH_3$, and $CH_4$. The idea of treating the lone pairs on a par with bonds has been encouraged by Hall, Lennard-Jones, and co-workers [83].

Starting from the canonical Hartree–Fock molecular orbitals, we are free to take linear combinations which may correspond to such intuitive notions, without, of course, changing the many-particle determinantal wave function. Going over to new linear combinations may have consequences for consideration of excitation or ionization (Koopmans's theorem). It is possible to derive, from the canonical Hartree–Fock equations, self-consistent field equations for localized (e.g., lone pair) orbitals. Peters [84] has investigated these equations and analyzed molecular SCF wave functions in this way, with particular attention being paid to expressions for ionization potentials.

## 5. Restricted Hartree–Fock Functions

The restricted Hartree–Fock function (RHF) is a determinant constructed from spin orbitals which are chosen to minimize the energy of the determinant while maintaining certain restrictions [other than Eq. (4)] on the form of the spin orbitals. The restrictions are necessary, for example, to ensure that the determinant is an eigenfunction of $S^2$. The resulting spin orbitals may not necessarily be solutions to the unrestricted Hartree–Fock equations, (1)–(2). In this section, we discuss the use of RHF functions and their stability properties.

For most diatomic systems with an even number of electrons, the ground state is a singlet. A possible solution to the unrestricted Hartree–Fock equations which is an eigenfunction of $S^2$ has $n = \frac{1}{2}N$ doubly occupied spatial orbitals. We put $\lambda_1 = \phi_1\alpha$, $\lambda_2 = \phi_1\beta$, $\lambda_3 = \phi_2\alpha$, $\lambda_4 = \phi_2\beta$, ..., $\lambda_{N-1} = \phi_n\alpha$, $\lambda_N = \phi_n\beta$. For this case, we get an effective Hamiltonian which operates on spatial functions only, by carrying out the spin integrations in Eq. (2).

$$h^{\mathrm{HF}}(\mathbf{r}_i) = h(\mathbf{r}_i) + \sum_{m=1}^{n} \left( 2 \int d\mathbf{r}_j \mid \phi_m(\mathbf{r}_j) \mid^2 \mid \mathbf{r}_i - \mathbf{r}_j \mid^{-1} \right.$$
$$\left. - \int d\mathbf{r}_j \phi_m{}^*(\mathbf{r}_j) \mid \mathbf{r}_i - \mathbf{r}_j \mid^{-1} \mathscr{P}_{ij} \phi_m(\mathbf{r}_j) \right) \tag{28}$$

We have used the fact that $\langle \lambda_m \lambda_n \mid g \mid \lambda_p \lambda_q \rangle$ vanishes when $\lambda_m$ and $\lambda_p$ (or $\lambda_n$ and $\lambda_q$) have different spins. The orbital energies $\tilde{\varepsilon}_i$, defined by

$$h^{\mathrm{HF}}\phi_i = \tilde{\varepsilon}_i\phi_i, \tag{29}$$

are identical in pairs. It is understood that the actual determinantal function is formed by associating each $\phi_i$ with both spins. Then the Hartree–Fock energy is

$$E^{\mathrm{HF}} = 2\sum_{i}^{n} \tilde{\varepsilon}_i - \frac{1}{2} \sum_{i<j=1}^{n} (2J_{ij} - K_{ij}) \tag{30}$$

The first sum is over the $n$ spatial orbitals and the second over the $\frac{1}{2}n(n-1)$ pairs of spatial orbitals. We have used the usual notation $J_{ij}$ and $K_{ij}$ for the Coulomb and exchange integrals between orbitals $i$ and $j$:

$$J_{ij} = \iint d\tau_1 \, d\tau_2 \mid \phi_i(\mathbf{r}_1) \mid^2 \mid \phi_j(\mathbf{r}_2) \mid^2 e^2/r_{12} \tag{31a}$$

$$K_{ij} = \iint d\tau_1 \, d\tau_2 \phi_i{}^*(\mathbf{r}_1)\phi_j{}^*(\mathbf{r}_2)\phi_j(\mathbf{r}_1)\phi_i(\mathbf{r}_2)e^2/r_{12} \tag{31b}$$

These integrals are over the spatial coordinates of electrons 1 and 2. Note that $J_{ii} = K_{ii}$.

The restricted Hartree–Fock procedure derives Eqs. (28)–(31) by considering a determinantal wave function with $n$ doubly occupied spatial orbitals, and varying these orbitals maintaining orthonormality as was done in the important article of Roothaan [13]. The energy of such a determinant is given by [Volume I, Chapter III, Eq. (62)]:

$$\langle H \rangle = 2 \sum_{i=1}^{n} \langle \phi_i(i) \mid h(i) \mid \phi_i(i) \rangle + \sum_{i<j=1}^{n} (4J_{ij} - 2K_{ij}) + \sum_{i=1}^{n} J_{ii}. \tag{32}$$

Then minimizing $\langle H \rangle$ with the constraints

$$\langle \phi_i \mid \phi_j \rangle = \delta_{ij}, \qquad i, j = 1, \ldots, n$$

leads to Eq. (29), when reasoning analogous to that of Volume I, Chapter III, Section A.5 is used. Since the use of doubly occupied spatial orbitals is a restriction on the wave function beyond the single-determinant form, one might expect that Brillouin's theorem does not hold, and the accuracy of one-electron properties would be lost. However, the equivalence between electrons of $\alpha$ and $\beta$ spin in this case means that a determinant of $n$ doubly occupied orbitals can exactly satisfy (in a self-consistent way) the unrestricted Hartree–Fock equations (1) and (2). Such a determinant thus gives a stationary energy even in the absence of restrictions, and Brillouin's theorem holds.

We may contrast this with the case of a system with an odd number $N$ of electrons. It is reasonable to doubly occupy $\frac{1}{2}(N - 1)$ spatial orbitals, leaving one unpaired electron. This determinant is an eigenfunction of $S^2$, but the orbitals will not satisfy the Hartree–Fock equations; spin orbitals of different spin would be subject to different exchange potentials, so they should have different spatial parts.

Explicitly, we write our spin-paired function as:

$$\Phi = \mathscr{A}\{\phi_1(1)\alpha(1)\phi_1(2)\beta(2)\phi_2(3)\alpha(3) \cdots \phi_n(N-1)\alpha(N-1)\phi_n(N)\beta(N)\} \tag{33}$$

The double occupation assures that the determinant be an eigenfunction of total spin with eigenvalue zero, but additional restrictions may be needed to give the desired spatial symmetry properties. As we have shown, if only closed shells are present, the unrestricted Hartree–Fock equations may be satisfied. For the diatomic molecule, this means that the two functions (25) differing only in the sign of $m$ must appear an equal number of times in the determinant. If we demand double occupation, this means four electrons for each orbital of $m \neq 0$ and two electrons for each $\sigma$ orbital.

Roothaan [13] showed in general that, if the determinant represents closed shells, so that all partners in a representation appear when any one appears, the determinant is invariant to all symmetry operations, no matter what the symmetry group. As already mentioned, the symmetry operator acts on all electrons simultaneously, so

$$R\Phi = \mathscr{A}\{[R\phi_i(1)\alpha(1)][R\phi_1(2)\beta(2)] \cdots [R\phi_{N/2}(N)\beta(N)]\} \qquad (34)$$

The spin is not affected. Here $R\phi_i$ is a linear combination of orbitals [Eq. (26)] which already appear in $\Phi$. If $\phi_i$ belongs to the $l_i$ row of the $\mu_i$ representation,

$$R\Phi = \sum_{j_1=1}^{d_1} [D^{(\mu_1)}(R)]_{j_1 l_1} \sum_{k_1=1}^{d_1} [D^{(\mu_1)}R]_{k_1 l_1} \cdots \sum_{k_n=1}^{d_n} [D^{(\mu_1)}(R)]_{k_n l_n}$$

$$\times \mathscr{A}\{\phi_{j_1}(1)\alpha(1)\phi_{k_1}(2)\beta(2) \cdots \phi_{k_n}(N)\beta(N)\}$$

Only products of orbitals in which each appears only once with each spin, can survive the antisymmetrization. A detailed analysis shows that the determinantal wave function will be multiplied by the product of the determinants of the representation matrices, each of which appears twice. If the orbitals are real, each determinant is $\pm 1$, and the product then is unity. The case where inherently complex irreducible representations appear was treated by Roothaan [13].

A review of the formalism involved in the computation of closed-shell restricted Hartree–Fock functions was given by McLean and Yoshimine [85]. They also explained the construction of automatic computer programs for carrying out such computations with Slater basis functions, including flow charts, linkage of subroutines, and schemes for automatically optimizing parameters and integral evaluation. Their method for the systematic evaluation of large numbers of two-electron integrals is explained in detail. A supplement to this paper tabulates wave functions obtained by their programs for a large number of diatomic and linear molecules.

Only for a closed-shell function can one satisfy symmetry and self-consistency conditions with a single determinant. Otherwise, one can sometimes write a function of desired spin and symmetry properties as a single determinant by demanding that the orbitals obey certain restrictions, but one cannot derive equations to determine the best orbitals to use without at some point averaging over the occupations of orbitals in the open shell.

The determinant formed from the resulting orbitals does not really make the energy stationary, and the Brillouin, virial, Hellmann–Feynman, and other theorems do not hold. There also exist open-shell states for which one cannot even write down a single-determinant wave function of the desired symmetry; sums of determinants must be used. One may use either a sum of determinants as the variation function, or correct a single determinant function, such as a UHF function, to give proper symmetry behavior, by turning it into a sum of determinants. The latter alternative will be discussed in this subsection, the former in Section B.

We can imagine the UHF function to be expanded in the *N*-electron spin eigenfunctions. This expansion may be done in a unique way [86]. As mentioned earlier, a determinant constructed from eigenfunctions of $S_z$, such as the $\lambda_i$ we use here, is also an eigenfunction of $S_z$, with eigenvalue $\sigma$ equal to half the excess of orbitals with $\alpha$ spin over those with $\beta$ spin. Then only the spin functions for this value of $\sigma$, and different eigenvalues of $S^2$, will appear in the expansion.

$$\Psi^{\mathrm{UHF}} = \sum_S \sum_n f_{Sn} g_{S\sigma n} \tag{35}$$

where $g_{S\sigma n}$ is a spin eigenfunction and $f_{Sn}$ is a spatial function. For one electron, we have $g_{\frac{1}{2},\frac{1}{2},1} = \alpha$ and $g_{\frac{1}{2},-\frac{1}{2},1} = \beta$. For two electrons, the four spin functions are just the familiar singlet and triplet. The index $n$ is needed because, in general, there can be several eigenfunctions of spin with the same quantum numbers for $S^2$ and $S_z$. Thus, for three electrons one could put $g_{\frac{1}{2}\frac{1}{2}1} = 3^{-1/2}(\alpha\beta\alpha + \alpha\alpha\beta - 2\beta\alpha\alpha)$ and $g_{\frac{1}{2}\frac{1}{2}2} = 2^{-1/2}(\alpha\beta\alpha - \alpha\alpha\beta)$, where $\alpha\beta\alpha$ means $\alpha(1)\beta(2)\alpha(3)$, and so on. These are orthonormal. We must have $S \geq \sigma$ and for $N$ electrons $S \leq \frac{1}{2}N$. (A tighter restriction may obtain if closed shells are used to describe some of the electrons: Then the number of electrons in open shells replaces $N$.) The problem is to eliminate the contamination of unwanted spin functions from the determinant.

Suppose we operate on $\Psi^{\mathrm{UHF}}$ with the operator $P_{S'} = S^2 - S'(S' + 1)$. $P_{S'}$ annihilates the functions $g_{S'\sigma n'}$ and multiplies $g_{S\sigma n}$ for $S \neq S'$ by $[S(S + 1) - S'(S' + 1)]$. Forming the product of such operators for all $S'$ except the one desired and renormalizing, we have the operator

$$\mathscr{O}_{S''} = \prod_{S' \neq S''} [S''(S'' + 1) - S'(S' + 1)]^{-1}[S^2 - S'(S' + 1)]. \tag{36}$$

$\mathscr{O}_{S''}$ annihilates all terms in (35) except for $S = S''$, and leaves the terms for $S = S''$ invariant. It may be referred to as a normalized spin projection operator, since it produces from any eigenfunction of $S_z$, a function which is an eigenfunction of $S^2$ as well. Since only $S^2$ and numerical constants appear in $\mathscr{O}_{S''}$ and $S^2$ is symmetric in the electrons, $\mathscr{O}_{S''}$ commutes with the antisymmetrizer. Thus $\mathscr{O}_{S''}\Psi^{\mathrm{UHF}}$ is antisymmetric. This shows that one can construct from a determinant an antisymmetric function which is an eigenfunction of the spin operators $S^2$ and $S_z$. It may also be seen, from the discussion following Eq. (22), that the result of operating $\mathscr{O}_{S''}$ on the determinantal wave function is a linear combination of determinants. All these determinants belong to the same configuration. In a sense, we have a generalization of the process of going from the Hartree product to the Slater determinant using the antisymmetrizer $\mathscr{A}$, since the determinant is a sum of products generated from the Hartree product by permutation operators [86].

The expectation value of $H$ must be recalculated with the projected function, although most of the integrals over orbitals required will already have been evaluated in carrying out the UHF calculation. It is not hard to see that at least one of the components which may be projected out has a lower energy than the starting function itself. Since the Hamiltonian is spin-independent and the $g_{S\sigma n}$ are orthonormal, the expectation value of (35) is

$$\frac{\langle \Psi \mid H \mid \Psi \rangle}{\langle \Psi \mid \Psi \rangle} = \frac{\sum_S \sum_n \langle f_{Sn} \mid H \mid f_{Sn} \rangle}{\sum_S \sum_n \langle f_{Sn} \mid f_{Sn} \rangle}$$

$$= \sum_S \left[ \frac{\sum_n \langle f_{Sn} \mid H \mid f_{Sn} \rangle}{\sum_n \langle f_{Sn} \mid f_{Sn} \rangle} \right]\left[ \frac{\sum_n \langle f_{Sn} \mid f_{Sn} \rangle}{\sum_{S'} \sum_n \langle f_{S'n} \mid f_{S'n} \rangle} \right] \qquad (37)$$

The first square bracket is the energy of a projected component, so that the expectation value of Eq. (37) is a weighted sum of these energies. The weighting factors all being positive, not all these energies can lie above $\langle H \rangle$.

The projection process is tractable if there are only a small number of unwanted spin components in the original determinantal wave function. However, the product of the projectors $P_{S'}$ gets quite unwieldy if it contains many factors. Sometimes, it is sufficient to use a single one, to purify the determinantal function of its most significant unwanted component. Thus, for a ground state we generally seek a function with the lowest possible spin ($S'' = 0$ or $\frac{1}{2}$) and the most important unwanted component is that with spin $S = S'' + 1$. This conforms to the general idea that energy lowering is associated with maximal occupation of low-lying orbitals consistent with reduction of interelectronic repulsion, this tending to promote spin pairing. This case was worked out by Amos and Hall [1] and Amos and Snyder [2], who gave expectation values of various operators after spin annihilation. It was pointed out that other components, if their contribution is not too large, make only a negligibly small contribution to certain important properties.

The projection operator technique may also be used for obtaining a desired spatial symmetry. From the operators of the spatial symmetry group, one can construct an operator that projects out unwanted representations, as does $P_{S''}$. Spin and spatial symmetry are handled together by using a product of space and spin projection operators (they commute). As can be seen in any text on group theory, the projection operators need not be written in the form of Eq. (36). For more details on the use of projection operators, the article by Löwdin [86] is recommended.

Thus, one can carry out an unrestricted Hartree–Fock calculation, ignoring all or some of the spin and symmetry restrictions on the orbitals, to produce a single-determinant wave function which optimizes the energy. Then one can project out the part of the resulting function which has the proper spin and spatial symmetry. This is sometimes referred to as a projected Hartree–Fock scheme, and produces a multideterminant function. The procedure may also be used on closed-shell states to produce a function that is free from restrictions like double occupation. Since double occupation leads to a large repulsion energy between the two electrons (of opposite spin) in the same spatial orbital, projection can give energies below that of the usual Hartree–Fock function. Slater's idea of different spatial orbitals for different spins [87], for instance, can be treated this way. It has been [86] stated that by projecting an unrestricted Hartree–Fock function, 95% of the energy error in the *restricted* Hartree–Fock (relative to the exact nonrelativistic energy) can be recovered.

The matrix elements between projected determinants may be simplified by use of the properties of the projection operators. Letting $\mathscr{O}$ represent any spinless operator totally symmetric in the electron coordinates, we have

$$\langle \mathscr{O}_S \Phi_J | \mathscr{O} | \mathscr{O}_S \Phi_K \rangle = \langle \Phi_J | \mathscr{O}_S \mathscr{O} | \mathscr{O}_S \Phi_K \rangle = \langle \Phi_J | \mathscr{O} | \mathscr{O}_S \Phi_K \rangle$$

where $\Phi_J$ and $\Phi_K$ are single determinants and $\mathscr{O}_S$ the spin projection operator. The idempotency of $\mathscr{O}_S$ has been employed. If $\mathscr{O}_S \Phi_k$ is expressed as a sum of determinants, the matrix element is a single sum of matrix elements between determinants.

Projection operators need not be used to give correct symmetry behavior. Other schemes, which take advantage of the properties of the spin and other operators, or of the characteristics of the permutation group, are available to convert a single determinant into a multideterminantal function of desired symmetry. We mention some of these in Section B.1.

Since the multideterminant function produced from the UHF is energy-optimized before projection rather than after, the projected determinant is not the best which can be written in this form. We should rather write a multideterminant function by projecting or otherwise correcting a single determinant, and vary the orbitals to minimize the expectation value of $H$ over *this* function. Such methods will be discussed in Section B.

In some cases it may be possible to use orbitals determined from a closed-shell calculation to construct an open-shell wave function, as was done by Nesbet [88] for the case of TiO. The ground state is $^3\Delta$, and arises from single occupation of the lowest $\delta$-orbital and the $9\sigma$ orbital. Nesbet constructed a determinantal wave function for this state from orbitals obtained from a Hartree–Fock calculation on a closed-shell (excited) state whose configuration is almost identical to that of the ground state. In this excited state, the $1\delta$ orbital is unoccupied and the $9\sigma$ doubly filled. Although the ground state function does not make the energy stationary, it can be expected that the optimum orbitals would not differ much from those actually used. This procedure is especially useful when one is interested in several electronic states. The problem in using it is to find a configuration or state for which a single-determinant calculation can legitimately be carried out, and which resembles the open shell state of interest to a high degree.

Even if a restricted Hartree–Fock determinant satisfies the UHF equation, there is no guarantee that another solution to these equations cannot be found which gives a lower energy. In addition there is the problem of stability to small variations of the spin orbitals which destroy the spin eigenfunction property or the single-determinant form. For the ground state of the two-electron atom, the energy is known to be a saddle point rather than a true minimum, and to be unstable with respect to variations leading to an open-shell (two-determinant) wave function [89].

Hibbert and Coulson [90] showed that, for a wide class of Hartree–Fock functions corresponding to open and closed shells, the energy may be lowered by splitting a pair of orbitals. These functions may be sums of determinants (extended Hartree–Fock: see Section B), constructed to give desired spin and symmetry properties, with the orbitals varied to make the energy stationary. If there are $Q$ determinants,

$$\Psi^{\mathrm{HF}} = \sum_{i=1}^{Q} c_i D_i \qquad (38)$$

where, in each $D_i$, the orbitals $\phi_i$, $i = 1 \cdots N$, are doubly occupied. The $D_i$ differ among themselves in the occupation of orbitals belonging to the open shell, denoted by $\psi_i$. Let one doubly occupied orbital, say $\phi_1$, be split by replacing $\phi_1(1)\phi_1(2)$ in all the $D_i$ by

$$u_1(1)v_1(2) + u_1(2)v_1(1)$$

where

$$u_1 = \phi_1 + \varepsilon\chi_1, \qquad v_1 = \phi_1 - \varepsilon\chi_1. \tag{39}$$

Here, $\varepsilon$ is small and $\chi_1$ is normalized and orthogonal to all the orbitals. It was noted [90] that the proper spin symmetry is maintained because

$$u_1 v_1 + v_1 u_1 = \tfrac{1}{2}(u_1 + v_1)(u_1 + v_1) + \tfrac{1}{2}(u_1 - v_1)(u_1 - v_1)$$

so that each $D_i$ becomes a sum of two determinants, one of which derives from $D_i$ by replacement of $\phi_1$ by $2^{-1/2}(u_1 + v_1)$, and the other by replacement of $\phi_1$ by $2^{-1/2}(u_1 - v_1)$. Each has the spin property of $D_i$, so $\sum c_i D_i$ becomes a sum of two functions of the same spin. Since the $\phi_i$ and $\psi_i$ have been chosen to make the energy stationary, the term proportional to $\varepsilon$ in the energy expression for the split-shell function vanishes. It could be shown [90] that, within a positive normalizing constant, the change in the energy would be

$$\Delta E = -\varepsilon^2 \sum_i c_i' \langle D_i \,|\, H - E^{\mathrm{HF}} \,|\, D_i' \rangle \tag{40}$$

where $c_i'$ differs from $c_i$ by a normalizing constant, and $D_i'$ is obtained from $D_i$ by replacement of $\phi_1\phi_1$ by $\chi_1\chi_1$. The orthogonality properties of the $\chi_1$ mean that no cross-terms between different $D_i$ survive. Now $\langle D_i \,|\, H - E^{\mathrm{HF}} \,|\, D_i \rangle$ becomes $\langle \phi_i\phi_i \,|\, r_{12}^{-1} \,|\, \chi_i\chi_i \rangle$ which is easily shown to be positive if the orbitals are real. Thus, this variation decreases the energy. A more detailed discussion [90] was given for the single-determinant wave function in which all the open-shell orbitals have the same spin. This is a pure spin state and remains so on replacement of $\phi_i\phi_i$ by $u_i v_i + v_i u_i$. Hibbert and Coulson discussed the choice of $\chi_1$ and the open-shell orbitals for the greatest energy lowering.

The formalisms of Adams [6] and Čížek and Paldus [8] (see Subsection 1) have been used to investigate the stability of restricted Hartree–Fock functions. To put spin properties into Adams's formalism, we note that, since all the spin orbitals are products of spatial orbitals with $\alpha$ or $\beta$ spin functions, the density matrix [Eq. (11)] is

$$\varrho(i, j) = \sum_k^{\text{(spin up)}} \lambda_k(i)\lambda_k^*(j) + \sum_k^{\text{(spin down)}} \lambda_k(i)\lambda_k^*(j)$$

and factoring off spin functions gives

$$\rho = \alpha\alpha^\dagger\rho_+ + \beta\beta^\dagger\rho_-. \tag{41}$$

The double occupancy condition corresponds to demanding equality of $\rho_+$ and $\rho_-$. Restrictions other than double occupancy may be treated similarly.

Since the restricted Hartree–Fock function seems often to work well for ground states of diatomic molecules near the equilibrium internuclear distance, Adams [6] investigated its stability with respect to the unrestricted Hartree–Fock. For LiH, he investigated particular variations of the orbitals which break the double occupation condition, using a limited-basis approximate SCF calculation. The restricted Hartree–Fock was shown to be unstable above $R = 4\,a_0$, but it was not proved that it is stable for smaller $R$. It was concluded that the stability condition is difficult to apply. Only instability can be demonstrated. This is done by actually displaying the variation of the orbitals which produces an energy lowering in second order.

Čížek and Paldus [8] also specialized their formalism to doubly-occupied states. Four coefficients $c_i^m$ correspond to each pair of orbitals. If one works in terms of the spatial functions $\phi_k$ ($k = 1, \ldots, \tfrac{1}{2}N$) one can promote from $\phi_j$ ($j \le N$), with spin $\alpha$ or $\beta$, to a virtual orbital with either spin. Suppose the coefficients corresponding to the cases $\alpha \to \alpha$, $\alpha \to \beta$, $\beta \to \alpha$, and $\beta \to \beta$, respectively are $c_i^m$, $c_i^{m'}$, $c_{i'}^m$, and $c_{i'}^{m'}$. Čížek and Paldus [8] defined

$$d_j^{l(s)} = 2^{-1/2}(c_i^m + c_{j'}^{m'}) \tag{42a}$$

$$d_j^{l(t_1)} = 2^{-1/2}(c_i^m - c_{j'}^{m'}) \tag{42b}$$

$$d_j^{l(t_2)} = 2^{-1/2}(c_j^{m'} + c_i^m) \tag{42c}$$

$$d_j^{l(t_3)} = 2^{-1/2}(c_j^{m'} - c_{i'}^m) \tag{42d}$$

The coefficients in previous expressions were written in terms of the new coefficients, and inspection of $\langle \Phi_1 | \tilde{H} | \Phi_1 \rangle$ and $\langle \Phi_0 | \tilde{H} | \Phi_2 \rangle$ showed that each divides into four groups of terms, one for each kind of $d_j^l$. Then the matrix $\mathbf{E}$ of Eq. (15) is block-diagonal, the four blocks corresponding to $(s)$, $(t_1)$, $(t_2)$, and $(t_3)$.

For example, in $\langle \Phi_1 | \tilde{H} | \Phi_1 \rangle$, the product of coefficients $d_i^{m(s)*}d_j^{n(s)}$ multiplies the matrix element of $\tilde{H}$ between

$$2^{-1/2}[\Phi^{i\alpha \to m\alpha} + \Phi^{i\beta \to m\beta}] \quad \text{and} \quad 2^{-1/2}[\Phi^{j\alpha \to n\alpha} + \Phi^{j\beta \to n\beta}].$$

These two-determinant functions represent spin singlets, and these terms are referred to as singlet excitations. The others are called triplet excitations because of the form of the matrix elements, but the functions involved are not necessarily spin eigenfunctions. The terms $(t_1)$, the terms $(t_2)$ and the terms $(t_3)$ involve identical matrix elements. The matrix elements have been given in a form convenient for this computation by Čížek [91]. The contributions of $\langle \Phi_0 \mid \tilde{H} \mid \Phi_2 \rangle$ and $\langle \Phi_2 \mid \tilde{H} \mid \Phi_0 \rangle$ behaving similarly, one must consider two eigenvalue problems, corresponding to the singlet and triplet excitations.

The applications given by Čížek in this article [91] are to $\pi$-electronic systems, the simplest being ethylene, but the general ideas are applicable: We must consider the eigenvalues of $\mathbf{E}$. The Hartree–Fock function may be triplet unstable (the submatrix corresponding to triplet excitations has a negative eigenvalue) or singlet unstable (the submatrix corresponding to singlet excitations has a negative eigenvalue).

If there is a triplet instability, one can construct an unrestricted Hartree–Fock function of lower energy than the original determinant, but the new determinant will not be a spin singlet. In general, we expect that an unrestricted determinant can give a lower energy than the restricted Hartree–Fock, which demands double occupancy, since a constraint on the variation is lifted. Indeed, it may be shown [92] that starting from a closed-shell doubly occupied determinant one can always obtain a lower energy by replacing the identical spatial parts of two spin orbitals with different functions and projecting the resulting determinant to give a pure singlet. The determinant without projection may not have a lowered energy, but one projected component will [see Eq. (37)]. In fact, if we allow variation of spin parts as well, we can lower the energy by using different axes of spin quantization for different pairs. This possibility is not included in our discussion.

If we find singlet instability, there exists a true singlet wave function that has lower energy than the original single determinant. Although this function has correct spin symmetry, projection or another technique might have to be used to give its desired behavior with respect to the molecular spatial symmetry group. Furthermore, the new function is constructed from the orbitals which are eigenfunctions of the restricted Hartree–Fock Hamiltonian. These orbitals will not be eigenfunctions of the Hartree–Fock Hamiltonian formed from the new set of occupied orbitals. Then the new function does not make the energy stationary, and a further energy lowering may be possible.

The extension of this discussion of instabilities to the open-shell case where there is a single electron outside closed shells, giving a doublet, has recently been given by Paldus and Čížek [93, 94]. The use of the creation and annihilation operators, together with diagrammatic perturbation theory, is found convenient here to deal with the complicated formulas which arise. Again, an instability which destroys the spin eigenfunction property (nondoublet) is always possible. There are also doublet instabilities, which lead to a pure spin state. For these, the problem of generating the doublet function of lower energy from the original one was discussed.

### 6. Current Hartree–Fock Calculations

Although a number of workers are considering stability and other basic problems of the method, a great effort continues to produce better Hartree–Fock calculations for larger and larger systems. These are usually restricted Hartree–Fock calculations, and almost invariably expand the spin orbitals in a basis set. In Section C, we discuss the results of such calculations for $U(R)$, and Chapter IV gives bibliographies. However, we will cite some results here to show the accuracy obtainable and the range of molecules that can be treated.

O'Hare and Wahl [95] treated SF, SeF, and their positive and negative ions, all at the experimental equilibrium internuclear distances for the neutral molecules. The compound SeF has 43 electrons, and there are 10 $\sigma$ orbitals, 5 $\pi$ orbitals, and 1 $\delta$ orbital whose expansions in a basis set must be determined. Both SF and SeF have a $\pi^3$ configuration for the electrons outside closed shells, and hence $^2\Pi$ ground states. A single determinant function may be used to represent such states. The negative ions have closed outer shells and $^1\Sigma^+$ terms; the positive ions have $^3\Sigma^-$ terms, resulting from a $\pi^2$ configuration, which also allows representation as a single determinant.

Basis functions determined for the constituent atoms were used for the molecular calculation [95], augmented by functions designed to describe the polarization of each atomic density by the other in the molecule. On the basis of previous calculations, O'Hare and Wahl estimated an error in the energy of 0.005 a.u. due to the incompleteness of the basis set. The difference between the Hartree–Fock energy of the atoms and the Hartree–Fock energy of the molecule was 0.0532 a.u. for SF and 0.0449 a.u. for SeF. The correction for different correlation energies in atoms and molecule (molecular extra correlation energy; see Section C) was estimated, from the known value of $0.105 \pm 0.007$ a.u. for the related molecule OF, as 0.103 a.u. for SF and 0.101 a.u. for SeF. Thus, the corrected dissocia-

energies were $0.156 \pm 0.008$ a.u. and $0.146 \pm 0.008$ a.u. for SF and SeF, respectively. Expectation values of a large number of operators for all six species were given. Koopmans's theorem (Volume I, Chapter III, Section A.5) ionization potentials and electron affinities of SF and SeF were compared to values for these quantities obtained from Hartree–Fock energy differences.

Wahl et al. [11] have given a "state-of-the-art" review of near-Hartree–Fock calculations (within 0.1 eV of the Hartree–Fock energy) for diatomic molecules (1967), while Clementi [96] has presented some extrapolations for the future. Wahl et al. found the accuracy of $D_e$, $\omega_e$, and other properties of $U(R)$ from such calculations discouraging, except for pairs of rare gas atoms, highly ionic molecules, hydrides, and certain homopolar molecule-ions. For the homopolar molecule–ions, the accuracy involved a fortuitous cancellation of errors; the other systems either dissociate correctly or seem to (because the exact wave function resembles ions rather than atoms out

**Table II**

*Results of near Hartree–Fock Calculations*[a]

| Molecule | State | $R_e$ $(a_0)$ | | $\omega_e$ $(cm^{-1})$ | | $D_e$ $(eV)$ | |
|---|---|---|---|---|---|---|---|
| | | Calc. | Expt. | Calc. | Expt. | Calc. | Expt. |
| $H_2$ | $^1\Sigma_g^+$ | 1.385 | 1.400 | 4561 | 4400 | 3.65 | 4.75 |
| $Li_2$ | $^1\Sigma_g^+$ | 5.25 | 5.051 | 326 | 351 | 0.17 | 1.05 |
| $N_2^+$ | $^2\Sigma_g^+$ | 2.041 | 2.113 | 2571 | 2207 | 3.24 | 8.85 |
| $N_2$ | $^1\Sigma_g^+$ | 2.013 | 2.075 | 2730 | 2358 | 5.27 | 9.91 |
| $F_2$ | $^1\Sigma_g^+$ | 2.51 | 2.725 | 1257 | 892 | −1.63 | 1.65 |
| LiF | $^1\Sigma^+$ | 2.941 | 2.955 | 1033 | 964 | 4.08 | 5.95 |
| BF | $^1\Sigma^+$ | 2.355 | 2.391 | 1496 | 1402 | 6.18 | 8.58 |
| CO | $^1\Sigma^+$ | 2.08 | 2.132 | 2431 | 2170 | 7.84 | 11.23 |
| $Cl_2$ | $^1\Sigma_g^+$ | 3.78 | 3.76 | 577 | 559 | 0.87 | 2.51 |
| $He_2^+$ | $^2\Sigma_g^+$ | 2.0 | — | 1792 | — | 2.7 | — |
| $F_2^-$ | $^2\Sigma_g^+$ | 3.6 | — | 510 | — | 1.66 | — |
| $Ne_2^+$ | $^2\Sigma_g^+$ | 3.2 | — | 660 | — | 1.65 | — |
| $Cl_2^-$ | $^2\Sigma_g^+$ | 5.0 | — | 260 | — | 1.28 | — |
| $Ar_2^+$ | $^2\Sigma_g^+$ | 4.6 | — | 300 | — | 1.25 | — |
| NaH | $^1\Sigma^+$ | 3.62 | 3.57 | 1187 | 1172 | 0.932 | (2.3) |
| SiH | $^2\Pi$ | 2.86 | 2.87 | 2144 | 2042 | 2.23 | (3.3) |
| HCl | $^1\Sigma^+$ | 2.39 | 2.41 | 3181 | 2989 | 3.48 | 4.616 |
| NaF | $^1\Sigma^+$ | 3.65 | 3.64 | 590 | 536 | 3.08 | 4.49 |

[a] From Wahl et al. [11].

to $R \gg R_e$). We reproduce their table of typical results of near-Hartree–Fock calculations and comparisons with experiment, and refer to Wahl *et al.* [11] for references (see Table II). They also list all such calculations for systems involving first and second row atoms.

Clementi [96] analyzed the status of molecular calculations in 1969 and attempted to extrapolate to the future. For "small molecules," which include diatomics with up to 40 or 50 electrons, he anticipated that near Hartree–Fock computations, and calculations giving a good part of the correlation energy, should be available. From these, calculated dissociation energies should be accurate to a few thousand calories and vibrational frequencies correct to close to the experimental accuracy. For "medium-sized molecules," which include diatomics with 100 or fewer electrons, near Hartree–Fock calculations should be available.

## REFERENCES

1. T. Amos and G. G. Hall, *Proc. Roy. Soc. Ser. A* **263**, 483 (1961).
1a. H. F. King, R. E. Stanton, H. Kim, R. E. Wyatt, and R. G. Parr, *J. Chem. Phys.* **47**, 1936 (1967).
2. T. Amos and L. C. Snyder, *J. Chem. Phys.* **41**, 1773 (1964).
3. C. A. Coulson and I. Fischer, *Phil. Mag.* **40**, 386 (1949).
4. H. F. King and R. E. Stanton, *J. Chem. Phys.* **50**, 3789 (1969).
5. R. E. Stanton, *J. Chem. Phys.* **48**, 257 (1968).
6. W. H. Adams, *Phys. Rev.* **127**, 1650 (1962).
7. R. E. Watson, *Phys. Rev.* **118**, 1036; **119**, 1934 (1960); D. J. Thouless, "The Quantum Mechanics of Many-Body Systems." Academic Press, New York, 1961.
8. J. Čížek and J. Paldus, *J. Chem. Phys.* **47**, 3976 (1967).
9. T. A. Kaplan and W. H. Kleiner, *Phys. Rev.* **156**, 1 (1967).
10. D. R. Hartree, "The Calculation of Atomic Structures." Wiley, New York, 1957.
11. A. C. Wahl, P. J. Bertoncini, B. Das, and T. L. Gilbert, *Int. J. Quantum Chem. Symp.* **1**, 123 (1967).
12. J. A. Keefer, J. K. Su Fu, and R. L. Belford, *J. Chem. Phys.* **50**, 160 (1969).
13. C. C. J. Roothaan, *Rev. Mod. Phys.* **23**, 69 (1951).
14. C. C. J. Roothaan and P. S. Bagus, *Methods Comp. Phys.* **2**, 62 (1963).
15. S. Huzinaga, *Progr. Theor. Phys. Suppl.* **40**, 52 (1967).
16. L. C. Allen and A. M. Karo, *Rev. Mod. Phys.* **32**, 275 (1960).
17. H. Eyring, J. Walter, and J. Kimball, "Quantum Chemistry," Sect. 9k. McGraw-Hill, New York, 1944.
18. T. Vladimiroff, *Int. J. Quantum Chem.* **4**, 89 (1970); *J. Chem. Phys.* **54**, 2292 (1971).
19. D. K. Harriss, C. W. Mitchell, and B. Musulin, *Theor. Chim. Acta* **1**, 93 (1963); C. M. Wang and B. Musulin, *Can. J. Chem.* **49**, 3406 (1971).
20. R. S. Mulliken, *J. Chem. Phys.* **36**, 3428 (1962).
21. M. Weissman and V. Tolmachev, *Phys. Rev. A* **3**, 1291 (1971).
22. H. O. Pritchard and F. H. Skinner, *J. Phys. Chem.* **65**, 641 (1966).

23. D. D. Ebbing, *J. Chem. Phys.* **36**, 1361 (1962).
24. M. Kotani, A. Amemiya, E. Ishiguro, and T. Kimura, "Table of Molecular Integrals." Maruzen, Tokyo, 1955, 2nd ed., 1963.
25. H. Preuss, "Integraltafeln zur Quantenchemie," Vols. 1–4. Springer-Verlag, Berlin and New York (1956–1961).
26. C. A. Coulson, *Proc. Cambridge Phil. Soc.* **38**, 210 (1941).
27. K. Ruedenberg, K. O-Ohata, and D. G. Wilson, *J. Math. Phys.* (*N.Y.*) **7**, 539 (1966).
28. K. Ruedenberg, *J. Chem. Phys.* **19**, 1459 (1951).
29. A. C. Wahl, P. E. Cade, and C. C. J. Roothaan, *J. Chem. Phys.* **41**, 2578 (1964).
30. M. P. Barnett and C. A. Coulson, *Phil. Trans. Roy. Soc. London* **243**, 221 (1951).
30a. F. E. Harris and H. H. Michels, *J. Chem. Phys.* **43**, S165 (1965); **45**, 116 (1966).
31. M. P. Barnett, *Methods Comp. Phys.* **2**, 95 (1963).
32. A. C. Wahl and R. H. Land, *Int. J. Quantum Chem. Symp.* **1**, 375 (1967); *J. Chem. Phys.* **50**, 4725 (1969).
33. R. A. Bonham, J. L. Peacher, and H. L. Cox, Jr., *J. Chem. Phys.* **40**, 3083 (1964); R. A. Bonham, *J. Mol. Spectrosc.* **15**, 112 (1965); K. Ruedenberg, *Theor. Chim. Acta* **7**, 359 (1967).
34. K. O-Ohata and K. Ruedenberg, *J. Math. Phys.* (*N.Y.*) **7**, 547 (1966).
35. R. E. Christoffersen and K. Ruedenberg, **49**, 4285 (1968).
36. F. T. Ormand and F. A. Matsen, *J. Chem. Phys.* **29**, 100 (1958). J. C. Browne and F. A. Matsen, *Phys. Rev. A* **135**, 1227 (1964).
37. H. M. James and A. S. Coolidge, *J. Chem. Phys.* **1**, 825 (1933).
37a. J. K. Knipp, *J. Chem. Phys.* **4**, 300 (1936).
38. F. Harris and H. S. Taylor, *J. Chem. Phys.* **32**, 3 (1960); **38**, 2591 (1963); *Mol. Phys.* **6**, 183 (1963); *Physica* (*Utrecht*) **30**, 105 (1964).
39. I. Shavitt, *Methods Comp. Phys.* **2**, 1 (1963).
40. S. F. Boys, *Proc. Roy. Soc. Ser. A* **200**, 542 (1942).
41. J. C. Browne and R. D. Poshusta, *J. Chem. Phys.* **36**, 1933 (1962).
42. I. Shavitt and M. Karplus, *J. Chem. Phys.* **36**, 550 (1962); **43**, 398 (1965); C. W. Kern and M. Karplus, *Ibid.* **43**, 415 (1965).
43. S. Huzinaga, *J. Chem. Phys.* **42**, 1293 (1965).
44. A. Veillard, *Theor. Chim. Acta* **12**, 405 (1968).
45. B. Roos, A. Veillard, and G. Vinot, *Theor. Chim. Acta* **20**, 1 (1971).
46. T. H. Dunning, Jr., *J. Chem. Phys.* **55**, 716 (1971).
47. L. C. Allen, *J. Chem. Phys.* **37**, 200 (1962).
48. T. H. Dunning, Jr., *J. Chem. Phys.* **55**, 3958 (1971).
49. D. M. Bishop, *J. Chem. Phys.* **40**, 1322 (1964); **48**, 291 (1968).
50. L. C. Allen, *J. Chem. Phys.* **37**, 200 (1962).
51. R. Ditchfield, D. P. Miller, and J. A. Pople, *J. Chem. Phys.* **53**, 613 (1970).
52. S. Huzinaga and Y. Sakai, *J. Chem. Phys.* **50**, 1371 (1969).
53. S. Huzinaga and C. Arnau, *J. Chem. Phys.* **52**, 2224 (1970).
54. T. A. Claxton and N. A. Smith, *Theor. Chim. Acta* **22**, 378 (1971).
55. W. J. Hehre, R. Ditchfield, R. F. Stewart, and J. A. Pople, *J. Chem. Phys.* **52**, 2769 (1970).
56. A. A. Frost, B. H. Prentice, III, and R. A. Rouse, *J. Amer. Chem. Soc.* **89**, 3064 (1967); and subsequent articles up to S. Y. Chu and A. A. Frost, *J. Chem. Phys.* **54**, 764 (1971).

57. M. E. Schwartz and L. J. Schaad, *J. Chem. Phys.* **46**, 4112 (1967); **47**, 5325 (1967).
58. J. L. Whitten, *J. Chem. Phys.* **44**, 359 (1966).
59. J. L. Whitten and L. C. Allen, *J. Chem. Phys.* **43**, S170 (1965).
60. C. P. Yue and R. L. Somorjai, *J. Chem. Phys.* **55**, 4594 (1971).
61. J. S. Alper, *J. Chem. Phys.* **55**, 3770 (1971).
62. J. Goodisman, *J. Chem. Phys.* **43**, 3037 (1965); **45**, 1515 (1966); H. F. Schaefer, III, *Ibid.* **52**, 6241 (1970); **54**, 2207 (1971); *J. Comput. Phys.* **6**, 142 (1971).
63. S. Hagstrom and H. Shull, *J. Chem. Phys.* **30**, 1314 (1959).
64. A. S. Coolidge, *Phys. Rev.* **42**, 189 (1932); R. Landshoff, *Z. Phys.* **102**, 201 (1936).
65. D. M. Bishop, *Advan. Quantum Chem.* **3**, 25 (1967).
66. E. F. Hayes and R. G. Parr, *Progr. Theor. Phys. Suppl.* **40**, 78 (1967).
67. D. M. Bishop, *J. Chem. Phys.* **43**, 3052 (1965).
68. R. Moccia, *J. Chem. Phys.* **40**, 2164 (1964).
69. A. G. Turner, B. H. Honig, R. G. Parr, and J. R. Hoyland, *J. Chem. Phys.* **40**, 3216 (1964).
70. F. E. Harris and H. H. Michels, *J. Chem. Phys.* **43**, S165 (1965).
71. R. B. Hake and K. E. Banyard, *J. Chem. Phys.* **45**, 3199 (1966).
72. K. E. Banyard and R. B. Hake, *J. Chem. Phys.* **41**, 3221 (1964); **43**, 2684 (1965); **44**, 3523 (1966).
73. R. Moccia, *J. Chem. Phys.* **40**, 2164, 2176, 2186 (1964).
74. P.-O. Löwdin, *Advan. Chem. Phys.* **2**, 227–231 (1959).
75. J. C. Slater, "Quantum Theory of Molecules and Solids," Vol. 1, Appendix VII. McGraw-Hill, New York, 1963.
76. J. Lennard-Jones, *Proc. Roy. Soc. Ser. A* **198**, 14 (1949).
77. R. K. Nesbet, *Rev. Mod. Phys.* **33**, 28 (1961); *Advan. Chem. Phys.* **9**, 321 (1965).
78. K. D. Carlson and D. R. Whitman, *Int. J. Quantum Chem. Symp.* **1**, 81 (1967).
79. E. R. Davidson, *in* "Physical Chemistry: an Advanced Treatise," Vol. III, p. 154. Academic Press, New York, 1969.
80. F. W. Birss and S. Fraga, *J. Chem. Phys.* **38**, 2552 (1963); R. McWeeny and B. T. Sutcliffe, "Methods of Molecular Quantum Mechanics," Sects. 5.1, 5.2. Academic Press, New York and London, 1969; H. Margenau and N. R. Kestner, "Theory of Intermolecular Forces," p. 77. Pergamon, Oxford, 1969; G. Berthier, *in* "Molecular Orbitals in Chemistry, Physics, and Biology (P.-O. Löwdin and B. Pullman, eds.), p. 57. Academic Press, New York, 1964; M. Simonetta and E. Gianinetti, *Ibid.* p. 83.
81. P.-O. Löwdin, *Rev. Mod. Phys.* **35**, 496 (1963).
82. R. K. Nesbet, *Rev. Mod. Phys.* **35**, 498 (1963).
83. G. G. Hall and J. Lennard-Jones, *Proc. Roy. Soc. Ser. A* **202**, 155 (1950); **205**, 357 (1951).
84. D. Peters, *J. Chem. Soc.*, pp. 2901, 2908, 2916 (1964); *J. Chem. Phys.* **45**, 3474 (1966).
85. A. D. McLean and M. Yoshimine, *IBM J. Res. Develop.* **12**, 206 (1968).
86. P.-O. Löwdin, *in* "Quantum Theory of Atoms, Molecules, and the Solid State," p. 601. Academic Press, New York, 1966; H. Nakatsuji, H. Kato, and T. Yonezawa, *J. Chem. Phys.* **51**, 3175 (1969).
87. J. C. Slater, *Phys. Rev.* **82**, 538 (1951); J. A. Pople and R. K. Nesbet, *J. Chem. Phys.* **22**, 571 (1954).
88. K. D. Carlson and R. K. Nesbet, *J. Chem. Phys.* **41**, 1051 (1964).
89. A. Hibbert and C. A. Coulson, *Proc. Phys. Soc. London* **92**, 17 (1967).
90. A. Hibbert and C. A. Coulson, *J. Phys. B* **2**, 458 (1969).

91. J. Čížek, *Theor. Chim. Acta* **6**, 292 (1966).
92. E. G. Larson, *Int. J. Quantum Chem. Symp.* **2**, 83 (1968).
93. J. Paldus and J. Čížek, *J. Chem. Phys.* **52**, 2919 (1970).
94. J. Paldus and J. Čížek, *Chem. Phys. Lett.* **3**, 1 (1969).
95. P. A. G. O'Hare and A. C. Wahl, *J. Chem. Phys.* **53**, 2834 (1970).
96. E. Clementi, *Int. J. Quantum Chem. Symp.* **3**, 179 (1969).

## B. Multideterminant Wave Functions

We may distinguish two reasons for using wave functions that are linear combinations of determinants. First, for certain open-shell configurations, one cannot write a wave function of desired space and spin symmetry as a single determinant. Second, use of many determinants permits a lower (better) energy and a closer approach to the exact wave function. In the former case, all the determinants belong to the same configuration. The proper determinants to use and the way to combine them are fixed by the symmetry itself, once the orbitals to be employed are chosen, and no secular equation need be solved. (A secular equation would automatically produce the correct results, of course.) We will discuss methods of generating linear combinations of determinants for states of desired symmetry species in Subsection 1 of this section. It is conceivable that a single configuration could yield several states of a desired symmetry species, in which case one would have to go to a secular equation, but this rarely occurs, particularly for ground states. The second case just referred to represents a true configuration interaction. The linear combination coefficients for the determinants are determined by the physics of the situation, not by symmetry, so they are found by solving the secular equation, although the order of the secular equation can often be decreased by using symmetry properties. Approaches to the approximate solution of a secular equation have been discussed in Volume I, Chapter III, Section A.

In either case, we have the problem of determining the best orbitals to use in construction of the determinants. While it is possible to use a fixed set of orbitals from any desired source, we obviously obtain better energies by varying the orbitals to optimize the energy of the multideterminant function. This is an extension of the Hartree–Fock problem to sums of determinants. We will call the determination of the best orbitals extended Hartree–Fock (Subsection 2), when the linear combinations of determinants are determined by symmetry, and multiconfiguration SCF (Subsection 3), where there is a true configuration interaction. These terms are sometimes used with different meanings in the literature. Some discussion of multi-configuration SCF was given in Volume I, Chapter III, Section A. Of course, whenever the orbitals are expanded in a basis set, problems related to the choice of basis functions, discussed in Section A, arise.

### 1. *Formation of Spin and Symmetry Functions*

The projection operator technique, introduced in the previous section, produces a linear combination of determinants having a desired symmetry from a single determinant. We will consider the formation of spin eigenfunctions first. It must be noted at the outset that, with a spin-free Hamiltonian, the spin of a wave function affects the energy only because, due to the Pauli principle, it implies the permutation symmetry of the spatial wave function. Several authors [1, 2] have emphasized that one can develop a "spin-free" quantum chemistry. All properties may be calculated, as we will see, as integrals over the coordinate (space) part of the wave function only.

Let the spin projection operator be written as $\mathcal{O}_S$, and the determinant on which it acts as $\mathcal{A}\{\Phi\}$, where

$$\Phi = F(\mathbf{r})G(S) \tag{43}$$

and $F$ is a function of spatial coordinates and $G$ of spin coordinates only. For a single determinant, both $F$ and $G$ are simple products (independent particle model), but generalizations are possible in which $F$ includes inter-electronic coordinates as well (correlated wave function). $\mathcal{O}_S$ will operate on $G$, a product of $\alpha$ and $\beta$ spin functions, to give a sum of such products, each having the same number of $\alpha$'s and $\beta$'s (since $S^2$ and $S_z$ commute) but in different order. Each, multiplied by $F(\mathbf{r})$, gives a determinant.

To calculate the energy of the projected function, we need expectation values of operators $Q$ which are symmetric to interchange of coordinates. Since both $Q$ and $\mathcal{O}_S$ commute with $\mathcal{A}$,

$$\langle \mathcal{O}_S\mathcal{A}\{\Phi\} \mid Q \mid \mathcal{O}_S\mathcal{A}\{\Phi\}\rangle = \langle \mathcal{O}_S\Phi \mid Q \mid \mathcal{O}_S\mathcal{A}\{\Phi\}\rangle \tag{44}$$

Several properties of $\mathcal{O}_S$ follow from its definition and can be seen from the explicit form, Eq. (36). Like any projection operator, $\mathcal{O}_S$ is idempotent; $\mathcal{O}_S^2 = \mathcal{O}_S$. Let $\Psi$, a function of space and spin, be expanded in the $N$-electron spin functions $g_{SMn}$, as in Eq. (35):

$$\Psi = \sum_{S,M,n} f_{SMn}g_{SMn} \tag{45}$$

Then

$$\mathscr{O}_S{}^2\!\left(\sum_S \sum_n f_{Sn} g_{SMn}\right) = \mathscr{O}_S\!\left(\sum_n f_{Sn} g_{SMn}\right) = \sum_n f_{Sn} g_{SMn}$$

Also, $\mathscr{O}_S$ is Hermitian. Let $\Psi^{(1)}$ and $\Psi^{(2)}$, eigenfunctions of $S_z$ with eigenvalues $M'$ and $M''$, respectively, be expanded as in Eq. (45). Then

$$\langle \Psi^{(1)} \mid \mathscr{O}_S \mid \Psi^{(2)}\rangle = \sum_{S'} \sum_{n'} \sum_{S''} \sum_{n''} \int f_{S'n'}^{(1)*} f_{S''n''}^{(2)}$$

$$\times d\tau \langle g_{S'M'N'} \mid \mathscr{O}_S \mid g_{S''M''n''}\rangle \qquad (46)$$

If $M' \neq M''$, $\langle \Psi^{(1)} \mid \mathscr{O}_S \mid \Psi^{(2)}\rangle$ vanishes, as does $\langle \Psi^{(2)} \mid \mathscr{O}_S \mid \Psi^{(1)}\rangle$. If $M' = M''$, (46) becomes, using the fact that the different spin functions for a particular $S$ and $M'$ are orthonormal,

$$\langle \Psi^{(1)} \mid \mathscr{O}_S \mid \Psi^{(2)}\rangle = \sum_{n'} \int f_{Sn'}^{(1)*} f_{Sn'}^{(2)} \, d\tau$$

By a similar calculation,

$$\langle \Psi^{(2)} \mid \mathscr{O}_S \mid \Psi^{(1)}\rangle = \sum_{n'} \int f_{Sn'}^{(2)*} f_{Sn'}^{(1)} \, d\tau = \langle \Psi^{(1)} \mid \mathscr{O}_S \mid \Psi^{(2)}\rangle^*$$

which shows $\mathscr{O}_S$ is Hermitian. Using the Hermiticity and idempotence in Eq. (44) and assuming $Q$ to be spin-independent (so that it commutes with $\mathscr{O}_S$) we have

$$\langle \mathscr{O}_S \mathscr{A}\{\varPhi\} \mid Q \mid \mathscr{O}_S \mathscr{A}\{\varPhi\}\rangle = \left\langle \varPhi \mid Q \mid \mathscr{O}_S \sum_P (-1)^{\delta_P} P\varPhi \right\rangle. \qquad (47)$$

The permutation $P$ permutes space and spin coordinates, i.e., it affects both factors in $\varPhi$, Eq. (43). Then (47) is written,

$$\sum_P (-1)^{\delta_P} \langle F \mid QP \mid F\rangle \langle G \mid P\mathscr{O}_S \mid G\rangle. \qquad (48)$$

Now we consider the evaluation of the spin factors $\langle G \mid P\mathscr{O}_S \mid G\rangle$.

A simplification is possible. The determinant, originally written as $\mathscr{A}\{\varPhi\}$, could just as well be written as $\mathscr{A}\{P\varPhi\}$, where $P$ is any permutation operator. Suppose we choose the permutation $P_G$ that produces, from $G$, a product of $n_\alpha$ $\alpha$ spins followed by $n_\beta$ $\beta$ spins. We denote this product by $G_0$. Then instead of (48) we have

$$\langle \mathscr{O}_S \mathscr{A}\{\varPhi\} \mid Q \mid \mathscr{O}_S \mathscr{A}\{\varPhi\}\rangle = \sum_P (-1)^{\delta_P} \langle P_G F \mid QP \mid P_G F\rangle$$

$$\times \langle G_0 \mid P\mathscr{O}_S \mid G_0\rangle \qquad (49)$$

and only the effects of permutations on $\mathscr{O}_S G_0$ need be considered. A permutation $P'$ which permutes only the first $n_\alpha$ and the last $n_\beta$ spins among themselves does not affect $G_0$. Since $\mathscr{O}_S P' = P' \mathscr{O}_S$, $P'$ must also leave $\mathscr{O}_S G_0$ unaffected. Therefore all products in $\mathscr{O}_S G_0$ which differ by permutations among the first $n_\alpha$ and the last $n_\beta$ factors appear with the same coefficient. We write $\mathscr{O}_S G_0$ as

$$\mathscr{O}_S\{\alpha(1) \cdots \alpha(n)\beta(n_\alpha + 1) \cdots \beta(n_\alpha + n_\beta)\}$$
$$= \sum_j C_j(S, M, N)[\alpha^{n_\alpha - j}\beta^j][\alpha^j \beta^{n_\beta - j}] \qquad (50)$$

where $[\alpha^p \beta^q]$ is the symmetric sum of all permutations of $p$ $\alpha$ functions and $q$ $\beta$ functions. Here, $N$ is the number of particles. Then for the second factor in (49) we have

$$\langle G_0 \mid P \mid \mathscr{O}_S G_0 \rangle = \sum_j C_j(S, M, N)$$
$$\times \langle \alpha(1) \cdots \alpha(n_\alpha)\beta(n_\alpha + 1) \cdots \beta(n_\alpha + n_\beta) \mid P \mid [\alpha^{n_\alpha - j}\beta^j][\alpha^j \beta^{n_\beta - j}] \rangle$$

There will be no more than one nonvanishing term on the right because of the scalar product. Thus all one really needs are the coefficients $C_j(S, M, N)$. These have been extensively studied, and their properties are derived and discussed by Smith and Harris [3]. Many of the properties were previously treated by Löwdin [4]. Manne [5] has calculated and tabulated the coefficients for systems of up to 20 electrons.

For example, a recursion relation between the $C_j(S, M, N)$ is obtained [3] by operating on (50) with $S^2$, which must simply multiply it by $S(S+1)$. Here, it is convenient to have a representation of $S^2$ in terms of spin permutation operators. In Section A, we wrote [Eq. (22) and subsequent discussion]

$$S^2 = S_z{}^2 + \mathscr{S} - S_z$$

and noted that the operator $\mathscr{S}$ produced, from a product of spin functions, $n_\alpha(n_\beta + 1)$ products. Of these, $n_\alpha$ were identical to the original product, and the rest were obtained from it by exchanges of opposite spins. Then $S^2$, operating on a product of spin functions, may be replaced by

$$[\tfrac{1}{2}(n_\alpha - n_\beta)]^2 + n_\alpha + \sum' \mathscr{S}_{ij}^{(s)} - \tfrac{1}{2}(n_\alpha - n_\beta)$$

where the sum is over interchanges of opposite spins only. Since any function of spin may be written as a sum of products of spin functions, we can use this to get a general formula for $S^2$. If we replace the sum by a sum over *all* spin interchanges, we will be including $\frac{1}{2}n_\alpha(n_\alpha - 1) + \frac{1}{2}n_\beta(n_\beta - 1)$ additional terms, corresponding to interchanges of identical spins, which change nothing, so that

$$
\begin{aligned}
S^2 &= [\tfrac{1}{2}(n_\alpha - n_\beta)]^2 + n_\alpha - \tfrac{1}{2}(n_\alpha - n_\beta) \\
&\quad - \tfrac{1}{2}n_\alpha(n_\alpha - 1) - \tfrac{1}{2}n_\beta(n_\beta - 1) + \sum_{i,k}^{(i \neq k)} \mathscr{I}_{ik}^{(s)} \\
&= n_\alpha + n_\beta - \tfrac{1}{4}(n_\alpha + n_\beta)^2 + \sum_{i,k}^{(i \neq k)} \mathscr{I}_{ik}^{(s)} \\
&= \tfrac{1}{4}N(4 - N) + \sum_{i,k}^{(i < k)} \mathscr{I}_{ik}^{(s)}
\end{aligned}
\tag{51}
$$

When we operate $S^2$, Eq. (51), on the right member of (50), the effect of the spin interchanges on $[\alpha^{n_\alpha - j}\beta^j][\alpha^j\beta^{n_\beta - j}]$ is easy to compute: The result is a linear combination of $[\alpha^{n_\alpha - j}\beta^j][\alpha^j\beta^{n_\beta - j}]$, $[\alpha^{n_\alpha - j + 1}\beta^{j-1}][\alpha^{j-1}\beta^{n_\beta - j + 1}]$, and $[\alpha^{n_\alpha - j - 1}\beta^{j+1}][\alpha^{j+1}\beta^{n_\beta - j - 1}]$. Since $S^2$ operating on the right member of Eq. (50) must give

$$
S(S + 1) \sum_j C_j(S, M, N)[\alpha^{n_\alpha - j}\beta^j][\alpha^j\beta^{n_\beta - j}],
$$

we obtain a recursion relation between the $C_j(S, M, N)$ [4]. For example, if $S_z = 0$, i.e., $n_\alpha = n_\beta = \frac{1}{2}N$, we have [4] from the coefficient of $[\alpha^{n_\alpha - j}\beta^j]$ $\times [\alpha^j\beta^{n_\beta - j}]$:

$$
\begin{aligned}
S(S+1)C_j(S, M, N) &= [\tfrac{1}{2}n_\alpha(4 - 2n_\alpha) + n_\alpha(n_\alpha - 1) + j(2n_\alpha - 2j)]C_j(S, M, N) \\
&\quad + [(n_\alpha - j)^2]C_{j+1}(S, M, N) + [j^2]C_{j-1}(S, M, N)
\end{aligned}
$$

a three-term recursion relation. Successive application of this relation can give the ratios of all $C_j(S, M, N)$ to $C_0(S, M, N)$. For $j = 0$, we find

$$
C_1(S, M, N)/C_0(S, M, N) = n_\alpha^{-1}[S(S + 1) - n_\alpha] \tag{52}
$$

and so on.

As an example of the use of these formulas, the expectation value of the Hamiltonian over a projected single determinant formed from spin orbitals with orthonormal spatial parts is

$$
\begin{aligned}
\langle H \rangle &= \frac{\langle \mathscr{O}_S \mathscr{A}\{\Phi\} \mid H \mid \mathscr{O}_S \mathscr{A}\{\Phi\} \rangle}{\langle \mathscr{O}_S \mathscr{A}\{\Phi\} \mid \mathscr{O}_S \mathscr{A}\{\Phi\} \rangle} \\
&= \frac{\sum_p (-1)^{\delta_p}\langle F \mid HP \mid F \rangle\langle G_0 \mid P \mid \mathscr{O}_S G_0 \rangle}{\sum_p (-1)^{\delta_p}\langle F \mid P \mid F \rangle\langle G_0 \mid P \mid \mathscr{O}_S G_0 \rangle}.
\end{aligned}
\tag{53}
$$

Here $F$ is a product of orbitals with $PF$ orthogonal to $F$. Then only the identity permutation contributes to the denominator, and only the identity and single interchanges to the numerator.

$$\langle H\rangle = \frac{\langle F\,|\,H\,|\,F\rangle\langle G_0\,|\,\mathscr{O}_S\,|\,G_0\rangle - \sum_{i<j}\langle F\,|\,H\,|\,\mathscr{T}_{ij}F\rangle\langle G_0\,|\,\mathscr{T}_{ij}\,|\,\mathscr{O}_S G_0\rangle}{\langle G_0\,|\,\mathscr{O}_S\,|\,G_0\rangle}$$

$$= \langle F\,|\,H\,|\,F\rangle - \sum_{i<j}\langle F\,|\,H\,|\,\mathscr{T}_{ij}F\rangle C_1(S, 0, N)/C_0(S, 0, N).$$

The first term is a sum of one-electron and two-electron integrals over the orbitals, including exchange integrals over pairs of orbitals with parallel spin. In the summation, only the two-electron part of $H$ enters:

$$\sum_{i<j}\langle F\,|\,H\,|\,\mathscr{T}_{ij}F\rangle = \sum_{i<j}\langle \phi_i(1)\phi_j(2)\,|\,g_{ij}\,|\,\phi_j(1)\phi_i(2)\rangle,$$

which is a positive quantity. Using Eq. (52) for the ratio

$$C_1(S, 0, N)/C_0(S, 0, N)$$

shows that the state of highest spin multiplicity will have the lowest energy, an extension of Hund's rule [4].

Secrest and Holm [6] considered several problems relating to spin projection of a Slater determinant using the operator $\mathscr{O}_S$. They derived formulas for the coefficients of the Slater determinants appearing in the projected function, like those discussed in Eqs. (50) *et seq.* In addition, they pointed out that, starting from a set of linearly independent Slater determinants, the projected functions need not be linearly independent. Secrest and Holm gave rules [6] for choosing the original set to ensure linear independence of the projection.

The coefficients $C_j(S, M, N)$ express the properties of the permutation group (symmetric group) and its representations. One can produce the projection operator explicitly in terms of permutations by considering the properties of the representations of this group. The orthonormal spin functions $g_{S'Mn}$, where $g_{S'Mn}$ is an eigenfunction of $S^2$ with eigenvalue $S'$ and of $S_z$ with eigenvalue $M$, as earlier, form bases for the representations. From these functions, the matrices of the representations can be obtained, and these matrices suffice to give the spin projection operators. This approach is explained in detail in the book of Kotani *et al.* [7]. There are several methods for generating the spin eigenfunctions $g_{S'Mn}$. In addition to the reference just cited, discussion and comparisons of these methods are found in the books of McWeeny and Sutcliffe [8] and Pauncz [9].

Let $d_{S'}$ be the number of spin functions $g_{SMn}$ $(n = 1, \ldots, d_{S'})$ for $S = S'$, and denote the number of electrons by $N$. Since $S^2$ and $S_z$ commute with all permutations $P$, we can easily see that $Pg_{S'Mn}$ is, like $g_{S'Mn}$, an eigenfunction of $S^2$ with eigenvalue $S'(S' + 1)$ and of $S_z$ with eigenvalue $M$. Therefore, it may be expanded in the spin functions corresponding to the same values of $S$ and $M$.

$$Pg_{S'Mn} = \sum_{k=1}^{d_{S'}} U_{kn}^{(S')}(P)\, g_{S'Mk} \tag{54}$$

The coefficients in the expansion clearly depend on $P$ and on $S'$, and we have written them as elements of a matrix. By performing several permutations one after another, one sees that the matrices $\mathbf{U}^{(S')}(P)$ form a representation of the permutation group. This representation is unitary and irreducible [7]. From well-known formulas of group theory, it can be shown that the projection operator $\mathcal{O}_{S'}$ can be written in terms of these representation matrices:

$$\mathcal{O}_{S'} = (d_{S'}/N!) \sum_{P} \sum_{k=1}^{d_{S'}} [U_{kk}^{(S')}(P)]^* \, P \tag{55}$$

Only the traces of the $U^{(S)}(P)$ are needed here. The matrices are given explicitly, with much further discussion, in the book of Kotani et al. [7]. The projection operators

$$(d_S/N!)^{1/2} \sum_{P} U_{nk}^{(S)}(P)\, P, \qquad n = 1, \ldots, d_S, \quad (k \text{ fixed}) \tag{56}$$

can be used to generate the individual partners in a representation characterized by $S$.

Now we can use the matrices $\mathbf{U}^{(S)}(P)$ to construct an antisymmetric $N$-electron function of space and spin which is an eigenfunction of $S^2$ and $S_z$. We imagine this function, denoted by $\psi$, to be expanded as in Eq. (45). Only spin eigenfunctions corresponding to one value of $S$ and one value of $M$ appear in the expansion, and the coefficients $f_{SM}$ will be independent of $M$.

$$\psi = \sum_{n=1}^{d_S} f_{Sn} g_{SMn} \tag{57}$$

To show the properties the $f_{Sn}$ must have so that $\psi$ be antisymmetric, operate on $\psi$ with a permutation $P$ (of space and spin coordinates simultaneously):

$$P\psi = \sum_{n=1}^{d_S} (Pf_{Sn}) \sum_{k=1}^{d_S} U_{kn}^{(S)}(P) g_{SMk}$$

using (54). We can show that this will equal $(-1)^{\delta_P}$ times $\psi$ if the $f_{Sn}$ transform under permutation operators as follows:

$$Pf_{Sn} = \sum_{k=1}^{d_S} V_{kn}^{(S)}(P)f_{Sk} \tag{58}$$

where

$$V_{kn}^{(S)}(P) = (-1)^{\delta_P} U_{nk}^{(S)}(P^{-1})^* \tag{59}$$

i.e., $\mathbf{V}(P) = (-1)^{\delta_P}\mathbf{U}^+(P)$. It is not difficult to show that the $\mathbf{V}^{(S)}(P)$ form a representation of the group of permutations [7]. From any function of spatial coordinates $F$ one can project a function transforming according to Eq. (58), i.e., a basis for the representation of the $\mathbf{V}(P)$, by standard group theoretical methods. The projection operator here is a sum of permutation operators which operate on the spatial variables. For example, starting with a "primitive function" $F$,

$$f_{Sn} = (d_S/N!)^{1/2}\sum_{P} V_{nk}^{(S)}(P)PF; \qquad n = 1, \ldots, d_S \tag{60}$$

where $k$ is arbitrary. If the primitive function $F$ already has some symmetry, the functions $PF$ may not be linearly independent, and care must be taken [7].

To calculate overlaps and (spin independent) Hamiltonian matrix elements for a function like (57), we require the matrix $\mathbf{V}$. Because of the orthonormality of the spin functions $g_{SMn}$,

$$\langle \psi \mid H \mid \psi \rangle = \sum_{n=1}^{d_S} \int f_{Sn}^* Hf_{Sn}\, d\tau$$

$$= \sum_{n=1}^{d_S} (d_S/N!)\sum_{P,Q} V_{nk}^{(S)}(P)^* V_{nk}^{(S)}(Q)\int (P^*F^*)HQF\, d\tau. \tag{61}$$

The only integrals needed are over the primitive functions. All reference to spin is gone, except as a label for representations of the group of permutations.

Starting from any product function of space and spin coordinates, $FG$, where $G$ is already an eigenfunction of $S_z$, one can project out a function $f_{Sn}g_{SMn}$ using a product of two projection operators. One is for space, say Eq. (60), and one for spin, say Eq. (56). To obtain the antisymmetric function (57) one can use the formula

$$(d_S/N!)\sum_{n}\sum_{P}\sum_{Q} V_{nk}^{(s)}(P)U_{nk}^{(s)}(Q)P_r Q_s(FG). \tag{62}$$

Here, $P_r$ permutes spatial coordinates only and $Q_s$ spin coordinates only. If we are interested in projecting single determinants, we take $F$ as a simple product of spatial orbitals. But the formalism is more general, since any function can be written as a sum of products $FG$. Thus, starting with any function of space and spin coordinates (assumed to be eigenfunction of $S_z$), we can use (62) or an equivalent operator to produce a function which is (*a*) antisymmetric to simultaneous permutations of space and spin coordinates and (*b*) an eigenfunction of $S^2$ as well as $S_z$.

There are other procedures for doing this in addition to those we have discussed. Goddard [10] compared several of these and gave additional valuable discussion. He showed, for instance, that the exact wave function can be written as the result of operation of such a space–spin projection operator on a spatial function multiplied by a product of $\alpha$ and $\beta$ spin functions. He developed a formalism for projecting out functions of desired symmetry which we shall discuss further later. Gallup [11] has used the Young operator formalism for this purpose and compared different methods.

For many applications in molecular calculations, the spin projection or equivalent formalism is unnecessary. If there is only one open shell and if a single configuration for space and spin is expected to be valid, the function

$$\Phi_1^{Sn}(1, \ldots, k) = \left[ \prod_{j=1}^{k} \phi_j(j) \right] g_{SMn}^{(k)}(1, \ldots, j) \tag{63}$$

(the superscript gives the number of electrons) describes the electrons of the open shell. If $\Phi_1^{Sn}$ is multiplied by

$$\Phi_2 = \prod_{j=k+1}^{l} [\phi_j(2j-1)\phi_j(2j)\alpha(2j-1)\beta(2j)], \tag{64}$$

a paired-electron description of the closed-shell electrons, and the result is antisymmetrized, one will have a suitable wave function for the system. Since the $g_{SMn}^{(k)}$ are sums of products of $\alpha$'s and $\beta$'s, the wave function will be a sum of determinants.

Functions having the correct spatial symmetry can be generated from a determinant by the projection operator technique. Instead of the permutation operators in (60), the operators of the molecular symmetry group would appear in the projection operator. For diatomic systems, it is usual to assume the orbitals from which we construct determinantal wave functions are of the form

$$f(r_A, r_B)e^{im\varphi}$$

with $f(r_B, r_A) = \pm f(r_A, r_B)$ for a homonuclear molecule [see the discussion after Eq. (25)]. Then each spin orbital is an eigenfunction of all the symmetry operators except, for $m \neq 0$, the vertical reflection (in a plane containing the symmetry axis). Because of this, it is usually possible to generate the symmetry-adapted functions by inspection, without the full projection operator formalism. The number of different cases that can arise is limited, especially for ground states, and the wave functions for many of them have been tabulated [12].

## 2. Extended Hartree–Fock Procedures

In Section A we mentioned the possibility of deriving a wave function for an open-shell state for which a single determinant cannot have the proper spin and symmetry behavior, by optimizing the orbitals of a single determinant without regard to the spin and symmetry problem, and then projecting out from the determinant a function with the desired properties. The projected Hartree–Fock procedure may be relatively easy to use, but it is evident that it does not really optimize the orbitals for the desired state. Rather, it gives the best orbitals for some average of all the states that can be projected from the unsymmetrized determinant. One really should perform the optimization on the projected function, i.e., find the best spin orbitals to use in the multideterminant function which is an eigenfunction of $S^2$ and symmetry operators. To carry out such an extended Hartree–Fock procedure, one requires a convenient expression for the energy of this function. Each of the methods for generating such a function provides such an expression [see Eqs. (53) and (61), for instance). Ruedenberg [13] has recently reviewed various approaches and given some new formulations. In this subsection, we consider some methods for choosing the best orbitals in a many-determinant open-shell wave function. References to methods other than those we mention are found in the articles cited. Formally, we have to minimize

$$\langle \Phi \mid \mathcal{O}H\mathcal{O} \mid \Phi \rangle / \langle \Phi \mid \mathcal{O} \mid \Phi \rangle$$

where $\Phi$ is the single determinant and $\mathcal{O}$ the projector, which is much more complex than $\langle H \rangle_\Phi$ itself. But it is in fact possible, in many cases, to derive equations for the orbitals.

Roothaan [12, 14] considered some of the important open-shell cases that arise for molecules. He displayed wave functions, as sums of determinants, for all the states arising from a number of configurations likely to arise, and gave explicit expressions for the energies of each. The energy actually was most easily calculated as the average energy for all the degenerate total wave functions of a given state. Because the average energy expression was minimized, the orbitals determined here were not the best for representation of any one of the degenerate wave functions corresponding to the state of interest, but the best for representation of all of them. It was shown [12] that the orbitals could be taken as symmetry orbitals in this treatment, but not if one considered just one state of the degenerate set. In each case, the energy consisted of a contribution of the orbitals of the closed shell, which are the same in all the determinants, plus a sum of one-electron and two-electron integrals involving the orbitals of the open shell. Roothaan defined $f$ as the fractional occupation of the orbitals of the open shell: $f$ equals the number of fully occupied open shell orbitals divided by the number of orbitals available in the open shell set. Then the one-electron operators contributed $2f \sum_m H_m$ to the energy, where $H_m$ represents the expectation value of the kinetic energy and nuclear attraction energy over such an orbital. The contribution to the interelectronic repulsion energy consisted of the interaction of the open-shell orbitals with the closed shell, $2f \sum_{k,m} (2J_{km} - K_{km})$, and the interactions of the open-shell orbitals among themselves, $f^2 \sum_{mn} (2aJ_{mn} - bK_{mn})$. Here, $k$ refers to orbitals of the closed shell, $m$ and $n$ refer to orbitals of the open shell, and $J_{ij}$ ($K_{ij}$) are the usual Coulomb (exchange) integrals between orbitals $i$ and $j$. The values of the constants $a$ and $b$ depend on the specific case.

Roothaan then minimized the energy with respect to variations in the orbitals, keeping them orthonormal by use of Lagrangian multipliers. The manipulations followed those for the single-determinant closed-shell case, up to the point at which the matrix of Lagrangian multipliers was to be diagonalized, to give a pseudoeigenvalue equation for each orbital. In the open shell case, one may not carry out a unitary transformation on the orbitals to accomplish this. The total wave function is invariant to a unitary transformation of the closed-shell orbitals among themselves, since this leaves all determinants unchanged; it is also invariant to a transformation of the open-shell orbitals among themselves, which transforms one degenerate wave function into another. Many off-diagonal Lagrangian multipliers can be removed by such transformations. However, one cannot carry out a transformation to eliminate the Lagrangian multipliers which couple open-shell and closed-shell orbitals. An exception is the case where the closed- and open-shell orbitals have no common symmetry, so are automatically orthogonal. One cannot reduce the open-shell problem to pseudoeigenvalue equations for individual orbitals by simply carrying out a linear transformation.

Roothaan [12] did succeed, in another way, in deriving pseudoeigenvalue equations for both closed- and open-shell orbitals. It was even possible to arrange things so that both closed- and open-shell orbitals were eigenfunctions of the same operator (which, of course, involved the orbitals themselves), although the eigenvalues could not necessarily be interpreted as orbital energies. His method is applicable to half-filled shells (all spins parallel in the open shell) and to all the states arising from $\pi^n$, $\delta^n$, ... of a linear molecule. The values of $f$, $a$, and $b$ for a number of such cases were tabulated [12].

A treatment like Roothaan's, for the wave function described by Eqs. (21)–(22), was given by Davidson [15]. The wave function

$$\psi_{Sn} = (N!)^{-1/2} \sum_{P} \delta_P P \{\Phi_1^{Sn} \quad \Phi_2\} \tag{65}$$

is easily seen to be normalized if the orbitals used for the open shell are orthogonal to those of the closed shell (which is no restriction on the orbitals) and to each other (which is). Here, $N = 2l - k$, where $l$ is the total number of orbitals and $k$ is the number of orbitals in the open shell. The average energy for the spin states from the single configuration was shown by Davidson [15] to be $\mathcal{N}^{-1} \sum_{M,P} H_{PP}^M$, where $\mathcal{N}$ is the number of spin states and $H_{PP}^M$ is the expectation value of $H$ over the Slater determinant

$$\Psi_P^M = \mathscr{A} \left\{ \prod_{j=1}^{k} [\phi_j(j)] \prod_{j=k+1}^{l} [\phi_j(2j - 1)\phi_j(2j)] \right.$$
$$\left. \times P[\alpha(1) \cdots \beta(k)]\alpha(k + 1)\beta(k + 2)\alpha(k + 3)\beta(k + 4) \cdots \right\}.$$

$P$ is one of the $k!$ permutations of $k$ electrons and acts on any product of $k$ $\alpha$'s and $\beta$'s.

We may use Eq. (62) of Volume I, Chapter III, to evaluate the terms $H_{PP}^M$, rewriting the two-electron part, for convenience, as

$$S = \tfrac{1}{2} \sum_{p,q} (\langle \lambda_p \lambda_q \mid g \mid \lambda_p \lambda_q \rangle - \langle \lambda_p \lambda_q \mid g \mid \lambda_q \lambda_p \rangle).$$

We have, for each $H_{PP}^M$, the same contribution from the one-electron Hamiltonian, from the Coulomb integrals, and from the exchange integrals between orbitals of the closed shell and between closed- and open-shell orbitals. Only the contribution of the open-shell exchange integrals is not the same for all determinants. By inspection, we can derive the coefficient of each $K_{ij}$ in the sum of $H_{PP}^M$. The result is that the average energy for the configuration is

$$\sum_{j=1}^{k} H_j + 2 \sum_{j=k+1}^{l} H_j + \sum_{i=k+1}^{l} \sum_{j=k+1}^{l} (2J_{ij} - K_{ij})$$

$$+ \sum_{i=k+1}^{l} \sum_{j=1}^{k} (2J_{ij} - K_{ij}) + \sum_{i=1}^{k} \sum_{j=1}^{k} (\tfrac{1}{2}J_{ij} - \tfrac{1}{4}K_{ij}[1 + \delta_{ij}]).$$

This may be minimized with respect to the orbitals.

Starting from the energy expression (53), with the function $F$ written as a product of orbitals, Löwdin [4, 16] derived one-particle eigenvalue equations for the best spatial orbitals. Then there is a single-particle model, like the Hartree–Fock, with the possibility of simple physical explanations. We refer to his papers for discussion of these extended Hartree–Fock equations.

Goddard [10] derived his own version of the extended Hartree–Fock equations, by defining operators $G_i{}^\mu$ such that $G_i{}^\mu \Phi$ is antisymmetric and an eigenfunction of $S^2$ for $\Phi$ an arbitrary function of space and spin, and minimizing the expectation value of the Hamiltonian over

$$\psi = G_i{}^\mu \left\{ \prod_k \phi_k(k) s_k(k) \right\}$$

(where $s_k$ is $\alpha$ or $\beta$) while maintaining all the $\phi_k$ normalized. The variational problem gave equations which could be written as:

$$H_k^{\mu i} \phi_k = \varepsilon_{kk} \phi_k \tag{66}$$

where $\varepsilon_{kk}$ is related to the Lagrange multiplier associated with normalization of $\phi_k$. The independent-particle picture is preserved. The operator $H_k^{\mu i}$, which may be different for different $k$, contains the usual one-electron operators in the Hamiltonian plus terms due to electrons in the orbitals other than $\phi_k$. The operator $H_k^{\mu i}$ is nonlocal like the Hartree–Fock one-electron Hamiltonian, but generally considerably more complicated. The fact that not all the $\phi_k$ are eigenfunctions of the same Hamiltonian operator means that a symmetry operation does not always transform the occupied orbitals among themselves. One cannot always choose them to transform according to irreducible representations of the symmetry group. In particular, for a homonuclear diatomic molecule at large $R$ one can have singly occupied orbitals localized on each atom and still have a many-electron function symmetric to interchange, as in the unrestricted Hartree–Fock function. Similarly, for a singlet state one does not have to have doubly occupied orbitals as in restricted Hartree–Fock methods, where double occupation leads to incorrect dissociation (see Section C.2). The restricted Hartree–Fock method appears as a special case, when equality of certain $\phi_k$ is demanded. Goddard [10] also showed the equivalence between his method and the extended Hartree–Fock methods in which the trial function is a spin-projected determinant, already mentioned, and gave comparisons with other methods. Calculations were carried out, by expanding each orbital in a suitable basis set, for $H_2$ and LiH [10].

Gallup [11] has discussed a projected Hartree–Fock method and compared it to configuration interaction calculations. McWeeny and Sutcliffe [8, Sect. 5.4–5.6] have derived extended Hartree–Fock equations for open shells within the density matrix formalism, using basis set expansions, and proposed a method of solution. They give a "master equation" and show how other formalisms appear as special cases of it.

The relations between various self-consistent field schemes (Hartree, Hartree–Fock, extended Hartree–Fock, etc.), which all seek to conserve the independent-particle model, were discussed in a very interesting article by Löwdin [17]. The extension of the model developed by Brueckner and others [18–20] for the problem of interacting nuclei was also included, and an "exact self-consistent field theory," which extends to the exact wave function the idea of individual particles moving in the average field of the others, was developed. Löwdin [17] analyzed and compared the self-consistent field schemes by writing the Hamiltonian $H$ as a zero-order part,

$$H_0 = \sum_i (h(i) + u(i))$$

and a perturbation,

$$V = \sum_{i<j} g(i, j) - \sum_i u(i).$$

Here, $h_i$ is the one-electron part of the Hamiltonian and $u_i$ is a potential chosen according to the particular theory. The form of $u_i$ was derived [17] for the Hartree, Hartree–Fock, and Brueckner schemes. As shown in the partitioning-projection operator formalism (Volume I, Chapter III, Eqs. (174) $et\ seq.$; Volume II, Chapter I, Eqs. (151) etc.], the exact energy may be written as

$$E = \langle \varphi_0 \mid H_0 + t \mid \varphi_0 \rangle$$

where $H_0 \varphi_0 = E_0 \varphi_0$. The exact SCF theory may be derived [19] by writing $E$ as

$$E = \left\langle \varphi_0 \middle| \sum_i h_i + \tau \middle| \varphi_0 \right\rangle,$$

regarding $\tau$ fixed (it depends on the exact wave function), and varying $\varphi_0$ to minimize the expression (which then yields the exact energy). Other derivations are possible [17].

Kobe [21] has recently reviewed Brueckner's and related schemes, and further developed Löwdin's exact SCF theory. He showed Löwdin's theory to be equivalent to imposing on the spin orbitals the criterion that the mean-square deviation of $\varphi_0$ from the exact wave function be minimized. This means that the overlap between $\varphi_0$ and the exact wave function will be maximized (see Volume I, Chapter III, Sections A.2, A.5, E.3). The exact SCF theory has not been employed in molecular calculations.

### 3. *Configuration Interaction*

Now we consider wave functions which are linear combinations of determinants belonging to several configurations. The number of determinants that can be formed by choosing $N$ orbitals ($N =$ number of electrons) from a set of $n$, even if there are spin and symmetry conditions to fulfill, quickly becomes very large as $n$ increases relative to $N$. In choosing those that will be most useful in lowering the energy, it is usual to draw on previous experience, and to invoke theories of various degrees of complexity (Volume I, Chapter III, Section A; Volume II, Chapter I, Section B) which indicate the physical effects that must be described by the excited configurations. To some extent, the number of configurations may be reduced if the molecular orbitals from which they are constructed are optimized. Clearly, the best orbitals in this respect are the natural orbitals, and this is exploited in the iterative NO–CI approach (Volume I, Chapter III, Section A.7) [22]. Other discussion of the choice of orbitals is found in Volume I, Chapter III, Section A. It is generally not possible to optimize all the parameters in the molecular orbitals, since they enter the energy in a nonlinear fashion. Most commonly, the orbitals are fixed and a large number of configurations used. Methods for dealing with large secular equations have been developed (see Volume I, Chapter III, Section A.7). Where one has an idea of what physical effects are important and how to describe them with a few configurations, one may hope to optimize all the orbitals which appear (multiconfiguration SCF). The molecular orbitals are generally expanded in a basis set. For the choice of basis functions, the considerations of Section A.2 are important. It is not common to reoptimize the basis in an extensive configuration interaction (CI) calculation, and a basis set generally is obtained from other calculations on the same or related systems.

We will first mention some possibilities which have been proposed to simplify CI calculations. Then we will consider the problem of describing interelectronic correlation as it relates to CI. Theories that are used to choose configurations are discussed next, illustrated by references to LiH calculations. We then discuss multiconfiguration SCF formalisms, and close with projections for CI work in the future.

Advanced techniques of group theory may be combined with physical reasoning to give simplifications in the calculation of a wave function by configuration interaction [23, 24]. The evaluation of the electronic repulsion integrals may also be simplified. The order of the secular equation is already much reduced when symmetry species, which do not interact, are formed from the determinants. Additional simplifications are obtained [24] when the orbitals are chosen to form bases for representations of subgroups o- the molecular symmetry group, and the configurations are classified act cording to these representations. Determinants belonging to differenf representations do not interact.

As in single-determinant calculations, we can use one-center expansions in configuration interaction calculations. This simplifies the evaluation of integrals, but makes it necessary to use a larger number of basis functions than would be necessary for a multicenter basis. Formation, from the one-center functions, of configurations and states which transform according to the molecular symmetry group becomes a problem in vector-coupling coefficients.

The convergence of a single-center configuration interaction calculation for $H_2$ was carefully considered by Hagstrom and Shull [25]. They gave an analysis which essentially isolated the terms giving the different kinds of correlation (see page 200). Using configurations constructed from s orbitals only, they obtained the "spherical limit" energy. They also found the best energy which could be obtained using functions of cylindrical symmetry. This permits left–right or in–out, but not angular correlation. The problem of poor representation of the cusps at the nuclei arises exactly as it does for the Hartree–Fock calculation, and leads to slowed convergences with increase in the size of the basis sets.

Increased flexibility in one-center calculations is gained if Slater orbitals with nonintegral principal quantum numbers are used. Hebert and Ludwig [26] calculated energies for He–He at $R = 1.0$, 0.75, and 0.5 $a_0$ using three configurations built from such orbitals. At 0.5 $a_0$, their energies were only 3.6 eV above those from two-center calculations.

For two-electron diatomic molecules at least, a powerful one-center approach is available [27]. It involves expansion of the exact wave function in eigenfunctions of orbital angular momentum and $z$-component of orbital angular momentum. The expansion coefficients, which are functions of $r_1$, $r_2$, and $r_{12}$, are determined from coupled differential equations, derived by substitution of the expansion into the Schrödinger equation.

Since the superiority of CI over single-configuration (restricted Hartree–Fock) calculations is in the treatment of interelectronic correlations, theories of the correlation energy are useful guides in setting up configurations to be used in CI. Many of the pair theories mentioned in Volume I, Chapter III, Section A actually produce wave functions of CI form, since they employ expansions of the pair functions in spin orbitals. The equations determining these spin orbitals may then be used as a guide in setting up configurations for CI.

Nesbet [28] reformulated the configuration interaction problem to give equations like those of the Brueckner theory for nuclear matter. The configurations were the Hartree–Fock plus single and double excitations from it. Then the CI equations could be written in terms of an effective two-particle Hamiltonian which represents the complete interaction. A recent review article by Nesbet [29] gives further references to Brueckner-like and other theories of correlation energy, whose ideas are useful in configuration interaction calculations. As discussed in Volume I, Chapter III, Section A, a configuration interaction function which includes a configuration resembling the Hartree–Fock plus others designed to give the effect of pair correlations should give good results. The pairing of electrons in a restricted Hartree–Fock calculation tends to favor the approach of electrons of opposite spins (electrons of the same spin are automatically kept apart by the antisymmetry of the wave function) and raises their contribution to the interelectronic repulsion energy. There are also long-range effects of correlation between electrons, which are analogous to the dispersion forces between atoms [30], but the short-range, intrapair effects are most important. The ideas of Hurley *et al.* [31] which suggest a trial function as a product of correlated pair functions, make for difficult calculations, but use of a CI function which simulates such a trial function can be fruitful.

In a diatomic molecule, we can identify several kinds of correlation within a pair:

(*i*) In–out correlation refers to the fact that, when one electron is close to the internuclear axis, the other tends to be far from it.

(*ii*) Left–right correlation means that electrons tend to stay on opposite ends of the molecule.

(*iii*) Angular correlation refers to electrons keeping on opposite ends of an imaginary stick perpendicular to the internuclear axis.

In–out correlation corresponds to radial correlation in an atom and the others to angular correlation. In a CI wave function, one should include configurations which describe all three effects. The familiar valence bond function, which often provides a good first approximation to the molecular wave function for $R$ near $R_e$, provides left–right correlation for the electrons in the bond. It is a sum of two determinants, with the second differing from the first by a spin exchange between the atomic orbitals (on different atoms) which form the bond. Strictly speaking, these two determinants correspond to the same configuration in an atomic orbital basis, but in terms of molecular orbitals the function is a configuration interaction.

It may be argued that, while correlation between electrons on the same atom, i.e., in atomic orbitals, is important, it changes only slightly as $R$ changes. Then providing such correlation in the wave function may not be necessary if $U(R)$, and not energies on an absolute scale, is of interest. The invariance of part of the correlation energy to molecule formation is at the basis of the atoms-in-molecules theory, developed by Moffitt, which we discuss in Chapter III, Section A.3 and of the method of optimized valence configurations, discussed later in this section and in Section C. The inaccuracies due to errors in atomic correlation energy can be corrected from empirical knowledge about the atoms. Correlation between electrons in orbitals on different atoms, not involved in the bond, should be small. Thus, correlation between electrons in strongly overlapping atomic orbitals, i.e., those forming the bond, is of primary importance.

The molecule for $R$ near $R_e$ still resembles in some ways the separated atoms, and the united atom. Thus, one should consider, in molecular calculations, the splittings between the terms generated from ground state configurations in both limits. The determinantal wavefunctions in the configuration interaction should include all the determinants necessary to construct the states corresponding to these terms.

The work on Li and LiH by Harris and Taylor [32] is a good example of how to go about choosing configurations to provide for particular kinds of correlation. It also illustrates the construction of wave functions as combinations of determinants to give the functions of proper symmetry corresponding to each configuration. The orbitals used were two-center functions, in ellipsoidal coordinates, which allow for great flexibility [32].

While most CI calculations build configurations from molecular orbitals, Matsen and Browne [33] argued the advantages, for small molecules, of forming configurations from atomic orbitals instead. For LiH, a 20-term wave function constructed in this way gave results better than those of other calculations at the time, but subsequent CI work on this system (see Section C) has given much better agreement with experiment.

As we have mentioned, the wave functions of the valence bond theory are most conveniently represented as configurations built from atomic orbitals. Since the simple covalent and ionic valence bond wave functions often work well, the valence bond theory can be used to set up configurations in CI calculations. In modern usage, however, a valence bond wave function often shares with the original valence bond theory only the spin-pairing structure (see Chapter I, Section B), without necessarily being constructed from atomic orbitals. This may be referred to as a generalized valence bond configuration. Some authors have hopes that the simple valence bond theory may become useful in the future [8, Sect. 3.3, Chapter 6].

However, such calculations are found mainly in the earlier literature. For example, in the $^2\Sigma^+$ ground state of LiH, the wavefunction would be expected to be dominated by the covalent function describing a bond between the H 1s and the Li 2s, plus an ionic function corresponding to $Li^+H^-$. However, the fact that the 2s–2p splitting is small in Li means that configurations involving excitation to the 2p orbitals must be included. In excited configurations, the 2p$\pi$ orbitals must be either doubly filled or empty to give $^1\Sigma$ terms. Miller et al. [34] considered only configurations in which the 2p$\pi$ orbitals were not used, and still came up with 20 possible configurations, each corresponding to ions or to a single bond. There are 12 ionic states. Letting orbitals with no subscript refer to Li, the configurations of the ionic states are: $(1s)^2(1s_H)^2$, $(1s)^2(2s)^2$, $(1s)^2(2p\sigma)^2$, $(2s)^2(1s_H)^2$, $(1s)^22s2p\sigma$, $(1s_H)^21s2s$, $(1s_H)^21s2p\sigma$, $(1s_H)^22s2p\sigma$, $1s2s(2p\sigma)^2$, $1s(2s)^22p\sigma$, $(2s)^2(2p\sigma)^2$, $(2p\sigma)^2(1s_H)^2$. Single-bond functions were considered for the 2s–1s$_H$ and the 2p$\sigma$–1s$_H$ bonds. In both cases, the two remaining electrons may be in the Li 1s orbital, or excited to the 2-shell. A bond between 1s$_H$ and the Li 1s orbital (the other electrons being both either in the 2s or the 2p orbital) was also considered. Miller et al. [34] gave results for 1, 2, 3, 4, 5, 6, 10, and 20 configurations, the final energy for LiH being $-217.02$ eV. Including only the ground state for $Li^+H^-$ together with the 2s and 2p bond states gave $-216.95$ eV. States in which the Li inner-shell electrons were promoted were particularly unimportant (0.01 eV energy gain). The energy of $-217.02$ eV compares to $-219.71$ derived from experimental data. The binding energy is reasonable only relative to SCF atomic energies. Computations on $BeH^+$, which is isoelectronic with LiH, gave similar results [34].

Browne and Matsen [35], in a subsequent configuration interaction calculation on LiH (1964), used a mixed ellipsoidal and atomic basis set. With 28 generalized valence bond configurations, they obtained the lowest energy for the molecule available at that time. The dissociation energy, from calculated molecular and atomic energies, was 2.34 eV (experiment, 2.516 eV). The value of $R_e$ was calculated as 3.046 $a_0$ (experiment, 3.015 $a_0$), $\omega_e$ as 1438 cm$^{-1}$ (experiment, 1406 cm$^{-1}$), and $\omega_e x_e$ as 86 cm$^{-1}$ (experiment, 23 cm$^{-1}$). A check of

$$\Delta_A \equiv k/Z_A - q_A$$

[see Volume I, Chapter III, Eq. 364 and discussion following] was made. Here, $\Delta_A$ would vanish for a spherical charge density on nucleus A which followed the motion of A. This was not found to be the case.

By using wave functions derived from the valence bond theory, one includes left–right correlation, which is extremely important in determining $U(R)$. The failure to include left–right correlation is responsible for the "incorrect dissociation" of the restricted Hartree–Fock function (Section A.2 and C.1). On the other hand, the molecular orbital description is superior to the valence bond for small $R$. The method of optimized valence configurations [36, 37] attempts to get accurate potential curves by building into a configuration interaction wave function the good features of both the molecular orbital and valence bond descriptions, hopefully with only a few configurations, and also describe the correlation of the electrons in the bond. Das and Wahl [36, 37] wrote the wave functions for diatomic molecules such as $H_2$, $Li_2$, and $F_2$ as a sum of configurations, of which one was the restricted Hartree–Fock function. The others were formed from it by excitation of the two electrons in $\phi_\sigma$, the highest filled $\sigma$ orbital, into excited orbitals, or into a two-electron function written as a sum of products of one-electron functions. In the determinant, $\phi_\sigma(1)\phi_\sigma(2)$ was replaced by $\phi_A(1)\phi_B(2) + \phi_B(1)\phi_A(2) + \Sigma\phi_A(1)\phi_A(2) + \Sigma\phi_B(1)\phi_B(2)$ where $\phi_A$ and $\phi_B$ denote atomic orbitals on atoms A and B. This gave correlation within this pair of valence electrons, and also allowed for proper dissociation. The "core" was left unchanged through all the configurations.

Then it was possible [36, 37] to write the energy in terms of core and valence contributions, and minimize the energy with respect to the core and valence orbitals themselves as well as with respect to the configuration mixing coefficients. The solution of the resulting equations was by expansion of the orbitals in a basis set. (References to previous related methods were given [36, 37].) For $H_2$, $Li_2$, and $F_2$, addition of a single configuration to the restricted Hartree–Fock function, corresponding to replacing $(n\sigma_g)^2$

by $(n\sigma_u)^2$, sufficed to give correct dissociation to Hartree–Fock atoms. Other configurations accounted for the three types of correlation: left–right (closely related to correct dissociation), angular, and in–out. The optimized valence configuration wave functions for all three molecules had much lower energy than did the corresponding Hartree–Fock functions. More important, the shapes of the $U(R)$ curves were significantly changed. Unlike those derived from the Hartree–Fock, the curves now appeared quite parallel to those derived from spectroscopic data. This is illustrated by the results for $Li_2$ in Table III. For $F_2$, there is no binding in Hartree–Fock $(D_e = -1.37 \text{ eV})$, but a binding energy of 0.54 eV (experimental value, 1.68 eV) was obtained from two configurations. Here, of course, there are three valence orbitals, and excitations from all are required to get accurate binding energies.

**Table III**

*Optimized Valence Configuration Method for $Li_2{}^a$*

|  | $R_e$ ($a_0$) | $D_e$ (eV) | $\omega_e$ (cm$^{-1}$) |
|---|---|---|---|
| Hartree–Fock | 5.26 | 0.17 | 326 |
| Two configurations | 5.43 | 0.46 | 344 |
| Optimized valence configuration | 5.09 | 0.93 | 345 |
| Experimental | 5.05 | 1.05 | 351 |

$^a$ See Das and Wahl [36], Wahl et al. [37].

No attempt was made [36, 37] to improve the description of the core electrons, which certainly make a large contribution to the correlation energy. This contribution might be expected to be unchanged in going from separated atoms to molecule and to affect the potential curve $U(R)$ only for very small $R$, whereas the method of optimized valence configurations was specifically designed to describe chemical binding and the interatomic potential. The results show that this can be done by describing accurately the valence electrons.

In general, multiconfiguration SCF calculations, in which parameters entering the molecular orbitals from which the configurations are built are optimized simultaneously with the linear parameters for mixing of configurations, are extremely difficult. Some of the work in developing such formalisms was mentioned in Volume I, Chapter III, Section A. It is not surprising that such calculations for molecules have appeared only recently.

McWeeny [38] has given a formalism for multiconfiguration SCF and commented on problems connected with its computer implementation. The functions he started with were products of orbitals multiplied by one of the linearly independent spin functions. These were antisymmetrized and normalized. In choosing the set of spin functions, a correspondence could be made with the bonding structures of valence bond theory. Then formulas for matrix elements of the Hamiltonian between spin eigenfunctions could be taken from those derived by Cooper and McWeeny [39] for the valence bond problem. The secular equation was set up and the equations for the optimum orbitals were derived. A procedure for iterative adjustment of the orbitals themselves was outlined.

Mukherjee and McWeeny [40] reviewed other multiconfiguration SCF treatments, and carried out calculations on LiH using McWeeny's formalism [38]. Numerical methods for solution of the equations were tested, and results compared with those of other workers. Seven configurations sufficed to give a total energy as good as that of much larger CI calculations (but still only halfway between restricted Hartree–Fock and the iterative natural orbital–CI results).

Hinze and Roothaan [40] also derived a multiconfiguration SCF formalism. The energy of a configuration interaction wave function was expressed in terms of expansion coefficients of the orbitals in a basis set and configuration mixing coefficients. The variation principle led to equations for the orbital expansion coefficients which resembled the Hartree–Fock equations. A method of solution analogous to a Newton–Raphson iteration in a multidimensional space was outlined. Veillard and Clementi [42] have also developed and implemented a multiconfiguration SCF scheme.

It seems likely that multiconfiguration SCF formalisms will play an increasing role in molecular calculations in the future [43]. At present, however, most CI calculations employ configurations constructed from a fixed set of orbitals, and vary only the parameters describing the combination of configurations (linear variation). Only for certain special cases [44], such as those treated by the optimized valence configuration method, does one attempt to optimize the orbitals simultaneously with the linear combination coefficients.

McLean and Yoshimine [45] estimated that, for systems with up to 20 electrons, accuracy comparable to that of the James–Coolidge $H_2$ calculations will be attainable by the mid-1970s. However, since the time required for CI and related calculations increases as the fourth or fifth power of the size of the basis set, they suggested that new techniques of a different kind should be investigated. These might be inefficient for small systems, but increase slowly enough in complexity for larger systems to be better than conventional methods. Clementi [46] also discussed the future of CI calculations, projecting that one will be able to calculate a good part of the correlation energy for diatomic molecules with up to about 50 electrons. He noted, however, that while one can now get about 70% of the correlation energy for a large number of systems (see Section C), it seems very difficult to go much beyond this. For $U(R)$, we do not require accurate calculation of *all* the correlation energy, so the situation is more hopeful.

## REFERENCES

1. F. A. Matsen, *Advan. Quantum Chem.* **1**, 60 (1964); J. I. Musher, *J. Phys. (Paris)* **31**, C4-51 (1970); D. J. Klein and B. R. Junker, *J. Chem. Phys.* **54**, 4290 (1971).
2. I. G. Kaplan, *Theor. Exp. Chem.* **3**, 83 (1967); *Teor. Eksp. Khim.* **3**, 150 (1967); and preceding papers.
3. V. H. Smith and F. E. Harris, *J. Math. Phys. (N.Y.)* **10**, 771 (1969).
4. P. O. Löwdin, *Phys. Rev.* **97**, 1509 (1955).
5. R. Manne, *Theor. Chim. Acta* **6**, 116 (1966).
6. D. Secrest and L. M. Holm, *J. Math. Phys. (N.Y.)* **5**, 738 (1964).
7. M. Kotani, A. Amemiya, E. Ishiguro, and T. Kimura, "Table of Molecular Integrals," Chapter 1. Maruzen, Tokyo, 1955.
8. R. McWeeny and B. T. Sutcliffe, "Methods of Molecular Quantum Mechanics," Sect. 3.6. Academic Press, New York, 1971.
9. R. Pauncz, "The Alternant Molecular Orbital Method," Chapter 2. Saunders, Philadelphia, Pennsylvania, 1967.
10. W. A. Goddard, III, *Phys. Rev.* **157**, 73, 81 (1967).
11. G. A. Gallup, *J. Chem. Phys.* **48**, 1752 (1968); **50**, 1206 (1969).
12. C. C. J. Roothaan, *Rev. Mod. Phys.* **32**, 179 (1960).
13. K. Ruedenberg, *Phys. Rev. Lett.* **27**, 1105 (1971); K. Ruedenberg and R. D. Poshusta, *Advan. Quantum Chem.* **7**, 1 (1972).
14. C. C. J. Roothaan and P. Bagus, *Methods Comp. Phys.* **2**, 47 (1963).
15. E. R. Davidson, *in* "Physical Chemistry: An Advanced Treatise" (D. Henderson, ed.), Vol. III, p. 154. Academic Press, New York, 1969.
16. P. O. Löwdin, *Rev. Mod. Phys.* **32**, 328 (1960).
17. P. O. Löwdin, *J. Math. Phys. (N.Y.)* **3**, 1171 (1962).
18. K. A. Brueckner, C. A. Levinson, and H. M. Mahmoud, *Phys. Rev.* **95**, 217 (1954).
19. H. Bethe, *Phys. Rev.* **103**, 1353 (1956).
20. K. A. Brueckner, A. M. Lockett, and M. Rotenberg, *Phys. Rev.* **121**, 255 (1961).
21. D. H. Kobe, *Phys. Rev. A* **3**, 417 (1971).
22. C. F. Bender and E. R. Davidson, *J. Phys. Chem.* **70**, 2675 (1966); *J. Chem. Phys.* **47**, 360 (1967).
23. J. R. Gabriel, *J. Chem. Phys.* **51**, 3713 (1969).

24. R. J. Buenker and S. D. Peyerimhoff, *Theor. Chim. Acta* **12**, 183 (1968).
25. S. Hagstrom and H. Shull, *J. Chem. Phys.* **30**, 1314 (1959).
26. C. J. Hebert and O. G. Ludwig, *J. Chem. Phys.* **47**, 3086 (1967).
27. A. K. Bhatia and A. Temkin, *Rev. Mod. Phys.* **36**, 1050 (1964); A. Temkin and A. K. Bhatia, *J. Chem. Phys.* **42**, 644 (1965); A. K. Bhatia and A. Temkin, *J. Chem. Phys.* **44**, 3656 (1966).
28. R. K. Nesbet, *Phys. Rev.* **109**, 1632 (1958).
29. R. K. Nesbet, *Int. J. Quantum Chem. Symp.* **4**, 117 (1970).
30. R. K. Nesbet, *Advan. Chem. Phys.* **9**, 321 (1965).
31. A. C. Hurley, J. E. Lennard-Jones, and J. A. Pople, *Proc. Roy. Soc. Ser. A* **220**, 446 (1953).
32. F. E. Harris and H. S. Taylor, *J. Chem. Phys.* **32**, 3 (1960); *Mol. Phys.* **6**, 183 (1963); *J. Chem. Phys.* **38**, 2591 (1963); **39**, 1012 (1963); *Physica (Utrecht)* **30**, 105 (1964).
33. F. A. Matsen and J. C. Browne, *J. Phys. Chem.* **66**, 2332 (1962).
34. J. Miller, R. H. Friedman, R. P. Hurst, and F. A. Matsen, *J. Chem. Phys.* **27**, 1385 (1957).
35. J. C. Browne and F. A. Matsen, *Phys. Rev. A* **135**, 1227 (1964).
36. G. Das and A. C. Wahl, *J. Chem. Phys.* **44**, 87 (1966).
37. A. C. Wahl, P. J. Bertoncini, G. Das, and T. L. Gilbert, *Int. J. Quantum Chem. Symp.* **1**, 123 (1967).
38. R. McWeeny, *Proc. Roy. Soc. Ser. A* **232**, 114 (1955); *Symp. Faraday Soc.* **2**, 7 (1968).
39. I. L. Cooper and R. McWeeny, *J. Chem. Phys.* **45**, 226 (1966); **49**, 3223 (1968).
40. N. G. Mukherjee and R. McWeeny, *Int. J. Quantum Chem.* **4**, 97 (1970).
41. J. Hinze and C. C. J. Roothaan, *Progr. Theor. Phys. Suppl.* **40**, 37 (1967).
42. A. Veillard and E. Clementi, *Theor. Chim. Acta* **7**, 133 (1967).
43. H. F. Schaefer, III, "The Electronic Structure of Atoms and Molecules," pp. 40–41. Addison-Wesley, Reading, Massachusetts, 1972.
44. A. Veillard, *Theor. Chim. Acta* **4**, 22 (1966).
45. A. D. McLean and M. Yoshimine, *IBM J. Res. Develop.* **12**, 206 (1968).
46. E. Clementi, *Int. J. Quantum Chem. Symp.* **3**, 179 (1969).

## C. $U(R)$ Derived from Calculations

In this section, we are concerned with the derivation of $U(R)$ curves from calculations like those described in the preceding chapters. From a set of values of $R$ and corresponding energies, one can obtain parameters to use with a chosen functional form for $U(R)$ (see Volume I, Chapter II), or resort to some curve-fitting technique. How one does this may affect the accuracy of the derived results. Problems related to the generation of $U(R)$ from a set of energies are discussed in Subsection 1. When restricted Hartree–Fock calculations are used to get the energies, there may be serious errors, but much is known about the source of the errors and how they can be corrected (Subsection 2). In Subsections 3 and 4, we cite the results of calculations by Hartree–Fock (Subsection 3) and other methods (Subsection 4) of $U(R)$ for diatomic ground states.

## 1. Derivation of $U(R)$ from Calculated Energies

In Volume I, Chapter II, Section D, we have reviewed many of the analytical forms for $U(R)$ which have been proposed on theoretical or other grounds. The number of parameters specific to a particular molecule varies from one up to rather large numbers. The possibility of a universal function, to be parametrized for each molecule, is discussed in Chapter III.

From energies for $n$ values of $R$, one can determine an $n$-parameter function for $U(R)$. Use of a function with fewer than $n$ parameters would mean a loss of information only if the functional form is known to be exactly right. Since this is never the case, some redundancy [use of $U(R)$ with fewer than $n$ parameters] is necessary to check the validity of the functional form. The fitting of the parameters is generally done by least-squares. For instance, Meckler [1] fitted CI energies for $O_2$ at $R = 1.5, 2.0, 2.5$ and $4.0\ a_0$ to a Morse curve and found good agreement with experiment for $R_e$, $\omega_e$, and the dissociation energy.

Expansions of $U(R)$ as polynomials in $R$ have this advantage over most other functional forms: The number of parameters can be increased as much as desired. However, the value of any one coefficient in the expansion depends on the number of terms used. The behavior of the coefficient of $R^i$ as terms in $R^j$ ($j > i$) are added can indicate the confidence we should place in the derived values for potential constants. Beckel and co-workers [2] have investigated the power series expansions of several forms of $U(R)$, mostly for $H_2$, and used them to determine spectroscopic constants. The power series were about $R_e$, which was determined by fitting a polynomial of $n$th order to $n + 1$ points close to $R_e$. It was noted [2] that a narrow range of $R$ values helped to ensure rapid convergence of the series and independence of the first few coefficients to the number of terms used in the series. However, the same choice of $R$-values made higher coefficients inaccurate or indeterminate. The effect of inaccuracies of the calculated energies on the derived potential constants was studied.

Nesbet [3] proposed the use of expansions in Chebyshev polynomials [4], and subsequent conversion to power series, to obtain potential constants by differentiation. One advantage of the Chebyshev expansions is that they allow an estimation of the error in the expansion coefficients due to errors in the calculated energies used. Furthermore [3, 4], if the $R$-values for which energies are computed are chosen as the roots of Chebyshev polynomials, the error due to truncation is as uniform as possible over the interval of $R$ considered.

Since $U(R)$ becomes infinite at $R = 0$, it may be preferable to use a polynomial expansion on the electronic energy,

$$U^e(R) = U(R) - Z_A Z_B/R, \qquad (67)$$

which approaches a constant (the united atom energy—see Chapter I, Section C) as $R$ approaches zero. The use of power series for $U^e(R)$ rather than $U(R)$ also seemed recommended by McLean's experience with his scaling method [5], which we have discussed in Volume I, Chapter III [Eqs. (293) *et seq.*]. Yoshimine [6] compared power series expansions of $U(R)$ and $U^e(R)$; the differences between the values of quantities derived from the two were taken as a measure of the uncertainty in the fitting procedure. An alternative [6] is to work with the nonsingular function $R\,U(R)$. If this is fit to a polynomial, the derivatives of $U(R)$, which give the spectroscopic constants, are easily obtained by differentiation.

Nesbet [3], as well as Schwendeman [7], investigated the errors introduced by polynomial fitting and differentiation of the polynomial. They carried out this procedure for the analytical Morse curve (Volume I, Chapter II, Section D), and compared the derived derivatives with the correct values (obtained analytically). When five points were used, the error in the fourth derivative could be several percent. Seven points were needed to get reliable values of the third and fourth derivatives. More reliable values from fewer points could be obtained by carrying out polynomial fitting on $U^e(R)$, Eq. (67), as well as on $U(R)$, and averaging the results.

Silver *et al.* [8], in connection with their separated pair calculations, least-square fitted a series of energies to a polynomial in $(R - R_e)/R_e$. They considered polynomials of various orders and monitored the change in the coefficients. Their conclusion was that the number of points must be much larger than the order of the polynomial. Then it was possible to have higher coefficients remain small and the first few coefficients remain constant as the order was changed. Under these circumstances, one could be sure of the adequacy of the fit to a truncated power series, and the values for the first few coefficients could be relied on.

Schwendeman [7] discussed a possible large error which could enter the derivation of force constants from a power series fitted to calculated energies.

He compared the power series for the exact electronic energy, $U^e(R)$, and $V^e(R)$, the electronic energy corresponding to an approximate wavefunction.

$$U^e(R) = U_e + U_e'(R - R_e) + \tfrac{1}{2}U_e''(R - R_e)^2 + \cdots \qquad ,(68\text{a})$$

$$V^e(R) = V_e + V_e'(R - R_e) + \tfrac{1}{2}V_e''(R - R_e)^2 + \cdots \qquad (68\text{b})$$

The subscript e means evaluation at $R_e$, the true equilibrium internuclear distance, and primes are differentiations with respect to $R$. Letting $V_N$ represent the nuclear repulsion energy, we have [7] that $U_e' + (dV_N/dR)_e$ vanishes. The quantity $R_V$, the value for $R_e$ predicted by the approximate energy $V^e(R)$, is the value of $R$ which makes $d(V^e + V_N)/dR$ vanish. From (68b),

$$V_e' + V_e''(R^V - R_e) + (dV_N/dR)_V = 0$$

If $V_N$ is also expanded in a power series about $R_e$,

$$(dV_N/dR)_V = (dV_N/dR)_e + V_{Ne}''(R_V - R_e) = -U_e' + V_{Ne}''(R_V - R_e)$$

so that

$$R_V - R_e = (U_e' - V_e')/(V_e'' + V_{Ne}'').$$

If the approximate function is a Hartree–Fock function, so that $U^e(R)$ and $V^e(R)$ differ by a second-order quantity (see Volume I, Chapter III, Section A.5) this shows that the error in the predicted equilibrium internuclear distance is formally second order, so that typically, $U_e' - V_e'$ is of the order 1% of $U_e$ [7, 9]. We have already presented the argument (Volume I, Chapter III, Section D.3) that the error in $R_e$ should be second order when a Hartree–Fock function is used. However, Schwendeman [7] showed that $V_e''$ and $V_{Ne}''$, which have opposite signs, typically cancel to 90%, so that $V_e'' + V_{Ne}''$ should be considered as a pseudofirst-order term, and $R_e$ may be off by more than typical expectation values. Furthermore, the pseudofirst-order error in $R_e$ can give errors in derivatives of $U^e(R)$, which are larger than typical second-order errors. In particular, a change in the origin of the expansion (68b), from $R_e$ to the predicted equilibrium internuclear distance, gave large errors in the force constants. It was suggested [7] that the experimental $R_e$ in (68b) be used when possible, to get more accurate results.

From the power series representation of $U(R)$, we can derive the spectroscopic constants $\omega_e$, $\omega_e x_e$, etc. by the Dunham analyses [Volume I, Chapter I, Section C, Eqs. (70)–(73), and succeeding discussion]. This is often the procedure employed to calculate spectroscopic constants and/or vibrational–rotational energy levels from calculated $U(R)$. The energy levels are given in terms of the spectroscopic constants by:

$$E_{VJ} = \omega_e(v + \tfrac{1}{2}) - \omega_e x_e(v + \tfrac{1}{2})^2 + \omega_e y_e(v + \tfrac{1}{2})^3 + \cdots$$
$$+ B_e J(J + 1) - \alpha_e(v + \tfrac{1}{2})J(J + 1) + \cdots$$
$$+ D_e J^2(J + 1)^2 + \beta_e(v + \tfrac{1}{2})J^2(J + \tfrac{1}{2})^2 + \cdots \quad (69)$$

Sometimes, the $Y_{lm}$ [Volume I, Chapter I, Eq. (73)] are computed. However, because of difficulties with truncated expansions and other problems associated with the Dunham analysis, some authors are beginning to compute the vibrational–rotational energy levels numerically, directly from the computed energies. If desired, the spectroscopic constants may then be obtained by fitting the energy levels to Eq. (69). If the energy levels are to be computed accurately from a set of internuclear distances $R$ and associated energies, the values of $R$ must be closely spaced, and a wide range of $R$ must be investigated. This means a large number of energy calculations are necessary.

For instance, Sahni et al. [10] computed electronic energies for LiH at 52 internuclear distances from $R = 1.7\,a_0$ to $R = 9\,a_0$. They carried out Hartree–Fock Roothaan (HFR) calculations and a limited (41 configurations) configuration interaction (CI). Since a minimal basis was employed (1s, 2s, and 2p orbitals on Li), it was possible to optimize all orbital exponents at each $R$. The first 14 vibrational levels were computed and compared to experiment. Their results for the first four, including minimum and maximum $R$-values corresponding to the semiclassical technique used [10] to calculate these levels, are shown in Table IV. Here, $B_v$ is proportional to the average value of $R^{-2}$ over the vibrational wave function; in terms of the constants of Eq. (69),

$$B_v = B_e - \alpha_e(v + \tfrac{1}{2})$$

From the vibrational levels, Sahni et al. derived the spectroscopic constants and compared with experiment (see Table V). It appeared that going from HFR to CI gave most improvement for large values of $R$.

Brown and Shull [11] computed energies for LiH by CI (discussed further on page 271) for 34 values of $R$, and interpolated to obtain additional points on $U(R)$, which was then used for direct calculation of vibrational–rotational energies by a method due to Davidson [12]. The energy $E_{vJ}$ was determined up to $v = 18$, $J = 4$ for LiH and LiD, and from the values

## Table IV

### Vibrational-Rotational Energy Levels for LiH

| | HFR Results | | | CI Results | | | | Experimental Results | | |
|---|---|---|---|---|---|---|---|---|---|---|
| $v$ | $\varepsilon$ (cm$^{-1}$) | $R_{\min}$ (Å) | $R_{\max}$ (Å) | $B_v$ (cm$^{-1}$) | $\varepsilon$ (cm$^{-1}$) | $R_{\min}$ (Å) | $R_{\max}$ (Å) | $B_v$ (cm$^{-1}$) | $\varepsilon$ (cm$^{-1}$) | $R_{\min}$ (Å) | $R_{\max}$ (Å) |
| 0 | 717.65 | 1.459 | 1.786 | 7.3286 | 687.62 | 1.453 | 1.789 | 7.3723 | 698 | 1.447 | 1.779 |
| 1 | 2114.48 | 1.367 | 1.942 | 7.1360 | 2012.02 | 1.365 | 1.955 | 7.1077 | 2058 | 1.354 | 1.938 |
| 2 | 3465.64 | 1.310 | 2.064 | 6.9395 | 3285.03 | 1.310 | 2.085 | 6.83:0 | 3376 | 1.297 | 2.061 |
| 3 | 4775.81 | 1.268 | 2.171 | 6.7658 | 4514.62 | 1.269 | 2.201 | 6.5923 | 4643 | 1.255 | 2.173 |

**Table V**

*Spectroscopic Constants for* LiH *Derived from Energy Levels*

|  | HFR | CI | Experiment |
|---|---|---|---|
| $\omega_e$ (cm$^{-1}$) | 1440.78 | 1376.07 | 1405.65 |
| $\omega_e x_e$ (cm$^{-1}$) | 21.277 | 23.628 | 23.20 |
| $B_e$ (cm$^{-1}$) | 7.3911 | 7.3935 | 7.5131 |
| $\alpha_e$ (cm$^{-1}$) | 0.176 | 0.2109 | 0.2131 |

Brown and Shull derived (for LiH) $\omega_e = 1374.3$ cm$^{-1}$, $\omega_e x_e = 22.5$ cm$^{-1}$, $B_e = 7.2$ cm$^{-1}$. Davidson's method [12] involved solution of the differential equations for the vibrational wave function with the potential (energy as a function of $R$) approximated by a step function. Its accuracy was tested on a Morse potential and on values for $U(R)$ for the hydrogen molecule.

As an alternative to treating energy, one can get all information about the shape of $U(R)$ from its derivative with respect to $R$. According to the Hellmann–Feynman theorem (Volume I, Chapter III, Section D), $dU/dR$ at any value of $R$ may be calculated as the expectation value of a variety of force operators, corresponding to different combinations of displacements of the two nuclei which have the effect of changing $R$ by $dR$. For a wave function not satisfying the theorem, the value $dU/dR$ calculated in this way may not be identical to that derived by differentiation (e.g., numerically) of $U(R)$. It may be in better or worse agreement with experiment, but the latter seems more often to be the case. According to the virial theorem (Volume I, Chapter III, Section C), $dU/dR$ may also be computed from expectation values of kinetic and potential energies. In general, calculated wave functions satisfy the virial theorem more closely than they do the Hellmann–Feynman theorem; parameters evaluated by way of the former tend to agree with the values obtained from $U(R)$ itself.

As we have mentioned in Chapter II, Section A.2, force constants may be obtained readily from configuration interaction calculations at a single value of $R$ when a one-center expansion is used [13, 14]. Suppose that a configuration interaction wavefunction,

$$\Psi_i = \sum_s c_{is} \Phi_s$$

is used in which the same basis functions $\Phi_s$ are employed for all $R$. The label $i$ refers to the different solutions of the secular equation. If there are

$n$ basis functions, $n$ solutions (sets of coefficients) will be found when $\mathbf{H}$ is diagonalized. The energy of the function $\Psi_i$ as a function of the internuclear distance $R$ is

$$U_i(R) = \sum_{s,t} c_{is}^* c_{is} H_{st} \qquad (70)$$

where $H_{st} = \langle \Phi_s \mid H \mid \Phi_t \rangle$. The function $\Psi_i$ is taken as normalized to unity independent of $R$ and the $\{\Phi_s\}$ have been assumed orthonormal. We will also assume, for simplicity, that all basis functions and coefficients are real. These restrictions may be relaxed without changing the argument.

Since the $\{c_{is}\}$ are obtained from a secular equation, Eq. (332) of Volume I, Chapter III shows that

$$dU_i/dX = \sum_{s,t} c_{is} c_{is} (dH_{st}/dX)$$

where $X$ is any nuclear coordinate. Then, if $Y$ is a nuclear coordinate (which may be the same as $X$),

$$d^2U_i/dXdY = \sum_{s,t} c_{is} c_{it} (d^2 H_{st}/dXdY)$$
$$+ 2 \sum_{s,t} (dc_{is}/dY) c_{it} (dH_{st}/dX) \qquad (71)$$

The coefficients and energies obey

$$\sum_t H_{st} c_{kt} = U_k c_{ks} .$$

Differentiating this, multiplying by $c_{is}$, and summing over $s$ yields

$$\sum_{s,t} c_{is} (dH_{st}/dY) c_{kt} + U_i \sum_t c_{it} (dc_{kt}/dY)$$
$$= (dU_k/dY) \delta_{ik} + U_k \sum_s c_{is} (dc_{ks}/dY) \qquad (72)$$

The eigenvalue equation (71) has been used in the second term on the left and the orthogonality of the eigenvectors has been used in the first term on the right. For $k \neq i$, this may be rearranged to

$$\sum_t c_{it} (dc_{kt}/dY) = \left[ \sum_{s,t} c_{is} (dH_{st}/dY) c_{kt} \right] \Big/ [U_k - U_i] \qquad (73)$$

while, for $k = i$,

$$\sum_t c_{it} (dc_{kt}/dY) = \tfrac{1}{2} d \left( \sum_t c_{it} c_{it} \right) \Big/ dY = 0 .$$

Therefore, we may multiply (73) by $c_{iu}$ and sum over $i$ to obtain

$$
\sum_i^{(i \ne k)} c_{iu}(U_k - U_i)^{-1} \sum_{s,t} c_{is}(dH_{st}/dY)c_{kt}
$$
$$
= \sum_{i,s} c_{iu}c_{is}(dc_{ks}/dY)
$$
$$
= \sum_s \delta_{us}(dc_{ks}/dY) = dc_{ku}/dY.
$$

This expression for the derivative of the coefficient may be substituted into Eq. (71) to give the second derivative of $U_i$ in terms of derivatives of the matrix elements $H_{st}$, and the coefficients (for a single value of $R$).

Therefore [13] one can obtain force constants from a single calculation if one can conveniently evaluate the derivatives of $H_{st}$. If the one-electron basis functions are $R$-independent, as they could be in a single-center calculation, the configurations constructed from them will also be $R$-independent. Then the derivative of $H_{st}$ with respect to $X$ is just $\langle \Phi_s \mid dH/dX \mid \Phi_t \rangle$, which is not difficult to calculate. Bratož and Allavena [13] have used this formalism for several molecules. The situation is much more complicated for a multicenter calculation, where the basis functions depend on $R$ [see Volume I, Chapter III, Eq. (265)]. Bishop and Randić [14] argued that this direct method for calculating force constants was more economical, more accurate, and gave better physical insight into the terms contributing to force constants, than the traditional method (using a series of calculations at different $R$).

They also extended the formulas [14] to include the effects of changes in nonlinear parameters in the basis functions with $R$. The second derivative of the energy function $U$ which depends on parameters $\varrho_1, \ldots, \varrho_n$ as well as on nuclear coordinates $X$ and $Y$ is

$$
\frac{d^2U}{dX\,dY} = \frac{\partial^2 U}{\partial X\,\partial X} + \sum_i \left[ \left( \frac{\partial}{\partial \varrho_i} \frac{\partial U}{\partial X} \right) \frac{d\varrho_i}{dY} + \frac{d}{dY} \left( \frac{\partial U}{\partial \varrho_i} \right) \frac{d\varrho_i}{dX} \right.
$$
$$
\left. + \frac{\partial U}{\partial \varrho_i} \frac{d^2\varrho_i}{dY\,dX} \right]
$$

The values of $\varrho_i$ are determined according to the variation principle for each value of the nuclear coordinates so that $\partial U/\partial \varrho_i$ and $d(\partial U/\partial \varrho_i)/dY$ vanish. The first term is the expectation value of $\partial^2 H/\partial X\,\partial Y$ and may be calculated from the wavefunction for a single configuration. For the remaining term, Bishop and Randić used

$$
\frac{d}{dX} \frac{\partial U}{\partial \varrho_i} = \frac{\partial^2 U}{\partial X\,\partial \varrho_i} + \sum_k \frac{\partial^2 U}{\partial \varrho_i\,\partial \varrho_k} \frac{d\varrho_k}{dX} = 0
$$

Thus, the second derivative with respect to nuclear coordinates is:

$$\frac{d^2U}{dX\,dY} = \frac{\partial^2 U}{\partial X\,\partial Y} - \sum_{i,k} \left(\frac{\partial^2 U}{\partial \varrho_i\,\partial \varrho_k}\right)\left(\frac{d\varrho_i}{dY}\right)\left(\frac{d\varrho_k}{dX}\right)$$

We need to know how the energy, for a single nuclear configuration, depends on the parameters $\varrho_i$. The calculations needed to establish this would normally be carried out in determining the optimum values for these parameters. The quantities $\partial \varrho_k/\partial X$ may be obtained [14] from

$$\frac{d}{dX}\left(\frac{\partial U}{\partial \varrho_i}\right) = 0 = \frac{\partial^2 U}{\partial X\,\partial \varrho_i} + \sum_{k=1}^{n}\left(\frac{\partial^2 U}{\partial \varrho_i\,\partial \varrho_k}\right)\left(\frac{d\varrho_k}{dX}\right). \qquad (74)$$

All other quantities being known, Eqs. (74) are simultaneous linear equations for the derivatives of the optimum values of the parameters with respect to nuclear coordinates. In principle, multicenter calculations could be treated with this formalism but "floating" (see Volume I, Chapter III, Section D.2) would have to be considered as a possible variation. Recently, Bishop and Macias [15] extended the analytic method to calculate force constants to wave functions given in terms of natural orbital expansions. Illustrative calculations were done for $H_2$.

The formalism of Gerratt and Mills [16], discussed in Volume I, Chapter III [Eqs. (249) *et seq.*], also makes possible direct calculations of force constants from Roothaan-type wave functions. The changes in the basis functions with $R$ (they are centered on nuclear positions) are taken into account by the occurrence of derivatives of orbitals with respect to $R$. The force constant was given by

$$k = \frac{2Z_A Z_B}{R^3} - Z_B\left\langle \Psi \left| \sum_{i=1}^{N} \left(\frac{3\cos^2 \theta_{Bi} - 1}{r_{Bi}^3} - \frac{4\pi}{3}\,\delta(\mathbf{r}_{Bi})\right) \right| \Psi \right\rangle$$

$$+ Z_B\left[\left\langle \Psi \left| \sum_{i=1}^{N} \left(\frac{\cos \theta_{Bi}}{r_{Bi}^2}\right) \right| \frac{\partial \Psi}{\partial R} \right\rangle + \left\langle \frac{\partial \Psi}{\partial R} \left| \sum_{i=1}^{N} \left(\frac{\cos \theta_{Bi}}{r_{Bi}^2}\right) \right| \Psi \right\rangle\right] \qquad (75)$$

assuming nucleus B moves with nucleus A remaining fixed. The test calculations on LiH, BH, and HF [16] gave only rough agreement with experimental force constants. This was ascribed to the inadequacy of the basis sets employed, since, when the same wave functions were used to calculate $k$ by fitting energies to a polynomial and differentiating, quite different results were obtained. Equation (75) assumes that the Hellmann–Feynman theorem is obeyed by the wave function, and the different values of $k$ obtained from the two methods show that this was not the case. If the Hellmann–Feynman theorem were not invoked, the equation for $k$ corresponding to Eq. (75) would involve second derivatives of atomic orbitals with nuclear coordinates.

Using a CI calculation for $H_2$ with a Gaussian basis, Bishop and Macias [17] sought to prove that, for force constants, direct analytical differentiation of the energy expression [Eqs. 70 *et seq.*] is superior to other methods in use from the viewpoint of accuracy as well as ease. Only calculations at a single value of $R$ are required; the derivatives of nonlinear parameters with respect to nuclear displacements may be determined from Eq. (74). If these derivatives are known, the force may be computed analytically for a series of values of $R$ and differentiated numerically, but this is more difficult and possesses no advantages over direct double differentiation of the energy [17]. The most common method, analysis of a series of calculated energies, requires optimization of parameters for several values of $R$. Only the virial theorem,

$$dU/dR = -(2\langle T \rangle + \langle V \rangle)/R$$

was found to be a reasonable alternative to direct differentiation, since accurate values of $dU/dR$ can be obtained without full optimization of a large number of parameters.

For a potential curve $U(R)$ which exhibits an energy minimum, the important measurable parameters are the equilibrium internuclear distance $R_e$, the dissociation energy from the lowest vibrational state, the equilibrium vibrational frequency $\omega_e$, and the anharmonicity constants of Eq. (69). The dissociation energy of the lowest vibrational state, $D_0$, differs by the zero-point energy of vibration from $D_e$, the dissociation energy from the potential minimum, which is just the energy difference between atoms and the molecule with $R = R_e$. If $\omega_e$ is 2000 cm⁻¹, this zero-point energy is 1000 cm⁻¹, or 0.12 eV which is small compared to typical dissociation energies of a few electron volts. In order to calculate accurately the parameters just mentioned, we require an approximate $U(R)$ which is parallel to the exact one for all $R$; an $R$-independent error will be of no consequence. For all the parameters except $D_e$, it suffices to have the approximate $U(R)$ parallel to the exact one for $R$ near $R_e$. The range for which the curves are parallel must be larger for $\omega_e x_e$ than $\omega_e$, and still larger for $\omega_e y_e$. Understanding of the sources of errors in calculated electronic energies can lead to predictions of the errors in derived values of the parameters, and to methods for correcting them to accurate values. If $D_e$, for instance, is computed as the difference of energies calculated for $R = \infty$ and $R = R_e$, one would like to take full advantage of cancellation of errors. This suggests that both calculations be performed with wave functions of the same complexity. The good dissociation energy obtained by Meckler (see above) was a difference between poor atomic and molecular energies. If, however, there is reason to expect an error entering for large $R$ which is not present for $R$ near $R_e$, another route to $D_e$ can be suggested that should be more accurate (see Subsection 2).

Klein *et al.* [18] considered the semiempirical correction of potential curves derived from *ab initio* calculations. The error $\Delta U(R)$, defined as the difference between calculated and true electronic energies (the internuclear repulsion is calculated exactly), was to be estimated. $\Delta U$ for $R = 0$ and $R = \infty$ was assumed to be known from spectroscopic data on atoms. Then the calculated $U(R)$ was to be corrected by subtracting

$$\Delta E_A(R) = \Delta U(\infty) + [\Delta U(0) - \Delta U(\infty)] f(R) \tag{76}$$

with the form of $f(R)$ to be investigated. The function $f(R)$ must obey

$$[\Delta U(\infty) - \Delta U(0)] f(R) \leq \Delta U(\infty)$$

for all $R$ so that $\Delta E_A(R)$ always be negative, and must approach unity for $R \to 0$ and zero for $R \to \infty$. Furthermore since any reasonably accurate calculation will obey the exact result, $dU/dR = 0$ for $R = 0$ (Volume I, Chapter III, Section D and Chapter I, Section C of this volume), it was required [18] that $df/dR$ vanish at $R = 0$. For large $R$ (Chapter I, Section A), the energy will be a series in $R^{-1}$, and, on dissociation to neutral atoms, the first few powers of $R^{-1}$ would vanish for the true $U(R)$ as well as for the calculated $U(R)$. Thus, Klein *et al.* required $\lim_{R \to \infty} R^3 f(R) = 0$. Several choices for $f(R)$, obeying these conditions, and guaranteeing a correction for intermediate $R$ lying between $\Delta U(0)$ and $\Delta U(\infty)$, were investigated and tested, for an excited state of $He_2$ with a stable minimum. The "pseudo-kinetic energy correction,"

$$f(R) = [\tilde{T}(R) - \tilde{T}(\infty)]/[\tilde{T}(0) - \tilde{T}(\infty)]$$

was found useful and employed [19] on the $He_2$ ground state potential curve. The pseudokinetic energy was defined as

$$\tilde{T}(R) = -U(R) - R\, dU/dR$$

and equals the electronic kinetic energy for a wave function satisfying the virial theorem [Volume I, Chapter III, Eq. (299)].

## 2. *Errors in Restricted Hartree–Fock Calculations*

In the usual restricted Hartree–Fock calculation for a diatomic molecule, expansions in a necessarily limited basis are employed, so the true energy of the restricted Hartree–Fock function (minimum energy for a determinant of specified form) is not obtained. Because of the restrictions on the forms of the molecular orbitals, this restricted Hartree–Fock energy is higher than that of the best determinant, the unrestricted Hartree–Fock. Furthermore, the exact energy can never be obtained from a single-determinant wave function. The improvement in going from restricted to unrestricted Hartree–Fock has rarely been investigated (see the end of this subsection). With respect to the error ("expansion error") due to limitations of the basis set, it is possible to estimate it by experimenting with extensions of the basis, or by carrying out calculations for the separated atoms, for which one can approach the restricted Hartree–Fock energy by other calculations, with the same basis set. In many cases, much of the error is from sources, such as the inner shells, which are not very different in molecule and atoms, and hence do not affect $U(R)$. Results for TiO by Nesbet [20] showed that, sometimes at least, an approximate Hartree–Fock binding energy may be obtained with a basis set that is quite inadequate to approach the restricted Hartree–Fock energies for atoms or molecule. It is now feasible to reduce the expansion error, for all but large diatomics, to 0.001 a.u., which is negligible compared to errors inherent in the restricted Hartree–Fock model.

Because of the widespread use of restricted Hartree–Fock calculations, for diatomic molecules, an understanding of the error referred to as "correlation energy" is important. Following a suggestion of Löwdin [21, 22], the correlation energy is defined as the difference between the restricted Hartree–Fock energy and the exact nonrelativistic energy. The correlation energy is associated with the fact that, in the Hartree–Fock model, the correlation between electrons is not taken into account, except for the Fermi hole. The main problem in the restricted Hartree–Fock is with electrons in the same spatial orbital, since they have opposite spins, and their mutual interaction is largest. If a diatomic molecule with a closed-shell configuration dissociates to atoms with open shells, the correlation energy will be different for the molecule and the atoms. Then the dissociation energy, if calculated as a difference of Hartree–Fock energies for atoms and molecule, will be inaccurate.

Stanton [23] considered the variation of correlation energy with $R$ using the fact that the Hartree–Fock function satisfied the Hellmann–Feynman theorem [Volume I, Chapter III, page 226 ]. This implies (with Brillouin's theorem) that the correlation energy is, to a first approximation, unchanged on change in a parameter in the one-electron part of the Hamiltonian, such as $R$ (see Volume I, Chapter III, Section D.3). Therefore Hartree–Fock and exact potential curves should be close to parallel over small ranges of $R$. [On the other hand, an arbitrary configuration interaction wave function, which gives better energies at each $R$, may not do as well: The energy improvement may vary sharply with $R$, and approximate and exact $U(R)$ curves will no longer be parallel.] Equilibrium internuclear distances and force constants should then be well-predicted by Hartree–Fock functions, but not necessarily dissociation energies, for which the change in $R$ involved is not small. Note that only for a closed shell is the restricted Hartree–Fock function a solution to the Hartree–Fock equations (Section A.4) (but see the end of this subsection).

Some calculations by Wahl [24] for homonuclear diatomics bear this out. For example, he found for $F_2$ $\omega_e = 1257$ cm$^{-1}$ (919), $\omega_e x_e = 9.85$ cm$^{-1}$ (13.6), $B_e = 1.003$ cm$^{-1}$ (0.8901), $\alpha_e = 0.0108$ cm$^{-1}$ (0.0146), $R_e = 1.33$ Å (1.42). Values in parentheses are experimental values. The calculated dissociation energy was $-0.060$ a.u. (molecular energy higher than atomic energy) instead of 0.062 a.u. Apparently, the calculated $U(R)$ rises too steeply for $R > R_e$. This molecule dissociates incorrectly (see pages 222, 227). There is clearly a large decrease in correlation energy between $F_2$ near its equilibrium internuclear distance and a pair of F atoms. (It may also be noted that errors in quantities other than the dissociation energy are much larger than 1%, which is the size of a typical second-order error [7, 9]. The discussion of Schwendeman, presented in the previous section, may be relevant here.)

Since a number of arguments indicate that the biggest problem with restricted Hartree–Fock functions is the failure to describe the motions of the electrons paired in the same spatial orbital, there should be a correlation between the number of such pairs and the correlation energy. The total correlation energy of a molecular system may be expressed exactly as a sum of contributions of independent localized pairs of orbitals [6]. There are large contributions from a pair corresponding to a doubly occupied orbital, referred to as a strong pair. It appears that the correlation energy contribution for each such pair is between 1 and 2 eV [21, 22, 25], and is roughly independent of shape and size. The correlation interaction between electrons in different strong pairs is like the London dispersion force, and contributes less to the correlation energy. Of course, there are more such interactions than there are strong pairs.

If the number of strong pairs in a molecule did not change much with $R$, correlation energy would be roughly $R$-independent. However, electron pairing is closely connected with bond formation. Thus the correlation energy could be quite different for the molecule and the separated atoms, but would more likely be the same for the molecule and the united atom. Even if $dE^{\text{corr}}/dR$ is small, it is not surprising that the correlation energy can be very different for $R = R_{\text{e}}$ and $R = \infty$ (separated atoms); however, it should not change much between $R_{\text{e}}$ and 0 (united atom). As we shall see later, we can use the correlation energy of the united atom to correct the restricted Hartree–Fock energy of the molecule near $R_{\text{e}}$, for calculation of the dissociation energy. In addition to possible formation of new pairs, decreasing $R$ from $R = \infty$ changes the shape of the orbitals, and, in general, brings all electrons closer together. Hopefully, these effects are less important energetically than pair formation. Now we shall take up some of these points in more detail.

Much of our understanding of correlation energies for molecules comes from studies of correlation energies in atoms. Clementi [26, 27] discussed the correlation energies for neutral atoms and positive ions (note that the Hellmann–Feynman theorem implies that $E^{\text{corr}}$ should be about the same for a positive ion as for the neutral atom with the same number of electrons). It was found that $E^{\text{corr}}$ depends only weakly on $Z$, except for states like $1s^2 2s^2\,{}^1S$, which becomes degenerate with $1s^2 2p^2\,{}^1S$ as $Z \to \infty$. At least for the first few periods, the correlation energy for neutral atoms was more linear than quadratic with the number of electrons $N$. This reflects the fact that the contributions of most of the $\frac{1}{2}N(N-1)$ pairs are small compared to those of the $\frac{1}{2}N$ strong pairs. The value of 0.03–0.04 a.u., found by Clementi [26] for the average ratio $k = E^{\text{corr}}/N$, corresponds to a correlation energy contribution of 1.6–2.2 eV per strong pair. As shown by Clementi, the deviations of the individual points from the line $E^{\text{corr}} = kN$ can be explained in terms of pair formation. With addition of an electron, $E^{\text{corr}}$ increases by about 0.02 a.u. if a new shell is started, by about 0.05 a.u. if a new pair is formed, and by something in between if no new pair is formed, but electrons with the same principal quantum number as that added are already present.

When the number of electrons became large, Clementi [27] found that these simple ideas held less well. The distinction between strong and weak pairs became less useful. The number of weak pairs, as well as the importance of each one to $E^{\text{corr}}$, grew, and the degeneracies and near degeneracies became more numerous. As a result, Clementi found that the correlation energy became more proportional to the total number of pairs. He also noted that the relativistic energy exceeded the correlation energy for $Z > 10$ [this is probably not a problem for discussion of $U(R)$] and that the configurations, and hence the pair structure, were less well-defined. The latter problem is associated with the fact that the magnetic interactions correspond to $J–J$ rather than $L–S$ coupling. Linderberg and Shull [28] tested correlation energies of three and four-electron systems, concluding that simple additivity of pair correlation energies did not always hold. As a function of $Z$, the correlation energy was roughly constant except when there were degeneracies for infinite $Z$ ("bare nucleus" or coulombic field). For molecules, they referred to the inadequacy of the restricted Hartree–Fock at large $R$, which we shall refer to in what follows as "exchange correlation" and emphasized its importance as a kind of correlation energy.

Comparing the molecule with the separated atoms, we usually have an increased correlation energy because one or more additional electron pairs are formed, and because the electrons are closer together on the average. The nature of all the pairs changes as $R$ decreases from infinity, but inner-shell pairs are very little affected by molecule formation. Most important is the formation of the new electron pair(s). When there are fewer pairs for the separated atoms than for the molecule, there will be an important error in subtracting Hartree–Fock energies to get a dissociation energy.

Furthermore, under these circumstances, the restricted Hartree–Fock functions which we use for the molecule cannot correctly describe the separated atoms. It is often said that such Hartree–Fock functions "dissociate incorrectly." This occurs for $H_2$, as we have discussed: The simple molecular orbital function becomes a mixture of H atoms and $H^+$ and $H^-$ ions for large $R$. Other examples are $F_2$, mentioned earlier, and LiF, for which the molecular orbital function represents a mixture of covalent and ionic states of the fragments at large $R$. The restricted Hartree–Fock function for BeO dissociates to neutral atoms, but the Oxygen atom is represented as being in a mixture of $^1D$ and $^1S$ states (ground is $^3P$) as pointed out by Yoshimine [6]. Incorrect dissociation leads to a correlation energy, defined as the difference between restricted Hartree–Fock and exact energies, which increases with increasing $R$. The actual energy of electron correlation probably increases [6] with decreasing $R$, since inter-electronic distances decrease. In addition to cases of incorrect dissociation, the correlation energy may change rapidly with $R$ where there are avoided crossings between curves for different configurations.

We will consider first the correction to more accurate values of dissociation energies, obtained by subtracting SCF energies for the molecule from those for the atomic dissociation products. Clementi [29] has discussed the procedure for doing this. First, we generally must correct the "SCF" energy obtained from a calculation using linear combinations of atomic-like orbitals (LCAO) for the incompleteness of the basis set. This is done by determining the effect of additional basis functions on the atomic calculations and considering the role of various basis orbitals in the atomic and molecular cases. We then have an estimate of the molecular SCF energy. Now by comparing the experimental and calculated atomic energies, we get the sum of relativistic and correlation energies for the atoms. An estimate of the former leads to values for atomic correlation energies. What is important for $D_e$, however, is what Clementi has called [30] "molecular extra correlation energy" (MECE), which is the correlation energy of the molecule compared to that of the separated atoms.

To a first approximation, the molecular correlation energy is the same as the correlation energy of the united atom, since the number of pairs of electrons is the same. Thus, LiF has as united atom Mg ($^1S$), for which $E^{corr} = 0.444$ a.u. The separated atoms are Li ($^2S$) and F ($^2P$), with correlation energies of 0.046 a.u. and 0.327 a.u. Then MECE for LiF is 0.071 a.u. Since a large part of MECE comes from formation of an electron pair which resembles that formed in going from F to F$^-$, Clementi [30] also considered the correlation energy of F$^-$ ($^1S$) compared to that of F. The correlation energy $E^{corr}$ for F$^-$ is 0.401 a.u., giving MECE for LiF equal to 0.074 a.u. A third route to MECE used the observation that it should be equal to MECE for HF, since the extra inner-shell electrons are the same for LiF and the separated atoms. The MECE for HF $= 0.074 \pm 0.002$ a.u. Relativistic energies need not be considered, since they are generally the same for the atoms as for the molecule. [Matcha [31] (see Section C.3) has included the change in relativistic energies, which amounts to about 0.002 a.u., in his calculations on the alkali halides, while Cade and Huo [32] (see page 238) considered several ways of estimating the variation of relativistic energy with $R$ for the hydrides.] Clementi noted [30] that the largest uncertainty in the calculation is in the restricted Hartree–Fock energy of the molecule. This uncertainty corresponds to the error due to the finite basis set.

In the above example, MECE was obtained from three sources: (1) the united atom, (2) a related isoelectronic system, (3) a related molecule. The use of isoelectronic systems is justified by the Hellmann–Feynman theorem. Often, a convenient isoelectronic system is a pair of atomic systems that resemble the electronic structure of the restricted Hartree–Fock function more closely than the actual dissociation products of the molecule. For $H_2$, the pair structure suggests $H^+ + H^-$; adding the $H^-$ correlation energy of 0.040 a.u. to the dissociation energy of 0.134 a.u. (from restricted Hartree–Fock) we come close to the experimental 0.175 a.u. Useful "related molecules" may sometimes be negative or positive ions for which the additional or missing electron is in a shell by itself. Other related molecules are those related through the periodic table. For example, MECE is expected to decrease gradually in a series of molecules for which one atom is the same and the other moves down the periodic table in the same group (e.g. OF, SF, SeF). For the correlation energy of a molecule–ion, we can use information on the parent neutral molecule. Since the energy of a molecule–ion with a single positive charge (or a singly charged atomic ion) is the energy of the parent molecule (or atom) plus the ionization potential $I$, MECE for the molecule–ion may be obtained [33, 33a] from MECE for the molecule and the ionization potentials.

$$\begin{aligned}
\text{MECE(mol} - \text{ion)} = {} & [E^{\mathrm{HF}}(\text{mol} - \text{ion}) - E^{\mathrm{exact}}(\text{mol} - \text{ion})] \\
& - [E^{\mathrm{HF}}(\text{atom} + \text{ion}) - E^{\mathrm{exact}}(\text{atom} + \text{ion})] \\
= {} & E^{\mathrm{HF}}(\text{mol} - \text{ion}) - E^{\mathrm{HF}}(\text{atom} + \text{ion}) \\
& - I(\text{molecule}) + I(\text{atoms}) - E^{\mathrm{HF}}(\text{molecule}) \\
& + E^{\mathrm{HF}}(\text{atoms}) + \text{MECE(molecule)}.
\end{aligned}$$

Here, "HF" refers to Hartree–Fock energy and "atom + ion" to the dissociation products of the molecule–ion.

For BeO, Yoshimine [6] estimated $E^{\mathrm{corr}}$ from Mg, the united atom. This gave $E^{\mathrm{corr}} = 0.444$ a.u. (as for LiF). He also used the value for $Be^{+2} + O^{-2}$; 0.450 a.u. A more accurate value should be: $E^{\mathrm{corr}}(Be^+) + E^{\mathrm{corr}}(O^-) +$ contribution of the pair formed from the $Be^+$ 2s orbital and the $O^-$ 2p orbital. Obtaining the last quantity from the results for the LiF and HF bonds, he found $E^{\mathrm{corr}}$ (BeO) 0.048 a.u. + 0.332 a.u. + 0.074 a.u. = 0.454 a.u., which agrees well with the other values. Note that the uncertainty of 0.01 a.u. is a fraction of an electron volt, which is a few percent of the dissociation energy.

Nesbet [3] explored a related approach to accurate dissociation energies. He considered the ion pairs obtained by analyzing the molecular orbital configuration at large $R$. For the molecules $N_2$, CO, and BF, the configuration is

$$(1\sigma)^2(2\sigma)^2(3\sigma)^2(4\sigma)^2(5\sigma)^2(1\pi)^4$$

and the ground state is $^1\Sigma^+$. For $R$ large, the first eight electrons are effectively represented by $(1s_A)^2(1s_B)^2(2s_A)^2(2s_B)^2$. The last six, for $N_2$, have the configuration $(2p\sigma_g)^2(2p\pi_u)^4$, which is a mixture of $N^{+3}N^{-3}$, $N^{+2}N^{-2}$,

$N^{+1}N^{-1}$, and NN, mostly in excited configurations. The average correlation energy for this situation could be approximated as that for $N^{+3}$ and $N^{-3}$ in their ground states [3]. This is because MECE is due to the formation of three electron pairs in the molecule, as in $N + N \rightarrow N^{+3} + N^{-3}$. Similarly, $C^{+2} + O^{-2}$ was used for CO and $B^{+1}F^{-1}$ for BF. The resulting dissociation energies were correct to $\sim$1 eV and in the order $CO > N_2 > BF$, which agrees with experiment. With no correction, the order was wrong and the errors ranged from 2 to 5 eV. The fact that $D_e$, corrected, was not more accurate was ascribed to interpair contributions to $E^{corr}$ and to the 2s–2p near-degeneracy in the atoms. For the $^1\Sigma^+$ state of TiO (not the ground state) Nesbet [20] calculated MECE as the difference in $E^{corr}$ for $O^{-2}$ ($^1S$) and O ($^3P$) plus the difference for $Ti^{+2}$ ($^1S$) and Ti ($^3F$). The state of $Ti^{+2}$ used corresponds to double occupation of the 4s orbital, to which the highest valence orbital of the molecule correlates. For the ground state of the molecule the configuration correlates with the 4s3d configuration of $Ti^{++}$, so the value of $E^{corr}$ for the (4s3d) $^3D$ state of $Ti^{++}$ was used. The MECE added 1.39 eV to the calculated value of $D_e$, 4.22 eV. The difference of 5.61 eV from the experimental value, $6.86 \pm 0.30$ eV, was ascribed to the inadequacies of the basis set. The error in molecular or atomic energies from this source is about 50 eV, but most of this cancels between molecule and atom.

Hollister and Sinanoğlu [34] also discussed the correction of binding energies for MECE. They suggested that the use of the united atom correlation energy to approximate the molecular correlation energy may lead to unsatisfactory results because of the difference in the inner-shell cores between molecule and united atom. Instead of the united atom, they proposed a "shrunk-core model," in which the core of the united atom was replaced by an atomic ion whose net charge was equal to the sum of the net charges of the cores in the molecule, but whose number of electrons was the same as for the cores in the molecule. Then $E^{corr}$ for the valence electrons was taken as $E^{corr}$ for the electrons moving around the shrunk core. The correlation energy $E^{corr}$ for the core electrons of the molecule was calculated as a sum of correlation energies of atomic ions. For example, $E^{corr}$ for $C_2$ was approximated as $E^{corr}(Ne) - E^{corr}(Ne^{+8}) + 2E^{corr}$ ($C^{+4}$). A second method estimated $E^{corr}$ by performing a population analysis on a simple MO function to get the number of electrons in each atomic orbital and hence in each kind of pair [34]. Then atomic data on pair correlation energies were used. Both methods gave MECE to a few tenths of an electron volt on the average, with the second doing a bit better than the first.

We now turn to other features of $U(R)$. It appears that potential curves derived from Hartree–Fock energies give good values for spectroscopic constants like $B_e$ (rotational constant, proportional to $R_e^{-2}$), $\omega_e$ (equilibrium vibrational frequency—depends on second derivative of $U$ near $R_e$), $\omega_e x_e$ (anharmonicity—depends on third derivative of $U$ near $R_e$). This is consistent with Stanton's ideas [23] on the constancy of correlation energy with $R$. Most commonly, the Hartree–Fock energy curve $U^{HF}(R)$ rises too steeply for $R > R_e$, since the restricted Hartree–Fock wave function does not describe the system correctly for $R \to \infty$, and hence is more and more in error [6]. This would make predicted values of $\omega_e$ too high, and those for $\omega_e x_e$ (anharmonicity) too low. Also, $R_0$, if it is calculated as the value of $R$ for which $U^{HF}(R)$ goes through a minimum, should be too small. The size of errors in $R_e$ has been discussed from a theoretical standpoint (see Volume I, Chapter III, Section D, and Subsection 2 of this section). Elimination of some of the error in $\omega_e$ and other quantities depending on derivatives of $U(R)$, by the use of the experimental, rather than the calculated, $R_e$ as the point at which derivatives are evaluated, was suggested by Schwendeman [7].

McLean [35] discussed the errors in the restricted Hartree–Fock potential curves for $H_2$ and LiF in some detail. He compared results of a Dunham analysis of the calculated $U(R)$ with experiment, and also derived a curve of the error in $U(R)$ as a function of $R$. It was expected that these cases would be typical of larger molecules. The main errors could be traced [35] to the incorrect dissociation of the restricted Hartree–Fock function. The results are summarized in Table VI. Note that $R_e$ is too small (as reflected in $B_e$) and $\omega_e$ is too large in both cases. For $H_2$, $\omega_e x_e$ is too small, but $\omega_e x_e$ for LiF is not in accord with our predictions (see the next paragraph, with reference to LiH). Here, $D_e$ is not the dissociation energy, but the coefficient of $[J(J + 1)]^2$ in Eq. (3), and equal to $4B_e^3/\omega_e^2$.

Cade [36] discussed the results of restricted Hartree–Fock (RHF) calculations for the diatomic hydrides of the first two rows and their ions. In general, $\omega_e$ and $B_e$ were high by about 5%, except for $B_e$ of LiH and $\omega_e$ of NaH, while $R_e$ was 1–2% low except for LiH and NaH. For first-row hydrides, $\alpha_e$ was about 10% low. Highly ionic LiH is an exception to the general rules, because the wave function resembles the structure $Li^+H^-$ rather than the structure LiH out to large $R$, so that the correct $U(R)$ looks like that for $Li^+H^-$ over a sizable region of $R$ ($R > R_e$). The avoided crossing point, $R_c$, between LiH and $Li^+H^-$ may be estimated from

$$e^2/R_c = I(\text{Li}) - A(\text{H}).$$

**Table VI**

*Restricted Hartree–Fock Results for Two Molecules*

|  | | $H_2$ | | LiF | |
| --- | --- | --- | --- | --- | --- |
|  | | RHF | Expt. | RHF | Expt. |
| $\omega_e$ | (cm$^{-1}$) | 4581.7 | 4400.39 | 1134.96 | 964.07 |
| $\omega_e x_e$ | (cm$^{-1}$) | 108.17 | 120.82 | 9.457 | 8.895 |
| $\omega_e y_e$ | (cm$^{-1}$) | 2.048 | 0.724 | 0.0518 | — |
| $B_e$ | (cm$^{-1}$) | 62.162 | 60.864 | 1.5805 | 1.5087 |
| $\alpha_e$ | (cm$^{-1}$) | 2.719 | 3.076 | 0.0182 | 0.02409 |
| $D_e$ | (cm$^{-1}$) | 0.046 | 0.046 | $1.225 \times 10^{-5}$ | $1.478 \times 10^{-5}$ |
| $\beta_e$ | (cm$^{-1}$) | $-0.0016$ | $-0.0016$ | $6.4 \times 10^{-8}$ | $-1.87 \times 10^{-7}$ |

The ionization potential of Li is 5.4 eV and the electron affinity $A$ of H is 0.8 eV, so $R_c \sim 1/(0.17 \text{ a.u.})$ or $6 a_0$. Cade [36] found that the RHF potential curves for AH and AH$^-$ were quite similar, except for vertical displacement, meaning that the added electron is in an orbital which is relatively unchanging with $R$. However, $E^{\text{corr}}$ differs for A and A$^-$, so there will be errors if Hartree–Fock energies are used in electron affinity calculations. The change in $E^{\text{corr}}$ was found [36] to be comparable in size to the electron affinity itself (electron affinities are typically much smaller than ionization potentials).

Green [37] has recently discussed the problems of incorrect dissociation of the restricted Hartree–Fock in some detail, particularly as applied to the ground state of CO. A large number of expectation values for different values of $R$ were considered to illustrate the effect of the large-$R$ behavior. The molecular orbital configuration of the $^1\Sigma^+$ ground state of CO for $R$ near $R_e$ is

$$(1\sigma)^2(2\sigma)^2(3\sigma)^2(4\sigma)^2(5\sigma)^2(1\pi^4).$$

Where $R$ becomes large, $1\sigma$ and $2\sigma$ correlate with the 1s orbitals of O and C, and $3\sigma$ and $4\sigma$ with the 2s orbitals of these atoms. The $5\sigma$ orbital becomes the C $2p_0$ orbital and the $1\pi$ the O $2p_+$ and $2p_-$ orbitals, so the wave function yields neutral atoms in singlet states. These states are not the correct atomic ground states, which are triplets; in fact, they are not eigenfunctions of orbital angular momentum. For C, the configuration is $1s^2 2s^2 p_0^2$, and the state [37] is $(2/3)^{1/2}\,^1D - (1/3)^{1/2}\,^1S$; for O, the configuration is $1s^2 2s^2 2p_+^2 2p_-^2$ and the state $(2/3)^{1/2}\,^1D - (1/3)^{1/2}\,^1S$. Green demonstrated that the orbital energies approached the $\frac{2}{3}$, $\frac{1}{3}$ linear combination of atomic values for large

$R$ ($R = 10$), while the total energy approached $\frac{2}{3}$ the sum of $^1D$ energies plus $\frac{1}{3}$ the sum of $^1S$ energies of C and O. This energy, $-112.29385$ a.u., is considerably above the sum of $^3P$ energies, $-112.49797$ a.u.

In order for the molecular wave function to dissociate properly to ground state atoms, two additional configurations must be included [37]. In one, $(1\pi)^4$ is replaced by $(1\pi)^2(2\pi)^2$; in the other, $(5\sigma)^2(1\pi)^4$ is replaced by $(5\sigma)(6\sigma)(1\pi)^3(2\pi)$. Both are doubly excited with respect to the ground state configuration, and so should not be very important in changing the energy for $R$ near $R_e$. Green [37] also discussed the configurations needed for proper dissociation in other molecules. For BeO, the $(1\pi)^4$ part of the ground configuration must be replaced by doubly excited $(5\sigma)^2(1\pi)^2$ ($5\sigma$ is unoccupied in the ground state). For the hydrides BH, CH, NH, and OH, whose restricted Hartree–Fock functions dissociate to ions, the additional configuration needed is singly excited [37].

The problem of incorrect dissociation is avoided in many cases when the unrestricted Hartree–Fock (UHF) theory is used (see Section A). However, the wave functions are not properly eigenfunctions of spin and orbital angular momentum. Estimating the errors for atoms and molecule, relative to the exact wave functions, also presents problems, since the solutions to the unrestricted Hartree–Fock equations have not been sufficiently studied. Thus, there are few diatomic molecule calculations employing the unrestricted Hartree–Fock method, with or without subsequent projection to a desired pure spin state.

Near the equilibrium internuclear distance, it is reasonable to expect that little improvement in the energy would be obtained. Green [37] argued that this would be the case on the basis of an analysis of the unrestricted Hartree–Fock as the restricted Hartree–Fock plus singly excited determinants. By Brillouin's theorem, the singly excited determinants would enter in the second order of perturbation theory and make little contribution to the energy, except in special cases. It does not seem that, for any diatomic molecule with a $^1\Sigma^+$ ground state, an unrestricted Hartree–Fock has given an improvement for intermediate $R$-values over the restricted Hartree–Fock [38]. For large $R$, where there are more degeneracies and near-degeneracies, things are different, and unrestricted Hartree–Fock functions can lead to proper dissociation products where the restricted Hartree–Fock does not (see p. 143).

Some of the problems in using unrestricted Hartree–Fock have been discussed in Section A.1. The wave function is a single determinant which is not an eigenfunction of total spin, and must be "purified" of unwanted components. Because of the nonlinear character of the Hartree–

Fock equations, the lowest energy UHF function may undergo non-continuous changes with $R$, as we go from one solution to the UHF equation to another whose $U(R)$ curve crosses the first. Then $U(R)$, calculated using the lowest energy for each $R$, may have discontinuities in slope. These difficulties are exemplified in one of the rare UHF calculations for molecules, by Salotto and Burnelle [38]. Although primarily interested in small polyatomic molecules, these authors reported UHF calculations for $U(R)$ of LiH and $H_2$. Using a Gaussian basis set, they computed energy and other properties for a series of $R$-values in the RHF and UHF approximations.

The plots of $\langle S^2 \rangle_{\mathrm{UHF}}$ as a function of $R$ show an interesting shape for both molecules. $\langle S^2 \rangle_{\mathrm{UHF}}$ was identically 0 out to a critical value of $R$ ($\sim 2.3\ a_0$ for $H_2$, $\sim 4.4\ a_0$ for LiH), then rose with a large slope and smoothly leveled off to a value near 1 (at $\sim 4.0\ a_0$ for $H_2$, $\sim 6.5\ a_0$ for LiH). Near $R_e$, the UHF was identical to the RHF in both cases; there was no energy lowering. At the point where the spin became nonzero, the energies of RHF and UHF diverged, with the latter lying lower. The actual energy of interest is the energy of the UHF function after spin projection. For $H_2$, the only unwanted component is the triplet; for LiH, Salotto and Burnelle [38] considered the effect of a single annihilation to remove triplet contamination, and also complete annihilation of all unwanted components. The projected $H_2$ energy followed the experimental potential curve more closely than did the energy of RHF. For LiH, the effect of projection on the energy was small at large $R$, but near $4\frac{1}{2}\ a_0$ the projected UHF had an energy several hundredths of an atomic unit below that of the unprojected function.

Some LiH calculations using "different orbitals for different spins," which is equivalent to UHF, were reported by Lindner [39], but only for $R = 3.015\ a_0$. The molecular orbitals were constructed from the minimal set with exponential parameters optimized for the atoms, and the effects on the energy of (a) opening the inner shell, keeping the valence orbital doubly occupied, and (b) opening the valence shell, keeping the inner shell closed, were investigated. The effect of (b), which would allow correct dissociation of the molecule at large $R$, was less than 0.015 a.u.

In connection with semiempirical calculations on the bonding electrons of the hydrogen halides, Harris and Pohl [40] employed the generalized valence bond wave function

$$\Psi_0 = \tfrac{1}{2}\{[\psi_1(1)\psi_2(2) + \psi_2(1)\psi_1(2)][\alpha(1)\beta(2) - \beta(1)\alpha(2)]\}.$$

Energies were compared with those from

$$\Psi_c = 2^{-1/2}\psi_3(1)\psi_3(2)[\alpha(1)\beta(2) - \beta(1)\alpha(2)].$$

The molecular orbitals $\psi_1$, $\psi_2$, and $\psi_3$ were written as a linear combination of atomic orbitals, one on each atom, with coefficients chosen to minimize the energy. All integrals over atomic orbitals were evaluated according to the same recipe. It is interesting that, in some cases, Harris and Pohl [40] found that the optimum coefficients in $\psi_1$ and $\psi_2$ were *complex*, making $\psi_1$ and $\psi_2$ complex conjugates of each other.

Another problem with UHF, involving convergence of the iterative methods employed, was recently discussed by Claxton and Smith [41]. Neither of the two best-known algorithms, Roothaan's iterative pseudo-eigenvalue method or McWeeny's steepest descent energy minimization, is wholly satisfactory. Both require a good starting guess for convergence to be obtained. The convergence problem also arises for open-shell calculations. As a result of experiences with calculations on $OH^+$, $OH$, and $OH^-$, Sleeman [42] suggested that the related closed-shell system should be done first to obtain a starting approximation to the orbitals for the open-shell calculation. Only then can either Roothaan's or McWeeny's method work successfully. A discussion was given of the possible methods available and the results of other workers on the convergence problem discussed. The various methods were compared in numerical detail by Claxton [43].

In addition to convergence problems encountered in finding the solution, the method by which the open shell calculation is set up may be of importance, different methods involving different approximations and yielding different solutions. Because an RHF treatment cannot satisfy the UHF equations (see Section A), some of the advantages of Brillouin's theorem may be lost. As a result, RHF calculations on open-shell ground states may give different kinds of errors in $U(R)$ from the closed-shell states we have been discussing. However, Walker and Richards [44] have concluded from several calculations that the singly excited states, which could enter in first order of perturbation theory and thus contribute to relatively large errors in calculated properties, have small Hamiltonian matrix elements with the ground state. For HeH, the errors in the energy due to such contributions seemed negligible (0.1 eV).

Huo [45] noted, similarly, that errors in many properties from restricted Hartree–Fock functions for open-shell atomic systems were not much worse than for closed-shell systems, despite the fact that Brillouin's theorem is satisfied by RHF functions in the latter and not the former case. According to the discussion at the end of Volume I, Chapter III, Section A.5 [between Eqs. (73) and (74)], the first-order error for the operator $Q$ depends on matrix elements like $\langle \Psi^{RHF} \mid Q \mid \Psi^{i \to p} \rangle$, with $\Psi^{i \to p}$ a determinant differing from $\Psi^{RHF}$ by the substitution of spin orbital $p$ for spin orbital $i$. It is clear, for molecules as well as for atoms, that the error will be more serious for some operators than others.

Huo gave a detailed analysis for the force operator. Using the fact that the spin orbitals obey the Hartree–Fock equations, she showed that

$$-\nabla_A E^{RHF} = \langle \Psi^{RHF} | -\nabla_A H | \Psi^{RHF} \rangle - \sum_{k,m} 2d_{km} \langle \Psi^{RHF} | H | \Psi^{k\to m} \rangle \quad (77)$$

where the coefficient $d_{km}$ is defined by

$$\nabla_A \phi_k = \sum_{m \neq k} d_{km} \phi_m .$$

The presence of the sum in (77) means that the Hellmann–Feynman theorem is not satisfied (as for restricted Hartree–Fock functions in open shell situations). The validity of the Hellmann–Feynman theorem is closely linked to that of Brillouin's theorem [Volume I, Chapter III, discussion preceding Eq. (353)]. Huo pointed out that, if the RHF and UHF energies are not very different, $\langle \Psi^{RHF} | H | \Psi^{k\to m} \rangle$ is likely to be small, so that the Hellmann–Feynman theorem is close to being obeyed. Furthermore, Huo [45] argued that, when a limited basis expansion is used, the expansion error may be comparable to the error in using restricted Hartree–Fock, and $\Psi^{RHF}$ and $\Psi^{UHF}$ may satisfy the Hellmann–Feynman theorem equally well.

A self-consistent field open-shell calculation on ScO by Carlson et al. [46] had, as principal goal, the determination of the symmetry of the ground state. The basis set consisted of the minimal basis orbitals for the separated atoms, plus basis functions required to represent the distortion and polarization due to binding. Computations were performed to optimize the molecular orbitals for the states $(8\sigma)^2(3\pi)^4(1\delta)\ ^2\Delta$ and $(8\sigma)^2(3\pi)^4(9\sigma)\ ^2\Sigma^+$. Then the energies of various states formed from each of the resulting sets of molecular orbitals were compared. In both cases, $^2\Sigma^+$ lay lowest. Since the molecular orbital configuration for the $^2\Sigma^+$ state dissociates incorrectly, the computed $\omega_e$ (1373 cm$^{-1}$) should be too high and the computed $R_e$ (1.615 $a_0$) too low. The experimental values, $\omega_e = 972$ cm$^{-1}$ and $R_e = 1.668$ $a_0$, were, for the ground state, of unknown symmetry. The dissociation energy was calculated as 3.0 eV, and it was expected [46] that MECE for this molecule would be somewhat larger than 3.5 eV, found [20] for TiO.

The open shell configuration $(1\sigma)^2(2\sigma)^2(3\sigma)^2(4\sigma)^2(5\sigma)^2(1\pi)^4(2\pi)$ may be represented by a single determinant and represents the NO ground state. In the treatment by Brion et al. [47] the off-diagonal Lagrangian multipliers coupling the $\pi$ orbitals of the closed and open shells were treated by a special method. It involved correcting the $\pi$ orbitals after each iteration on the $\sigma$ orbitals, which were determined by SCF equations of the usual sort. The computed binding energy was negative unless some configuration interaction was included. Other open-shell calculations are discussed in the next Subsections.

### 3. *Results of Hartree–Fock Calculations*

In this Subsection, we review the restricted Hartree–Fock calculations of $U(R)$ for the ground states of diatomic systems. The few unrestricted Hartree–Fock calculations known to us were mentioned in the previous section. Subsection 4 reviews results of configuration interaction and other calculations more elaborate than Hartree–Fock. We hope to touch on most recent work (except for one- and two-electron systems), but it seems impossible to make the list of calculations mentioned complete. Chapter IV gives more extensive bibliographies. The work discussed here and in Subsection 4 should illustrate the discussions of previous sections and give an idea of the successes of modern techniques in obtaining accurate *ab initio* results for a wide range of interesting systems.

This subsection is divided into third-level sections, by the kinds of systems. In (*a*) we consider pairs of rare gas atoms, in (*b*) hydrides, and in (*c*) homonuclear molecules other than those of (*a*). Other neutral species formed from the first 18 elements are discussed in (*d*), and ionic species in (*e*). Level (*f*) considers alkali halides and systems larger than those of levels (*a*)–(*e*).

The ground states of the homonuclear molecules of the first two rows of the periodic table were discussed from the viewpoint of molecular orbital theory in some early work of Lennard-Jones [48], without quantitative calculations. Lennard-Jones derived the general characteristics of the molecular orbitals (including their dependence on $R$ and $Z$) by considering the one-electron two-center problem for nuclei of charge $Z$. The ground state of each molecule was obtained by placing electrons into the molecular orbitals in order of increasing energy. The process of filling was assumed to occur in steps, in which two electrons were added and each nuclear charge increased by unity. The symmetries and bonding properties of the many-electron molecular states were derived from the symmetries and bonding properties of the individual orbitals. The behavior of the molecules on dissociation was discussed [48] assuming that one electron from a doubly occuped molecular orbital went to each atom. Where several pairs of electrons were involved in bonds, all the electron spins going to each atom, outside of closed shells, were supposed to remain parallel. The extension and elaboration of these ideas, particularly by Mulliken [49], led to a full-fledged semiempirical molecular orbital theory of diatomic molecules, which included discussion of excitation and ionization energies as well as of the shapes of $U(R)$ curves.

After Roothaan [50] had given a systematic treatment of the molecular orbital (MO) theory in the linear combination of atomic orbital (LCAO) approximation, the developing and improving computers were used for *a priori* MO calculations on a wide variety of molecular systems. We mention several of these earlier calculations, but emphasize those of the last decade. Allen and Karo [51] summarized the molecular orbital calculations on diatomic molecules performed up to 1960. More accurate calculations appearing since then, within and without the single determinant framework, have improved the possibilities for accurate prediction of molecular properties.

### (a) Rare Gas Atoms

For two rare gas atoms, $U(R)$ is a monotonically decreasing function, except for the van der Waals attraction, which it seems cannot be described by a Hartree–Fock function. The simplest (LCAO–MO) function is equivalent to the simplest valence bond function for the homonuclear pairs, so that the two methods produce the same $U(R)$.

An early calculation on the $He_2$ system by Griffing and Wehner [52] obtained good agreement with previous calculations by other methods, and at the same time, demonstrated the relative simplicity of the molecular orbital approach. The MO wave function for $He_2$ is a single determinant, formed by double occupation of two $\sigma$ orbitals. In the minimal basis set calculation, the orbitals were symmetric and antisymmetric combinations of Slater orbitals with an exponential parameter of 27/16, which is appropriate for free He atoms. Variation of the exponential parameter gave little improvement in the energy [52]. The interaction potential $U(R)$ was computed and compared to experiment.

For $He_2$, we now have single-configuration calculations, done by expansion techniques, which approach the true restricted Hartree–Fock energy to a small fraction of an electron volt. Somewhat less accurate, but still very satisfactory, calculations have been performed for other homonuclear pairs of rare gas atoms. For these systems, there is no problem of incorrect dissociation; the single-configuration wave function should be a good approximation at large $R$. It is for small $R$ that difficulties arise because the lowest molecular orbital configuration does not correlate with the lowest atomic orbital configuration of the united atom (see Chapter I, Section C.3).

Gilbert and Wahl [53] carried out single-configuration calculations for $He_2$, $Ne_2$, and $Ar_2$ using different basis sets, in order to discuss ways of estimating the "expansion error," due to the incompleteness of the basis set. They suggested that by studying, in a systematic and prescribed way, the changes in a calculated property as the basis set was improved, such an estimate could be obtained. For $He_2$, the expansion error was estimated to be 0.001 eV. It rose to 0.098 eV for $Ne_2$ and 0.293 eV for $Ar_2$, but it was expected [53] that much of the error for the latter two molecules would not affect $U(R)$, since it involved the inner-shell electrons. In any case, Gilbert and Wahl estimated the error due to use of a single configuration ("correlation error") to be large compared to the expansion error in all cases. Since no bond and no new electron pairs are formed when two rare gas atoms approach each other, the correlation energy for $He_2$, $Ne_2$, etc. should be roughly independent of $R$ (MECE $\sim 0$). Indeed, the correlation energy for Be, the united atom of He + He, is 2.56 eV, while $E^{corr} = 2.29$ eV for He + He. For large $R$, there are no low-lying electronic states which could cause changes in $E^{corr}$ due to near-degeneracy. The situation for Ne + Ne is at first sight less satisfactory, since $E^{corr} = 21.4$ eV, while $E^{corr}$ for the united atom, Ca, is estimated at about 25.8 eV. However, it was argued [53] that most of the change in $E^{corr}$ occurs at very small $R$, where the inner shells start to overlap.

The maximum error in $U(R)$ for $He_2$, compared to the true Hartree–Fock, was given [53] as 0.01 eV for $R \geq 2.5\,a_0$. Somewhat larger errors were assigned to $U(R)$ for $Ne_2$ and $Ar_2$. Comparisons with other calculations and with experiment were given. Kestner [54, 54a] subsequently extended the SCF calculations for larger $R$ with a bigger basis set. He was particularly interested in determining whether an energy less than that of the separated atoms could be found, which would mean van der Waals attraction appearing in a single configuration wave function. It was not.

Margenau and Kestner [54a] review the approximate Hartree–Fock calculations for $U(R)$ of $He_2$. By giving exponential fits to different approximate results, they point up the differences due to the different basis sets used. They also analyze the difference between Hartree–Fock molecular and atomic energies into atomic orbital contributions.

For the diatom formed from two rare gas molecules, the wave function can be written down immediately if we use a minimal basis. Similarly, if it is assumed that distortions of the atomic orbitals as $R$ decreases from infinity are unimportant, the wave function may be given as a single determinant formed by doubly occupying all the atomic orbitals of the two atoms. In this case, the atomic orbitals used might themselves be linear combinations of orbitals of some basis set. The difference between the energy of such a wave function and the energy of the SCF function formed from the same orbitals as are used to construct the atomic orbitals was referred to by Gilbert and Wahl [53] as the "distortion error." It turned out to be generally small. This suggested to Butler and Kestner [55] the possibility of obtaining accurate potential curves for these systems without performing SCF calculations. The wave function built from the atomic orbitals would be corrected by a limited configuration interaction in a basis of localized orbitals. The configurations added would include configurations having equal numbers of orbitals centered on the two atoms ("covalent") and configurations having more orbitals on one atom than the other ("ionic" or "charge transfer"). From analysis of the Hartree–Fock function, it was argued that the latter would be most important in improving the energy for He–He and Ne–Ne at large $R$. For $R \leq 2.5$ in the former case, the distortions of the atoms became so large that the analysis was no longer useful.

It seems that, for most of the range of $R$, accurate $U(R)$ can be obtained by superposing a calculated van der Waals interaction on $U(R)$ from Hartree–Fock calculations. However, at small $R$, interaction of configurations becomes important.

### (b) Hydride Molecules

We will not discuss calculations on $H_2$, because of their great number and because one can do many things for this molecule which cannot be extended to larger ones. Then the simplest neutral hydride is HeH. The best calculation for the neutral HeH, especially at small $R$, is that by Fischer and Kemmey [56]. This was an open-shell SCF calculation employing a basis of 26 s and p Gaussian orbitals. Since the same basis set was used over a wide range of $R$, it was expected [56] that the error would be less than the difference of the He and Li correlation energies, 0.0032 a.u., i.e., the error due to the incomplete basis set should cancel. A single-configuration wave function sufficed to accurately describe the system. Comparisons to other theoretical and experimental $U(R)$ curves were given. Other recent calculations of the HeH potential curve are those of Das and Ray, and Miller and Schaefer [56a].

A great deal of work has been done toward obtaining the restricted Hartree–Fock energies for LiH. Lithium hydride has attracted a large number of calculations because it has only four electrons, includes an inner shell and a bond, and possesses a large number of properties whose values are accurately known.

An early calculation on this molecule is that by Fischer [57]. This was a minimal basis set calculation for several values of $R$, and led to $R_e = 2.91\ a_0$

(experiment, 3.02 $a_0$), $\omega_e = 1.8 \times 10^3$ (1433), and a dissociation energy of 0.0722 a.u. (0.0967 a.u.) from the lowest vibrational state. The ion BeH$^+$, which is isoelectronic to LiH, was also treated. We have already cited (Subsection 1) the results for spectroscopic constants of LiH obtained from minimal basis set calculations by Sahni *et al.* [10]. The good agreement with experiment they obtained ($\omega_e = 1440.78$ cm$^{-1}$, $R_e = 3.037\,a_0$, $B_e = 7.39$ cm$^{-1}$ versus 7.51 from experiment, etc.) shows that the "expansion error" or error due to the limited basis is quite $R$-independent.

An earlier matrix Hartree–Fock calculation, by Kahalas and Nesbet [58], used 13 carefully selected basis functions with optimized exponents. Configuration interaction, carried out by perturbation theory formulas, was required to obtain a reasonable dissociation energy. The best Hartree–Fock calculations are those of Cade and Huo [32], discussed on pp. 238 *et seq.*, along with those of the other hydrides.

One-center MO calculations have not been very successful for this system, except for those of Keefer *et al.* [59]. They expanded each molecular orbital in spherical harmonics with Li at the center, and similarly expanded $1/r_{12}$ and other operators in the Hamiltonian. As a result, all integrals over angles could be performed explicitly to derive coupled differential equations in one dimension for the MO expansion coefficients, which were radial functions. After truncation of the expansions for the molecular orbitals, the coupled equations were solved numerically by methods like those used by Hartree for atomic SCF calculations. The resulting energy for LiH was 1.4 eV higher than that of Cade and Huo, but 0.3 a.u. better than that of previous one-center expansions. The predicted value of $B_e$ was 3.02 $a_0$, fortuitously closer to the experimental value than the Hartree–Fock result, and considerably better than the previous one-center result (3.70 $a_0$). Computations of this kind were also done for HeH$^+$, where an energy only 0.015 a.u. higher than Peyerimhoff's (see page 251) was obtained, and for H$_2$.

Another hydride for which a large number of calculations has been carried out is HF. The hydrides LiH, BH, and HF are the closed-shell cases of the first row, and much less is known about BH than about the other two. Restricted Hartree–Fock calculations for HF successfully predict $R_e$ and $\omega_e$. Thus, a calculation by Harrison [60], using 18 Gaussian functions in the basis, gave $R_e$ correct to 0.01 $a_0$. It was claimed that the energy for $R = R_e$ was only about 0.03 a.u. from the true Hartree–Fock energy.

Clementi's approximate restricted Hartree–Fock calculations [33] for HF and HF$^+$ used a "double-zeta" Slater basis set, i.e., two atomic orbitals of each kind, with different exponential parameters. The optimization was performed by finding the best minimal basis set and then replacing each Slater orbital by two orbitals with exponential parameters split symmetrically from that for the minimal set. The same basis functions, 11 of $\sigma$ symmetry and 5 of $\pi$ symmetry, were used for the ground state of HF$^+$ ($^2\Sigma$) as for HF ($^1\Sigma$). For HF, $R_e$ was predicted as 1.74 $a_0$ (experiment, 1.733), and $D_e$ (difference of Hartree–Fock energies for atoms and molecule) as 0.149 a.u. Since the experimental value is 0.2235 a.u., MECE is 0.074 a.u. = 2.026 eV, corresponding to the formation of one additional electron pair.

Nesbet's calculation [33a] of about the same date yielded a better value of $D_e$ (4.438 eV = 0.1632 a.u.) but a poorer value of $R_e$ (1.6865 $a_0$). A good value for $\omega_e$, 4055 cm$^{-1}$ (experimental value, 4138.5), was also obtained. The importance of inclusion of basis orbitals to describe distortion and polarization of the atomic electron clouds was emphasized, particularly for accurate calculation of $D_e$. It will be noted, on referring to Table VII, that neither Nesbet's nor Clementi's calculation really approached the true Hartree–Fock. Nesbet estimated MECE as the difference of correlation energies of F$^-$ and F, 0.082 a.u. = 2.2 eV. The corrected dissociation energy of 6.6 eV was 0.5 eV higher than the experimental value and would be further increased by improvements in the basis set.

Moccia [61] carried out one-center single-configuration calculations on HF and HI, all basis orbitals being centered on the heavy atom in each case. He found that addition of basis functions with higher angular momentum was more important in improving results than optimization of nonlinear parameters. The predicted $R_e$ for HF of 0.915 Å (1.729 $a_0$) compares favorably with the results just cited. Similarly, $R_e$ for HCl was calculated [61] as 1.272 Å and the experimental value is 1.274 Å. Again, Table VII (below) shows that the excellent agreement between calculated and experimental values of $R_e$ is partly fortuitous; it would be decreased by improvements allowing a closer approach to the true RHF function.

A restricted Hartree–Fock calculation using 17 Slater basis functions for HCl by Nesbet [62] predicted $R_e = 1.333$ Å and $\omega_e = 2498$ cm$^{-1}$ (experimental value, 2989.74 cm$^{-1}$) from computations at three values of $R$ near $R_e$. Since $\omega_e$ was too low and $R_e$ too high, opposite to what is expected from the known errors of the RHF function, Nesbet suspected (correctly: see Table VII) that the basis set was insufficient to give results close to the true RHF and that further optimization of parameters was necessary. The calculated $D_e$ of 2.78 eV relative to SCF atoms was corrected for MECE to give 3.76 eV (experimental, 4.615 eV); a better basis set would raise the value appreciably. For estimation of MECE, Nesbet assumed $E^{corr}$ for HCl approximated $E^{corr}$ for Cl$^-$, and then since $E^{corr}$ (Cl$^-$) was unavailable, approximated $E^{corr}(Cl) - E^{corr}(Cl^-)$ by $E^{corr}(Ar^+) - E^{corr}(Ar)$. A more recent calculation [63] of the same type, using a smaller basis set, was aimed at obtaining the proper exponent for the Cl 3d orbital for use in semiempirical calculations.

The study of the ground states of the first- and second-row hydrides by Cade and Huo [32] is extremely important for several reasons. (Earlier systematic work on the hydrides by Cade [36] has already been mentioned.) A large number of accurate calculations were presented. The ground states of the first- and second-row hydride molecules were computed in the Hartree–Fock approximation: LiH ($^1\Sigma^+$), BeH ($^2\Sigma^+$), BH ($^1\Sigma^+$), CH ($^2\Pi$), NH ($^3\Sigma^-$), OH ($^2\Pi$), HF ($^1\Sigma^+$), NaH ($^1\Sigma^+$), MgH ($^2\Sigma^+$), AlH ($^1\Sigma^+$), SiH ($^2\Pi$), PH ($^3\Sigma^-$), SH ($^2\Pi$), HCl ($^1\Sigma^+$). A comprehensive bibliography of the many calculations on the first-row hydrides was given (only a few had been done for the second row). The accuracy (closeness of approach to the true Hartree–Fock) of the new calculations, which employed expansions in Slater orbitals, was assessed from several points of view. Detailed discussion of the choice of basis functions and the effects of adding functions to the basis, was given. The errors in the spectroscopic constants derived from the calculated energies were carefully discussed. Cade and Huo also considered the difference between the Hartree–Fock and exact energies as a function of $R$ and of molecule. They attempted to separate the relativistic and correlation contributions to this energy difference, and to elucidate the effects of incorrect dissociation. Finally, the possibility of correcting properties calculated from Hartree–Fock calculations, to give accurate predictions of spectroscopic and other properties, was considered. Here, we cannot touch on all the points raised in their discussions.

With respect to accuracy, Cade and Huo [32] concluded that the accuracy to which their calculation for the hydride AH approached the Hartree–Fock was comparable to that available for the corresponding atom A. All evidence indicated that the quality of the second-row wave functions was comparable to that of the first row in most respects. The Hellmann–Feynman forces as $R_e$, which should vanish, were calculated to be about 0.02 a.u./$a_0$ for first-row hydrides, but this quantity is extremely sensitive to small errors in the wave function (it was estimated [32] that a force of this magnitude could correspond to an energy error of 0.0001 $a_0$). The virial theorem ratio, $\langle V \rangle / \langle T \rangle$, was checked to be within 0.00005 of the theoretical value of $-2$ at the calculated equilibrium internuclear distance.

The orbital exponents were optimized for a single value of $R$, near $R_e$, but the effect of reoptimization at other internuclear distances was checked for a few cases: It amounted to less than 0.0005 a.u. in each one. For

**Table VII**

*Spectroscopic Constants[a] for First and Second Row Hydrides[b]*

| Molecule | $B_e$ (cm$^{-1}$) | $\alpha_e$ (cm$^{-1}$) | $\omega_e$ (cm$^{-1}$) | $\omega_e x_e$ (cm$^{-1}$) | $R_e$ ($a_0$) | $k$ ($10^5$ dynes/cm) |
|---|---|---|---|---|---|---|
| LiH | 7.426 | 0.1945 | 1433 | 23.26 | 3.034 | 1.066 |
|  | 7.513 | 0.213 | 1405.6 | 23.20 | 3.015 | 1.026 |
| BeH | 10.392 | 0.2647 | 2147 | 34.60 | 2.528 | 2.461 |
|  | 10.308 | 0.300 | 2058.6 | 35.5 | 2.538 | 2.263 |
| BH | 12.273 | 0.3726 | 2499 | 49.04 | 2.305 | 3.397 |
|  | 12.016 | 0.408 | 2367.5 | (49) | 2.336 | 3.044 |
| CH | 14.882 | 0.4712 | 3053 | 55.50 | 2.086 | 5.106 |
|  | 14.448 | 0.530 | 2868.5 | 64.4 | 2.124 | 4.506 |
| NH | 17.319 | 0.5715 | 3556 | 66.78 | 1.923 | 7.003 |
|  | 16.668 | 0.646 | (3125.6) | — | 1.9614 | (5.410) |
| OH | 19.712 | 0.6501 | 4062 | 74.54 | 1.795 | 9.216 |
|  | 18.871 | 0.714 | 3735.2 | 82.81 | 1.8342 | 7.791 |
| FH | 21.868 | 0.7693 | 4469 | 80.34 | 1.696 | 11.261 |
|  | 20.949 | 0.797 | 4139.0 | 90.44 | 1.7328 | 9.657 |
| NaH | 4.766 | 0.1151 | 1187 | 14.71 | 3.617 | 0.8020 |
|  | 4.901 | 0.135 | 1172.2 | 19.72 | 3.566 | 0.781 |
| MgH | 5.859 | 0.1315 | 1598 | 21.79 | 3.259 | 1.454 |
|  | 5.818 | 0.167 | 1495.7 | 31.5 | 3.271 | 1.274 |
| AlH | 6.394 | 0.1719 | 1741 | 26.86 | 3.113 | 1.735 |
|  | 6.391 | 0.186 | 1682.6 | 29.09 | 3.114 | 1.620 |
| SiH | 7.558 | 0.1894 | 2144 | 30.91 | 2.861 | 2.635 |
|  | 7.498 | 0.214 | 2042.5 | 35.67 | 2.874 | 2.390 |
| PH | 8.664 | 0.2145 | 2515 | 36.37 | 2.668 | 3.636 |
|  | (8.412) | — | (2380) | — | 2.708 | (3.257) |
| SH | 9.756 | 0.2494 | 2860 | 42.32 | 2.513 | 4.707 |
|  | (9.461) | — | (2702) | — | 2.551 | (4.201) |
| ClH | 10.767 | 0.2682 | 3181 | 47.76 | 2.389 | 5.838 |
|  | 10.591 | 0.302 | 2989.7 | 52.05 | 2.4087 | 5.157 |

[a] Upper value is calculated; lower value is experimental.
[b] See Cade and Huo [32].

$R = 0$ and $R = \infty$, atomic Hartree–Fock calculations were used, and molecular calculations were carried out for about 15 other values of $R$. Using nine or ten of these, Dunham analysis was performed to obtain the spectroscopic constants. Table VII compares the results to experiment.

It was noted [32] that the percentage errors in the spectroscopic constants are generally smaller than what was obtained by previous calculations of this kind. In particular, $\omega_e x_e$ is now in fair agreement for all molecules. Except for $\omega_e x_e$, agreement with experiment is about as good for the second-row molecules as for those of the first row. It seems that the incorrect dissociation of the restricted Hartree–Fock wave function is less harmful to the calculated spectroscopic constants for the hydrides than for other molecules. Furthermore, the errors in a given property for different molecules follow almost uniform trends, making it possible to derive corrections to particular results for the hydrides and for the ions $AH^+$ and $AH^-$. The dissociation energies, shown in Table VIII, exhibit the usual errors due to MECE.

The error in $U(R)$ is due to variation of the correlation and relativistic energies with $R$. In order to discuss the former, an estimate of the latter

**Table VIII**

*Dissociation Energies for Hydrides$^{a,b}$*

| Molecule | $D_e$ (calc) | $D_e{}'$ | $D_e$ (expt) |
|----------|--------------|----------|--------------|
| LiH | 1.49 | 2.83 | 2.52 |
| BeH | 2.18 | 3.01 | (2.6) |
| BH | 2.78 | 3.95 | 3.58 |
| CH | 2.47 | 3.81 | 3.65 |
| NH | 2.10 | 4.01 | (3.80) |
| OH | 3.03 | 4.79 | 4.63 |
| FH | 4.38 | 6.23 | 6.12 |
| NaH | 0.932 | 2.19 | (2.3) |
| MgH | 1.15 | 2.10 | (2.3) |
| AlH | 2.36 | 3.71 | $3.01 \pm 0.05$ |
| SiH | 2.23 | 3.61 | $3.32 \pm 0.25$ |
| PH | 2.03 | 4.19 | (3.34) |
| SH | 2.63 | 4.75 | $3.70 \pm 0.12$ |
| ClH | 3.48 | 5.40 | 4.616 |

$^a$ All values are in electron volts.
$^b$ From Cade and Huo [32].

is necessary. Cade and Huo [32] first plotted, as a function of the atomic number of the heavy atom A, the difference ($\Delta E$) between Hartree–Fock and experimental energies for the united atom ($R = 0$), the molecule at $R_e$, and the separated atoms ($R = \infty$). The plots were fairly parallel, with $\Delta E$ ($R = \infty$) lying about 1 eV above $\Delta E$ ($R = R_e$), which lay above $\Delta E$ ($R = 0$). Two approximations were considered to obtain the relativistic energy for the molecule at $R = R_e$ from atomic data: either using (I) the relativistic energy of the separated atoms or (II) the mean of the relativistic energies of separated and united atoms. Method I is justified if $E^{rel}$ arises mostly from the 1s electrons, since it should then be constant from the separated atom limit down to values of $R$ comparable to the radius of the 1s orbital, which is less than $R_e$. For smaller values of $R$, $E^{rel}$ should drop rapidly to the united atom value. For the first row atoms, there is only about 0.01 a.u. difference between the estimates I and II. When either correction for $E^{rel}$ is made to $\Delta E$ ($R = R_e$), the result (which represents the correlation energy of the molecule) was quite close to $E^{corr}$ ($R = 0$), and much less close to $E^{corr}$ ($R = \infty$). For second-row hydrides, the larger size of $E^{rel}$ relative to $E^{corr}$ made differences between estimates I and II apparent. When the latter was used, unreasonable correlation energies were obtained [32].

To obtain the exact energy as a function of $R$ for comparison with calculated $U(R)$, Cade and Huo [32] used the "experimental" RKR curve; a Morse curve parametrized with experimental values of $D_e$, $R_e$, and $k_e$; and a Hulburt–Hirschfelder curve parametrized with experimental $D_e$, $R_e$, $\omega_e$, $\alpha_e$, and $\omega_e x_e$. For molecules for which the RKR curve was available, the Hulburt–Hirschfelder curve lay very close to it. No RKR curves were available for the second row hydrides. Cade and Huo plotted $\Delta(R)$ $= E(R)_{expt} - E(R)_{calc}$ versus $R$, and also $\Delta(R) - \Delta(\infty)$ and $\Delta(R) - \Delta(0)$ versus $R$. In order to derive $E^{corr}(R)$ from $\Delta(R)$, the dependence of the relativistic energy with $R$ was required, but it is not known. The change in $E^{rel}$ between $R = \infty$ and $R = 0$ was shown to be large enough and of the right sign to account for all of $\Delta(R)$ in all cases. Thus, it was impossible to derive $E^{corr}(R)$ to any degree of accuracy, although it was felt [32] that it was probably fairly flat, possibly going through a maximum. $\Delta(R)$ approached $\Delta(0)$ for $R \to 0$ except in the case of NH. For large $R$, Cade and Huo felt that $\Delta(R)$ was no longer meaningful, since it reflected the incorrect dissociation due to the form of the wave function and no physical effect.

Turning to dissociation energies, Cade and Huo [32] found the use of $E^{corr}$ for the united atom in calculating MECE worked well. They pointed out, however, that the resulting formula for corrected dissociation energies,

$$-D_e' = E^{\mathrm{HF}}(R_e) - E^{\mathrm{corr}}(R = 0) - E^{\mathrm{nonrel}}(R = \infty)$$

originally proposed by Stanton [23], actually assumes that $E^{\mathrm{rel}}$ is unchanged between molecule and separated atoms. This assumption was supported by the considerations relative to approximation I and II. The calculated and corrected dissociation energies are tabulated in Table VIII.

For the first-row hydrides, the errors in $D_e'$ are several tenths of an electron volt; those in $D_e$ (with no correlation energy correction) were as high as 1.7 eV. However, for the second-row hydrides, agreement of $D_e'$ with experiment is less good. Except for NaH and MgH, where the experimental values are questionable, $D_e'$ is always much too high. This was consistent [32] with Cade and Huo's deduction that $E^{\mathrm{corr}}$ was less at $R_e$ than at $R = 0$. Except for AlH in the second row and BeH in the first, $D_e' - D_e(\mathrm{expt})$ varied slowly in going across the period; this was used [32] to choose between rival experimental values, i.e., 2.2 eV or 3.34 eV for PH.

### (c) Homonuclear Diatomics

Some results of restricted Hartree–Fock calculations for homonuclear diatomics of the first two rows have been given in Table I of Section A.6. There, $\omega_e$ and $D_e$ were given for $\mathrm{Li}_2$, $\mathrm{N}_2$, and $\mathrm{F}_2$ and compared with experiment. The formally triple-bonded $\mathrm{N}_2$ has attracted most interest, and we will discuss calculations on $\mathrm{N}_2$ in more detail.

The early calculation of Scherr [64] used a minimum basis with exponents chosen by Slater's rules. The difference between the energy calculated for the molecule at $R_e$, $-108.574$ a.u., and twice the energy calculated for a nitrogen atom with the same basis, $-108.530$ a.u., was only about a tenth of the experimental dissociation energy. While, as subsequent work showed, the single-configuration energy of $\mathrm{N}_2$ can be improved by over 0.4 a.u. by using a bigger basis set, a single-determinant calculation cannot give anything like the correct dissociation energy without a correction for MECE.

Richardson [65] doubled the number of 2s and 2p basis functions, but took the exponential parameters for 2s and 2p orbitals as identical, leaving three nonlinear parameters. They were chosen on the basis of results of atomic calculations. Richardson found that the improvement in the basis was more important for the molecule than for the atoms, since the dissociation energy of 3.48 eV was considerably better than that of previous calculations (Scherr found 1.20 eV). A scheme for estimation of MECE was set up [65] on the basis of the correlation energies for $N^{5+}$ (1.79 eV), $N^{4+}$ (1.90 eV), $N^{3+}$ (4.23 eV), and $N^0$ (5.48 eV). $N^{5+}$ is $(1s)^2$, so that the 1s–1s correlation energy was estimated as 1.79 eV. The difference, 1.90 eV — 1.79 eV, was ascribed to the correlation of a 1s–2s pair with opposite spins, so that the 2s–2s pair correlation energy was estimated as 4.23 eV — 1.79 eV — 2(0.11 eV) = 2.22 eV. The additional correlation energy of $N^0$ (1.25 eV) represents three 1s–2p pairs plus three 2s–2p pairs (only electrons of opposite spins are considered); assigning 0.11 eV to the former means the latter are 0.31 eV. For $N_2$, Richardson [65] assigned correlation energies between molecular orbitals as follows: 1.79 eV for the $1\sigma_g$ and $1\sigma_u$ pairs and 0.22 eV for each pair of molecular orbitals when one was inner shell and one valence shell. Intraorbital correlation energy for the valence shell was assigned 2.22 eV and interorbital correlation energy for the valence shell 0.62 eV per pair. The total was 22.5 eV, so that MECE was 22.5 eV — 2(5.48 eV) = 11.5 eV. Adding this to his computed dissociation energy gave fair agreement with experiment, but it was admitted that MECE was probably overestimated by this process.

Nesbet [3] performed calculations using the double-zeta basis set as part of a study of the isoelectronic $N_2$, CO, and BF (see Subsection *d*). He also included d orbitals. While the double-zeta set sufficed to describe "radial distortion" of the atoms in the molecule, the d orbital was very important in describing the "angular distortion," which corresponds to polarization of the atomic charge clouds from spherical symmetry. Addition of the d orbital had a large effect on the computed binding energy. The energy multiplied by $R$ was calculated at five values of $R$, chosen as the roots of the fifth-order Chebyshev polynomial. The results for spectroscopic constants, given in Table IX, reflect the incorrect dissociation which makes $R_e$ too small, $\omega_e$ too large, and $\omega_e x_e$ too small (for BF, there is an additional error [3, 4] because the true Hartree–Fock function is not closely approached).

The dissociation energies were not in very good agreement with experiment because of the large MECE; indeed, the order for the three molecules was incorrect. A crude correlation energy correction was made [3] as described in Section A, by using atomic data to estimate the effect of additional pairs formed on molecule formation. The corrected $D_e$'s were in the correct order (see Table IX); the remaining errors were explained [3] in terms of the residual s–p degeneracy of the atoms.

For $N_2$ and several states of $N_2^+$, Cade *et al.* [66] gave approximate Hartree–Fock calculations for a range of $R$, giving particular attention to the choice of basis functions and the values of $R$ for which calculations

**Table IX**

*Results of Approximate SCF Calculations for $N_2$, CO, and $BF^a$*

| | $R_e$ (calc) (Å) | $R_e$ (obs) (Å) | $\omega_e$ (calc) (cm$^{-1}$) | $\omega_e$ (obs) (cm$^{-1}$) | $\omega_e x_e$ (calc) (cm$^{-1}$) | $\omega_e x_e$ (obs) (cm$^{-1}$) | $B_e$ (calc) (cm$^{-1}$) | $B_e$ (obs) (cm$^{-1}$) | $D_e$ (calc) (eV) | $D_e$ (calc) +MECE (eV) | $D_e$ (obs) (eV) |
|---|---|---|---|---|---|---|---|---|---|---|---|
| $N_2$ | 1.072 | 1.098 | 2722.4 | 2358.1 | 10.9 | 14.2 | 2.094 | 1.999 | 4.842 | 9.79 | 9.90 |
| CO | 1.119 | 1.128 | 2357.2 | 2170.2 | 11.1 | 13.5 | 1.965 | 1.931 | 7.312 | 10.49 | 11.24 |
| BF | 1.272 | 1.262 | 1413.8 | 1399.8 | 9.9 | 11.3 | 1.495 | 1.518 | 5.825 | 7.49 | 8.58 |

$^a$ See Nesbet [3].

were made. It was concluded that the choice of *R*-values must be "reasonable" to get consistent results from a Dunham analysis. The basis set was large enough so that reoptimization of orbital exponents at each value of *R* was not necessary. In the case of $N_2$, $U(R)$ from the calculations was parallel to that derived by the RKR method for $R < R_e$, but shifted to smaller *R*. It rose too sharply, compared to the RKR curve, for $R > R_e$. The difference, $E^{HF} - E^{RKR}$, was plotted as a function of *R*, showing clearly the deterioration of the results for large *R*. The slope of $U(R)$ near $R_e$ (which is the derivative of $E^{corr} + E^{rel}$ with *R*) was about 0.1 a.u./$a_0$. The energy error near $R_e$, about half an atomic unit, was mostly due to relativistic and correlation energy of the inner shells, so was also present for the separated atoms. Cade *et al.* [66] attempted to analyze the errors in $U(R)$ and associate them with specific physical effects. The results of a Dunham analysis for the $^1\Sigma_g^+$ state of $N_2$ were: $R_e = 1.065$ Å, $\omega_e = 2729.6$ cm$^{-1}$, $\omega_e x_e = 8.38$ cm$^{-1}$, $B_e = 2.121$ cm$^{-1}$, $\alpha_e = 0.01347$ cm$^{-1}$ (experiment, 0.01781), and $\omega_e y_e = -0.4745$ cm$^{-1}$ (exp't., $-0.0124$). The improved calculation [66] gives worse agreement than Nesbet's [3].

Diatomic carbon is known to have a $^1\Sigma^+$ ground state, but the lowest observed state for BN, which is isoelectronic with BeO and $C_2$, is $^3\Pi$. Verhaegen *et al.* [67, 68] performed open-shell calculations, of the kind performed for BeO (see Subsection *d*) for $C_2$ and BN. After the correction for differences in correlation energy, the $^3\Pi$ state of BN, corresponding to $(1\sigma)^2(2\sigma)^2(3\sigma)^2(4\sigma)^2(5\sigma)(1\pi)^3$, lay 0.42 eV below the $^1\Sigma^+$ closed shell state. For $C_2$, the $^1\Sigma^+$ state was pushed below the $^3\Pi$ by a perturbation of a low-lying excited $^1\Sigma^+$ state, whose effect on the energy could be estimated. From calculations at three different values of *R*, equilibrium internuclear distances and vibrational frequencies were derived [67, 68] for the ground and excited states of the molecules and compared with experiment. Some ground state results are presented in Table X.

**Table X**

*SCF Calculations for 12-Electron Molecules*[a]

| Molecule | $R_e$ (calc) (Å) | $R_e$ (expt) (Å) | $\omega_e$ (calc) (cm$^{-1}$) | $\omega_e$ (expt) (cm$^{-1}$) |
|----------|----------|----------|----------|----------|
| BeO $^1\Sigma^+$ | 1.300 | 1.331 | 1732 | 1487 |
| BN $^3\Pi$ | 1.305 | 1.281 | 1750 | 1515 |
| $C_2$ $^1\Sigma_g^+$ | 1.255 | 1.242 | 1970 | 1855 |

[a] See Verhaegen [67, 68].

Verhaegen [69] subsequently pointed out that the correlation energy corrections neglected possible irregularities associated with near degeneracies. In some calculations for $C_2$ and $C_2{}^+$, he derived approximate values for the near-degeneracy contributions from limited configuration interactions. Then a semiempirical correction was superposed on the near-degeneracy contribution: $E^{corr}$ for $C_2$ was taken as $E^{corr}(C^{2+}) + E^{corr}(C^{2-})$, and $E^{corr}$ for $C_2{}^+$ as $E^{corr}(C^{2+}) + E^{corr}(C^-)$, in accordance with the pairing structures determined from population analyses of the molecular wave functions. For the ground state of $C_2$, $R_e$ was 1.252 Å, $\omega_e$ 1946 cm$^{-1}$, and $D_e$ 6.16 eV (experiment, 6.25). For the $^4\Sigma_g{}^-$ state of $C_2{}^+$, determined to be the ground state by a set of open-shell calculations, $R_e$ was 1.412 Å, $\omega_e$ 1560 cm$^{-1}$, and $D_e$ 4.9 eV (experiment, 5.2).

Sahni [70] calculated single-configuration wavefunctions for various states of NF and $O_2$. He found it necessary to go beyond the minimal Slater set for these molecules, unlike others he treated, to get accurate results. $R_e$ values, and excitation and ionization energies, were reported. For the ground states, the results are shown in Table XI. It appears from more accurate calculations [71] that $R_e$ for the $O_2$ ground state in the RHF limit is more like 2.19 $a_0$. The dissociation energy is 1.41 eV (experiment, 5.21 eV), $\omega_e$ is too high, and $\omega_e x_e$ too small. The RHF wave function dissociates incorrectly, and all the errors are in the expected directions.

**Table XI**

|  | $R_e$ (calc) | $R_e$ (obs) |
|---|---|---|
| $O_2$ ($^3\Sigma_g$) | 2.28 $a_0$ | 2.2817 $a_0$ |
| $O_2{}^+$ ($^2\Pi_g$) | 2.19 | 2.126 |
| NF ($^3\Sigma$) | 2.44 | 2.456 |
| NF$^+$ ($^2\Pi$) | 2.30 | — |

The RHF results for $F_2$ by Wahl [24] have been discussed in Section C.2. The incorrect dissociation makes $\omega_e$ much too large and $\omega_e x_e$ and $R_e$ too small, while the effect of MECE is to give a negative dissociation energy. Apparently, MECE is about 3.3 eV, which is twice $D_e$.

Gilbert and Wahl [72] performed limited-basis Hartree–Fock calculations for $Cl_2$ in connection with work on related ions (see Subsection e for details). The derived values for $D_e$, $R_e$, $\omega_e$ and $B_e$ were 0.87 eV, 3.8 $a_0$, 577 cm$^{-1}$, and 0.24 cm$^{-1}$. The poor value for $D_e$ compared to the experimental 2.51 eV is due to the large MECE, as for F$_2$. In both cases, an accurate dissociation energy can be obtained from

$$D_e(X_2) = D_e(X_2^-) + \text{electron affinity}(X) - \text{electron affinity }(X_2)$$

using calculated dissociation energies for the molecule–ions (which seem correct to a fraction of an electron volt) and experimental electron affinities.

### (d) Other Molecules of First Two Rows

The isoelectronic molecules BF, CO, and N$_2$ have attracted a lot of interest, as already mentioned. Chemical binding and other properties have been discussed [73] in terms of the molecular charge densities and the Hellmann–Feynman forces on the nuclei. (The same was done for the isoelectronic set C$_2$–BeO–LiF.)

Nesbet's results [3] for the potential constants of CO and BF were given in Table IX. Subsequently, Huo [74] gave near Hartree–Fock calculations for CO and BF. Her results are given in Table XII.

Sahni et al. [75] derived $U(R)$ for CO by calculating energies at 43 values of $R$ using a minimal basis set with exponential parameters optimized at a single $R$, and also using the Slater rules for these parameters. The value

**Table XII**

| | | Calculated | Observed |
|---|---|---|---|
| $\omega_e$ | CO | 2431 cm$^{-1}$ | 2169.8 |
| | BF | 1496 | 1402.1 |
| $\omega_e x_e$ | CO | 11.69 cm$^{-1}$ | 13.295 |
| | BF | 12.07 | 11.8 |
| $B_e$ | CO | 2.027 cm$^{-1}$ | 1.9313 |
| | BF | 1.1559 | 1.510 |
| $\alpha_e$ | CO | 0.01525 cm$^{-1}$ | 0.017535 |
| | BF | 0.01851 | 0.016 |
| $R_e$ | CO | 2.081 $a_0$ | 2.132 |
| | BF | 2.354 | 2.391 |
| $k$ | CO | 23.864 × 10$^5$ dynes/cm | 19.02 |
| | BF | 9.189 | 8.071 |

$D_e$, calculated as $E(R = 6.5\, a_0) - E(R = 2.132\, a_0)$ (the potential curve is level at 6.5, so it is equivalent to infinity), was 12.496 eV in the first case, 11.949 in the second. These are much closer to the experimental 11.242 eV than values for $D_e$ obtained by other workers, but this reflects a cancellation between MECE and the energy error due to incorrect dissociation. Vibrational–rotational levels were calculated by the WKB method and by direct integration; from various numbers of these, the rotational constants were derived. It was claimed [75] that this procedure was more "fundamental" than the Dunham analysis, which checks $U(R)$ around $R_e$. Nesbet [3] and Huo [74] used larger basis sets than Sahni et al. for their matrix Hartree–Fock calculations, and a different method for deriving the spectroscopic constants (see Tables IX and XII for their results).

Like $N_2$, CO, and BF, the isoelectronic molecules $C_2$, BeO, and LiF have closed shell $^1\Sigma^+$ ground states, and there is evidence [67, 68] that the ground state of BN, isoelectronic to the last three, is $^1\Sigma^+$ as well. We have given the results of Verhaegen et al. [67, 68] for BeO and BN in Table X. We now discuss other work for BeO and LiF.

The matrix Hartree–Fock calculations on BeO by Yoshimine [6] have already been mentioned. The energies of this Roothaan Hartree–Fock closed-shell calculation were estimated to be within 0.1 eV of the Hartree–Fock energies. The basis set was built from a double-zeta basis, derived from atomic calculations, by adding functions to describe polarization within the molecule and optimizing their exponents. For one value of $R$, a further enlarged basis set was used to check the possible energy improvement. The molecular Hartree–Fock energy was estimated, by subtracting correlation and relativistic energies from the exact energy, as $-89.456 \pm 0.01$ a.u.; the result of calculation was $-89.450 \pm 0.002$ a.u. The potential curve was derived by McLean's scaling method [5] (see Volume I, Chapter III, Section C) and a Dunham analysis performed on it. An idea of the uncertainty due to the fitting procedure was obtained by using two functional forms for $U(R)$: a power series, and a power series plus $R^{-1}$. In Table XIII, we cite an average of the two results.

Yoshimine's calculations on BeO were for the $^1\Sigma^+$ state, assumed to be the ground state, with configuration $(1\sigma)^2(2\sigma)^2(3\sigma)^2(4\sigma)^2(1\pi)^4$. According to the Wigner–Witmer rules, this state cannot dissociate into atoms in their ground states; Be ($^1S$) and O ($^3P$) can combine to form triplets only. Since a ground state molecule which correlates with excited atoms is rare, the possibility that the $^1\Sigma^+$ state is not the ground state, but some hitherto unobserved triplet state, was investigated by Verhaegen and Richards [67]. They calculated energies for the $^1\Sigma^+$ state, for the $^3\Pi$ state formed from the

**Table XIII**

$^9Be^{16}O$ *Hartree–Fock Calculation*

| | $\omega_e$ | $\omega_e x_e$ | $\omega_e y_e$ | $\omega_e z_e$ | $B_e$ | $\alpha_e$ | $\gamma_e \times 10^3$ | $\delta_e \times 10^4$ | $D_e \times 10^5$ | $\beta_e \times 10^8$ |
|---|---|---|---|---|---|---|---|---|---|---|
| Calc. | 1807.6 | 10.65 | −0.143 | −0.004 | 1.75746 | 0.0151 | −0.118 | −0.102 | 0.6646 | 0.69 |
| Expt. | 1487.3 | 11.830 | 0.022 | — | 1.6510 | 0.0190 | — | — | — | — |

[a] All values in reciprocal centimeters.

configuration $(4\sigma)^2(1\pi)^3(5\sigma)$, and for the $^3\Sigma^+$ state formed from the configuration $(4\sigma)(5\sigma)(1\pi)^4$. It was shown that CI using determinants formed from virtual orbitals from the $^1\Sigma^+$ calculation led to poor results. Thus, open-shell variational calculations, with symmetry and equivalence restrictions on the molecular orbitals, were carried out for the two triplet states, at various values of $R$. The one-electron Hamiltonians were displayed [67, 68]. Both the $^3\Pi$ state and the $^3\Sigma^+$ state may be written as single determinants. The computed energies had to be corrected for differences in correlation energy associated with the change in the pair structure. Verhaegen and Richards did this by performing population analyses of the molecular wave functions and using atomic correlation energies. The result was that $^3\Pi$ and $^3\Sigma^+$ were higher in energy than $^1\Sigma^+$, by 1.47 and 2.96 eV, respectively. It was concluded that the ground state BeO was $^1\Sigma^+$. Similar calculations were done for MgO [76]. The corrected energies of $^3\Pi$ and $^3\Sigma^+$ were 1.43 and 1.13 eV above $^1\Sigma^+$, so the ground state was again concluded to be $^1\Sigma^+$. For SrO and BaO, for which accurate calculations of this kind are not possible, experimental evidence indicates a $^1\Sigma^+$ ground state.

The calculations of Huo *et al.* [77] were also aimed at establishing whether the $^1\Sigma$ state of BeO was the ground state. Using 24 Slater basis functions, they performed SCF calculations for the lowest $^1\Pi$ and $^3\Pi$ states at several values of $R$. Like Verhaegen and Richards [67, 68] they concluded that $^1\Sigma$ was the ground state, but the better basis set used led to a much smaller $^1\Pi$–$^3\Pi$ splitting. There was only a slight improvement for the ground state compared to Yoshimine's [6] calculations.

Although we will discuss the alkali halides, for which single-configuration calculations by Matcha [31] have been appearing, later, we want to mention the earlier (1963) work on LiF by McLean [78]. This work obtained an energy at $R_e$ within 0.007 a.u. (0.2 eV) of the true Hartree–Fock energy, and also computed $U(R)$, using a slightly less accurate wave function, for $1.6 < R < 4.85\ a_0$. This calculation and that of Yoshimine for BeO [6] were carried out by the same procedures, so that comparisons between the two isoelectronic systems could be made. McLean [5, 78] discussed the procedure for obtaining the best values of all nonlinear parameters. He used a functional form for the energy in terms of the parameters, and found the minimum by differentiation. Many expectation values for various values of $R$ were tabulated. The distance $R_e$ was found to be 2.8877 $a_0$ (experimental $R_e = 2.955\ a_0$) and $D_e$ (uncorrected difference of Hartree–Fock energies) 0.1466 a.u. Adding to this the estimated MECE of 0.076 a.u. and the change in relativistic energy (assuming LiF had a relativistic energy equal to the sum of those for $Li^+$ and $F^-$) of $-0.001$ a.u., gave $D_e = 0.220$ a.u., which agreed to 0.001 a.u. with experiment.

Some results for BF were given in Table II of Section A, and for NO [47] at the end of Subsection 2. The NF results of Sahni [70] were cited along with his results for $O_2$ in subsection $c$. It should be noted that the calculated value for $R_e$ was found before the value was experimentally obtained; they agree to 0.02 $a_0$. For the $^2\Sigma^+$ ground state of BeF, calculations were performed by Walker and Richards [44], in connection with a discussion of the nature of the first excited ($^2\Pi$) state. In the Slater orbital basis, Be exponents were taken from Yoshimine's BeO work [6] and F exponents from Nesbet's (1964) BF work [3]. Three energies were fit to a quadratic, to give $R_e$ and $\omega_e$.

Hartree–Fock calculations for ClO, ClO$^+$, and ClO$^-$ have recently been reported [79]. The configuration of ClO is known to be $KK$ $(3\sigma)^2(4\sigma)^2(5\sigma)^2$ $(6\sigma)^2(1\pi)^4(2\pi)^3$; ClO$^-$ ($^1\Sigma$) was assumed to have an extra $2\pi$ electron, and ClO$^+$ ($^3\Sigma$) a $2\pi$ electron missing. O'Hare and Wahl [79] performed all their calculations for 1.570 Å, which is $R_e$ for the neutral molecule. Ionization potentials and electron affinities were calculated by Koopmans's theorem and by energy differences. The value $D_0$ for ClO was calculated as 0.0019 a.u. Since MECE for FO was 0.1045 a.u., $D_0$ was predicted to be slightly less than 0.1064 a.u. $= 2.86$ eV; the experimental value is 2.75 eV. These authors also calculated [80] spectroscopic properties for CF and its ions (see Subsection $e$). LiHe and NaHe are discussed together with their ions, in Subsection $e$.

### (e) Ionic Species of the First Two Rows

The two-electron HeH$^+$ is the simplest molecule after $H_2$, and so has been the subject of many calculations. Most are configuration interactions, but Peyerimhoff [81], in connection with some work on NeH$^+$, presented the results of Hartree–Fock–Roothaan calculations for HeH$^+$. The restricted Hartree–Fock wave function for the $^1\Sigma^+$ ground states dissociates correctly to He $+$ H$^+$. Since there is no change in pair structure with $R$, the correlation energy should be close to independent of $R$. The correlation energy of the united atom Li$^+$ is 0.043 a.u., that of the separated atoms (He $+$ H$^+$) is 0.042 a.u. Furthermore, the energy of He$^+$ $+$ H is so much higher than that of He $+$ H$^+$ that the single configuration should represent the situation well. All in all, HeH$^+$ is a favorable case for restricted Hartree–Fock calculations. (It is also a favorable case for one-center calculations, because it is a one-center system for large $R$ as well as for small $R$.)

Indeed, Peyerimhoff found [81] that $U(R)$, as derived from the single-configuration calculations, agreed closely with that derived from configuration interaction. The correlation energy (using Anex's configuration inter-action energies [82] instead of the exact energy) was 0.0406, 0.0416, 0.0410, 0.0392 a.u. at $R = 1.0$, 1.4, 1.8, 2.2 $a_0$, i.e., quite constant. According to $E^{corr}$ of He and $Li^+$, these numbers should be increased by about 0.002 a.u. $R_e$ and other potential constants, as well as the dissociation energy, were also in good agreement with results of configuration interaction (see Table XXII, page 268). These constants were derived in several different ways: by the Dunham treatment, by the virial theorem ($dE/dR$ computed from kinetic and potential energies), and by the Hellmann–Feynman theorem (force curves). Calculations were done [81] for 17 values of $R$ between 1 and 4.5 $a_0$, using various basis sets of Slater orbitals. Only for large basis sets with carefully optimized orbital exponents was calculation of forces a reliable way of deriving potential constants. Some of Peyerimhoff's results are given in Table XXII on page 268. The Dunham analysis employed various sets of 9, 10, or 11 points; the results given are averages of all the sets. The dissociation energy $D_e$ was computed relative to Hartree–Fock atoms. The only experimentally known constant was $D_e = 1.79$ eV.

The results of a one-center numerical calculation by Keefer et al. [59] (see discussion of their LiH results in subsection b) were fitted to a Morse curve. From this, $D_0$ was derived as 1.60 eV [corresponding to $D_e = 1.60 + \frac{1}{2}(0.38$ eV$) = 1.79$ eV], $R_e$ as 1.480 $a_0$, and $\omega_e$ as 3036 cm$^{-1}$.

We will consider larger hydride ions before going to homonuclear molecule–ions and other species. A single-determinant calculation for the negative species $BeH^-$ was performed by Kaufman and Sachs [83]. A Gaus-sian basis set was used, with 21 basis functions on each nuclear center. Orbital exponents were optimized at $R = 2.72$ $a_0$ and calculations were performed for 1.75 (0.25) 3.50 $a_0$ and for some larger $R$ values. The ne-cessity, in calculations on negative ions, for reoptimization of certain ex-ponents obtained from calculations on neutral species, even for large basis sets, was noted; other workers also have made this point. The minimum energy was at 2.655 $a_0$. Since the first virtual orbital energy was negative by about 0.016 a.u. for some values of $R$, it was speculated [83] that $BeH^{2-}$ is stable.

A calculation by Cade [36] for spectroscopic and other properties of the $OH^-$ and $SH^-$ ground states illustrates how accumulated experience on such calculations can be used to correct calculated values for properties to give reliable predictions. The wave functions for the $^2\Pi$ ground states of OH and SH were used to get the starting basis set for the $^1\Sigma^+$ states of $OH^-$ and $SH^-$, but exponential parameters were extensively reoptimized. In additional to spectroscopic properties, Cade computed electronic forces,

**Table XIV**

*Near Hartree–Fock Results[a] for OH⁻ and SH⁻*

| | OH⁻ | | | | | SH⁻ | | | | |
|---|---|---|---|---|---|---|---|---|---|---|
| | $B_e$ (cm⁻¹) | $\alpha_e$ (cm⁻¹) | $\omega_e$ (cm⁻¹) | $\omega_e x_e$ (cm⁻¹) | $R_e$ ($a_0$) | $B_e$ (cm⁻¹) | $\alpha_e$ (cm⁻¹) | $\omega_e$ (cm⁻¹) | $\omega_e x_e$ (cm⁻¹) | $R_e$ ($a_0$) |
| Calc. | 20.023 | 0.722 | 4087.9 | 87.82 | 1.781 | 9.734 | 0.2471 | 2827 | 43.84 | 2.516 |
| Est percentage error | +3–+5 | ~−10 | +7–+9.5 | ~−10 | −1.5–−2.5 | +2–+4 | ~−10 | +5–+7 | ~−10 | −1–−2 |
| Predicted | 19.4–19.1 | ~0.8 | 3820–3733 | ~98 | 1.81–1.83 | 9.58–9.36 | ~0.27 | 2692–2642 | ~49 | 2.54–2.57 |
| Experiment | 18.87 | — | 3735 | — | 1.834 | — | — | — | — | — |

[a] See Cade [36].

charge moments, and other properties as functions of $R$. To correct the spectroscopic properties [from a Dunham analysis of $U(R)$ derived from energy calculations at nine or ten $R$-values], Cade [36] plotted percentage errors in $\omega_e$, $B_e$, $R_e$ and $\alpha_e$ for the first- and second-row diatomic hydrides. The errors, which range from 0 to 10%, varied systematically enough for the first three properties to estimate corrections to calculated properties for OH$^-$ and SH$^-$. For example, the predicted $R_e$ of 1.781 $a_0$ for OH$^-$ was estimated to be 1.5–2.5% too low, so $R$ was predicted to be between 1.81 and 1.83 $a_0$, which compared well with the experimental value, 1.834 $a_0$. The results obtained in this way are summarized in Table XIV. The results for OH$^-$ give confidence to the predictions for SH$^-$.

The potential curves for OH and OH$^-$ were roughly parallel, as were those for SH and SH$^-$, which seemed to be in accordance with experiment. However, OH$^-$ lay above OH, so that the predicted electron affinity is negative in the Hartree–Fock approximation. The correction of electron affinities obtained from Hartree–Fock calculations is very similar to the use of MECE for dissociation energies: One must add the extra correlation energy of the molecular ion relative to the neutral molecule. Cade estimated this in several ways: (a) from the correlation energies of HF and HCl (iso-electronic with OH$^-$ and SH$^-$) relative to OH and SH, and (b) from the changes in correlation energy for the united atom systems: F $\rightarrow$ F$^-$ and Cl $\rightarrow$ Cl$^-$. Estimate (a) relies on the relative invariance of correlation energy to nuclear charge, estimate (b) to its invariance to internuclear distance. Various calculated electron affinities are given in Table XV and compared with experiment.

Peyerimhoff's Hartree–Fock–Roothaan calculations [81] on the NeH$^+$ ground state ($^1\Sigma^+$ symmetry) employed a Slater orbital basis set with optimization of exponential parameters at several values of $R$ (not at all those

### Table XV

*Electron Affinities for OH and SH*

|  | Hartree–Fock energy difference | Energy of highest ion orbital (Koopmans's theorem) | H–F E.A.$+\Delta E^{corr}$ from (a) | H–F E.A.$+\Delta E^{corr}$ from (b) | Experiment |
|---|---|---|---|---|---|
| OH | −0.10 eV | 2.90 eV | 1.40 eV | 1.99 eV | 1.83 eV |
| SH | 1.21 eV | 2.42 eV | 2.11 eV | 2.25 eV | (2.3) |

for which calculations were carried out). There were 11 $\sigma$-functions and $\pi$-functions centered on Ne, and 5 $\sigma$- and 2 $\pi$-functions centered on H. The calculated $U(R)$ closely resembled that computed for $HeH^+$ (see page 251). As for $HeH^+$, it was expected that the correlation energy would be roughly $R$-independent, and that $U(R)$ from the single-configuration calculation would be accurate. Table XVI gives some calculated potential constants derived by a Dunham analysis of energies calculated using exponential parameters optimized for a single $R$-value near $R_e$, and optimized for a number of $R$-values. The latter procedure improves energies more away from $R_e$, so $\omega_e$ is lowered and the anharmonicity increased. The use of force curves and the virial theorem to derive potential constants was discussed [81] and the results used to assess errors in the wave function. With the singly optimized basis, the virial method gave $R_e = 1.833\ a_0$ and the force curves gave $1.81$–$1.82\ a_0$. The only experimental data available were $D_e = 2.15 \pm 0.25$ eV.

### Table XVI

#### Near Hartree–Fock Results[a] for $NeH^+$

|  | $R_e\ (a_0)$ | $B_e\ (cm^{-1})$ | $\omega_e\ (cm^{-1})$ | $\omega_e x_e\ (cm^{-1})$ | $\alpha_e\ (cm^{-1})$ | $D_e\ (eV)$ |
|---|---|---|---|---|---|---|
| Single-opt. | 1.836 | 18.52 | 3069 | 122 | 1.13 | 2.022 |
| Multi-opt. | 1.836 | 18.52 | 3053 | 141 | 1.21 | 2.023 |

[a] From Peyerimhoff [81].

The rare gas pairs with a single electron removed are stable. From their minimal basis $He_2$ results, Griffing and Wehner [52] derived a potential curve for $He_2^+$ by invoking Koopmans's theorem. Simply subtracting the orbital energy of the second ($\sigma_u$) molecular orbital from the total energy at each value of $R$, they obtained a $U(R)$ which agreed well with available experimental data.

Mulliken [84] has given a qualitative and semiquantitative discussion of $U(R)$ for ground and excited states of rare gas molecules and ions. Dissociation energies were estimated by considering isoelectronic systems, by examining the effect of removal of an electron on $U(R)$ for the neutral system, and by invoking analogies to related molecules. For instance, it was expected that the ratio of dissociation energies for $Xe_2^+$ and $I_2$ would be the same as for $He_2^+$ and $H_2$. This reasoning leads to $D_e = 0.78,\ 1.21,\ 0.92,$

and 0.65 eV for $Ne_2^+$, $Ar_2^+$, $Kr_2^+$, and $Xe_2^+$, respectively. Experimental data indicate values a few tenths of an electron volt higher in each case. Similar reasoning was used for other properties of $U(R)$, such as force constants. It was also pointed out [84] that $U(R)$ for the ground state of $A_2^+$ should resemble in shape $U(R)$ for Rydberg states of $A_2$.

Limited basis Hartree–Fock calculations for $He_2^+$, $Ne_2^+$, and $Ar_2^+$ were recently presented by Gilbert and Wahl [72]. The ions $F_2^-$ and $Cl_2^-$ (isoelectronic to the last two) were also calculated, as well as $Cl_2$. Spectroscopic properties were derived from energy calculations at a series of $R$-values, giving the results shown in Table XVII. There are experimental values only for $D_e$, which seem quite well calculated (experimental values here are uncertain) despite the importance of left–right correlation. This result was unexpected and difficult to justify theoretically, although some analysis of the energy was attempted.

Table XVII

Results[a] for Rare-gas Molecule-Ions

|  | $D_e$ calc. (eV) | $D_e$ expt. | $R_e$ ($a_0$) | $\omega_e$ (cm$^{-1}$) | $B_e$ (cm$^{-1}$) |
|---|---|---|---|---|---|
| $He_2^+$ | 2.67 | 2.49 | 2.0 | 1790 | 7.4 |
| $Ne_2^+$ | 1.65 | 1.35 | 3.2 | 660 | 0.59 |
| $Ar_2^+$ | 1.25 | 1.6 | 4.6 | 300 | 0.139 |
| $F_2^-$ | 1.66 | 1.29 | 3.6 | 510 | 0.50 |
| $Cl_2^-$ | 1.28 | 1.26 | 5.0 | 260 | 0.136 |

[a] See Gilbert and Wahl [72].

The accuracy of the calculations, particularly with respect to the insufficiency of the basis set used, was discussed [72]. From experience with basis sets constructed in the same way, the energy error relative to the true Hartree–Fock was estimated as 0.01 a.u. and only slightly worse errors were anticipated for $U(R)$. Further arguments were presented to show that the expansion error in $D_e$ was only about 0.1 eV and that the calculated $U(R)$ was parallel to the true Hartree–Fock potential curve to 0.1 eV for $R \gtrsim R_e$.

For large $R$, a direct calculation of $U(R)$ is possible since the interaction is between a point charge and a polarizable charge distribution [72]. In the case of the noble gas molecule–ions,

$$U(R) \rightarrow -\tfrac{1}{2}\alpha_0 R^{-4} - (C + \tfrac{3}{4}\beta_0)R^{-6}$$

where $\alpha_0$ and $\beta_0$ are dipole and quadrupole polarizabilities and $C$ the van der Waals constant, estimated, by the London formula, in terms of the polarizabilities and ionization potentials for the neutral and positively charged atomic species. In the case of the halogen molecule–ions, a mono-pole–quadrupole interaction, going as $R^{-3}$, had to be added to the above form.

The ion $Li_2^+$ was studied in 1935 by James [85] to test various methods of calculation. More recently, SCF–MO calculations were done by Fischer and Kemmey [56] using 14 s and p Gaussians on each nucleus. They determined $D_e = 1.24$ eV and $R_e = 5.7\,a_0$, compared to James's results [85] of 1.234 eV and 5.96 $a_0$, although their energies were about 2 eV above James's for the range of $R$ studied.

The ions $N_2^+$ and $CO^+$ have been studied, generally in conjunction with the neutral molecules [65, 66, 86]. Schaefer [87] has emphasized that the accuracy of the open-shell calculations is insufficient to predict the right order of states of $N_2^+$. Koopmans's theorem, which means in this case consideration of the highest filled orbital energies of $N_2$, also gives an incorrect result.

Some light is thrown on the failure of Koopmans's theorem by the calculations of Sahni and Sawhney [86] for CO and $CO^+$. It will be remembered that Koopmans's theorem assumes that, on removal of an electron, the distribution of the remaining electrons does not change. A calculation of $U(R)$ for $CO^+$ in the Hartree–Fock approximation [86] compared results with two basis sets: first, using exponents taken from CO calculations and second, using exponents reoptimized for $CO^+$. Sahni and Sawhney [86] calculated molecular constants by finding vibration–rotation energy levels (using WKB or integration of the radial Schrödinger equation). The large improvement in the results on change of basis implied that Koopmans's theorem was unreliable in generating information on $CO^+$ from CO calculations. The results with reoptimized basis orbitals were: $R_e = 2.081\,a_0$ (experiment, 2.107), $D_e = 8.287$ eV (8.482), $\omega_e = 2439$ cm$^{-1}$ (2214), $\alpha_e = 0.017$ cm$^{-1}$ (0.0189). The value of $\omega_e x_e$ depended strongly on the number of energy levels used in deriving values for the constants. The dissociation energy $D_e$ was calculated as $E(6.5\,a_0) - E(R_e)$.

For $N_2$ and the lowest $^2\Sigma_u^+$, $^2\Pi_u$, and $^2\Sigma_g^+$ states of $N_2^+$ over a range of $R$, Cade et al. [66] gave approximate open-shell Hartree–Fock calculations, which we discussed briefly in Subsection c. The three $N_2^+$ states are those produced from $N_2$ by removal of an electron from the three highest $N_2$ orbitals. Errors for $N_2^+$ were more serious than for $N_2$, since the computed $U(R)$ for the $^2\Pi_u$ state lay below that for the $^2\Sigma_g^+$ state when $R$ exceeded about 1.9 $a_0$. Experimentally, the $^2\Sigma_g^+$ is below the $^2\Pi_u$ by about 0.05 a.u. for 1.8 $a_0 < R < 2.2\,a_0$, and approaches it near 2.6 $a_0$. The errors were ascribed [66] to the inadequacies of the Hartree–Fock approximation.

The difference, $E^{HF} - E^{RKR}$, was plotted as a function of $R$ for the states considered. This showed clearly the deterioration of the results for large $R$. The energy error for the three states, near the equilibrium internuclear distance of each, was greater than 0.5 a.u., but most of this was due to relativistic and correlation energy of the inner shells, so was also present for the separated atoms. Cade *et al.* [66] attempted to analyze the errors in $U(R)$ and associate them with specific physical effects. The results of a Dunham analysis for the $^2\Sigma_g^+$ state of $N_2^+$ were given and were comparable in accuracy to those for $N_2^+$.

Other ions for which we have cited results are the positive and negative ions of SF and SeF (dissociation energies, Section A.6), NF$^+$ and $C_2^+$ (Subsection c) and ClO$^+$ and ClO$^-$ (Subsection d). We conclude with another recent example.

Values for CF and its ions have been calculated by O'Hare and Wahl [80] at several internuclear distances with matrix SCF wave functions, and SiF and its ions have been calculated at the equilibrium internuclear distance of SiF. The potential constants derived for CF, CF$^+$, and CF$^-$ are given in Table XVIII. In the hope of deducing accurate dissociation energies in the absence of experiment for related molecules, O'Hare and Wahl considered various routes to MECE. The Hartree–Fock dissociation energy of CF was 0.128 a.u. or 3.48 eV and the experimental value 0.207 a.u. or 5.63 eV. Estimating $E^{corr}$ for CF as the mean of the values for BF and NF gives the dissociation energy correct to 0.03 eV, and good results are also obtained using $E^{corr}$ for the united atom, P ($^2P$). Furthermore, $E^{corr}$ for the first-row monofluorides from BeF (0.438 a.u.) to $F_2$ (0.740 a.u.) varies almost linearly with atomic number (slope 0.0060 $\pm$ 0.005 a.u.) as does MECE, from 0.028 a.u. for BeF to 0.108 a.u. for $F_2$. The experimental

**Table XVIII$^a$**

| | $R_e$ (Å) | $\omega_e$ (cm$^{-1}$) | $\omega_e x_e$ (cm$^{-1}$) | $B_e$ (cm$^{-1}$) | $\alpha_e$ (cm$^{-1}$) |
|---|---|---|---|---|---|
| CF ($^2\Pi$) | 1.259(1.2667) | 1388(1308.1) | 10.59(11.10) | 1.433(1.4172) | 0.017(0.0184) |
| CF$^+$ ($^1\Sigma$) | 1.135 | 1963 | 13.05 | 1.764 | 0.017 |
| CF$^-$ ($^3\Sigma$) | 1.465 | 750 | 11.48 | 1.060 | 0.022 |

$^a$ Experimental results are given within the parentheses.

and Hartree–Fock dissociation energies are not monotonic, however. Similar results obtain [80] for the second-row fluorides, but less experimental data are available. The dissociation energies that follow, corrected for MECE, are expected to be accurate to $\frac{1}{2}$ eV.

|            | CF  | CF$^+$ | CF$^{2+}$ | SiF | SiF$^+$ | SiF$^{2+}$ |
|------------|-----|--------|-----------|-----|---------|------------|
| $D_e$ (eV) | 5.3 | 7.4    | 5.3       | 5.8 | 6.6     | 2.2        |

The rare-gas–alkali systems LiHe, NaHe, and their ions have been calculated recently in the restricted Hartree–Fock approximation by Krauss *et al.* [88]. The neutral molecules, which are open-shell systems, were treated [88] according to the open-shell formalism of Roothaan (Section A). These open-shell–closed-shell systems dissociate properly in this approximation, and correlation energy may be expected to be quite *R*-independent since no electron pairs are formed. Thus, the Hartree–Fock *U(R)* was anticipated [88] to be a good approximation except for large *R*. The interaction for large *R* (dispersion interaction) may be calculated by perturbation theory (Chapter I, Section A). Slater and Gaussian basis sets were used and results compared. The sets were constructed by adding "polarization functions," whose exponents were optimized for one large value of *R*, to atomic basis sets. Differences between Slater and Gaussian results were apparent at the shorter values of *R* considered.

While *U(R)* was purely repulsive, it possessed a marked curvature on a plot of log *U* versus *R*, which distinguished it from *U(R)* for rare-gas–rare-gas interactions (which are close to straight lines). For the ions, which are of $^1\Sigma^+$ symmetry, *U(R)* showed a minimum because of the long-range ion-induced dipole attraction. This occurred at 3.8 $a_0$, with a depth of 0.07 eV, for LiHe$^+$, and at 4.7 $a_0$, with a depth of 0.03 eV, for NaHe$^+$. The curve *U(R)* was fit by least squares to $U(R) = A_1 e^{-\alpha R} + A_2 R^{-4}$. The value for $A_2$ from the Slater calculations was much closer than $A_2$ from the Gaussian calculations to the theoretical value of half the dipole polarizability of He. The $R^{-4}$ law fitted *U(R)* to $10^{-5}$ a.u. for $R > 5\,a_0$. For $R = 10$, the $R^{-6}$ van der Waals attraction amounted to only 10% of *U(R)* and the $R^{-8}$ contribution was negligible; it is practical to compute these corrections separately and add them to *U(R)* from a Hartree–Fock calculation. Krauss *et al.* [88] compared their results with those of previous workers.

### ( *f* ) Alkali Halides and Larger Systems

In this subsection, we consider diatomic systems formed from heavier atoms. The alkali halides, even those formed from the first 18 elements, are included here because of the systematic work of Matcha [31] on a number of such compounds, including those involving elements further down in the periodic table. In addition to molecules discussed here, preceding sections have given results for TiO by Nesbet [20] and ScO by Carlson *et al.* [46].

A series of publications by Matcha [31] has been appearing over the past few years, reporting LCAO wave functions for the important alkali halide molecules. The basis sets consisted of orbitals used for atomic calculations plus polarization functions with suitable angular behavior to describe the distortions accompanying molecule formation. Changes in the energy and a large number of other properties with improvements in the basis set were studied. The final wave functions, hopefully, are close to the Hartree–Fock limit [31]. In the calculation of $U(R)$, orbital exponents were not reoptimized at each $R$, since such reoptimization was found to change computed spectroscopic constants by much less than defects in the Dunham analysis and arbitrariness connected with the choice of the set of $R$-values. A large number of spectroscopic and other properties were tabulated for different isotopic species. We present only a few results (Table XIX), including the coefficients in the Dunham expansion of the potential,

$$U(R) = a_0 \varrho^2 (1 + a_1 \varrho + a_2 \varrho^2 + \cdots), \qquad \varrho = (R - R_e)/R_e.$$

In general, the calculated $U(R)$ seemed [31] to parallel the exact $U(R)$ for $R < R_e$, but not for large $R$. Because the restricted Hartree–Fock functions for the alkali halides AX approach Hartree–Fock functions for the $A^+$ and $X^-$ ions, which lie higher in energy than the A and X atoms, the curvature of the calculated $U(R)$ near $R_e$ is too large, as reflected in $a_0$ and $\omega_e$. $R_e$ and $\omega_e x_e$ would be expected to be too low for the same reason but this is not always the case, so other $R$-dependent errors are entering.

Matcha [31] corrected the calculated dissociation energies (differences of SCF atomic and molecular energies) for relativistic and correlation effects and obtained very good agreement with experiment. The correlation correction (MECE) has been discussed in Section C.2. Since the restricted Hartree–Fock wave function for a highly ionic alkali halide molecule AX resembles that for the ions $A^+$ and $X^-$, relativistic and correlation energies for the ions were used for the molecular quantities. Thus, for LiCl, MECE was approximated by

$$E^{corr}(Li^+) + E^{corr}(Cl^-) - E^{corr}(Li) - E^{corr}(Cl)$$
$$= 0.0435 + 0.5790 - 0.0453 - 0.5410 \text{ a.u.} = 0.0362 \text{ a.u.}$$

**Table XIX**

Results of SCF–LCAO Calculations by Matcha[a] for Alkali Halides

| Molecule | | $R_e$ ($a_0$) | Coeffs. in Dunham exp. | | | $D_e$ (a.u.) | Other properties (cm$^{-1}$) | | | |
|---|---|---|---|---|---|---|---|---|---|---|
| | | | $a_0 \times 10^{-5}$ (cm$^{-1}$) | $a_1$ | $a_2$ | | $\omega_e \times 10^{-2}$ | $\omega_e x_e$ | $\omega_e y_e \times 10^2$ | $\omega_e z_e \times 10^3$ |
| $^7$Li$^{35}$Cl | calc. | 3.825 | 1.6003 | $-2.7147$ | 4.7888 | 0.1406 | 6.715 | 4.671 | 6.893 | 1.855 |
| | expt. | 3.819 | 1.45±0.01 | $-2.72$±0.01 | 5.3±0.4 | 0.1781 | 6.41±0.3 | 4.2±0.3 | — | — |
| $^{23}$Na$^{19}$F | calc. | 3.628 | 1.774 | $-3.257$ | 6.602 | 0.112 | 5.584 | 4.387 | 6.954 | 2.503 |
| | expt. | 3.639 | 1.644 | $-3.133$ | 6.433 | 0.182 | 5.361 | 3.830 | — | — |
| $^{23}$Na$^{35}$Cl | calc. | 4.485 | 1.660 | $-3.441$ | 6.771 | 0.117 | 3.783 | 2.589 | 1.294 | 5.158 |
| | expt. | 4.4609 | 1.524 | $-3.076$ | 6.470 | 0.155 | 3.646 | 2.05 | — | — |
| $^{39}$K$^{19}$F | calc. | 4.188 | 1.872 | $-3.219$ | 6.785 | 0.113 | 4.485 | 2.484 | 4.033 | 0.861 |
| | expt. | 4.1035 | 1.621 | $-3.116$ | 6.340 | 0.187 | 4.260 | 2.430 | — | — |
| $^{39}$K$^{35}$Cl | calc. | 5.015 | 1.619 | $-3.077$ | 9.028 | 0.125 | 2.824 | 1.166 | 3.379 | 2.433 |
| | expt. | 5.039 | 1.521 | $-3.226$ | 7.1 | 0.161 | 2.798 | 1.167 | — | — |
| $^7$Li$^{81}$Br | calc. | 4.0655 | 1.751 | $-2.546$ | 3.513 | 0.144 | 6.282 | 3.881 | 7.015 | 6.251 |
| | expt. | 4.102 | 1.429 | $-2.717$ | 5.0 | 0.159 | 5.610 | 3.860 | — | — |

[a] See Matcha [31].

The relativistic energy correction was much smaller in size than the correlation energy correction. Analogously to MECE, the extra relativistic energy was computed as

$$E^{\text{rel}}(\text{Li}^+) + E^{\text{rel}}(\text{Cl}^-) - E^{\text{rel}}(\text{Li}) - E^{\text{rel}}(\text{Cl}) = -0.0021 \text{ a.u.}$$

This is to be combined with MECE. Adding 0.0341 a.u. to the computed $D_e$ of 0.1406 a.u., one obtains a result close to that of experiment. The relativistic and correlation corrections given by Matcha [31] are presented in Table XX. The corrected dissociation energies tend to be low by about 0.01 a.u., which presumably represents the difference of the calculated molecular energy from the restricted Hartree–Fock energy.

**Table XX**[a]

*Dissociation Energies for Alkali Halides*[b]

| Molecule | $D_e$ (calc.) | MECE | Rel. Corr. | $D_e$ (corr.) |
|---|---|---|---|---|
| LiCl | 0.1406 | 0.0362 | −0.0021 | 0.1747 |
| NaF | 0.112 | 0.067 | −0.0015 | 0.178 |
| NaCl | 0.117 | 0.031 | −0.002 | 0.146 |
| KF | 0.113 | 0.65 | −0.001 | 0.176 |
| KCl | 0.125 | 0.029 | −0.002 | 0.152 |
| LiBr | 0.144 | 0.011 | −0.002 | 0.153 |

[a] Energies are given in atomic units.
[b] See Matcha [31].

Calculated forces for both atoms in each molecule were among the properties reported by Matcha [31]. The Hellmann–Feynman theorem was not always well satisfied. Thus, the force in atomic units on Li and Cl in LiCl as functions of $R$ could be represented by:

$$F_{\text{Li}} = 0.0518 + 0.3739\varrho - 1.7924\varrho^2; \qquad F_{\text{Cl}} = -0.0178 - 0.0603\varrho + 0.7223\varrho^2.$$

Only the former agrees with

$$dE/dR = 0.3812\varrho - 1.552\varrho^2,$$

derived by differentiating $U(R)$. The constant terms in $F_{\text{Li}}$ and $F_{\text{Cl}}$, 0.0518 and −0.0178, are the forces at the calculated equilibrium internuclear distance, which would vanish for a function satisfying the Hellmann–Feynman theorem. For other molecules, one or both of the force curves was generally poor compared to $dE/dR$ derived from $U(R)$, although the errors were not much greater for the heavier molecules than for LiCl.

The calculated forces were quite sensitive to improvements in the basis, as usually occurs (see Volume I, Chapter III, Section D). On the other hand, it was found [31] that the slope and curvature of a plot of $-\langle V \rangle / 2 \langle T \rangle$ versus $R$ was fairly insensitive to the quality of the basis set, which suggested that one could compute $U(R)$ from an accurate calculation of $E$ at one $R$ plus less accurate calculations at other $R$. From the latter, $dE/dR$ could be obtained by invoking the virial theorem, and the constant term needed to define the energy could be derived from the single accurate calculation.

Near Hartree–Fock calculations for SF and SeF at the experimental values of $R_e$ have been performed [89], giving $D_e$ equal to 0.0532 a.u. $\pm$ 0.005 and 0.0449 $\pm$ 0.005, respectively. The MECE, from the known value for OF (0.105 $\pm$ 0.009) was estimated as 0.103 and 0.101 a.u., respectively.

Of the rare-gas–halogen systems, KrF and its positive ion have been studied in SCF and CI approximations by Liu and Schaefer [90]. The basis orbitals were Slater functions, the set being built from the double-zeta atomic sets by the addition of polarization functions and other orbitals. For both theoretical and experimental reasons, the ground state of KrF was expected to be $^2\Sigma^+$ and that of KrF$^+$ $^1\Sigma^+$. Both dissociate properly in the single-determinant approximation. The error in the calculated energies, relative to the true Hartree–Fock, was estimated as 0.005 a.u. in this calculation. Some potential constants which were derived are found on p. 285. The molecule KrF was found to be not bound in the single-determinant approximation, and it appeared that the effect of correlation was not large enough to account for this.

### (g) Conclusions

It appears that one can perform approximate Hartree–Fock calculations leading to $U(R)$ for any diatomic system desired. The resulting potential constants and other features, however, will generally be reliable only qualitatively, unless one has an idea of (a) the expansion error—how closely the limited basis expansion approaches the true Hartree–Fock function and (b) the correlation error—in what respects the Hartree–Fock $U(R)$ resembles or differs from the true one. These questions can be answered when one has available similar calculations on a set of related molecules, for which exact Hartree–Fock and exact results are known. In many cases, analysis of the correlation error may make it possible to correct calculated potential constants to accurate values.

4. *Results of Configuration Interaction Calculations*

The energy of the restricted Hartree–Fock function can be approached fairly closely from above by the use of basis expansions for the molecular orbitals, as the results of the preceding sections demonstrate. In configuration interaction calculations, the lower limit of energy is the exact non-relativistic energy. Correlation energy ($E^{corr}$) being defined as the difference between the restricted Hartree–Fock energy and the nonrelativistic energy, the energy improvement of configuration interaction over the restricted Hartree Fock may be expressed as percentage of correlation energy recovered. The amount of computational effort needed for additional energy improvement rises rapidly as we approach the exact energy, and not many calculations (except for very small systems) can recover more than half the correlation energy. However, the accurate computation of $U(R)$ is still possible if the part of the correlation energy not accounted for is $R$-independent.

Similarly, if $U(R)$ rather than absolute values of energy is of interest, a simple wave function whose shortcomings are roughly $R$-independent is superior to more sophisticated work which may improve results preferentially for one region of $R$. For example, full valence-shell configuration interaction using a minimal basis, which is often practical, seems to give a good picture of the shape of the potential curve. The actual energies may be no better than what is afforded by a good Hartree–Fock calculation, but atoms and molecules are described equally well.

The discussions of configuration interaction calculations are divided into third-level subsections as follows: (*a*) rare-gas interactions, (*b*) hydrides, (*c*) other systems. Calculations by methods, such as separated pair theories, which differ formally from CI but which in fact reduce to CI calculations (see Volume I, Chapter III, Section A) are included. For further discussion of the accomplishments of CI, Schaefer's book [87] may be consulted. Complete bibliographies are found in Chapter IV.

(*a*) *Rare Gas Interactions*

Improved calculations of the repulsive interaction between two He atoms in their ground states have appeared concurrently with improved experimental measurements of the interaction potential. Margenau and Kestner [56a] discuss many of the He–He calculations and comparison of the results with experiment. They give analysis of many of the atomic and molecular wave functions to describe the different types of electronic correlation. The curve $U(R)$ is analyzed into Hartree–Fock, pair correlation, and many-electron effects (less than 15% of the interaction energy). By interpreting the effects of different contributions to $U(R)$ at various values of $R$, they point out shortcomings in some of the approximate calculations, and suggest how certain approximate results may be corrected to give $U(R)$ accurately.

As was discussed in Chapter I, Section C, there is an avoided curve crossing for small $R$. For $R < 0.6\,a_0$, the wave function is described in the single-configuration approximation as $(1\sigma_g)^2(2\sigma_g)^2$, which correlates with the $(1s)^2(2s)^2$ ground state configuration of the united atom, Be. For larger $R$, the configuration $(1\sigma_g)^2(1\sigma_u)^2$ has lower energy, since the $1\sigma_g$ and $1\sigma_u$ molecular orbitals correlate with the He 1s atomic orbitals. A configuration interaction calculation is thus needed for accurate small-$R$ energies. There is no problem for large $R$, since the restricted Hartree–Fock function dissociates correctly to Hartree–Fock He atoms.

Phillipson [91] carried out CI calculations for $He_2$ at small $R$ in 1962, using up to 64 configurations. Discrepancies of his final results with scattering data showed the necessity of reconsideration of the latter. Phillipson's CI [91] used 1s, 1s′ (different exponential parameter), 2s, 2p, and 2p$\pi$ orbitals on each center. So long as only s orbitals were used, $U(R)$ was only lowered, and not changed in shape, compared to the single-configuration results. With 64 configurations, about half of the estimated correlation energy was recovered. The energy error in the range 0.5–2 Å was estimated as less than 1.25 eV, while the error in $U(R)$ was estimated at half of this because of cancellation from one $R$ value to another.

Barnett [92] performed a limited CI study using $\sigma$-type orbitals in ellipsoidal coordinates. All the configurations used were constructed from a product of four such basis functions with one of the two singlet spin functions for four electrons. Each such product was antisymmetrized and made symmetric to inversion and vertical reflection using projection operators. Various choices of one- and two-configuration functions were investigated, using open shells and optimizing nonlinear parameters. Barnett concluded [92] that a very few configurations could suffice if chosen carefully (for $R \geq 2$ the two-configuration results were superior to those of Phillipson).

Matsumoto et al. [93] carried out more extensive CI calculations, using a basis set composed of 28 basis functions in ellipsoidal coordinates (10 $\sigma_g$, 4 $\sigma_u$, 8 $\pi_u$, 4 $\pi_g$, and 2 $\delta_g$ functions). They kept the number of configurations manageable by an iterative technique (see Volume I, Chapter III, Section A.7) in which natural orbitals were derived at each stage and used to construct configurations for a succeeding linear variation calculation. Their final wave function contained 50 configurations of natural orbitals. They also carried out SCF calculations for the three important configurations. The results, shown in Table XXI, indicated that the crossing between the $(1\sigma_g)^2(2\sigma_g)^2$ and $(1\sigma_g)^2(1\sigma_u)^2$ curves occurred for $R$ about 0.6 $a_0$, but that it did not lead to large distortions in the shape of $U(R)$ because the single-configuration curves intersect at a small angle. Matsumoto et al. [93] estimated that their $U(R)$, which lay about 0.02 a.u. below that of Phillipson's, was still about 0.02 a.u. above the exact $U(R)$. The estimated exact curve could be fit with $ae^{-bR}$, with $a = 237$ eV and $b = 4.23/\text{Å}$.

**Table XXI**

*SCF and CI Energies$^a$ for $He_2$ in a.u.*

| $R$ | $(1\sigma_g)^2(2\sigma_g)^2$ | $(1\sigma_g)^2(1\sigma_u)^2$ | $(1\sigma_g)^2(1\pi_u)^2$ | NO-CI |
|---|---|---|---|---|
| 0.25 | $-12.7256$ | $-12.206$ | $-12.488$ | $-12.79858$ |
| 0.50 | $-10.7393$ | $-10.591$ | $-10.273$ | $-10.81998$ |
| 0.75 | $-9.309$ | $-9.5026$ | $-9.212$ | $-9.57766$ |
| 1.00 | $-8.272$ | $-8.7928$ | $-8.210$ | $-8.85830$ |

$^a$ See Matsumoto et al. [93].

Several calculations have started from the Hartree–Fock wave functions and added in configurations to describe specific physical effects. Murrell and Shaw [94] used their perturbation formalism (see Chapter I, Section B.2 and B.3) to compute Coulomb and exchange energies for $He_2$ from approximate Hartree–Fock wave functions for He, simply calculating the expectation value of the perturbing operator over the antisymmetrized atomic wave functions. The dispersion energy from other calculations was added and the minimum in $U(R)$ was found [94] at 5.5 $a_0$, in agreement with results of other workers. The calculated dissociation energy was $3.3 \times 10^{-5}$ a.u. Kestner and Sinanoğlu [95] also started with the SCF potential curve and corrected it for the effects of pair correlations, finding $R_e = 5.75 a_0$ and $D_e = 1.3 \times 10^{-5}$ a.u., but incorrect long-range behavior [$C_0$ (van der Waals) $= 0.853$ instead of $1.471$].

It was convenient to go from the Hartree–Fock orbitals to orbitals localized on the two atoms. For the He–He interaction, the $n$-electron ($n > 2$) contribution to the correlation energy was necessarily zero for the separated atoms and only 0.02 eV for the united atom. Furthermore, the pair correlations were of two kinds: the intraorbital pair correlations (referring to the localized orbitals) are those for the He atoms, with some small distortion, while the interorbital pair correlations made an important contribution to $U(R)$. It was expected [95] that these could be well-described by a configuration interaction trial function.

An accurate calculation [19] for $He_2$ over the range $R = 0$ to $R = \infty$ was corrected by Klein *et al.* by the "pseudokinetic energy procedure" [18] (see Subsection 1). The resulting $U(R)$ curve was estimated to be only 1 eV distant from the true. The value for $U(R)$ as derived from scattering measurements was too low by about 3 eV. Fits were given [19] to the corrected $U(R)$, over limited ranges of $R$, to Born–Mayer forms, $Ae^{-bR}$.

Bertoncini *et al.* [96, 97] performed five-configuration calculations for $He_2$ for $R$ between 3.78 $a_0$ [$U(R)$ for shorter distances is accurately known from other CI work] and 7.5 $a_0$ [for $R > 7.5$, the $R^{-6}$ and $R^{-8}$ van der Waals terms accurately represent $U(R)$]. One of the goals was accurate knowledge of $U(R)$ near the van der Waals minimum at 5.7 $a_0$. They used pairs of excitations in a localized orbital basis, one on each atom (like those contributing to the van der Waals attraction), to form the four excited configurations. In the spirit of the optimized valence configuration approach (see Section B) intraatomic correlation energy was not calculated, since it would not affect the shape of $U(R)$, but only the absolute value of the energy. The resulting potential curve was tabulated and plotted, and comparisons made with other theoretical and experimental work.

In connection with the description of charge transfer reactions, Michels and Harris [98] calculated the ground state of $HeNe^+$ using a complete valence shell configuration interaction in a limited basis. There has been nothing like the extensive work for He–He for other rare gas interactions, CI becoming more difficult and the possibility of finding and interpreting new effects more slight.

*(b) Hydrides*

The two-electron $HeH^+$ system was the subject of an accurate calculation by Evett [99] in 1956. Evett used as many as 23 correlated basis functions, like those of James and Coolidge [100], in ellipsoidal coordinates:

$$\psi_i = \exp[-\alpha(\xi_1 + \xi_2)][\xi_1^{m_i}\xi_2^{n_i}\eta_1^{j_i}\eta_2^{k_i} + \xi_1^{n_i}\xi_2^{m_i}\eta_1^{k_i}\eta_2^{j_i}]r_{12}^{p_i}.$$

He calculated energies for six values of $R$ between 1.20 and 1.60 $a_0$, with $\alpha = 1.375$ (chosen from consideration of related systems). Some of his results are included in Table XXII. Because there was no attempt to op-

## Table XXII

### CI Results for HeH$^+$

| | $R_e$ ($a_0$) | $D_e$ (eV) | $\omega_e$ (cm$^{-1}$) | $\omega_e x_e$ (cm$^{-1}$) | $\alpha_e$ (cm$^{-1}$) | $D_e$ (cm$^{-1}$)[a] |
|---|---|---|---|---|---|---|
| SCF [81] | 1.4554 | 1.943 | 3233 | 176 | 2.82 | — |
| Evett [99] | 1.432 | 1.90 | 3600 | 201 | — | — |
| Anex [82] | 1.446 | 1.931 | — | 184 | — | — |
| Michels [102] | 1.44 | 1.85 | 3379 | 313.7 | 3.785 | 0.01615 |
| Stuart and Matsen [104] | 1.468 | 1.930 | 3184 | 255 | — | — |
| Wolniewicz [106] | 1.4632 | 2.039 | — | — | — | — |

[a] Not a dissociation energy.

timize $\alpha$ at different values of $R$, they are not as accurate as calculations performed later by, in principle, less efficient methods. For example, Anex [82] performed configuration interaction calculations using one-electron basis functions in ellipsoidal coordinates (no correlation factors), optimizing nonlinear parameters at each of four values of $R$. He considered up to eight orbitals of $\sigma$ symmetry, four of $\pi$ symmetry, and two of $\delta$ symmetry; the largest basis used $6\sigma + 3\pi + 2\delta$. Estimated limits for different contributions to the correlation energy were given for different values of $R$.

Michels [102] performed single configuration open shell calculations with a mixed ellipsoidal-atomic basis. For each of 12 values of $R$, values of two nonlinear parameters were optimized. There was good agreement in $U(R)$ with more complex calculations. Part of this, as explained by Michels, was due to cancellation of errors. The curve $U(R)$ for excited states was also computed; indeed, this was the primary interest of Michels's work [102]. The united atom and separated atom correlations of the states calculated were discussed. The local energy calculations of Conroy [103] (see Volume I, Chapter III, Section E.2) gave a dissociation energy $D_e$ of 2.28 eV, which seems too much higher than the values obtained by other workers (see Table XXII) to be due to the increased sophistication of his calculations.

The line labeled Stuart and Matsen in Table XXII refers to these authors' 30-configuration calculations [104], which employed a one-center expansion centered on He. Because HeH$^+$ is one-center (both electrons centered on one nucleus) at large $R$, good results were expected from a one-center calculation. A subsequent one-center calculation [105] on this system allowed for a shift of the expansion center from the He nucleus. Except for small $R$, this gave little improvement, and it was concluded that optimization of nonlinear parameters in the basis, other than the location of the expansion center, would be more profitable.

The results of Wolniewicz [106] probably are extremely close to the exact results, and are the standard against which we judge other calculations on this system. Wolniewicz used 64 correlated James–Coolidge basis functions, of the kind used by Evett [99], but including a factor of $\exp[-\beta(\eta_1 + \eta_2)]$. He optimized the energy with respect to the nonlinear parameters $\alpha$ and $\beta$. (Note that $\beta = 0$ is appropriate to a homonuclear molecule and $\beta = \pm\alpha$ to a one-center system.) Methods like the James–Coolidge expansion, with powers of $r_{12}$, cannot be used conveniently for systems with more than two electrons, so that we shall not encounter them again in this chapter.

For the neutral species HeH, Duparc and Buckingham [107] computed energies for $0 \leq R \leq 1\ a_0$ using two configurations constructed from four molecular orbitals. Energies were calculated for 16 values of $R$, but parameters were optimized only for a few $R$ values and values interpolated for others. A fit of the energy to a quartic in $R$ with no linear term (see Chapter I, Section C, Eq. 245) gave

$$U = -4.015 + 18.375R^2 - 47.488R^3 + 45.535R^4.$$

The term in $R^2$ agreed with Bingel's predictions for the united atom (see Chapter I, Section C) but not that in $R^3$. The corrected theory of Brown and Steiner gave much better agreement. According to Eq. (239) of Chapter I, the ratio of the coefficient of $R^3$ to that of $R^2$ in the united atom expansion should be $-3$. Duparc and Buckingham's calculations [107] for $He_2$ at small $R$ gave similar results with respect to the united atom theories.

For the first-row diatomic hydrides from LiH to HF, Bender and Davidson [108, 109] gave very extensive CI calculations at the equilibrium value of $R$ for each. The basis of Slater orbitals was large enough to give an error in the SCF energy of only 0.0003 a.u. for LiH and 0.003 a.u. for HF, compared to the results of Cade and Huo [32]. A large number of singly and doubly excited configurations (from the Hartree–Fock) was used for each molecule. The iterative natural orbital technique (Volume I, Chapter III, Section A.7) was employed, and 70% or more of the correlation energy was obtained in each case. Among the breakthroughs making calculations of this kind possible, Bender and Davidson included faster computers, the availability of accurate SCF functions for these systems [32], and improved algorithms for diagonalization of large matrices. Even with these advances, the large number of functions used made a series of calculations for different $R$ values, which would be needed to compute $U(R)$, prohibitive. Because of the large amount of work and the large number of configurations needed to get high accuracy, even if the configurations are optimized (as by using natural orbitals), the authors [108, 109] were pessimistic about the possibility of using CI for calculation of potential surfaces. With respect to pair calculations of the correlation energy, they emphasized that the pairs and pair energies depend on the choice of SCF orbitals: canonical, localized, etc.

Among the first-row hydrides, LiH (the smallest) has been to a great extent a testing ground for methods of calculation. We can only discuss a few of the calculations on LiH here (see Chapter IV for bibliographies). The results are tabulated in Table XXIII.

Ebbing's configuration interaction calculation [109a] for this molecule, carried out in 1962, gave the best energy available at that time. It employed 53 configurations built from basis functions in ellipsoidal coordinates. Ebbing [109a] started with a restricted Hartree–Fock calculation for each value of $R$, and used the Hartree–Fock orbitals to form the configurations. At $R = 3.00\,a_0$, the single-configuration calculation gave $-7.985$ a.u. and the best CI $-8.041$ a.u. The exact energy was estimated at $-8.0703$ a.u. Kahalas and Nesbet [58], also starting from a Roothaan–Hartree–Fock function, performed CI by perturbation theory. With 13 carefully selected Slater orbitals as basis functions, the best energy obtained was $-8.0171$ a.u. Using the calculated value for the energy of Li, this gave a dissociation energy of $0.0826$ a.u., which is almost $90\%$ of the experimental value.

The Matsen and Browne CI calculation [110] (1962) and that of Browne and Matsen [111] (1964), were discussed in Section B. Bender and Davidson [112] performed a limited CI for LiH with a double-zeta basis, aimed at generating potential curves for excited states as well as the ground state. Thus, parameters were not optimized for any particular molecular state and the ground state energy near $R_e$ was only $-8.0036$ a.u. Subsequently, these authors performed [113] an iterative natural orbital CI (see Volume I, Chapter II, Section A.7) for this molecule at the experimental $R_e$, using a basis of ellipsoidal orbitals. They obtained an energy of $-8.0606$ a.u., which represents $80\%$ of the correlation energy. Unfortunately, $U(R)$ could not be computed.

In addition to a Hartree–Fock–Roothaan calculation with a minimal basis, Sahni et al. [10] performed a CI calculation using the same basis, exponents being optimized at each of 52 $R$-values. The 41 configurations were the restricted Hartree–Fock plus doubly excited determinants. The RHF and CI results are compared in Table XXIII. $D_e$ was computed in each case as $E(R_\infty) - E(R_e)$, with $R_\infty = 9\ a_0$ and $R_e = 3.015\ a_0$, which is the experimental value. The actual energies at $R_e$ were $-7.972840$ and $-7.989836$ a.u.

**Table XXIII**

*Some CI Results for* LiH

| | $R_e\ (a_0)$ | $D_e$ (eV) | $\omega_e$ (cm$^{-1}$) | $\omega_e x_e$ (cm$^{-1}$) | $\alpha_e$ (cm$^{-1}$) | $B_e$ (cm$^{-1}$) |
|---|---|---|---|---|---|---|
| Matsen and Browne [110] | 3.075 | 1.793 | 1602 | 28.19 | 0.0867 | |
| Browne and Matsen [111] | 3.046 | 2.34 | 1438 | 86 | | |
| Sahni *et al.* SCF [10] | 3.03681 | 3.657 | 1440.78 | 21.277 | 0.176 | 7.3911 |
| Sahni *et al.* CI [10] | 3.03492 | 2.96 | 1376.07 | 23.628 | 0.2109 | 7.3935 |
| Brown and Shull [11] | 3.060 | 2.12 | 1374.3 | 22.5 | | 7.2 |
| Silver *et al.* [8] | 3.046 | 2.30 | 1483 | 22 | 0.2849 | 7.381 |
| Experiment | 3.015 | 2.52 | 1405.6 | 23.20 | 0.213 | 7.5131 |

Other CI calculations on LiH obtained considerably better energies near $R_e$: $-8.0561$ a.u. (Browne and Matsen [111], mixed ellipsoidal-atomic basis) and $-8.0606$ a.u. (Bender and Davidson [113], natural orbital CI) but the work of Sahni *et al.* was aimed at reproducing $U(R)$ for a large range of $R$. By using the same configurations, and reoptimizing exponential parameters for each $R$, it was hoped [10] to achieve "uniform" accuracy and obtain a potential curve parallel to the exact one. The improvement in the energy near $R_e$ afforded by CI was small compared to the improvement at large $R$, where the incorrect dissociation of the RHF was corrected.

Brown and Shull [11] carried out CI for the LiH ground state using a basis of ellipsoidal functions. Parameters were chosen to give a good description of the three configurations shown in previous calculations to dominate the ground state wave function: Li ($^2S$) H ($^2S$), Li ($^2P$) H ($^2S$), and Li$^+$ ($^1S$) $H^-$ ($^1S$). 69 configurations were used for calculations at 34 internuclear distances between 1 $a_0$ and 10 $a_0$, although the results were expected [11] to be accurate in the more limited range $2\,a_0 \leq R \leq 8\,a_0$. The lower energies obtained by other workers were attributed to more careful optimization of inner-shell orbitals. Results for properties of $U(R)$ are given in Table XXIII. We will discuss the last calculation, which involves pair functions, later.

The binding energy of LiH$^+$ was computed by Browne [114], using a generalized valence bond function in a mixed ellipsoidal–Slater basis set. Several different numerical potential curves were obtained. The difference of atomic and molecular energies gave $D_e = 0.104 \pm 0.016$ eV.

Brown's 38-configuration calculation [115] for BeH$^+$ at $R_e$ resembled the Brown and Shull [11] LiH calculation, and comparisons were made. There were four major configurations, Be$^+$ ($^2S$) H ($^2S$), Be$^+$ ($^2P$) H($^2S$), Be$^{2+}$ ($^1S$) H$^-$ ($^1S$), and Be ($^1S$) H$^+$. The correlation energy of 0.0934 a.u. was very close numerically to that of Be, but analyses into different kinds of correlation and into contributions of different shells showed important differences between the two systems.

The pair-correlation theories have often been tested on the hydrides. Ahlrichs and Kutzelnigg [116] used a wave function for LiH which was an antisymmetrized product of strongly orthogonal geminals and derived decoupled equations for the pairs by dropping certain terms (see Volume I, Chapter III, Section A.8). The correlation energy was a sum of pair contributions minus a correction which was small for this molecule. The basis of Gaussian functions was felt to be inadequate for the inner-shell pair, and only 78% of the overall correlation energy was obtained. It was estimated that another 15% was due to the inadequacy of the basis set. Between 2.7 and 3.3 $a_0$ ($R_e = 3.015\,a_0$) the correlation energy of the valence pair was roughly constant, although this resulted from an increase in left–right correlation and decrease in angular correlation with increasing $R$. If only the former were included, $R_e$ was predicted to be 3.10 $a_0$, in worse agreement with experiment than the Hartree–Fock, but inclusion of angular correlation as well brought $R_e$ back to a good value.

Also using separated pair wave functions, Silver et al. [8] calculated energies for a series of $R$-values for LiH and BH. The wave functions were antisymmetrized products of strongly orthogonal geminals, each expanded in their natural expansions. The energy was varied with respect to the geminal expansion coefficients and with respect to the coefficients of expansion of the natural orbitals in a set of basis functions (atomic orbitals). Over 80% of the correlation energy was obtained for Li and LiH but less than half of the correlation energy for BH. The BH energy was the lowest yet obtained, and the LiH energy within 0.006 a.u. of that obtained by

Bender and Davidson [113]. The energies were fit to Dunham polynomials,

$$U(R) = a_0 \varrho^2 \left( 1 + \sum_k a_k \varrho^k \right), \qquad \varrho = (R - R_e)/R$$

of various orders. For LiH, the derived values were (experimental values in parentheses): $a_0 = 0.3367$ (0.2995), $a_1 = -2.254$ ($-1.884$), $a_2 = 4.709$ (2.378), $R_e = 3.045\, a_0$ (3.015 $a_0$). The derived potential constants for LiH are found in Table XXIII; those for BH were $\omega_2 = 2928\ \text{cm}^{-1}$ (2367.5), $\omega_e x_e = 45.40\ \text{cm}^{-1}$ (49), $\alpha_e = 0.4887$ (0.408). It was emphasized [8] that careful optimization of all parameters at each $R$ was important, since results of pair calculations were more sensitive to variations in orbital exponents than were those of SCF calculations. Since there were about a hundred parameters, it was impossible to carry out optimization of all of them for each $R$, and values were interpolated from results of extensive calculations at a few $R$-values.

The recent calculations on BH by Gélus et al. [117], also using a separated electron pair wave function, accounted for a much greater part of the correlation energy, due to the relaxation of the strong orthogonality constraint. Each pair of localized orbitals derived from a Hartree–Fock calculation was replaced by singlet or triplet pair functions (see Volume I, Chapter III, Section A.8) in their natural orbital expansions. The energy, with correlation energy calculated as a sum of interpair and intrapair contributions, was not a true upper bound to the exact energy. It was argued [117] that the correction terms needed to guarantee an upper bound were roughly independent of $R$, so that they could be neglected in this work, whose aim was to study the effect of correlation energy on $R_e$ and $k$ (force constant). The calculation was performed in a basis of Gaussian-like functions.

The calculated potential curve was practically parallel to that derived by Cade and Huo [32] from SCF calculations, so that $R_e$ and $k_e$ were not changed much (see Table XXIV) by the effects of inclusion of pair correlations, but the changes seem to be in the right direction. This statement must be qualified, because the SCF calculations with the basis set of Gélus et al. [117] give results differing from those of the more accurate SCF calculations of Cade and Huo [32] by almost as much as they differ from the results of the pair calculation. Analysis of the pair calculations [117] showed that the inner shell correlations were unimportant in determining $R_e$ and $k$, but that interpair contributions, though small numerically relative to intrapair contributions, were of great importance here.

The "valence bond" calculation whose results are cited in Table XXIV is that of Harrison and Allen [118], which used Gaussian lobe functions.

**Table XXIV**

*Force Constants and Equilibrium Distances for BH*

| Method | Ref. | $R_e$ $(a_0)$ | $k$ $(10^5$ dynes/cm) |
|---|---|---|---|
| Accurate SCF | [32] | 2.305 | 3.397 |
| Approx. SCF | [117] | 2.311 | 3.330 |
| Valence bond | [118] | 2.536 | — |
| Strongly orth. pairs | [8] | 2.324 | 4.623 |
| Independent pairs | [117] | 2.316 | 3.190 |
| Experiment | [117] | 2.336 | 3.028 |

Valence bond functions corresponding to a full CI for the valence shell were used to calculate energies and potential curves for eight states of BH. For the $^1\Sigma^+$ state, which was found to lie lowest, the energy obtained was below any previously reported (Harrison and Allen [118] list and comment on previous calculations). The basis set gave an SCF energy 0.056 a.u. above that of Cade and Huo [32], but the SCF energies gave $R_e$ 0.3 $a_0$ greater. The value of $R_e$ from the CI (valence bond) calculation (see Table XXIV) is also high.

Blint *et al.* [119] computed energies for BH from a wave function of very general form, built from six different and nonorthogonal molecular orbitals. When certain restrictions were placed on these molecular orbitals, the general function reduced to the simple valence bond, restricted Hartree–Fock function, or extended Hartree–Fock functions previously proposed [120]. A particular coupling scheme for the electron spins, to give an overall spin singlet, was not imposed, as in the other methods; e.g., the generalized valence bond function pairs the 1s electrons on B, the 2s electrons on B, and the two electrons in the bond, to give three local singlets, which are then combined. By use of the properties of the permutation symmetry group (see Section B), the six-electron function was made into a $^1\Sigma^+$ state with full permutation symmetry. The dissociation energy $D_e$ from the most general function, computed as a difference of calculated energies for B+H and BH, was −25.04602 a.u. + 25.16640 a.u. = 3.275 eV. The generalized valence bond gave, correspondingly, −25.04558 a.u. + 25.16639 a.u. = 3.287 eV, while the difference of calculated SCF energies was −25.02905 a.u. + 25.13137 a.u. = 2.784 eV.

The CI calculations on NH by Kouba and Ohrn [121] attempted to obtain comparable accuracy for the ground and several excited states, with the correct ordering of levels, and spectroscopic constants in reasonable agreement with experiment. The core was frozen, the CI involving valence electrons only, and natural orbital analysis like that of Bender and Davidson [113] (Volume I, Chapter III, Section A.7) was used to keep the calculation manageable. The natural orbitals were used for new configurations and the process was repeated. For the ground state, of $^3\Sigma^-$ symmetry, the results are given in Table XXV.

**Table XXV**

*Calculated Properties of NH Ground State*

| | $D_e$ (eV) | $R_e$ (Å) | $\omega_e$ (cm$^{-1}$) | $\omega_e x_e$ (cm$^{-1}$) | $B_e$ (cm$^{-1}$) | $\alpha_e$ (cm$^{-1}$) |
|---|---|---|---|---|---|---|
| Kouba and Ohrn [121] | 2.42 | 1.12 | 3224 | 117.2 | 14.28 | 0.564 |
| Hartree-Fock [32] | 2.10 | 1.018 | 3556 | 66.7 | 17.32 | 0.572 |
| O'Neil and Schaefer [122] | 3.058 | 1.041 | 3300 | 120 | 16.56 | 0.760 |
| Experiment | 3.41±0.16 | 1.038 | 3125.6 | 78 | 16.65 | 0.646 |

O'Neil and Schaefer [122] used the Bender–Davidson iterative natural orbital scheme (see Volume I, Chapter III, Section A.7) to perform CI calculations for three states of NH, at different values of $R$. The basis was smaller than that used by Bender and Davidson [108, 109] in their work on the first-row hydrides; the series of calculations would otherwise have been prohibitively expensive. The basis was good enough to get an SCF energy only 0.003 a.u. above Cade and Huo's [32] for the ground $^3\Sigma^-$ state and comparable errors relative to their results for excited states. A least-square fit of a quartic through the energies at six values of $R$ gave the tabulated results for the ground state (see Table XXV).

Separated-pair wave functions (antisymmetrized products of geminals) were applied to NH by Mehler *et al.* [123]. For the $^3\Sigma^-$ ground state, the geminals used were: $K$-shell singlet, lone-pair singlet, bonding singlet, and triplet. Although $R_e$ was correct to 0.290, the binding energy was only $\frac{2}{3}$ the correct value. This was attributed to changes in intergeminal correlation energy within the valence shell on molecule formation, since intergeminal effects cannot be described by this sort of a wave function.

The separated pair calculations we have cited, together with the work of Bender and Davidson [124] on HF, show that the antisymmetrized product of strongly orthogonal geminals becomes increasingly inappropriate for systems more complicated than LiH. Interpair effects become more important. Force constants from the separated-pair calculations are quite unreliable, while $R_e$ and $D_e$ are hardly better than from restricted Hartree–Fock [125; 87, pp. 195 et seq.].

For HF and HF$^-$, Michels and Harris [98] generated $U(R)$ using a valence orbital configuration interaction. The same basis set was used for both systems, in order to get results useful for consideration of the electron detachment reaction. The HF$^-$ potential curve was found to be repulsive. This example serves as a reminder that, for the first few rows of the periodic table, potential curves good enough to be useful can be generated from CI.

## (c) Other Systems

Much of this subsection is devoted to the method of optimized valence configurations. We will consider homonuclear systems first.

Das and Wahl and others [126, 127] successfully applied their optimized valence configuration (OVC) approach to $H_2$, $Li_2$, and $F_2$. In addition to the SCF configuration, they included configurations formed by replacing the highest doubly occupied $\sigma$ orbital by a two-electron function, $\sum_\alpha \varphi_\alpha(1)\varphi_\alpha(2)$, expanded in the same basis functions as the SCF. The added configurations allowed correct dissociation (left–right correlation), and provided angular and in–out correlation. Their choice was made by trial. Results from the calculated $U(R)$ for $Li_2$ and $F_2$ are shown in Table XXVI. The optimized valence configuration results are compared with those from Hartree–Fock, those from the optimized two-configuration wave function (which permits proper dissociation) and with experiment.

### Table XXVI

*Optimized Valence Configuration Calculations[a]*

|                     | $\omega_e$ (cm$^{-1}$) | | $R_e$ ($a_0$) | | $D_e$ (eV) | |
|---------------------|-----------|-----------|-----------|-----------|-----------|-----------|
|                     | $Li_2$    | $F_2$     | $Li_2$    | $F_2$     | $Li_2$    | $F_2$     |
| Hartree–Fock        | 336       | 1257      | 5.26      | 2.50      | 0.17      | $-1.37$   |
| Two configurations  | 344       | 678       | 5.43      | 2.74      | 0.46      | 0.54      |
| Opt. val. con.      | 345       |           | 5.09      |           | 0.93      |           |
| Experiment          | 351       | 892       | 5.05      | 2.68      | 1.03      | 1.65      |

[a] See Das and Wahl [126], Wahl et al. [127].

A subsequent $Li_2$ calculation [128] by the OVC method, with seven configurations, gave a better value for $D_e$, reported other potential constants, and compared results with Hartree–Fock and experiment. The experimental values were derived from a Hurlburt–Hirschfelder potential function, parametrized to fit experiment. The effect on calculated potential constants of addition of configurations of different types was discussed; most properties are improved from their Hartree–Fock values by CI. Only a small fraction of $E^{corr}$ was obtained, but the shape of $U(R)$ was clearly improved, making it more parallel to the Hurlburt–Hirschfelder curve. The values of $R_e$, $\omega_e$, and $D_e$ were 5.09 $a_0$, 345.27 cm$^{-1}$, and 0.99 eV. The values of $\alpha_e$ and $\omega_e x_e$ were slightly improved compared to Hartree–Fock.

For $R > 5.3\, a_0$, a configuration interaction calculation [129] on $Li_2$ based on the Hartree–Fock function plus pair excitations in localized orbitals succeeded in obtaining the dispersion interaction (not as a series in $R^{-1}$, of course) as well as the exchange forces. Only the excitations of the valence electrons were considered, and the energy was minimized with respect to the linear CI parameters as well as parameters entering the orbitals used in building the configurations. Thus, the calculation is equivalent to an optimized valence configuration calculation. The basis functions were Gaussians, taken in groups; some parameters were not optimized for $Li_2$ but taken from atomic calculations. Kutzelnigg and Gélus [129] used the wave function.

$$\Psi = \mathscr{A}\{\phi_1(1)\alpha(1)\phi_1(2)\beta(2)\phi_2(3)\alpha(3)\phi_2(4)\beta(4)\omega_n(5,6)\mathscr{S}(5,6)\}$$

where the pair function was written in its natural expansion.

$$\omega_n(i,j) = \sum_{k=1}^{n} C_k \chi_k(i)\chi_k^*(j).$$

For $n = 2$, the pair function may be written in a form resembling the valence bond by putting

$$\chi_1 = [2(1+s)]^{-1/2}(u+v); \qquad \chi_2 = [2(1-s)]^{-1/2}(u-v);$$

then $\omega_2(i,j)$ is essentially $u(i)v(j) + u(j)v(i)$. Here, $u$ and $v$ are restricted Hartree–Fock atomic orbitals for the two Li atoms and are not orthogonal. The energy for one configuration (the Hartree–Fock), for two configurations (the valence bond), and for 12 configurations was tabulated [129] as a function of $R$. The difference in the last two, for large $R$, was identified as the dispersion interaction, and compared with calculated dispersion forces, since the exchange energy should vanish exponentially. This was consistent with the fact that energy for two configurations became $R$-independent for $R > 16$. Kutzelnigg and Gélus noted that the two-configuration wave function was still changing for large $R$, even though the energy did not reflect this, and that Gaussians seemed to work as well as Slater orbitals for the region of $R$ considered. For $R \geq 12$, they estimated [129] that their $U(R)$ was good to 10%; for smaller $R$, that the error could be several times this.

Later, further calculations using the method of optimized valence configurations were performed [130] for $F_2$. The possibility of accurate $U(R)$ curves exists because correlation within the $F_2$ core is independent of $R$, while intershell (core-valence) correlation is negligible to a good approximation. The procedure is:

(1) Augment the basis set of a restricted Hartree–Fock calculation.

(2) Obtain the restricted Hartree–Fock orbitals.

(3) Use them to construct the multiconfiguration wave function which permits correct dissociation.

(4) Add in configurations to describe the correlation of the valence electrons and carry out multiconfiguration SCF.

For $F_2$, the core, $1\sigma_g^2 1\sigma_u^2 2\sigma_g^2 2\sigma_u^2$, was invariant over all configurations. The wave function permitting correct dissociation [130] described the valence electrons as $1\pi_u^4 1\pi_g^4 [a(3\sigma_g^2) + b(3\sigma_u)^2]$. The remaining additional configurations described split-shell ($\sigma$–$\pi$) and intrashell ($\pi$–$\pi$) correlations, correlation associated with molecule formation, and single-center valence electron correlation modified by bonding. From the resulting $U(R)$ (tabulated [130]) the Dunham analysis gave $R_e = 2.59\, a_0$ (experiment, 2.68), $D_e = 1.57$ eV (experiment, 1.65), $\omega_e = 1021$ cm$^{-1}$ (experiment, 892).

Several CI calculations for $F_2$ have formed configurations from atomic orbitals. Harris and Michels [131, 132] performed CI calculations for $F_2$ (as well as HF), forming the functions by spin projection of open-shell determinants. This allowed for greater flexibility. Formulas for matrix elements of the Hamiltonian between spin-projected open-shell functions had previously been given [131, 132]. With a minimal basis of Slater orbitals, 25 configurations could be formed. Calculations were done with the minimal set and also with a set of double-zeta atomic orbitals. The effect of omitted configurations was considered. From calculations at a number of values of $R$ including $R = \infty$, potential curves were plotted. Near $R_e$, the CI energies using the Slater orbitals were almost 1 a.u. above the SCF energies, which in turn were less than 0.1 a.u. above the atomic orbital CI and Das and Wahl's optimized double configuration energies. Since the basis functions were optimized for the atoms, the error in the calculated energies should become worse for smaller $R$. It was concluded [131, 132] that CI using atomic orbitals could give good results but was risky; the basis orbitals should be optimized for the molecule. Extended CI calculations were carried out at two internuclear distances [133] using the minimal Slater set, symmetry molecular orbitals, and Hartree–Fock molecular orbitals constructed from them, and testing different methods of choosing configurations.

In Schaefer's CI calculations [134] on $F_2$ and other molecules, all integrals were computed numerically in ellipsoidal coordinates, using crossed Gauss quadratures, rather than by semianalytic procedures. This made it possible to use Hartree–Fock atomic orbitals, each of which is expressed as a sum over basis functions, to construct the configurations, since an integral over such functions was not expressed as a sum of a large number of individual integrals over basis functions. In his discussion of previous CI calculations on $F_2$, Schaefer noted [134] that dissociation energies, calculated as differences of CI atomic and molecular energies, generally decreased as the accuracy of the atomic orbitals used increased. Schaefer reported results of a number of CI calculations, with configurations built from Hartree–Fock atomic orbitals. Various sets of configurations were tried, the largest involving 318 configurations, but the highest value that could be obtained for $D_e$ was 0.80 eV. The complete CI from the Hartree–Fock atomic orbitals gave 0.32 eV. Similar calculations for $Cl_2$ led to $D_e = 0.71$ eV (experiment, 2.48 eV). The conclusion was that Hartree–Fock atomic orbitals are not a good set of basis functions for calculations of this type, even when augmented by polarization functions.

In particular, $R_e$ and $D_e$ were worse when the CI was built from the Hartree–Fock atomic orbitals than from the minimal Slater basis. Note that the set of Hartree–Fock atomic orbitals, if parameters internal to the orbitals are not varied, is a minimal set. Typically, overcontracted sets such as these give $D_e$ too small and $R_e$ too large [87, pp. 73, 74], because they are optimized for the atoms and are insufficiently flexible in the region of molecule formation.

In connection with atoms-in-molecules methods (Chapter III, Section A.3), Balint-Kurti and Karplus [135] presented multistructure valence bond calculations on $F_2$, $F_2^-$, and LiF. Gaussian representations of the atomic orbitals were used to construct the basis functions. Numerical and graphical potential energy curves were presented, with and without atoms-in-molecules corrections, and compared to experiment. For $F_2$, for which 18 ionic and covalent functions were used, the energy errors near $R_e$ were only a few hundredths of an atomic unit. The predicted value for $R_e$ was too high by about $0.15\,a_0$ and $D_e$ too low by 0.01 a.u. For both LiF and $F_2$, some experimentation with different many-electron functions in the basis was performed. No experimental data for $F_2^-$ were available. The *ab initio* calculation predicted $R_e = 3.28\,a_0$ and $D_e = 0.087$ a.u. The increased dissociation energy relative to $F_2$ (0.06 a.u.) was surprising.

Bender and Davidson's configuration interaction calculations [136] on the low-lying states of $B_2$ were aimed at describing a large number of states, obtaining their potential curves, and establishing their relative energies. Therefore, there was no exponent optimization for a particular state. The one electron basis was in ellipsoidal coordinates. An approximate calculation was carried out for the open-shell Hartree–Fock state corresponding to the configuration $(1\sigma_g)^2(1\sigma_u)^2(2\sigma_g)(2\sigma_u)(3\sigma_g)(3\sigma_u)(1\pi_g)(1\pi_u)$. A transformation was made to the Hartree–Fock orbitals, which became the basis from which the configurations were built. For 10 values of $R$, CI was carried out using configurations which maintained the inner shells closed and placed the other 6 electrons in the orbitals $2\sigma_g$, $2\sigma_u$, $3\sigma_g$, $3\sigma_u$, $1\pi_g$, and $1\pi_u$. The lowest state was found to be of $^5\Sigma_u^-$ symmetry and $R_e$ was computed as $2.95\,a_0$.

Meckler [1, 137] carried out CI calculations for $O_2$ in 1953, using a minimal Gaussian basis set without optimizing exponential parameters, and considering only configurations which kept the 1s and 2s orbitals doubly occupied. Many-electron functions of $^3\Sigma_g^-$ and $^1\Sigma_g^+$ symmetry were formed. For the lowest $^3\Sigma_g^-$, a Morse curve was fit to the energies for $R = 1.5$, 2.0, 2.5, and $4.0\,a_0$. From this curve, Meckler found [1] $R_e = 2.255\,a_0$, $\omega_e = 1503$ cm$^{-1}$, and $D_e = 0.183$ a.u. While these compared well with the experimental results ($2.28\,a_0$, 1580.4 cm$^{-1}$, and 0.187 a.u.), the Morse curve approached a value well above atomic energies for $R \to \infty$, indicating that a cancellation of errors was involved in the good agreement.

For $O_2$ and for CN, calculations allowing full valence-shell configuration interaction in a minimal Slater set have been used to generate $U(R)$ for about 60 low-lying molecular electronic states [138, 139]. States of various symmetries were considered, but only results for the ground states are cited here. It should be mentioned that the interest of this work [138, 139] was to generate $U(R)$ for a large number of excited states, for use in interpretation of molecular spectra. Calculations were performed for a dozen values of $R$ (see Table XXVII).

More recently, Schaefer [71] performed CI on $O_2$ with 128 configurations, using the iterative natural orbital procedure. Two-center integrals were computed purely numerically over the basis of Slater orbitals. The computed potential curve was in good accord with experiment: ($R_e = 2.30\,a_0$, $\omega_e$

**Table XXVII**

*Results of Valence-Shell Configuration Interaction*

|  | $D_e$ (eV) | $R_e$ (Å) | $\omega_e$ (cm$^{-1}$) | $\omega_e x_e$ (cm$^{-1}$) | $B_e$ (cm$^{-1}$) | $\alpha_e$ (cm$^{-1}$) |
|---|---|---|---|---|---|---|
| $O_2 \, {}^3\Sigma_g^-$ | 3.81 | 1.30 | 1582 | 14 | 1.25 | 0.0127 |
| Expt. | 5.12 | 1.21 | 1580 | 12 | 1.45 | 0.0158 |
| CN ${}^2\Sigma^+$ | 6.18 | 1.236 | 1939.2 | 14.54 | 1.610 | 0.0151 |
| Expt. | 7.75 | 1.172 | 2068.7 | 13.14 | 1.900 | 0.0173 |

$= 1614$ cm$^{-1}$, $D_e = 4.72$ eV). The potential curve calculated was compared graphically to the RKR. If $D_e$ was corrected for the deficiencies of the basis set by adding the difference between calculated and true SCF energies for the system, agreement with experiment was excellent.

We now turn to heteronuclear diatomics. The Bender–Davidson iterative natural orbital procedure (see Volume I, Chapter III, Section A.7) was used for CI calculations on NO by Kouba and Ohrn [140]. Although 269 configurations were used, the energy was still above that of a Hartree–Fock calculation using a better basis set.

The CI calculations on BC presented by Kouba and Ohrn [140, 141] produced potential curves for a large number of states of different symmetries. The potential curves were generated from the 54 lowest roots of the secular equation, and spectroscopic properties were calculated. The ground state, not experimentally observed for the diatomic species, was found to be ${}^4\Sigma^-$, and to have $R_e = 1.665$ Å, $\omega_e = 991$ cm$^{-1}$, $\omega_e x_e = 10.39$ cm$^{-1}$, $B_e = 1.0690$ cm$^{-1}$, $\alpha_e = 0.0128$ cm$^{-1}$. Its high spin might imply high reactivity [14], explaining the fact that it is experimentally unknown. The dissociation energy, to B $({}^2P)$ + C $({}^3P)$, was 3.046 eV.

The valence bond CI calculations for LiF of Balint-Kurti and Karplus [135] have been mentioned. Using 13 ionic and covalent functions, they obtained an energy about 0.1 a.u. above the experimental curve for $R$ near $R_e$, and a calculated binding energy too small by about a factor of 2. The predicted value for $R_e$ was low by about 0.1 $a_0$. The inadequacy of the results was ascribed [135] to the poor description of the F$^-$ ion afforded by the basis set.

Isoelectronic with LiF is the BeO molecule. It has a ${}^1\Sigma^+$ ground state, which cannot dissociate to atoms in their ground states, according to the Wigner–Witmer rules. Single determinant calculations, however, generally predict [142] that the ${}^1\Pi$ state lies lower than the ${}^1\Sigma^+$ for BeO, and configuration interaction is needed to get the correct ordering of energy levels. (The closeness of the states is also responsible for some difficulties in analyzing the spectra to get potential constants.) Furthermore, even though BeO is a closed-shell system, the Hartree–Fock wave function for the $^1\Sigma^+$ state, corresponding to the configuration $1\sigma^2 2\sigma^2 3\sigma^2 4\sigma^2 1\pi^4$, does not dissociate correctly to Be ($^1S$) and O ($^1D$), since the latter, corresponding to the configuration $1s^2 2s^2 2p^4$, needs two determinants to represent it. Schaefer [142] reported a CI calculation, using configurations which kept the core orbitals doubly filled but allowed single and double excitations from the SCF reference configuration, with the proviso that no more than one electron go to an orbital far removed in energy from the filled SCF orbitals. Even so, 157 configurations, corresponding to 569 determinants, were included. The Bender–Davidson iterative natural orbital procedure was used (see Volume I, Chapter III, Section A.7). Two-electron integrals were evaluated numerically. Potential constants evaluated with a near Hartree–Fock wave function [77] and with Schaefer's function are given in Table XXVIII. The dissociation energy obtained from an SCF calculation by Schaefer was 0.23 eV less than the Hartree–Fock value quoted in the table. This suggested that 0.23 eV be added to 6.58 eV to correct for deficiencies in the basis (a procedure which worked well for a similar calculation [71] on $O_2$, where the exact value for $D_e$ was known). The result compared well with 6.87 eV, a value derived by one author for $D_e$ from the spectrum.

### Table XXVIII

*Near Hartree–Fock and CI Results for* BeO

| | $D_e$ (eV) | $R_e$ (Å) | $\omega_e$ (cm$^{-1}$) | $\omega_e x_e$ (cm$^{-1}$) | $B_e$ (cm$^{-1}$) | $\alpha_e$ (cm$^{-1}$) |
|---|---|---|---|---|---|---|
| Near HF [77] | 4.13 | 1.29 | 1736 | 10.66 | 1.754 | 0.0157 |
| CI [142] | 6.58 | 1.313 | 1629 | 12.27 | 1.699 | 0.0174 |
| Experiment | 6.69±0.4 | 1.331 | 1487 | 11.83 | 1.651 | 0.0190 |

A configuration interaction calculation of the binding energy and other properties of CO was presented in 1960 by Hurley [143]. The choice of configurations which would be effective in improving the energy was made in part by an analysis into valence bond functions of a single determinant function, in which each molecular orbital was expanded in a minimal basis set of atomic orbitals. With a 48-configuration CI, a binding energy of

7.50 eV, several electron volts lower than that of preceding MO calculations, was obtained. The binding energy was calculated by subtracting the molecular energy at the experimental equilibrium internuclear distance from the sum of ground state C and O energies, calculated with the same basis functions. A 23-configuration function, corresponding to the inclusion of structures for CO, $C^-O^+$, $C^+O^-$, and $C^{+2}O^{-2}$, gave a binding energy only 0.05 eV smaller than 48 configurations.

Hurley invoked his theory of the interatomic correlation correction (see Chapter III, Section A.3) to correct matrix elements of the Hamiltonian using knowledge of atomic energies as follows: Let $\tilde{W}_i$ be the sum of calculated energies for the atomic states into which CO dissociates and $W_i$ be the experimental sum of atomic energies. The calculated Hamiltonian and overlap matrix elements between the orbital wave functions corresponding to states $i$ and $j$ are designated by $\tilde{H}_{ij}$ and $\tilde{S}_{ij}$. According to Hurley [143], $\tilde{H}_{ij}$ should be augmented by $\frac{1}{2}\tilde{S}_{ij}(W_i - \tilde{W}_i + W_j - \tilde{W}_j)$, and the linear variation calculation redone with the new matrix elements. The binding energy, calculated as the difference between the sum of experimental atomic energies, and the new calculated molecular energy, was now 11.00 eV, and should represent a lower bound to the true binding energy [143]. This enabled Hurley to choose between three possible values of this quantity suggested by experiment.

The multiconfiguration SCF procedure of Das and Wahl has been applied to alkali-noble gas systems [144], previously treated by Scheel and Griffing. The wave function for such a system consists of a function representing the Hartree–Fock configuration plus double-excitation terms corresponding to interatomic correlation, i.e., involving single excitations on each center. It is evident that the form of the wavefunction used makes this treatment capable of describing the dispersion interaction. The orbitals used were optimized simultaneously with the coefficients of the double-excitation terms. Intraatomic correlation was ignored because it is roughly independent of R and couples only slightly with interatomic correlation. For large $R$, Das and Wahl [144] started from the Hartree–Fock potential curve and evaluated separately the intraatomic correlation energy and the dispersion terms. In the range studied ($R \geq 6\,a_0$ for HeH and $R \geq 10\,a_0$ for LiHe), the intraatomic correlation was much less important than the dispersion interaction. Subsequent improved calculations on HeH [145], using five configurations, produced a potential curve expected to be accurate to 15% for $R > 2\,a_0$.

Bertoncini et al. [146] reported similar calculations for ground and excited states of NaLi and for the ground state of NaLi$^+$. Here, there are two or one electrons outside a closed shell. It was impossible to reoptimize all parameters for each $R$, so a basis set optimized for Hartree–Fock calculations on the atoms was used, augmented and modified somewhat. The results of calculations on the ground states are shown in Table XXIX. Here, "frozen" means that the orbitals of the SCF calculation were fixed and only the orbitals describing the correlation were varied. The SCF (restricted Hartree–Fock) for NaLi dissociates incorrectly to a mixture of $\{Na^+ (^1S) + Li^- (^1S)\}$, $\{Na^- (^1S) + Li^+ (^1S)\}$, and $\{Na (^2S) + Li (^2S)\}$. Bertoncini et al. [146] estimated the error in $D_e$ for NaLi as 1–1.5 thousandths of an atomic unit ($\sim$0.03 eV), and the expansion error, due to incompleteness of the basis set, as $\sim$0.0006 a.u. The curve $U(R)$ for NaLi$^+$ was estimated to be accurate to 0.006 a.u. Optimized valence configuration calculations have also been performed [127] for NaLi, as well as for NaF. The improvement in dissociation energies was most encouraging. For NaLi, the Hartree–Fock and OVC potential curves were quite parallel for $R \leq R_e$, but very different beyond about $6\frac{1}{2} a_0$.

**Table XXIX**

*Multiconfiguration SCF for* NaLi *and* NaLi$^+$

|            | $R_e$ ($a_0$) | $D_e$ (eV) | $\omega_e$ (cm$^{-1}$) | $B_e$ (cm$^{-1}$) |
|------------|---------------|------------|------------------------|-------------------|
| NaLi SCF   | 5.6473        | 0.0763     | 250.65                 | 0.34896           |
| NaLi frozen| 5.5550        | 0.8468     | 249.77                 | 0.36066           |
| NaLi relaxed| 5.5480       | 0.8526     | 248.48                 | 0.36160           |
| NaLi$^+$   | 6.5009        | 0.9194     | 186.62                 | 0.26335           |

Wahl et al. [127] discussed the kinds of configurations one must include in this method to properly describe different kinds of molecules. For ionic systems, a large number of double excitations must be included to get really accurate potential curves, but the Hartree–Fock curve should be parallel to the true $U(R)$ for internuclear distances considerably greater than $R_e$ (the molecule resembles ions out to the distance at which the difference of ionization potential and electron affinity exceeds the Coulomb attraction). According to these authors [127], OVC calculations are feasible for diatomic molecules composed from the first two rows of the periodic table.

Good results have also been obtained for $D_e$, $R_e$, and $\omega_e$ of homonuclear alkali molecules by this method [100, 128]. Thus the predictions for NaLi are probably reliable [87, p. 229].

We conclude with some CI results for a fairly large molecule. Liu and Schaefer [90] have reported $U(R)$ from SCF and CI calculations on the ground states of KrF and KrF$^+$. The compound KrF was not stable in the single-determinant approximation. The effect of a 158-configuration CI was to flatten out the potential curve, but still give no binding. On the basis of experience with comparable calculations on other molecules, it was concluded that $U(R)$ for KrF had no minimum, except for that due to the Van der Waals attraction, which could not be reproduced with the choice of configurations used [90]. For KrF$^+$ ($^1\Sigma^+$, isoelectronic to BrF), which dissociates to Kr$^+$ + F at large $R$ in the single determinant approximation, the potential constants given in Table XXX were calculated. The SCF dissociation energy is relative to SCF atoms. As in the case of halogen diatomics, MECE is large enough to give little or no bonding from the SCF function. Experimentally, $D_0{}^0 \geq 1.58$ eV.

**Table XXX**

*Potential Constants$^a$ for* KrF$^+$

|      | $R_e$ (Å) | $D_e$ (eV) | $\omega_e$ (cm$^{-1}$) | $\omega_e x_e$ (cm$^{-1}$) | $B_e$ (cm$^{-1}$) | $\alpha_e$ (cm$^{-1}$) |
|------|-----------|------------|------------------------|-----------------------------|--------------------|-------------------------|
| SCF  | 1.680     | −0.02      | 810                    | 6.3                         | 0.386              | 0.0029                  |
| CI   | 1.752     | 1.94       | 621                    | 8.3                         | 0.355              | 0.0044                  |

$^a$ See Liu and Schaefer [90].

*(d) Conclusions*

It is clear that one can now perform configuration interaction calculations, with hundreds of configurations, for fairly large systems. In contrast to the situation of the last section, where one could approach extremely closely to the true Hartree–Fock function, approaching the exact wave function by recovering a large fraction of the correlation energy does not seem a practical goal for most systems. More important than obtaining as low an energy as possible is finding out how to set up a calculation whose inaccuracies will be $R$-independent, so that $U(R)$ will be accurately calculated.

One would like to have a set of rules for constructing configurations, which could easily be implemented for a variety of new systems, and which yields a potential curve in which one can be confident. It does not appear from the preceding discussion that theoretical analysis suffices to derive such a method. Of course, one can improve on the Hartree–Fock by supplying configurations needed for proper dissociation, or correct for near-degeneracies, but no theory has been advanced to show how to set up a CI to give an energy above the exact by an *R*-independent amount. Confidence in the schemes which have been advanced grows out of the number of successful calculations based on them.

Many more calculations will be appearing. Unless only qualitative information about $U(R)$ suffices, an isolated CI calculation on a single system is no longer worth doing. What is required is a set of calculations to demonstrate the reliability of a proposed method. Schemes which are being tested include full valence shell CI with minimal basis, optimized valence configurations, and Schaefer's first-order CI (Hartree–Fock, near-degenerate configurations, and single and double excitations from them). The more complex the method, the harder it will be to get sufficient calculations to establish its reliability. It is encouraging that simple schemes often work as well as more complicated ones. It is likely that experience will lead to a reliable recipe (or recipes) for deriving $U(R)$ for any desired molecule.

## REFERENCES

1. A. Meckler, *J. Chem. Phys.* **21**, 1750 (1953).
2. C. L. Beckel, *J. Chem. Phys.* **33**, 1885 (1960); E. J. Finn and C. L. Beckel, *Ibid.* **33**, 1887 (1960); C. L. Beckel and J. P. Sattler, *Ibid.* **42**, 2620 (1965); *J. Mol. Spectrosc.* **20**, 153 (1966).
3. R. K. Nesbet, *J. Chem. Phys.* **40**, 3619 (1964).
4. R. K. Nesbet, *J. Chem. Phys.* **40**, 3619, Appendix A (1964); Z. Kopal, "Numerical Analysis," Appendix I. Wiley, New York, 1955.
5. A. D. McLean, *J. Chem. Phys.* **40**, 2774 (1964).
6. M. Yoshimine, *J. Chem. Phys.* **40**, 2970 (1964).
7. R. H. Schwendeman, *J. Chem. Phys.* **44**, 2115 (1966).
8. D. M. Silver, E. L. Mehler, and K. Ruedenberg, *J. Chem. Phys.* **52**, 1174, 1181 (1970).
9. J. Goodisman and W. A. Klemperer, *J. Chem. Phys.* **38**, 721 (1963).
10. R. C. Sahni, B. C. Sawhney, and M. J. Hanley, *J. Chem. Phys.* **51**, 539 (1969).
11. R. E. Brown and H. Shull, *Int. J. Quantum Chem.* **2**, 663 (1968).
12. E. R. Davidson, *J. Chem. Phys.* **34**, 1240 (1961).
13. S. Bratož and M. Allavena, *J. Chem. Phys.* **37**, 2138 (1962); M. Allavena and S. Bratož, *J. Chim. Phys.* **60**, 1199 (1963).
14. D. M. Bishop, *Mol. Phys.* **6**, 305 (1963); D. M. Bishop and M. Randić, *J. Chem. Phys.* **44**, 2480 (1966).
15. D. M. Bishop and A. Macias, *J. Chem. Phys.* **55**, 647 (1971).
16. J. Gerratt and I. M. Mills, *J. Chem. Phys.* **49**, 1719, 1730 (1968).
17. D. M. Bishop and A. Macias, *J. Chem. Phys.* **53**, 3515 (1970).
18. D. J. Klein, E. M. Greenawalt, and F. A. Matsen, *J. Chem. Phys.* **47**, 4820 (1967).

19. D. J. Klein, C. E. Rodriguez, J. C. Browne, and F. A. Matsen, *J. Chem. Phys.* **47**, 4862 (1967).
20. K. D. Carlson and R. K. Nesbet, *J. Chem. Phys.* **41**, 1051 (1964).
21. P.-O. Löwdin, *Advan. Phys.* **5**, 1 (1956).
22. P.-O. Löwdin, *Advan. Chem. Phys.* **2**, 207 (1959).
23. R. E. Stanton, *J. Chem. Phys.* **36**, 1298 (1962).
24. A. C. Wahl, *J. Chem. Phys.* **41**, 2600 (1964).
25. L. C. Allen, *J. Chem. Phys.* **34**, 1156 (1961); L. C. Allen, E. Clementi, and H. M. Gladney, *Rev. Mod. Phys.* **35**, 465 (1963).
26. E. Clementi, *J. Chem. Phys.* **38**, 2248 (1962); **39**, 175 (1963).
27. E. Clementi, *J. Chem. Phys.* **42**, 2783 (1965); E. Clementi and A. Veillard, *J. Chem. Phys.* **44**, 3050 (1966).
28. J. Linderberg and H. Shull, *J. Mol. Spectrosc.* **5**, 1 (1960).
29. E. Clementi, *J. Chem. Phys.* **38**, 2780 (1963); **39**, 487 (1963).
30. E. Clementi, *J. Chem. Phys.* **38**, 2781 (1963).
31. R. L. Matcha, *J. Chem. Phys.* **47**, 4595, 5295 (1967); **48**, 335; **49**, 1264 (1968); **53**, 485 (1970).
32. P. Cade and W. M. Huo, *J. Chem. Phys.* **47**, 614, 649 (1969).
33. E. Clementi, *J. Chem. Phys.* **36**, 33 (1962).
33a. R. K. Nesbet, *J. Chem. Phys.* **36**, 1518 (1962).
34. C. Hollister and O. Sinanoğlu, *J. Amer. Chem. Soc.* **88**, 13 (1966).
35. A. D. McLean, *J. Chem. Phys.* **40**, 243 (1964).
36. P. E. Cade, *J. Chem. Phys.* **47**, 2390 (1967).
37. S. Green, *J. Chem. Phys.* **52**, 3100 (1970).
38. A. W. Salotto and L. Burnelle, *J. Chem. Phys.* **52**, 2936 (1970).
39. P. Lindner, *Uppsala Quantum Chem. Group Preprint* # 166, January 1966.
40. F. E. Harris and H. A. Pohl, *J. Chem. Phys.* **42**, 3648 (1965).
41. T. A. Claxton and N. A. Smith, *Theor. Chim. Acta* **22**, 399 (1971).
42. D. H. Sleeman, *Theor. Chim. Acta* **11**, 135 (1968).
43. T. A. Claxton, *Chem. Phys. Lett.* **4**, 469 (1969).
44. T. E. H. Walker and W. G. Richards, *Proc. Phys. Soc. London* **92**, 285 (1967).
45. W. M. Huo, *J. Chem. Phys.* **45**, 1554 (1966).
46. K. D. Carlson, E. Ludeña, and C. Moser, *J. Chem. Phys.* **43**, 2408 (1965).
47. H. Brion, C. Moser, and T. Yamazaki, *J. Chem. Phys.* **30**, 673 (1959); **33**, 1871 (1960).ᐧ
48. J. E. Lennard-Jones, *Trans. Faraday Soc.* **25**, 668 (1929).
49. R. S. Mulliken, *Rev. Mod. Phys.* **4**, 1 (1932); *Phys. Rev.* **56**, 778 (1939); and numerous other papers.
50. C. C. J. Roothaan, *Rev. Mod. Phys.* **23**, 69 (1951).
51. L. C. Allen and A. M. Karo, *Rev. Mod. Phys.* **32**, 275 (1960).
52. V. Griffing and J. F. Wehner, *J. Chem. Phys.* **23**, 1024 (1955).
53. T. L. Gilbert and A. C. Wahl, *J. Chem. Phys.* **47**, 3425 (1967).
54. N. R. Kestner, *J. Chem. Phys.* **48**, 252 (1968).
54a. H. Margenau and N. R. Kestner, "Theory of Intermolecular Forces," Sect. 3.3. Pergamon, Oxford, 1969.
55. D. Butler and N. R. Kestner, *J. Chem. Phys.* **53**, 1704 (1970).
56. C. R. Fischer and P. J. Kemmey, *J. Chem. Phys.* **53**, 50 (1970).
56a. G. Das and S. Ray, *Phys. Rev. Lett.* **24**, 1391 (1970). W. H. Miller and H. F. Schaefer III, *J. Chem. Phys.* **53**, 1421 (1970).

57. I. Fischer, *Ark. Fys.* **5**, 349 (1952).
58. S. L. Kahalas and R. K. Nesbet, *J. Chem. Phys.* **39**, 529 (1963).
59. J. A. Keefer, J. K. Su Fu, and R. L. Belford, *J. Chem. Phys.* **50**, 160 (1969).
60. M. C. Harrison, *J. Chem. Phys.* **41**, 499 (1964).
61. R. Moccia, *J. Chem. Phys.* **37**, 910 (1962).
62. R. K. Nesbet, *J. Chem. Phys.* **41**, 100 (1964).
63. D. B. Boyd, *Theor. Chim. Acta* **20**, 273 (1971).
64. C. W. Scherr, *J. Chem. Phys.* **23**, 569 (1955).
65. J. W. Richardson, *J. Chem. Phys.* **35**, 1829 (1961).
66. P. E. Cade, K. D. Sales, and A. C. Wahl, *J. Chem. Phys.* **44**, 1973 (1966).
67. G. Verhaegen and W. G. Richards, *J. Chem. Phys.* **45**, 1828 (1966).
68. G. Verhaegen, W. G. Richards, and C. M. Moser, *J. Chem. Phys.* **46**, 160 (1967).
69. G. Verhaegen, *J. Chem. Phys.* **49**, 4696 (1968).
70. R. C. Sahni, *Trans. Faraday Soc.* **63**, 801 (1966).
71. H. F. Schaefer, III, *J. Chem. Phys.* **54**, 2207 (1971).
72. T. L. Gilbert and A. C. Wahl, *J. Chem. Phys.* **55**, 5297 (1971).
73. R. F. W. Bader and A. D. Bandrauk, *J. Chem. Phys.* **49**, 1653 (1968).
74. W. M. Huo, *J. Chem. Phys.* **43**, 624 (1965).
75. R. C. Sahni, C. D. La Budde, and B. C. Sawhney, *Trans. Faraday Soc.* **62**, 1993 (1966).
76. W. G. Richards, G. Verhaegen, and C. M. Moser, *J. Chem. Phys.* **45**, 3226 (1966).
77. W. M. Huo, K. F. Freed, and W. A. Klemperer, *J. Chem. Phys.* **46**, 3556 (1967).
78. A. D. McLean, *J. Chem. Phys.* **39**, 2653 (1963).
79. P. A. G. O'Hare and A. C. Wahl, *J. Chem. Phys.* **54**, 3770 (1971).
80. P. A. G. O'Hare and A. C. Wahl, *J. Chem. Phys.* **55**, 666 (1971).
81. S. Peyerimhoff, *J. Chem. Phys.* **43**, 998 (1965).
82. B. G. Anex, *J. Chem. Phys.* **38**, 1651 (1963).
83. J. J. Kaufman and L. M. Sachs, *J. Chem. Phys.* **53**, 446 (1970).
84. R. S. Mulliken, *J. Chem. Phys.* **52**, 5170 (1970).
85. H. M. James, *J. Chem. Phys.* **3**, 9 (1935).
86. R. C. Sahni and B. C. Sawhney, *Trans. Faraday Soc.* **63**, 1 (1967).
87. H. F. Schaefer, III, "The Electronic Structure of Atoms and Molecules," p. 150. Addison-Wesley, Reading, Massachusetts, 1972.
88. M. Krauss, P. Maldonado, and A. C. Wahl, *J. Chem. Phys.* **54**, 4944 (1971).
89. P. A. G. O'Hare and A. C. Wahl, *J. Chem. Phys.* **53**, 2834 (1970).
90. B. Liu and H. F. Schaefer, III, *J. Chem. Phys.* **55**, 2369 (1971).
91. P. E. Phillipson, *Phys. Rev.* **125**, 1981 (1962).
92. G. P. Barnett, *Can. J. Phys.* **45**, 137 (1967).
93. G. H. Matsumoto, C. F. Bender, and E. R. Davidson, *J. Chem. Phys.* **46**, 402 (1967).
94. J. N. Murrell and G. Shaw, *Mol. Phys.* **12**, 475 (1967).
95. N. R. Kestner and O. Sinanoğlu, *J. Chem. Phys.* **45**, 194 (1966).
96. P. Bertoncini and A. C. Wahl, *Phys. Rev. Lett.* **25**, 990 (1970).
97. P. Bertoncini, G. Das, and A. C. Wahl, *J. Chem. Phys.* **52**, 5112 (1970).
98. H. H. Michels and F. E. Harris, *Int. J. Quantum Chem. Symp.* **2**, 21 (1968).
99. A. A. Evett, *J. Chem. Phys.* **24**, 150 (1956).
100. H. M. James and A. S. Coolidge, *J. Chem. Phys.* **1**, 825 (1933).
101. N. Scheel and V. Griffing, *J. Chem. Phys.* **36**, 1453 (1962).

102. H. H. Michels, *J. Chem. Phys.* **44**, 3834 (1966).
103. H. Conroy, *J. Chem. Phys.* **41**, 1341 (1964).
104. J. D. Stuart and F. A. Matsen, *J. Chem. Phys.* **41**, 1646 (1964).
105. L. S. Combs and L. K. Runnels, *J. Chem. Phys.* **49**, 4216 (1968).
106. L. Wolniewicz, *J. Chem. Phys.* **43**, 1087 (1965).
107. D. M. Duparc and R. A. Buckingham, *Proc. Roy. Soc.* **83**, 731 (1964).
108. C. F. Bender and E. R. Davidson, *Phys. Rev.* **183**, 23 (1969).
109. C. F. Bender and E. R. Davidson, *Chem. Phys. Lett.* **3**, 33 (1969).
109a. D. D. Ebbing, *J. Chem. Phys.* **36**, 1361 (1962).
110. F. A. Matsen and J. C. Browne, *J. Phys. Chem.* **66**, 2332 (1962).
111. J. C. Browne and F. A. Matsen, *Phys. Rev. A* **135**, 1127 (1964).
112. C. F. Bender and E. R. Davidson, *J. Chem. Phys.* **49**, 4222 (1968).
113. C. F. Bender and E. R. Davidson, *J. Chem. Phys.* **47**, 360 (1967).
114. J. C. Browne, *J. Chem. Phys.* **41**, 3495 (1964).
115. R. E. Brown, *J. Chem. Phys.* **51**, 2879 (1969).
116. R. Ahlrichs and W. Kutzelnigg, *J. Chem. Phys.* **48**, 1819 (1968).
117. M. Gélus, R. Ahlrichs, Y. Staemmler, and W. Kutzelnigg, *Theor. Chim. Acta* **21**, 63 (1971).
118. J. F. Harrison and L. C. Allen, *J. Mol. Spectrosc.* **29**, 432 (1969).
119. R. J. Blint, W. A. Goddard, III, R. C. Ladner, and W. E. Palke, *Chem. Phys. Lett.* **5**, 302 (1970).
120. W. A. Goddard, III, *Phys. Rev.* **187**, 73, 81 (1967).
121. J. Kouba and Y. Öhrn, *J. Chem. Phys.* **52**, 5387 (1970).
122. S. V. O'Neil and H. F. Schaefer, III, *J. Chem. Phys.* **55**, 394 (1971).
123. E. L. Mehler, K. Ruedenberg, and D. M. Silver, *J. Chem. Phys.* **52**, 1181, 1206 (1970).
124. C. F. Bender and E. R. Davidson, *J. Chem. Phys.* **49**, 4222 (1968).
125. E. R. Davidson, *Rev. Mod. Phys.* **44**, 451 (1972).
126. G. Das and A. C. Wahl, *J. Chem. Phys.* **44**, 87 (1966).
127. A. C. Wahl, P. J. Bertoncini, G. Das, and T. L. Gilbert, *Int. J. Quantum Chem. Symp.* **1**, 123 (1967).
128. G. Das, *J. Chem. Phys.* **46**, 1568 (1967).
129. W. Kutzelnigg and M. Gélus, *Chem. Phys. Lett.* **7**, 296 (1970).
130. G. Das and A. C. Wahl, *Phys. Rev. Lett.* **24**, 440 (1970).
131. F. E. Harris and H. H. Michels, *Int. J. Quantum Chem. Symp.* **1**, 329 (1967).
132. F. E. Harris, *J. Chem. Phys.* **46**, 2769 (1967).
133. F. E. Harris and H. H. Michels, *Int. J. Quantum Chem. Symp.* **3**, 461 (1970).
134. H. F. Schaefer, III, *J. Chem. Phys.* **52**, 6241 (1970).
135. G. G. Balint-Kurti and M. Karplus, *J. Chem. Phys.* **50**, 478 (1969).
136. C. F. Bender and E. R. Davidson, *J. Chem. Phys.* **46**, 3313 (1967).
137. J. C. Slater, "Quantum Theory of Molecules and Solids," Vol. I, Sect. 6.5. McGraw-Hill, New York, 1963.
138. H. F. Schaefer, III and F. E. Harris, *J. Chem. Phys.* **48**, 4946 (1968).
139. H. F. Schaefer, III and T. G. Heil, *J. Chem. Phys.* **54**, 2573 (1971).
140. J. E. Kouba and Y. Öhrn, *Int. J. Quantum Chem.* **5**, 534 (1971).
141. J. E. Kouba and Y. Öhrn, *J. Chem. Phys.* **53**, 3923 (1970).
142. H. F. Schaefer, III, *J. Chem. Phys.* **55**, 176 (1971).
143. A. C. Hurley, *Rev. Mod. Phys.* **32**, 400 (1960).

144. G. Das and A. C. Wahl, *Phys. Rev. A* **4**, 825 (1971).
145. G. Das and S. Ray, *Phys. Rev. Lett.* **24**, 1391 (1970).
146. P. J. Bertoncini, G. Das, and A. C. Wahl, *J. Chem. Phys.* **52**, 5112 (1970).

## SUPPLEMENTARY BIBLIOGRAPHY

Below are listed, alphabetically by author, articles relevant to the subject of this chapter which came to our attention too late to be incorporated in the manuscript. We have given author, title, and, in some cases, additional information on the content of the work. The Section to which the work is relevant is given in parentheses in each case.

A. B. Anderson, On Evaluating Force Constants from LCAO–MO–SCF Electronic Change Densities for Diatomic Molecules. *J. Chem. Phys.* **58**, 381 (1973). (C1)

R. Albat and N. Gruen, Examples of Known SCF Procedures Which do not Satisfy All Necessary Conditions for the Energy to be Stationary. *Chem. Phys. Lett.* **18**, 572 (1973). (A4)

J. N. Bardsley, Potential Curves for $He_2^+$ and $Li_2^+$ (using configuration interaction). *Phys. Rev.* **A3**, 1317 (1971). (C4)

P. J. Bertoncini and A. C. Wahl, . . . He–He $^1\Sigma_g^+$ Potential (by Multiconfiguration SCF) at Intermediate to Large Separations II. Changes in Intraatomic Correlation Energy. *J. Chem. Phys.* **58**, 1259 (1973). (C4)

F. B. Billingsley, II, An Economic Storage and Processing Method for Two-Electron Integrals in LCAO–MO Calculations. *Int. J. Quantum Chem.* **6**, 617 (1972). (A3)

D. M. Bishop and A. Macias, Ab Initio Calculations of Harmonic Force Constants. VI. Application to SCF Wavefunctions. *J. Chem. Phys.* **56**, 999 (1972). (C1)

D. M. Bishop and J.-C. Leclerc, Unconventional Basis Sets in Quantum Mechanical Calculations. *Mol. Phys.* **24**, 929 (1972). (A5)

N. Björnao, Methods for Solving Constrained SCF–LCAO–MO Equations. Applications to . . . $N_2$. *Mol. Phys.* **24**, 1 (1972). (A, C3)

V. Bondybey, P. K. Pearson, and H. F. Schaefer III, Theoretical Potential Curves for OH, $HF^+$, HF, $HF^-$, $NeH^+$ and NeH by First-Order CI. *J. Chem. Phys.* **57**, 1123 (1972). (C4)

A. K. Chandra and R. Sundar, Vibrational Force Constant and Electron Relaxation in $H_2$ and $Li_2$ (using previously presented extended Hartree-Fock calculations). *Chem. Phys. Lett.* **14**, 577 (1972). (B2, C4)

G. Das and A. C. Wahl, New Techniques for the Computation of Multiconfiguration SCF Wavefunctions. *J. Chem. Phys.* **56**, 1769 (1972). (B)

G. Das and A. C. Wahl, Theoretical Study of the $F_2$ Molecule Using the Method of Optimized Valence Configurations. *J. Chem. Phys.* **56**, 3532 (1972). (B, C4)

K. K. Docken and J. Hinze, LiH Potential Curves and Wavefunctions for X $^1\Sigma^+$ . . . . *J. Chem. Phys.* **57**, 4928 (1972). (C)

V. Dyczmons, No $N^4$ Dependence in the Calculation of Large Molecules. *Theor. Chim. Acta* **28**, 307 (1973). (A5)

W. England, One-Center Coulomb, Two-Center Hybrid, and Two-Center Coulomb Integrals over STP Functions. *Int. J. Quantum Chem.* **6**, 509 (1972). (A3)

H. Fukutome, The Unrestricted Hartree-Fock Theory of Chemical Reactions...Homopolar Two-Center Two-Electron Systems. *Prog. Theor. Physics* **47**, 1156 (1972). (A)

S. Green, Calculated Properties for NO X $^2\Pi$ and A $^2\Sigma^+$ (by restricted Hartree–Fock). *Chem. Phys. Lett.* **13**, 552 (1972). (C3)

S. Green, P. S. Bagus, B. Liu, A. D. McLean and M. Yoshimine, Calculated Potential-Energy Curves for CH$^+$ (CI starting from Hartree–Fock). *Phys. Rev.* **5A**, 1614 (1972). (C4)

J. A. Hall and W. G. Richards, A Theoretical Study of the Spectroscopic States of the CF Molecule (LCAO–MO SCF). *Mol. Phys.* **23**, 331 (1972). (C3)

P. S. Julienne, M. Krauss, and A. C. Wahl, Hartree–Fock Energy Curves for X $^2\Pi$ and $^2\Sigma^+$ States of HF$^+$. *Chem. Phys. Lett.* **11**, 16 (1971). (C3)

J. J. Kaufman, LCAO–MO–SCF Calculations Using Gaussian Basis Functions X. $^2\Sigma^+$ States AlLi..., *J. Chem. Phys.* **58**, 1680 (1973). (A5, C3)

D. Kunik and U. Kaldor, Ground State of He$_2$ by the Spin-Optimized Method (optimizing both the orbitals and the spin-coupling scheme in the wavefunction). *J. Chem. Phys.* **56**, 1741 (1972). (B)

V. A. Kuprievich and O. V. Shramko, The Multiconfiguration SCF Theory—Method of One-Electron Hamiltonian. *Int. J. Quantum Chem.* **6**, 327 (1972). (B)

R. McWeeny and B. T. Sutcliffe, "Methods of Molecular Quantum Mechanics," Academic Press, New York, 1972. Generalized product functions and cluster expansions are discussed on pp. 175–192, configuration interaction on pp. 55–56, and in Sections 3.5 and 4.4. For Hartree–Fock procedures for open and closed shells, see Sections 5.1 and 5.2. (A, B, C)

R. Manne, Brillouin's Theorem in Roothaan's Open-Shell SCF Method. *Molec. Phys.* **24**, 935 (1972). (A3, B)

I. Mayer, Derivation of the Extended Hartree–Fock Equations. *Chem. Phys. Lett.* **11**, 397 (1972). (B)

M. B. Milleur and R. L. Matcha, Unified Treatment of Two-Center, Two-Electron Integrals...Over Slater Functions. *J. Chem. Phys.* **57**, 3029 (1972). (A3)

R. S. Mulliken, The Nitrogen Molecule Correlation Diagram (SCF calculations for $R \leq R_e$). *Chem. Phys. Lett.* **14**, 137 (1972). (C3)

N. S. Ostlund, Complex and Unrestricted Hartree–Fock Functions. *J. Chem. Phys.* **57**, 2994 (1972). (A)

D. Peters, Simple Procedure for Open Shell SCF Molecular Orbital Computations. *J. Chem. Phys.* **57**, 4351 (1972). (A)

S. D. Peyerimhoff and R. J. Buenker, Comparison of Various CI Treatments for the Description of Potential Curves for... O$_2$. *Chem. Phys. Lett.* **16**, 235 (1972). (C4)

B. Roos, A New Method for Large-Scale Configuration Interaction Calculations. *Chem. Phys. Lett.* **15**, 153 (1972). (B3)

J. Schamps and H. Lefebvre-Brion, SCF Calculations of the Electronic States of MgO (and MgO$^+$). *J. Chem. Phys.* **56**, 573 (1972). (C3)

L. L. Shipman and R. E. Christoffersen, A New Method for the Calculation of Vibrational Force Constants (from the energy). *Chem. Phys. Lett.* **11**, 101 (1971). (C1)

D. M. Silver, R. E. Christoffersen, E. L. Miller, W. England, and K. Ruedenberg, errata for previous publications on integrals over Slater orbitals. *J. Chem. Phys.* **57**, 3585 (1972). (A3)

V. Staemmler and M. Jungen, The Direct Determination of Brueckner Orbitals with Application to the H$_2$ Molecule (exact Hartree–Fock theories). *Theor. Chimica Acta* **24**, 152 (1972). (B)

E. W. Thulstrup and Y. Öhrn, Configuration Interaction Studies of NO and NO⁺ with Comparisons to Photoelectron Spectra. *J. Chem. Phys.* **57**, 3716 (1972).  (C4)

T. E. H. Walker, On the Calculation of Matrix Elements in Diatomic Molecules (using group theory). *Mol. Phys.* **23**, 489 (1972).  (B3)

T. E. H. Walker, On Open Shell Hartree–Fock Calculations. *Theor. Chimica Acta* **25**, 1 (1972).  (A)

B. Wirsam, Combined SCF and CI Calculations of the Spectrum of SiH. *Chem. Phys. Lett.* **10**, 180 (1971).  (C3, C4)

M. Yoshimine, Accurate Potential Curves and Properties for the X $^2\Pi$ and A $^2\Sigma$ States of LiO (by CI using pseudonatural orbitals). *J. Chem. Phys.* **57**, 1108 (1972).  (B3, C4)

W. T. Zemke, G. Das, and A. C. Wahl, Theoretical Determination of the...Binding Energy of $O_2{}^-$ (by Optimized Valence Configurations). *Chem. Phys. Lett.* **14**, 310 (1972).  (C4)

# Chapter III Semiempirical Calculations and Simple Models

By a semiempirical calculation, we mean one that substitutes, for all or part of the computation, information derived from experimental results. Semiempirical schemes often appear as quantum mechanical calculations with the insertion of measured data for quantities that would normally be calculated (Section A). However, this need not be the case. We have in mind simple models based on physical or intuitive reasoning, which give rise to relatively simple computations, but which cannot be rigorously justified from the Schrödinger equation (Section B). The validity and accuracy of predictions of such models are checked (and the formulas themselves sometimes adjusted) by comparison of their predictions with experiment for a number of cases before using them on others (sometimes called "calibration" or "curve-fitting," depending on one's viewpoint). This is where the semiempirical aspect comes in. Looking at things this way, many methods of calculation not normally referred to as semiempirical belong in this chapter, since we would not use them if they were not known to work; we shall not take so extreme a position.

Most semiempirical methods are aimed at more complex systems than diatomics, where the problems connected with purely theoretical solutions are more intractable, but a number of applications have been made to diatomic systems. Schemes for deriving properties of a molecule from known results on related systems must also be classed among semiempirical calculations, although they often consist only of a set of empirically determined rules, with theoretical justification added. We include a discussion of such algorithms in Section B. Some of these may be derived theoretically, such as the "Combination Rules" for van der Waals constants referred to in Section A, which give $C_{AB}$ in terms of $C_{AA}$ and $C_{BB}$.

## A. Semiempirical Calculations

The main motivation for semiempirical calculation is, of course, avoiding difficult computation, such as evaluation of certain integrals. If, for example, values of the two-electron integrals of the molecular orbital theories, or the integrals over products of atomic wave functions of the valence bond theories, were easily available, the remaining calculations would be very simple. At least two approaches are possible: Empirical formulas could be developed for these integrals (the measured data for calibration of the formulas in this case would be some exactly calculated values for the integrals), or some combination of the integrals could be associated to an experimentally accessible quantity. If all integrals are treated in the latter fashion, we have a theory relating one set of properties to another.

Semiempirical molecular orbital theories, designed to simplify MO calculations by avoiding integral evaluation, are discussed in Subsection 1. It appears that such calculations actually do more than this: Their results do not simply mimic those of the *ab initio* calculations they resemble. The pseudopotential calculations of Subsection 2 attempt to simplify MO calculations by reducing a many-electron problem to a few-electron one. This means simulating the effect of all the electrons except those of interest by a local potential. An examination of the Hartree–Fock equations shows this cannot be done rigorously; hence the need for empirical data in testing reasonable formulas. The atoms-in-molecules approach (Subsection 3) is closely related to the valence bond theory. Its principal use, however, seems to be to improve *ab initio* valence bond calculations using atomic data, rather than to permit valence bond calculations with a minimum of integral evaluation.

### 1. *Semiempirical MO Theories*

Many semiempirical calculations are in the framework of the self-consistent field–linear combination of atomic orbital approximation (SCF–LCAO), where the wavefunction is written as a determinant (or sum of determinants) of molecular orbitals (MO's) and each molecular orbital is written as a sum of atomic orbital (AO) basis functions. The big problem in carrying through such calculations *ab initio* is the evaluation of integrals. The empirical elements in semiempirical LCAO calculations usually are related to simplification or avoidance of integral evaluations. For instance, if we use as basis functions orbitals that actually describe the atoms from which the molecule is built, we can use information about atomic energies to evaluate some integrals. Alternatively, simple formulas for difficult integrals have been set up and parametrized to give reliable results in known cases. Since the evaluation of the two-electron integrals is the most difficult, we could attempt to parametrize these only, but many methods, such as those which approximate the matrix elements of an effective one-electron Hamiltonian, parametrize both one- and two-electron integrals.

Reviews of semiempirical MO theories by Jug [1], Jaffe [2], Pople [3], and Klopman and O'Leary [4] have recently appeared, as well as a book by Pople and Beveridge [5]. Whereas the last-named book deals almost exclusively with the study and development of semiempirical MO schemes carried out by Pople and co-workers, the book-length review of Klopman and O'Leary [4] attempts to discuss and compare a great variety of such calculations, mostly as applied to large organic molecules. It includes a history of their development, tables summarizing the various parameterization schemes and tables of results calculated with them. The different schemes are appraised for different properties. For instance, for prediction of equilibrium internuclear distances and force constants of diatomic molecules, Klopman and O'Leary [4] suggest that CNDO and MINDO (see page 310) would be the most reliable. Jaffe [2] attempts to compare different schemes within a general theoretical framework, and Sokolov [6] puts the Hückel and other schemes into the context of the goals of quantum chemistry. We shall give some general discussion, cite a number of simple calculations, and then turn to Pople's systematic work.

Semiempirical molecular orbital calculations go back to those of Hückel [6a] for the $\pi$ electrons of aromatic systems, where the effective Hamiltonian represents the effect of the nuclei and the electrons of the $\sigma$ core, which is assumed fixed. For diatomics, an analogous treatment could be a calculation for the valence electrons in the field of the fixed cores. We might argue that we have a good idea of the effect of the core electrons, since they are only slightly changed between the molecule and the atoms. With the core fixed, the field in which the valence electrons move is well-defined (although treating the core simply as a source of potential ignores the effect of the exclusion principle—see Subsection 2). If we neglect the interaction between the valence electrons, as in the simplest Hückel treatment, the problem is formally one-electron: the determination of the molecular orbitals for the valence electrons in the field of the core. (However, the successful parametrization schemes actually include part of the effect of the valence electron repulsion in the effective one-electron Hamiltonian.) As in some treatments of $\pi$ electrons, we can consider the interaction between valence electrons explicitly; then the many-electron problem has been reduced to a few-electron problem. Although it appears that some of the assumptions of the Hückel theory are incorrect, the theory gives qualitatively reasonable results because it reflects the molecular structure [6].

*(a) General Discussion*

As an example of work using the simplest approach, we cite Pearson's calculations [7] of bond energies, built in part on ideas of Pauling and Mulliken. For a heteropolar two-electron bond, Pearson assumed the valence electrons could be described by the molecular orbital

$$\psi_{AB} = a\phi_A + b\phi_B$$

where $\phi_A$ and $\phi_B$ are valence atomic orbitals on nuclei A and B. The energy corresponding to this wavefunction was calculated in terms of matrix elements for some effective Hamiltonian: $\bar{H}_{AA}$, $\bar{H}_{AB}$, and $\bar{H}_{BB}$. With neglect of overlap, the resulting formula for the energy was

$$\varepsilon_1 = \frac{a^2\bar{H}_{AA} + 2ab\bar{H}_{AB} + b^2\bar{H}_{BB}}{a^2 + b^2} \tag{1}$$

Minimizing $\varepsilon_1$ with respect to the values of $a$ and $b$, one gets the secular equations, which leads to

$$\varepsilon_1 = \tfrac{1}{2}(\bar{H}_{AA} + \bar{H}_{BB}) - [\tfrac{1}{4}(\bar{H}_{AA} - \bar{H}_{BB})^2 + \bar{H}_{AB}^2]^{1/2}$$

One might expect $\bar{H}_{AA}$ and $\bar{H}_{BB}$ to represent the energies of the valence electrons in the atoms. (Their difference may be related [7] to the difference of Pauling electronegativities.) If one further takes $\bar{H}_{AB}$ as the geometric mean of $\bar{H}_{AA}$ and $\bar{H}_{BB}$, one obtains $\varepsilon_1$ wholly in terms of atomic quantities. The dissociation energy of the molecule, calculated as $-2\varepsilon_1 + \bar{H}_{AA} + \bar{H}_{BB}$, gave results [7] in good qualitative agreement with reality for many molecules.

Despite the simplicity and reasonableness of this approach, it is open to several criticisms. Problems appear when the preceding formulas are compared with the formulas of a true limited-basis SCF calculation, as was done by Jug [1]. Let the $i$th molecular orbital, $\lambda_i$, be expanded as

$$\lambda_i = \sum_{j=1}^{N} c_j^{(i)}\phi_j, \qquad i = 1, \ldots, n \tag{2}$$

where the $\phi_j$ are members of a set of $N$ basis functions. According to the Hartree–Fock theory, the coefficients $c_j^{(i)}$ are obtained as solutions to a matrix pseudoeigenvalue equation; i.e., the coefficients are the eigenvector of a Hamiltonian matrix which requires the coefficients $c_j^{(k)}$ ($k = 1, \ldots, n$) for its construction, as well as one- and two-electron integrals. In the spirit of the Hückel theory, one assumes the effective Hamiltonian, represented by this matrix, is fixed, and attempts to approximate the matrix elements of the effective Hamiltonian over the basis functions $\phi_j$. We denote such a matrix element by $H_{ij}$, and the approximation to it by $\bar{H}_{ij}$. Then we must solve the matrix eigenvalue equation

$$\sum_{j=1}^{N} (\bar{H}_{ij} - \varepsilon_k S_{ij})c_j^{(k)} = 0, \qquad k = 1, \ldots, N. \tag{3}$$

where the $S_{ij}$ are the overlap integrals between the basis functions. The $\varepsilon_j$ are approximations to the orbital energies. If one really wants to mimic the Hartree–Fock equations, where the coefficients used in constructing the Hamiltonian must be the same as those obtained by solving the pseudo-eigenvalue equations (self-consistency), one requires a method for including the dependence of the $\bar{H}_{ij}$ on the coefficients.

Most modern methods make provision for such a dependence. In applying the scheme of Eq. (1) to metal chlorides, Pearson and Gray [8] introduced such a dependence. Letting A refer to metal and B to the chlorine, they defined the ionicity

$$x = (b^2 - a^2)/(b^2 + a^2)$$

and wrote the molecular orbital energy as

$$\varepsilon_1 = \tfrac{1}{2}(1 + x)q_{Cl} + (1 - x^2)^{1/2}\beta + \tfrac{1}{2}(1 - x)q_{M}.$$

The total energy of the electron pair was twice this. The "Coulomb integral" $q_M$ ($\bar{H}_{AA}$) was taken as the negative of an ionization potential, but $q_{Cl}$ ($\bar{H}_{BB}$) was supposed to depend on $x$, since for an ionic molecule it includes an electrostatic attraction between the negatively charged Cl atom (excess of electrons) and the positively charged (electron-deficient) metal atom. For the "exchange integral" $\beta$ ($\bar{H}_{AB}$), Pearson and Gray [8] used the geometric mean of the corresponding quantities for $M_2$ and $Cl_2$, obtained by comparing experimental binding energies with the formula just given. Then $x$ was determined by putting $\partial W/\partial x$ equal to zero.

Another problem with the one-electron scheme that appears on comparison with the limited-basis Hartree–Fock equations is as follows. Having calculated, by means of Eq. (3), values of orbital energies, we do not have the total electronic energy. In the Hartree–Fock theory, this is given by

$$E^{\mathrm{HF}} = \sum_{i=1} \varepsilon_i - \tfrac{1}{2} \sum_{i,j=1}^{N} \langle \lambda_i \lambda_j \,|\, 1/r_{12} \,|\, \lambda_i \lambda_j - \lambda_j \lambda_i \rangle \qquad (4)$$

Summing the orbital energies counts the interelectronic repulsion twice, which necessitates the extra terms in (4). But these are just the two-electron integrals we want to avoid calculating. An alternative formula,

$$E^{\mathrm{HF}} = \tfrac{1}{2} \sum_{i=1} \varepsilon_i + \tfrac{1}{2} \sum_{i=1} \langle \lambda_i \,|\, h_1 \,|\, \lambda_i \rangle \qquad (5)$$

where $h_1$ is the one-electron part of the Hamiltonian, is somewhat better, requiring that we evaluate only the one-electron integrals. If our goal is to avoid evaluation of all integrals, except overlaps, a recipe must be invented for obtaining an approximation to the Hartree–Fock energy from the sum of the orbital energies. It turns out that, for some purposes, one can use the sum of orbital energies for $E^{\mathrm{HF}}$ if one omits the internuclear repulsion (see below). As mentioned in Section C.3 of Chapter II, Lennard-Jones (before *ab initio* calculations were available) predicted bonding properties of diatomic molecules from consideration of the way orbital energies varied with $R$.

The extension of the Hückel method to saturated molecules (extended Hückel calculations) was developed by Hoffmann [9]. For a recapitulation of the scheme with discussion of implementation, see Kagan *et al.* [10]. Only valence electrons were considered and a minimal basis set was used. The energies reported by Hoffmann [9] were sums of valence electron orbital energies, so that correction terms such as those given in Eq. (4) were neglected together with the intercore repulsion. It was found that, from the behavior of these energies with changes in nuclear coordinates, equilibrium distances and even force constants could be predicted for many molecules. This could imply that the two neglected factors roughly cancel, at least near the equilibrium configuration, or that their difference is roughly independent of nuclear displacement in this region. Alternatively, we could contend that the method of guessing the $\bar{H}_{ij}$ also simulates the internuclear repulsion contribution to the energy. Note that $\bar{H}_{ij}/S_{ij}$, which we would expect to be $R$-dependent, is not so in this scheme [see Eq. (10)].

It is true that sums of Hartree–Fock orbital energies, as a function of nuclear displacement, mimic the behavior of the total energy in many cases. This is the basis of Walsh's rules [11, 11a] for shapes of polyatomic molecules. Equilibrium nuclear configurations, and $R_e$ values in particular, are sometimes correctly predicted by such sums of orbital energies. Binding energies are less successful. However, Allen [11a, 12] has shown that the sum of valence orbital energies is an unreliable guide to bond lengths, and that the sometimes good results of the extended Hückel theory are physically incorrect and misleading. The cancellation of the internuclear repulsion against the increase in interelectronic repulsion on molecule formation has been discussed by several authors.

Boer et al. [12a] gave an analysis in which the electrons were divided into core and valence electrons. The repulsion between the core electrons on atom A and the core electrons on atom B behaves, as a function of $R$, like the internuclear repulsion, while the repulsion between the core elec-trons on center A and the valence electrons behaves like the attraction of nucleus A for the valence electrons, except with opposite sign and with a different effective charge. If one can consider the valence electrons moving in the field of the nuclei and the cores, which are assumed not to deform, the situation is exactly that of a molecule with a very few electrons moving in the field of nuclei with small charges. Since the effective Hamiltonian of Hoffmann's Hückel method is just for the valence electrons, the cancellation required is between the effective internuclear repulsion and the change of interelectronic repulsion of the valence electrons only. These quantities are much smaller than the full internuclear repulsion $V_{NN}$ and tend to cancel as well. Boer et al. [12a] gave numerical data showing this cancella-tion.

This author [13] has noted that, formally, the approximate cancellation of $V_{NN}$ and the change of $V_{ee}$ is related to the isoelectronic principle. Let the binding energy be defined, for a diatomic molecule, as

$$D = -(E_e^m + V_{NN}) + E_e^A + E_e^B \qquad (6)$$

where $E_e$ is the energy of the electrons, and m, A, and B refer, respectively, to the molecule with $R = R_e$, atom A, and atom B. According to the iso-electronic principle, isoelectronic systems should have about the same binding energies; thus $\partial D/\partial Z_A = 0$, where $Z_A$ is the charge of nucleus A and the electron configuration is held fixed in taking the derivative of an energy with respect to a nuclear charge. By writing explicitly the expectation values in $E_e^m$, $E_e^A$, and $E_e^B$, and using the Hellmann–Feynman theorem (Volume I, Chapter III, Section D) in carrying out the differentiation, we get from (6)

$$\partial D/\partial Z_A = Z_A^{-1}(-V_{Ae}^m - V_{NN} + V_{Ae}^A). \tag{7}$$

Here, $V_{Ae}^m$ and $V_{Ae}^e$ are the expectation values of attraction of electrons to nucleus A, in the molecule and in the A atom. From (7) and the corresponding relation for B,

$$Z_A \, \partial D/\partial Z_A + Z_B \, \partial D/\partial Z_B = -2V_{NN} - V_{Ae}^m$$
$$- V_{Be}^m + V_{Ae}^A + V_{Ae}^B \tag{8}$$

From the virial theorem, which holds for the atoms and for the diatomic molecule at its equilibrium internuclear distance, and the definition of $D$, the right member of (8) is then just $-V_{NN} + \Delta V_{ee} + 2D$. Here, $\Delta V_{ee} = V_{ee}^m - V_{ee}^A - V_{ee}^B$. Let us denote by $\bar{D}$ the binding energy calculated using sums of orbital energies instead of total energies, so

$$\bar{D} = D + V_{NN} - \Delta V_{ee}.$$

Using Eq. (8),

$$\bar{D} = D - Z(\partial D/\partial Z_A) - Z_B(\partial D/\partial Z_B) + 2D. \tag{9}$$

If $\partial D/\partial Z_A$ and $\partial D/\partial Z_B$ vanished exactly, $\bar{D}$ would be three times the correct binding energy; the additional terms probably decrease the error. In fact, binding energies calculated by neglecting $V_{NN}$ and using differences of orbital energy sums tend to be high by about a factor of 2 [12a, 13]. Because $\partial D/\partial Z_A$ is small (isoelectronic principle), the error, $\bar{D} - D$, is of the size of $D$ itself. Dissociation energies, even if not accurately calculated, will not be wildly in error—as might be feared from the size of $V_{NN}$ relative to $D$. The preceding treatment could refer to the actual molecule, with $Z_A$ and $Z_B$ representing the true nuclear charges and all electrons being included in the calculation, or to the pseudomolecule, with $Z_A$ and $Z_B$ referring to the effective charges of the cores and only valence electrons included. In the latter case, the error associated with the core approximation may cancel the error due to the assumption that $V_{NN} = \Delta V_{ee}$.

### (b) Representative Calculations

We now consider the parametrization schemes which have been used in the extended Hückel and other methods. In Hoffmann's work [9], the basis functions were the atomic orbitals of the valence shells for all the atoms in a molecule. The overlap integrals were computed using SCF or Slater representations of these orbitals. It turned out that results were insensitive to the choice of orbitals, suggesting that the simpler Slater orbitals be used. The off-diagonal $\bar{H}_{ij}$, generally referred to as resonance integrals, were assumed proportional to the corresponding overlap integrals and to the average of the corresponding diagonal elements:

$$\bar{H}_{ij} = \tfrac{1}{2}K(\bar{H}_{ii} + \bar{H}_{jj})S_{ij} \qquad (10)$$

$K$ is sometimes referred to as a Wolfsberg–Helmholtz parameter [14]. The value 1.75 was used for $K$, although varying $K$ from 1.6 to 2.0 seemed to produce little change in the results. The success of this kind of approximation, which we will see is often used, may be associated with a favorable choice of diagonal matrix elements [15]. It is not a good approximation if the actual integrals are used. All overlaps (not just nearest neighbors, as in the simplest Hückel theory) were used. The diagonal elements $\bar{H}_{ii}$ ("Coulomb integrals") were assumed by Hoffmann [9] to depend on the atom and on the angular momentum type of the atomic orbitals. They were first taken as ionization potentials for the valence states of the atoms. Pople *et al.* [16] pointed out that the formula (10) violates certain invariance criteria with respect to rehybridization of basis orbitals (see page 310 below). Jug [17] has shown how formula (10) can be derived, and compared theories that employ it with others that are closer to *ab initio* calculations. He proposed the use of $2/(1 + S^2)$ instead of the constant $K$; other related suggestions had been made previously (see Jug's article [17] for details).

We have mentioned previously that schemes of this type cannot simulate *ab initio* SCF calculations, since they do not deal explicitly with the one- and two-electron parts of the SCF Hamiltonian, do not use a correct formula like Eq. (4) or Eq. (5) for the total energy, and do not provide for self-consistency between the matrix elements of the effective Hamiltonian and the solution. This is not to say that parametrization scheme considering the effective Hamiltonian as a whole cannot give accurate dissociation energies. Clearly there are interpolation formulas that predict the dissociation energy of a molecule from those of related molecules, so that a semiempirical calculation or parametrization scheme with parameters chosen to give correct results on several molecules can reasonably be expected to do well on related ones. However, a scheme in which the matrix elements of the effective one-electron Hamiltonian are parametrized, does not really approximate the SCF calculation. Some authors (see review articles) have abandoned the notion that the extended Hückel schemes are approximations to *ab initio* MO calculations. It is well known [2, 4, 15, 18] that the extended Hückel scheme sometimes gives good results even when individual approximations used in it are poor. Klopman and O'Leary [4] have pointed out that, when experimental atomic energies are employed in the parametrization of the matrix elements, some of the correlation energy is included in the semiempirical calculation, although it purports to be within the Hartree–Fock framework.

The introduction of self-consistency generally is done by making the $\bar{H}_{ij}$ in Eq. (3) depend on the coefficients $c_j^{(k)}$. If the effective Hamiltonian includes a Fock operator, it is obvious that such a dependence is needed. Physically, the energy of an electron in an atomic orbital must depend on the number of electrons in the orbital, which may be obtained from a population analysis of the molecular orbitals of Eq. (1). Indeed, extended Hückel calculations generally overestimate local charges in molecules, just because there is no mechanism for making a high occupation of any one atomic orbital energetically unfavorable. According to Jorgenson et al. [15] the proportionality of the off-diagonal $\bar{H}_{ij}$ to the average diagonal element and the overlap integral $S_{ij}$ does not properly represent the effect of ionicity. The ionicity should be more important for diagonal elements than for the off-diagonal elements, which are associated with covalent bonding. Semiempirical theories which include a dependence of $\bar{H}_{ii}$ on atomic orbital occupation numbers are discussed in an article by Cusachs and Reynolds [19]. One such method, by Pearson and Gray [8], has already been mentioned. Rein et al. [20] have described an iterative scheme in which populations are computed from the results of an extended Hückel calculation and used to calculate new values for $\bar{H}_{ii}$. With the new values, the extended Hückel calculation is repeated to give new populations and new $\bar{H}_{ii}$, and the process is iterated until the $\bar{H}_{ii}$ no longer change. This self-consistency method decreases the exaggerated charge separation found by the simple extended Hückel calculations.

Harris [21] has pointed out that simply iterating to self-consistency in this way does not necessarily lead to the minimum energy solution. One must start with the approximate energy expression in which the dependence of matrix elements on expansion coefficients is made explicit, and then differentiate to minimize it with respect to the coefficients. This is truly analogous to the Hartree–Fock procedure, and produces, instead of Eq. (3),

$$\sum_{j=1}^{N} (F_{ij} - \varepsilon_k S_{ij}) c_j^{(k)} = 0 \qquad (11)$$

where

$$F_{ij} = \bar{H}_{ij} + \sum_{m,n=1}^{N} P_{mn}(\partial H_{mn}/\partial P_{ij}). \qquad (12)$$

Here,

$$P_{ij} = 2 \sum_{k=1}^{n} c_i^{(k)*} c_j^{(k)}$$

is an element of the density matrix in the atomic orbital basis. In some cases, solution of (3) automatically gives a solution to (12) [21], but this is not always so.

It is customary to associate $\bar{H}_{ii}$ to the valence state ionization potential (VSIP), representing the energy per electron in atomic orbital $i$. According to Cusachs and Reynolds [19], the VSIP should be roughly a linear function of $q$, the occupation number of the orbital. The linear dependence of $\bar{H}_{ii}$ follows if we write the energy of the electrons in orbital $i$ as

$$E(q_i) = q_i U_i + \tfrac{1}{2} q_i(q_i - 1) W_{ii}. \qquad (13)$$

Here, $U_i$ is the orbital energy and $W_{ii}$ the repulsion between two electrons in the orbital $i$ ($U_i$ includes repulsions due to electrons in other orbitals). The VSIP or $\bar{H}_{ii}$ may be defined as $E(q_i - 1) - E(q_i)$, as $-\partial E/\partial q_i$, or as $-E/q_i$, making it linear in $q_i$. With one electron in orbital $i$, the ionization potential $I_i$ is just $-U_i$ and the electron affinity $A_i$ is $-(U_i + W_{ii})$. The approximation

$$W_{ii} = I_i - A_i \qquad (14)$$

(both quantities taken as positive) is often used in semiempirical calculations. Cusachs and Reynolds [19] went one step further by combining this formula with Eq. (13) and the definition, $\bar{H}_{ii} = -\partial E/\partial q_i$, to give

$$\bar{H}_{ii} = -I_i - \tfrac{1}{2}(I_i - A_i) + q_i(I_i - A_i).$$

Other such formulas were also discussed, and the importance of a dependence of $\bar{H}_{ii}$ on $q_i$ was emphasized [19].

If the diagonal element of the effective one-electron Hamiltonian is taken as a quadratic function of orbital occupation number, several schemes exist to relate the coefficients in the quadratic to experimental data. Jorgenson *et al.* [15] gave tables of values for these derived coefficients, and discussed their use in semiempirical calculations for diatomic and polyatomic molecules. In the simplest scheme, the electronic energy of the molecule MX relative to the atoms M and X can be taken as $I_M(\xi) - I_X(-\xi)$ where each ionization potential is a quadratic function of the net charge $\xi$. A Coulombic attraction of the charged species must be added. The minimum energy is obtained by setting the derivative with respect to $\xi$ equal to zero, which corresponds to an equalization of electronegativity principle. To relate this phenomenological formulation to quantum mechanics, Jorgenson *et al.* [15] considered that intraatomic interactions were included in $I_M$ and $I_X$, and interatomic interactions in the Coulomb or Madelung energy, proportional to $\xi^2$. This scheme leads to large ionicities, and predicts dissociation to ionic species. This is incorrect, but should not lead to too much trouble at small $R$ [15]. Since mutual penetration of the atoms is neglected, $R_e$ cannot be predicted without inclusion of additional terms in the energy, but the charge distribution for a given $R$ should be reasonably well described. A contradiction with the virial theorem (Volume I, Chapter III, Section C) was also discussed [15].

Schemes like those just mentioned are needed only when one adheres to the extended Hückel philosophy, i.e. parametrization of matrix elements of the effective Hamiltonian rather than the one- and two-electron integrals themselves. If one approximates the individual integrals, a Hartree–Fock calculation may be carried through using them [with energies being calculated correctly as in Eqs. (4) or (5)], and dependence of the Hartree–Fock matrix on populations is present. It does not follow, however, that such a procedure approximates an *ab initio* calculation, even if the same formulas are used; parametrization which yields good results for particular properties may, in fact, involve poor or unrealistic approximations to the individual integrals. For example, differential overlap between orbitals on different atoms is often neglected in some integrals, while the overlap integrals between the same orbitals are not set equal to zero. Such inconsistencies were discussed as part of a general treatment of semiempirical MO methods by Pople [3, 5] (see page 310). Before discussing this treatment and the methods which grew out of it, we mention briefly other semiempirical schemes which have been used for calculating $U(R)$.

Klopman [22, 23] considered only the valence electrons of single-bonded diatomic molecules, with a bond molecular orbital formed as a linear combination of atomic orbitals, one on each atom:

$$\psi = C_A \phi_A + C_B \phi_B \tag{15}$$

Because of the neglect of differential overlap, off-diagonal matrix elements of the one-electron Hamiltonian, $\langle \phi_A(1) \mid h(1) \mid \phi_B(1) \rangle$, were set equal to zero, while the only nonvanishing two-electron integrals were

$$W_{AA} = \int \phi_A(1)^* \phi_A(2)^* r_{12}^{-1} \phi_A(1) \phi_A(2) \, d\tau_1 \, d\tau_2,$$

$W_{BB}$, and

$$\Gamma_{AB} = \int \phi_A(1)^* \phi_B(2)^* r_{12}^{-1} \phi_A(1) \phi_B(2) \, d\tau_1 \, d\tau_2.$$

$W_{AA}$ was obtained from atomic energies, while $\Gamma_{AB}$ was assumed to have a Coulombic dependence:

$$\Gamma_{AB} = e^2 [R^2 + (\varrho_A + \varrho_B)^2]^{-1/2}$$

where $R$ is the internuclear distance and $\varrho_A = 0$ except for an $S$-orbital, where it equals the radius of the orbital. The diagonal one-electron integrals were divided into two parts:

$$\int \phi_A(1)^* h(1) \phi_A(1) \, d\tau_1 = \int \phi_A(1)^* (T + V_A) \phi_A(1) \, d\tau_1$$
$$+ \int \phi_A(1)^* V_B \phi_A(1) \, d\tau_1 \tag{16}$$

where $T$ is the kinetic energy operator; $V_A$ and $V_B$ are the potentials of the cores of A and B. The first part is a one-center contribution, whose value was derived [22, 23] from energies of atomic states, while the second part, denoted by $\beta_{AB}$, was chosen semiempirically. For the heteronuclear diatomic molecule AB, $\beta_{AB}$ was taken to be the geometric mean of the quantities $\beta_{AA}$ and $\beta_{BB}$, which refer to the homonuclear molecule and were derived by finding the values needed to give agreement between calculated and experimental bond dissociation energies for the homonuclear molecules. Thus $\beta_{AB}$ and $\beta_{BA}$ were identical. With all necessary parameters thus fixed, a matrix SCF calculation was performed to obtain $C_A$ and $C_B$, and bond energies were derived. They agreed with experiment to a few per cent, except for the fluorides, where errors were much greater. The calculations were subsequently extended to polyatomic systems [23].

Pohl *et al.* [24] performed semiempirical MO calculations on the hydrogen halides. Their parametrizations were generalizations of the Pariser–Parr–Pople treatment [25] for aromatic systems. As in Klopman's treatment [22, 23] only the valence electrons were treated explicitly, the others being subsumed into a nonpolarizable core, and the bond molecular orbital was represented by (15). For HX (X = halogen), $\phi_A$ was the hydrogen 1s orbital and $\phi_B$ the halogen $np$ orbital, for which a Slater orbital was used. The diagonal one-electron integral was written as (16), the first term taken as minus the valence state ionization potential $I_A$, and $\beta_{AB}$ as $R^{-1}$. The one-center core integral $\langle \phi_A | V_A | \phi_A \rangle$ was evaluated as $\langle \phi_A | 1/r_A | \phi_A \rangle$. For the two-electron integral $W_{AA}$, Pohl et al. [24] used Eq. (14), and $\Gamma_{AB}$ was approximated as $R^{-1}$. To evaluate other integrals, they put

$$\phi_A\phi_B \to \tfrac{1}{2}S_{AB}(\phi_A{}^2 + \phi_B{}^2).$$

(This is the basis for the so-called Mulliken approximation [26], used to evaluate molecular integrals before big computers became ubiquitous.) The energy of the system was the energy of the valence electrons plus the proton-core repulsion, an approximation to which could be obtained from the potential seen by a point charge penetrating the nondeforming halogen atomic charge cloud. The potential was taken from quantum statistical theories (Volume I, Chapter III, Section Γ, p. 277).

From calculations of total energy at several values of $R$, $R_e$ and the force constant $k$ were derived [24]. To derive the binding energy, the following hypothetical process was imagined: The atoms were promoted to their valence states and ionized, the ionic cores were brought to a distance $R_e$, and the valence electrons put back to form the molecule. The overall energy change was a sum of promotion energies and valence state ionization potentials, plus the core repulsion and the electronic energy. The results are included in Table I. The possibility of using such calculations to describe trends within series' of related molecules, perhaps predicting an unknown quantity for one molecule from calculated results plus experimental data on the other molecules, seemed promising [24]. It was possible to improve the calculated binding energies by multiplying off-diagonal matrix elements of the Fock Hamiltonian by an empirically determined parameter, like $K$ in Eq. (10). This also improved calculated equilibrium internuclear distances slightly, but made force constants worse.

**Table I**

*Semiempirical Calculations for Hydrogen Halides*

|      | $k$ ($10^5$ dynes/cm) | | | $R_e$ (Å) | | | $D_e$ (eV) | | |
| --- | --- | --- | --- | --- | --- | --- | --- | --- | --- |
|      | Ref.[24] | Ref.[28] | Expt. | Ref.[24] | Ref.[28] | Expt. | Ref.[24] | Ref.[28] | Expt. |
| HF  | 7.5 | 20.2 | 9.6 | 0.86 | 0.87 | 0.91 | 5.65 | 7.59 | 6.08 |
| HCl | 3.3 | 4.6 | 5.1 | 1.44 | 1.30 | 1.27 | 2.6 | 4.84 | 4.47 |
| HBr | 2.3 | 2.5 | 4.1 | 1.60 | 1.48 | 1.41 | 2.2 | 3.73 | 3.80 |
| HI  | 1.8 | 1.1 | 3.1 | 1.84 | 1.64 | 1.60 | 1.8 | 1.85 | 3.10 |

Subsequently, Harris and Pohl [27] considered a split shell wavefunction, which was a product of the singlet spin function with the spatial function

$$\psi_1(1)\psi_2(2) + \psi_2(1)\psi_1(2)$$

where each molecular orbital, $\psi_1$ and $\psi_2$, was of the form (15). This wave function behaves properly as R becomes infinite (see Chapter II, Section C). The same basis functions and integral approximations were used as in the previous calculation. Some improvement was found in binding energies, equilibrium distances, and force constants.

Ferreira and Bates [28] analyzed the relations between several semiempirical schemes as applied to the hydrogen halides. One of their aims was to show the relation of the parametrized SCF formalism to the method of differential ionization energy [22], which is the dependence of an ionization potential on the population of an atomic orbital, taken as a continuous variable. For a doubly occupied molecular orbital of the form (15), the energy is

$$\langle H \rangle_\varphi = c_A{}^2 H_{AA} + 2c_A c_B H_{AB} + c_B{}^2 H_{BB}.$$

If the effective Hamiltonian operator represents the effect of the nuclei and all other electrons (i.e., it simulates a Hartree–Fock Hamiltonian),

$$H_{AA} = \langle \phi_A \mid T + V_A + V_B + V_{el} \mid \phi_A \rangle$$

where $T$ is the kinetic energy operator, $V_A$ and $V_B$ are the effective potentials due to the cores, and $V_{el}$ is the electronic repulsion of the other electron in $\psi$. Most calculations approximate, as did those just discussed, the one-electron integral $\langle \phi_A \mid T + V_A \mid \phi_A \rangle$ as an atomic one-electron eigenvalue or the negative of an ionization potential, $I_A$. The integral $\langle \phi_A \mid V_{el} \mid \phi_A \rangle$ may be approximated by writing out the correct expression,

$$\langle \phi_A \mid V_{el} \mid \phi_A \rangle = \langle \phi_A(1)\psi(2) \mid 1/r_{12} \mid \phi_A(1)\psi(2) \rangle,$$

in terms of the atomic orbitals, and invoking zero differential overlap. (The electrons having opposite spins, there is no exchange term.) The result is

$$c_A{}^2 \langle \phi_A(1)\phi_A(2) \mid 1/r_{12} \mid \phi_A(1)\phi_A(2) \rangle + c_B{}^2 \langle \phi_A(1)\phi_B(2) \mid 1/r_{12} \mid \phi_A(1)\phi_B(2) \rangle.$$

Alternatively, the product $\phi_A(1)\phi_B(1)$ could be approximated by something proportional to $\phi_A(1)^2 + \phi_B(1)^2$ to get a similar expression. The first integral is often approximated (see page 303) by $I_A - A_A$. We denote the second by $\Gamma_{AB}$. The matrix elements $\bar{H}_{ij}$ now depend on the molecular orbital coefficients, just as in the true Hartree–Fock equations.

With the assumptions made so far, one has [28]

$$\bar{H}_{AA} = -I_A + c_A{}^2(I_A - A_A) + c_B{}^2\Gamma_{AB} + \beta_{AB}. \tag{17}$$

The "molecular terms" are $c_B{}^2\Gamma_{AB} + \beta_{AB}$ and the "atomic terms" are $-I_A + c_A{}^2(I_A - A_A)$. The contribution of the atomic terms to the energy of the bond molecular orbital is

$$2c_A{}^2(-I_A) + c_A{}^4(I_A - A_A) = -q_A I_A + \tfrac{1}{4}q_A{}^2(I_A - A_A) \tag{18}$$

where the occupation number or population $q_A = 2c_A{}^2$ has been introduced. The factor of two is from double occupation of the MO; the interelectronic repulsion must be counted only once. This differs from the energy of an isolated AO, Eq. (13). Ferreira and Bates [28] ascribed the difference to left–right correlation energy. The energy per electron may be taken as the derivative of (18) with respect to $q_A$. The approximations of Cusachs and Reynolds [19] [Eqs. (13) *et seq.*] yield expressions of similar form, but not exactly identical.

For computations on the hydrogen halide HX, Ferreira and Bates [28] assumed no overlap between the H and X AO's, so that $c_A{}^2 + c_B{}^2 = 1$. A point charge assumption was used for $\Gamma_{AB}$ and $\beta_{AB}$, and Eq. (10) was used for off-diagonal elements. The bond dissociation energy was calculated as the negative of twice the molecular orbital energy plus atomic promotion energies to valence states plus atomic valence state ionization potentials plus intercore repulsion (point charges assumed). With these formulas, a correspondence could be made with the method of differential ionization energies [22]. In this method, the ionicity is defined as $x = c_A{}^2 - c_B{}^2 = 2c_A{}^2 - 1 = 1 - 2c_B{}^2$, and the bond energy is

$$\varepsilon_1 = (1 - x^2)^{1/2}E_{cov} - x^2R^{-1} + Z_A Z_B R^{-1} + \varepsilon_A(x) + \varepsilon_B(x). \tag{19}$$

The differential ionization energy of atom A is a quadratic in $x$, chosen to give the known atomic energies for atomic populations of 0, 1, and 2 ($x = -1, 0, 1$); the coefficients depend on $I_A$ and $A_A$. The "ionic bond order" is $x^2$ and $(1 - x^2)^{1/2}$ is identified with the covalent bond order. The final formula for the energy could be made identical to that of the SCF theory.

The bond energies obtained by Ferreira and Bates [28], using the formulas we have discussed, were minimized with respect to the coefficients to yield pseudoeigenvalue equations. The input was the value of a Wolfsberg–Helmholtz parameter ( chosen as 0.4 by trial and error to get the best results), overlap integrals, atomic properties, and core–core repulsions (point charge model using charges estimated from atomic calculations). Some of the results were included in Table I. Potential curves were also displayed. In general, the results are better than those of Pohl et al. [24] but force constants are not well predicted. We refer to Ferreira and Bates's article [28] for many references to previous work on semiempirical calculations.

Earlier, Ferreira [29] had calculated accurate dissociation energies for 23 diatomic molecules with large ionic character, with a somewhat simpler formalism. The dissociation energy was written in terms of the ionicity as [see Eq. (19)]

$$D_{AB} = -2\varepsilon_1 + \bar{H}_{AA} + \bar{H}_{BB}$$
$$= -x\bar{H}_{BB} - 2(1 - x^2)^{1/2}\tilde{\beta}_{AB} + x\bar{H}_{AA} + (x^2 - Z_A Z_B)R^{-1}$$

For the homonuclear molecules A–A and B–B, where $x = 0$ and $Z_A Z_B$ is small, $-2\beta$ represents the dissociation energy, so if $\tilde{\beta}_{AB}$ is taken to be the mean of $\tilde{\beta}_{AA}$ and $\tilde{\beta}_{BB}$ one can write

$$-2(1 - x^2)^{1/2}\tilde{\beta}_{AB} = \tfrac{1}{2}(1 - x^2)^{1/2}(D_{AA} + D_{BB}).$$

Ferreira [29] introduced the linear dependence of $\bar{H}_{AA}$ and $\bar{H}_{BB}$ on $x$: for zero charge on an atom, $\bar{H}_{AA}$ was taken as the Mulliken electronegativity, $\tfrac{1}{2}(I_A + A_A)$, while $\partial \bar{H}_{AA}/\partial x$ was taken, as above, as $\tfrac{1}{2}(I_A - A_A)$. Then $D_{AB}$ was a quadratic in $x$. Ferreira fixed $R$ at the known equilibrium internuclear distance, differentiated with respect to $x$, and calculated $D_{AB}$. The average error in $D_{AB}$, which varied between 50 and 200 kcal for the molecules considered, was 3%.

The interesting halogen–halogen diatomics have been the subject of many semiempirical (and ab initio) calculations, as is evidenced by the list of references given by Cheesman et al. [30]. A number of the simple models, such as the free-electron and electrostatic models have also been employed. We cannot discuss all here, but merely summarize the conclusions of Cheesman et al. for the model they investigated. It appears that the bonding can be explained very well in terms of the $p\sigma$ orbitals alone, with the other orbitals of the valence shell contributing to the effective core potential. Since there is a single bond and one $p\sigma$ orbital per atom, one has a situation as simple as the Hückel $\pi$-electron theory. It was concluded that, to obtain good results for $R_e$, one must go beyond a simple treatment of the $p\sigma$ electrons, at least describing the core interaction. A theory that did this and resulted in good $R_e$ values was described; unfortunately, it also yielded poor charge distributions (dipole moments).

### (c) CNDO et seq.

A systematic study and development of semiempirical molecular orbital schemes has been given by Pople and co-workers [5]. They considered schemes to treat the valence electrons only, so that the core electrons gave rise to an effective potential, and the one-electron Hamiltonian included the attraction of the core. Pople *et al.* [16] began by considering what conditions had to be satisfied by an approximation scheme for the integrals of the Hamiltonian if the results of an SCF scheme using them were to be invariant to a transformation of the basis set (mixing basis orbitals among themselves). This should not change the results of the calculation. Such a transformation could rehybridize the orbitals on one atom, for instance. If the elements of the transformation matrix are given by $O_{mn}$, so that

$$t_m = \sum_{k=1}^{n} O_{km}\phi_k$$

expresses the $m$th new basis function in terms of the $n$ members of the original set, matrix elements of the two-electron Hamiltonian transform as (assuming the $O_{mn}$ are real)

$$
\begin{aligned}
g_{mnpq} &= \langle t_m(1)t_n(2) \mid g_{12} \mid t_p(1)t_q(2)\rangle \\
&= \sum_{i,j,k,l} O_{im}O_{jn}O_{kp}O_{lq}\langle \phi_i(1)\phi_j(2) \mid g_{12} \mid \phi_k(1)\phi_l(2)\rangle.
\end{aligned}
\tag{20}
$$

Pople *et al.* [16] showed that some schemes used led to contradictions. For instance, one cannot, as an approximation, simply assume zero differential overlap, so $\phi_m\phi_p = 0$ for $m \neq p$ and

$$\langle \phi_m\phi_n \mid g_{12} \mid \phi_p\phi_q\rangle = 0 \quad \text{unless} \quad m = p, \quad n = q. \tag{21}$$

If one takes new basis functions $t_a = \phi_m + \phi_p$ and $t_b = \phi_m - \phi_p$ [and take $\phi_n = \phi_q$ for simplicity in (21)] one would have

$$\langle t_a\phi_n \mid g_{12} \mid t_b\phi_n\rangle = \langle \phi_m\phi_n \mid g_{12} \mid \phi_m\phi_n\rangle - \langle \phi_p\phi_q \mid g_{12} \mid \phi_p\phi_n\rangle.$$

The rule (21) would make the left side zero, but not the right side. But suppose $\phi_m$ and $\phi_p$ are perpendicular 2p orbitals on the same atom, so

$t_a$ and $t_b$ are also perpendicular 2p orbitals. Evidently here it suffices to demand $\langle \phi_m \phi_n \mid g_{12} \mid \phi_m \phi_n \rangle$ and $\langle \phi_p \phi_n \mid g_{12} \mid \phi_p \phi_n \rangle$ equal to restore consistency. An often-used approximation, neglect of differential overlaps only between pairs of orbitals on different centers, is consistent provided that transformations which mix basis orbitals on different centers, leading to two-center basis functions, are ruled out.

Naturally the simplest approximation scheme is one that puts as many integrals as possible equal to zero, and the first scheme put forward by Pople *et al.* [16] involved complete neglect of differential overlap (CNDO) in the two-electron integrals. (If differential overlap were neglected everywhere, so that the one-electron integrals $\langle \phi_m \mid h \mid \phi_n \rangle$ were set equal to zero for $m \neq n$, the bonding effect of the overlap would be eliminated.) The overlap matrix was assumed diagonal. For orbitals on the same atom, taking the differential overlap equal to zero might be a bad approximation, but the overlap integrals between such orbitals vanish in any case. The integrals for the two-electron operator are now

$$\langle \phi_k \phi_l \mid g_{12} \mid \phi_m \phi_n \rangle = \delta_{km} \delta_{ln} \gamma_{kl} \tag{22}$$

This would lead to trouble (see page 310) unless some condition were imposed on the $\gamma_{kl}$. The simplest is to make $\gamma_{kl}$ the same for all $\phi_k$ on atom A and all $\phi_l$ on atom B, i.e., $\gamma_{kl}$ is replaced by $\gamma_{AB}$, depending on the centers involved only. For a diatomic molecule, there are only three parameters involved.

For the case of doubly occupied orbitals (closed shell) the matrix elements $H_{ij}$ of the Hartree–Fock Hamiltonian are

$$\langle \phi_i \mid h \mid \phi_j \rangle + \sum_{\alpha=1}^{n/2} \sum_{k,l=1}^{N} c_k^{(\alpha)} c_l^{(\alpha)} [2 \langle \phi_k \phi_i \mid g \mid \phi_l \phi_j \rangle - \langle \phi_k \phi_i \mid g \mid \phi_j \phi_l \rangle]$$

$$= h_{ij} + \sum_{\alpha=1}^{n/2} \sum_{k,l=1}^{N} c_k^{(\alpha)} c_l^{(\alpha)} [2 \delta_{kl} \delta_{ij} \gamma_{ki} - \delta_{kj} \delta_{il} \gamma_{ki}] \tag{23}$$

Pople *et al.* [16] introduced the atomic orbital population matrix

$$P_{ij} = 2 \sum_{\alpha=1}^{n/2} c_i^{(\alpha)} c_j^{(\alpha)} = P_{ji} \tag{24}$$

and rewrote (23) as

$$\bar{H}_{ij} = h_{ij} + \sum_k P_{kk} \delta_{ij} \gamma_{ki} - \tfrac{1}{2} P_{ij} \gamma_{ij}.$$

Note that $\sum_{\alpha,i} P_{ii} = n$, where $n$ is the number of valence electrons. Grouping together orbitals for each atom, and defining the total atomic populations

$$P_{AA} = \sum_{i}^{(i\ \mathrm{on}\ A)} P_{ii}$$

the diagonal elements become (assuming $\phi_i$ is on A)

$$\bar{H}_{ii} = h_{ii} + P_{AA}\gamma_{AA} + P_{BB}\gamma_{AB} - \tfrac{1}{2}P_{ii}\gamma_{AA} \qquad (25)$$

An off-diagonal element would be

$$\bar{H}_{ij} = h_{ij} - \tfrac{1}{2}P_{ij}\gamma_{ij} \qquad (26)$$

To parametrize the remaining integrals, Pople *et al.* [16] divided $h$ into the Hamiltonian for the core of atom A $(T + V_A)$ and that for the core of atom B $(V_B)$. Calling these parts $U^{(A)}$ and $-V_B$, they put $h_{ij} = U_{ij}^{(A)} - (V_B)_{ij}$, when $\phi_j$ was an orbital on atom A, and $h_{ij} = U_{ij}^{(B)} - (V_A)_{ij}$ when $\phi_j$ was an orbital on B. If both $\phi_i$ and $\phi_j$ were on A, $V_B$ included repulsion between core orbitals on B and the charge distribution $\phi_i(1)\phi_j(1)$ on A, represented by integrals like $\langle \phi_i\phi_k \mid g \mid \phi_j\phi_k \rangle$ ($\phi_k$ is in the core of B). Since similar integrals were dropped in (22), it was consistent to put $(V_B)_{ij}$ equal to zero when $\phi_i$ and $\phi_j$ were different orbitals on A. By a similar analogy, one should put $(V_B)_{ij} = V_{AB}\delta_{ij}$, with $V_{AB}$ independent of $i$ and $j$, but perhaps dependent on the separation between A and B. In addition, $\phi_i$ (on A) is roughly an eigenfunction of $U^{(A)}$, so $U_{ii}^{(A)}$ would be the eigenvalue (perhaps an atomic orbital energy) and $U_{ij}^{(A)}$ would vanish. Thus, for $\phi_i$ and $\phi_j$ on the same atom,

$$h_{ij} = (W_i - V_{AB})\delta_{ij} \qquad (27)$$

For $\phi_i$ and $\phi_j$ on different atoms, $h_{ij}$ should depend on the extent to which they overlap. Thus one can put [16]

$$h_{ij} = \beta_{AB}^0 S_{ij} \qquad (28)$$

where $\beta_{AB}^0$ depends on the atoms involved and perhaps on their separation, but not on the specific orbitals. This is necessary for proper invariance behavior. The value of $h_{ij}$ has roughly the proper dependence on interatomic separation even if $\beta_{AB}^0$ is assumed independent of $R_{AB}$. There is perhaps some inconsistency in neglecting overlap integrals elsewhere and not here; the $h_{ij}$ ("resonance integrals"), however, play a central role in bonding and cannot be neglected.

The parameters whose numerical values need to be specified for the diatomic A–B are

(a) $\gamma_{AA}$, $\gamma_{AB}$, and $\gamma_{BB}$ for the two-electron repulsion integrals.

(b) $W_\mu$ (essentially the energy of the orbital $\mu$ in the absence of the other core and other electrons),

(c) $V_{AB}$ (effect of core B on electrons in orbitals on A) and $V_{BA}$,

(d) the bonding parameters $\beta^0_{AB}$,

(e) the overlap integral $S_{ij}$ for use in (28).

Alternatively, if we do not calculate the overlaps, the quantities $\beta^0_{AB}S_{ij} = \beta_{ij}(R)$ can be taken as empirically assignable parameters, replacing (d) and (e). [A later paper [31, 32] reduced the number of parameters by putting $V_{AB} = Z_B\gamma_{AB}$, with $Z_B$ the charge on the core of B, instead of using (c), after it was found that this gave rise to better results for some simple cases.] The off-diagonal matrix elements of the Fock operator or effective potential were

$$\bar{H}_{ij} = \beta^0_{AB}S_{ij} - \tfrac{1}{2}P_{ij}\gamma_{AB} \qquad (29)$$

and the diagonal elements could be written (for orbital $i$ on atom A)

$$\bar{H}_{ii} = W_i + (P_{AA} - \tfrac{1}{2}P_{ii})\gamma_{AA} + P_{BB}\gamma_{AB} - V_{AB}.$$

The total energy of the molecule is given by:

$$\tfrac{1}{2}\sum_{i,j} P_{ij}(\bar{H}_{ij} + h_{ij}) + Z_A Z_B/R_{AB}$$

In the original application of the theory to diatomic molecules [33] (referred to as CNDO/1), the overlap integrals were calculated directly, while values for $\gamma_{AB}$ and $V_{AB}$ were chosen by calculating some representative integrals:

$$\gamma_{AB} = \iint \phi_{sA}(1)^2\phi_{sB}(2)^2 g_{12}\, d\tau_1\, d\tau_2$$

$$V_{AB} = \int \phi_{sA}(1)^2(Z_B/r_{1B})\, d\tau_1$$

($Z_B$ = charge of core B). The $W_\mu$ (for the 2s and 2p orbitals of the atoms considered) were derived from atomic energies. The bonding parameters $\beta^0_{AB}$ were written as averages of "atomic" quantities to reduce the number of parameters. Thus

$$\beta^0_{AB} = \tfrac{1}{2}(\beta_A^0 + \beta_B^0) \qquad (30)$$

with the $\{\beta_A{}^0\}$ chosen to get the best overall fit of CNDO results on certain diatomics to LCAO–SCF results. A calibration of this kind occurs in most semiempirical theories, and helps assure that reasonable results are obtained.

In general, dipole moments and charge distributions were grossly correct. Orbital energies, compared to vertical ionization potentials (Koopmans's theorem) were more reliable than those from extended Hückel calculations [34]. The errors in the method, however, increased for small $R$ in such a way as to make predicted $R_e$-values much too small and dissociation energies too large. Force constants were generally too high, and various modifications have been suggested to improve this situation [35]. The method was more useful for conformational changes in polyatomics, where bond lengths were not altered [31–33].

The CNDO/2 parametrization [31, 32] was introduced to correct these problems, which were traced to the "penetration integrals" $V_{AB}$. Instead of calculating $V_{AB}$ as an integral over a valence shell s orbital, Pople and Segal [31, 32] used

$$V_{AB} = Z_B\gamma_{AB}$$

without much theoretical justification, retaining $Z_A Z_B/R$ for the intercore repulsion. The method for estimation of the core matrix elements was also modified. For details and interpretations of the resulting formulas, see the book of Pople and Beveridge [5]. Improved predictions for $R_e$ resulted, with errors typically about 0.01 Å (with notable exceptions, such as $Li_2$ and $F_2$). For the heavier atoms, the parametrization of the CNDO/2 theory is made difficult because of a relative dearth of a priori calculations, necessary for calibration. Furthermore, predicted force constants were still poor [31, 32].

Various improved theories, such as intermediate neglect of differential overlap (INDO), which does not put one-center exchange integrals equal to zero (the values of such integrals are derived from Slater–Condon parameters), and neglect of diatomic differential overlap (NDDO), which retains dispersion-like integrals such as

$$\langle\phi_{sA}(1)\phi_{sB}(2) \mid g \mid \phi_{pA}(1)\phi_{pB}(2)\rangle,$$

have also been proposed [5, 33, 36]. They are improvements over CNDO largely with respect to electronic spectra, predictions of $R_e$, $k$, etc. for the ground state being little changed. Pople and Beveridge summarize predictions for CNDO and INDO. Table II is taken from theirs. For force con-

**Table II**

*Results of CNDO and INDO Calculations*

| Mole-cule | $R_e$ (Å) | | | $k$ ($10^5$ dynes/cm) | | | $D_e$ (eV) | | |
|---|---|---|---|---|---|---|---|---|---|
| | CNDO | INDO | Expt. | CNDO | INDO | Expt. | CNDO | INDO | Expt. |
| $H_2$ | 0.746 | 0.746 | 0.742 | 10.4 | 10.4 | 5.7 | 5.37 | 5.37 | 4.75 |
| $Li_2$ | 2.179 | 2.134 | 2.672 | 0.84 | 0.85 | 0.25 | 14.71 | 14.40 | 1.05 |
| $B_2$ | 1.278 | 1.278 | 1.589 | 17.83 | 17.93 | 3.5 | 24.43 | 24.68 | 3.66 |
| $C_2$ | 1.146 | 1.148 | 1.242 | 36.31 | 36.3 | 12.2 | 27.30 | 26.55 | 6.36 |
| $N_2^+$ | 1.127 | 1.129 | 1.116 | 50.3 | 50.3 | 20.1 | 2.40 | 2.21 | 8.84 |
| $N_2$ | 1.140 | 1.147 | 1.094 | 52.2 | 50.5 | 23.0 | 25.49 | 20.21 | 9.90 |
| $O_2^+$ | 1.095 | 1.100 | 1.123 | 66.3 | 64.8 | 16.6 | 2.32 | 4.72 | 6.76 |
| $O_2$ | 1.132 | 1.140 | 1.207 | 56.8 | 54.5 | 11.8 | 17.44 | 15.37 | 5.21 |
| $F_2$ | 1.119 | 1.128 | 1.435 | 56.7 | 53.8 | 3.60 | 14.62 | 12.85 | 1.64 |
| LiH | 1.573 | 1.572 | 1.595 | 1.95 | 1.94 | 1.03 | 5.90 | 5.71 | 2.52 |
| BeH | 1.324 | 1.323 | 1.343 | 5.02 | 5.03 | 2.26 | 7.07 | 7.43 | 2.62 |
| BH | 1.194 | 1.204 | 1.233 | 8.60 | 8.30 | 3.0 | 10.01 | 9.37 | 3.58 |
| CH | 1.108 | 1.118 | 1.120 | 12.07 | 11.65 | 4.5 | 9.61 | 8.62 | 3.64 |
| NH | 1.061 | 1.070 | 1.038 | 14.64 | 14.05 | 5.6 | 8.26 | 6.89 | 3.90 |
| OH | 1.026 | 1.033 | 0.971 | 17.02 | 16.39 | 7.8 | 7.36 | 6.30 | 4.56 |
| FH | 1.000 | 1.006 | 0.917 | 19.12 | 18.64 | 9.6 | 6.77 | 6.30 | 6.11 |
| BF | 1.404 | 1.408 | 1.262 | 17.01 | 17.28 | 7.94 | 12.25 | 11.57 | 4.38 |
| LiF | 2.161 | 2.162 | 1.51 | 1.97 | 1.94 | — | 1.85 | 1.83 | 5.99 |
| BeF | 1.671 | 1.670 | 1.361 | 7.50 | 7.38 | 5.8 | 35.94 | 35.49 | 5.48 |
| BeO | 1.463 | 1.474 | 1.331 | 28.92 | 17.30 | 7.5 | 9.52 | 6.11 | 4.69 |
| BO | 1.297 | 1.300 | 1.205 | 25.54 | 25.26 | 13.6 | 15.22 | 14.75 | 9.22 |
| CO | 1.191 | 1.196 | 1.128 | 41.34 | 40.46 | 19.0 | 22.17 | 19.82 | 11.22 |
| NO | 1.152 | 1.159 | 1.151 | 50.89 | 48.95 | 15.9 | 20.00 | 16.12 | 6.62 |
| BN | 1.268 | 1.269 | 1.281 | 26.55 | 26.70 | 8.3 | 17.11 | 13.36 | 5.09 |
| CN | 1.169 | 1.174 | 1.172 | 40.63 | 38.88 | 16.3 | 23.83 | 21.58 | 7.63 |

stants and dissociation energies, the agreement is not impressive. The average deviation of $R_e$ calculated from CNDO from $R_e$ (experimental) is 0.12 Å. For comparison, the statement that $R_e$ is 1.23 Å for all second-row diatomics has an average deviation of 0.16 Å.

Recently, Baetzold [37] performed extended Hückel and CNDO calculations for metallic diatomic molecules, deriving potential curves and other properties for $Ag_2$, $Cu_2$, $Pd_2$, $Au_2$, $Na_2$, $Cd_2$, and $Ca_2$. In the Hückel calculations, he tried 1.30 and 1.75 for the Wolfsberg–Helmholtz parameter $K$ [Eq. (10)], as well as $K = 2 - |S_{ij}|$, suggested by other workers. In the CNDO calculations, $\beta_A^0$ [Eq. (30)] was chosen empirically.

Baetzold concluded that one can get good results for $R_e$, $D_e$, and other properties, provided the values for the parameters ($K$, $\beta_A{}^0$, etc.) are properly chosen, i.e., one cannot use the same values for all calculations.

The semiempirical scheme of Dewar and Klopman [18] is closely related to Pople's NDDO. These authors also discussed the problem of choosing approximations such that calculated results would be invariant to the choice of basis functions. For their approximations, the invariance was not exact, but the integrals violating these criteria were expected to be small. Unlike other workers, Dewar and Klopman renounced the idea of attempting to reproduce the results of *ab initio* SCF calculations by semiempirical evaluation of integrals, and attempted only to predict molecular properties.

By contrast, in a recent publication Jug [38] invoked quantum mechanical commutator equations to discuss the choice of parameters for approximate MO theories. The commutator equations lead to relationships between exact matrix elements of various operators. In particular, the necessary dependence of certain two-center integrals on $R$ was investigated.

These examples illustrate two possible goals in seeking a semiempirical scheme for calculation of $U(R)$: (1) A simple procedure for getting roughly correct *results*; (2) A close approach to *ab initio* calculations with much less work. It seems that, in order to compete with the *ab initio* SCF calculations, which are becoming practical for almost all diatomics, one must introduce and evaluate so many parameters that semiempirical calculations have little use. It should be noted that many of the methods we have discussed are aimed at polyatomic molecules and deal with diatomics mainly for the purposes of testing and calibration. With respect to goal (1), semiempirical calculations have to compete with simple interpolation schemes. The more elaborate methods do not give better results, but, if they reproduce several different properties, they may lead to more useful rules and increased understanding.

## 2. *Pseudopotentials*

The semiempirical molecular orbital treatments, which consider valence electrons only, include the effect of the core as a "core potential." However, the core is not just an expanded nucleus, but consists of electrons, which are indistinguishable from the valence electrons and hence exchange with them. The Pauli principle, in an orbital picture such as the SCF, is satisfied when the orbitals describing the valence electrons are orthogonal to the orbitals describing the core. This orthogonality is assured if wave functions representing the valence orbitals are large only where those representing the core orbitals are small. If there is appreciable overlap between the core and valence electrons, the assumption of an "inert" core is invalid, and the effect of the Pauli exclusion principle associated with interpenetration of core and valence electrons must be taken into account. If a variational calculation on valence electrons only is performed without ensuring, in some way, that the trial function be orthogonal to the wave functions of the core, the energy of the valence electrons will converge down to the core energy ("nightmare of inner shells"). On the other hand, one feels that it should be possible to develop a variational method for treating the valence electrons without having to consider the core electrons explicitly; we should not have to know too much, say, about all 22 electrons of $Na_2$, to learn about the two bond electrons which determine, to a large extent, $U(R)$.

Since the Pauli principle tends to exclude the valence electrons from the core region, it has the effect of a repulsive potential. It can be shown formally (see page 318) that the orthogonality constraint can be replaced exactly by an additional term in the potential in the valence electron Hamiltonian, but the term is nonlocal, and not simply a multiplicative potential. A term that resembles a potential, i.e., a local, one-particle multiplicative operator, but that represents an effect which cannot really be described by a potential, is called a pseudopotential. As another example, consider the interaction of a point charge $q$ with a polarizable charge distribution of dipole polarizability $\alpha$. The interaction energy is $-\frac{1}{2}\alpha q^2 R^{-4}$, and we may choose to represent it as the interaction of $q$ with a potential $-\frac{1}{2}\alpha q R^{-4}$. This is an equivalent description only for values of $R$ such that penetration of the charge distribution by $q$ does not take place.

The introduction of a pseudopotential may provide a simple and easily visualizable description of a system. As in the examples above, it makes it possible to avoid the detailed description of all the particles except the one of interest. For purposes of computation, there is no simplification unless the pseudopotential can be represented by a local potential. In the first example, the nonlocal nature of the pseudopotential means that it depends on the wave function of the valence electron. This would not, of course, be the case for a true potential. It is impossible to represent exactly the effect of the Pauli principle by a potential. However, the pseudopotential can sometimes be represented approximately, or for certain well-defined applications, as a local potential. One knows that the nonlocal exchange operator can often be simulated [39, 39a] by a local operator: The effect of exchange may be approximated by a local exchange potential which is a function of electron density (see Volume I, Chapter III, Section E), and calculations on atoms and solids using such a formalism are becoming more and more widespread. Gombás and Kisdi [40] have discussed "occupation exclusion operators," related to the use of the Pauli principle in quantum statistical theories (which is responsible for the $\varrho^{5/3}$ dependence of the kinetic energy density—see Volume I, Chapter III, Section E).

A book on the subject of pseudopotentials has recently appeared [41], as well as a review article [42]. We direct our discussion toward the use of pseudopotentials for the problem of $U(R)$. Pseudopotentials have been used most for atomic and solid state calculations, but some applications to the ground state potential curves of diatomic systems have appeared. In some cases, the simple physical idea of a repulsive potential to exclude the valence electrons from the core region has been used without explicitly invoking the pseudopotential formalism. Sometimes, as in semiempirical MO calculations, this exclusion is assured by the orbitals used to describe the valence electrons.

We will first consider the use of pseudopotentials to allow the treatment of valence electrons only. The discussion we give here was presented by Phillips and Kleinman [43].

Let $H$ be the Hamiltonian for an electron in a molecule. It may include something like the Fock operator to represent the effect of the other electrons. The eigenvalue equation

$$H\psi = E\psi \qquad (31)$$

has solutions $\psi = \psi_c$ corresponding to the core electrons, in which we are not interested, and solutions $\psi = \psi_v$ corresponding to the valence electrons. Solutions for different eigenvalues are orthogonal. Within the subspace of functions orthogonal to the $\{\psi_c\}$, the $\{\psi_v\}$ correspond to the lowest eigenvalues of $H$. Then parameters in a trial function $\psi$ may be chosen to minimize the expectation value of $H$, provided that $\psi$ is maintained orthogonal to the $\{\psi_c\}$. We seek to introduce an additional term, denoted by $V'$, into $H$, such that the $\psi_v$ correspond to the lowest eigenvalues of the modified Hamiltonian. Letting $E_v$ be the eigenvalue for the valence electron, we want

$$(H + V')\psi = E\psi \qquad (32)$$

to have $E = E_v$ as the lowest eigenvalue, with $\psi_v$ the corresponding wave function. Formally, we can define $V'$ by its effect on an arbitrary function $f$:

$$V'f = \sum_c \psi_c \langle \psi_c \mid f \rangle (E_v - E_c) \qquad (33)$$

with the sum running over all core orbitals. Clearly, $\psi = \psi_v$ satisfies (32) with $E = E_v$, since $\langle \psi_c \mid \psi_v \rangle = 0$. On the other hand, so does any core orbital $\psi_c$, since

$$(H + V')\psi_c = E_c\psi_c + (E_v - E_c)\psi_c$$

Any eigenfunction of $H$ with eigenvalue $E$ is an eigenfunction of $H + V$, with the same eigenvalue, except for the core eigenfunctions, for which the eigenvalue is shifted. Thus, the lowest eigenvalue of $H + V'$ is $E_v$, and one can formally find $E_v$ by minimizing the expectation value of $H + V'$ over a trial function, with no need to impose orthogonality constraints. Note that the corresponding eigenfunction is not well-defined because any function of the form

$$\chi = \psi_v + \sum_c a_c\psi_c \tag{34}$$

(where $\{a_c\}$ is a set of arbitrary constants) is also an eigenfunction of $H + V'$ with eigenvalue $E_v$:

$$(H + V')\chi = E_v\psi_v + E_v \sum_c a_c\psi_c$$

$\chi$ may be referred to as a pseudoeigenfunction. A formulation of the modified Hamiltonian $H + V'$ in terms of projection operators is obviously possible. This, and generalizations of the pseudopotential just given, is discussed by Weeks *et al.* [42].

For purposes of computation, the formalism represents no simplification. However, despite the dependence of $V'$ on the function upon which it operates, it may be possible to replace it by an ordinary repulsive potential, at least if we use it only on functions which look like valence electron eigenfunctions. Furthermore, we might attempt, as suggested by Hellmann [44], to replace the sum of all true and pseudopotentials in $H + V'$ by one effective potential $V_{\mathrm{eff}}$, and investigate $V_{\mathrm{eff}}$'s form. Then we could consider the problem of finding eigenvalues of the one-electron operator

$$H_{\mathrm{eff}}(1) = T(1) + V_{\mathrm{eff}}(1) \tag{35}$$

where $T$ is the usual kinetic energy operator. For several valence electrons, we would use the Hamiltonian

$$H_{\mathrm{eff}} = \sum_i H_{\mathrm{eff}}(i)$$

and perhaps consider interactions between these electrons explicitly. The hope is that we can set up a simple model for $H_{eff}$ on physical or other grounds. Note that $V_{eff}$ is the "core potential" of the semiempirical MO theories.

Hellmann proposed a potential of the form

$$V_H = -ar^{-1} + be^{-2cr}r^{-1} \tag{36}$$

for atoms, where $a$, $b$, and $c$ were constants determined semiempirically [44]. The procedure for doing this was to calculate one-electron energy levels in this potential and choose $a$, $b$, and $c$ so that the calculated levels agreed with experiment. The first term in (36) is the Coulombic attraction to the core, while the second (in the form of a Yukawa potential) represents a short-range exchange repulsion due to penetration (see Chapter II, Section B). To the extent that the core is unaffected by excitations of the valence electrons, one can use the determined $V_H$ to compute excited energy levels. This approach has recently been carried through successfully by Rice and co-workers [42, 45], for both atoms and molecules. The extension of the idea to diatomic molecules seems simple: Derive a molecular pseudo-potential (perhaps as a sum of the atomic pseudopotentials) and use it to calculate wave functions and energies of the molecular valence electrons. However, if $R$-dependent properties are to be calculated, this requires the assumption that the cores are unaffected by the change in $R$. This is likely to be less true than the assumption that excitation of the valence electrons leaves the cores invariant. The core–core overlaps, which are strongly $R$-dependent, distort the core electron density and also change the effect of the exclusion principle. Like the semiempirical MO calculations of Subsection 1, pseudopotential models have had less success in calculating changes in energy with $R$ than in calculating properties at a single value of $R$.

Among the simplest diatomic systems for the use of a pseudopotential approach is $Li_2^+$, which has one electron outside two $k$-shell cores. McMillan [46] has treated this molecule as a one-electron problem, taking the potential energy as the sum of two core potentials. Each of these was of the form

$$V(r) = -(3/r) + (2/r)(1 - (1 + \alpha r)e^{-2\alpha r})$$

with $r$ the distance from the nucleus and $\alpha$ chosen as 1.655, which gives the correct atomic energy levels. The six lowest solutions to the molecular one-electron problem were used to derive potential curves for the six lowest energy levels of $Li_2^+$. The interatomic potential in each case was calculated according to

$$U(R) = E(R) + V_{ion-ion}(R) - E(\infty).$$

$E(\infty)$ is the energy of $Li^+ + Li$ for infinite separation, $E(R)$ is the eigenvalue for the valence electron, and $V_{ion-ion}$ consists of a sum of terms, representing the Coulomb repulsion between unit point charges, a van der Waals $R^{-6}$ attraction, and a repulsion between closed shells. The latter two terms were derived by scaling He–He interactions. Instead of finding the eigenvalues $E(R)$ directly or using variational approximations, McMillan obtained the energies for ground and excited states by a local energy method (Volume I, Chapter III, Section F). For fixed $E$, $F = \langle (H - E)^2 \rangle$ was minimized for a wave function of linear variation form. Then the value of $E$ was varied, with the wave function fixed, to minimize $F$, and the process repeated to self-consistency. The procedure is hard to justify theoretically, and there are no bounds on $E$ (see Volume I, Chapter III, Section F). For the ground state, McMillan found good agreement with Hartree–Fock–Roothaan results for certain properties. The $U(R)$ for the six states generated were used to discuss various scattering processes.

Cadioli et al. [47] treated the homonuclear two-valence-electron problem by considering two electrons in the field of two equal-point charges. Various configuration interaction wavefunctions built from 2s and 2p Slater orbitals of the two atoms were used, ignoring orthogonalization to the 1s orbitals. The collapse of the trial wave function to the inner shell was prohibited by the restrictions on the variational wave function (only orbitals of principal quantum number two were used) rather than by introduction of a repulsive potential of the Hellmann type. This is what is done in extended Hückel and related calculations. For $Li_2$, the effect of demanding orthogonality of the valence electron wavefunction to the cores was considered, using as effective potential a sum of hydrogenic potentials with nuclear charges of 1.30. This Hamiltonian, as well as the basis functions, was chosen on the basis of atomic calculations. Energies for various wave functions at different values of $R$ were reported, with and without orthogonalization to the core orbitals. Even in this favorable case, energies changed by as much as 30% due to introduction of orthogonalization. The relative effects of intraatomic and interatomic orthogonalization depended on the wave functions used and on $R$, as might be expected: intraatomic orthogonalization was less important at small $R$, while the effects of interatomic orthogonalization became almost negligible for large $R$. The values predicted for $D_e$ and $R_e$, 0.36 eV and 6.23 $a_0$, compare to the experimental 1.05 eV and 5.05 $a_0$, while $\omega_e$ was calculated correctly to 10%. It was noted that varying orbital exponents could improve the numbers: However, absurd results could also be obtained if no orthogonality constraints were imposed.

Actually, the derivation of the effective Hamiltonian where there are several valence electrons is not as straightforward as the derivation for a single valence electron. If we describe the pair by double occupation of a one-electron orbital, we can use the one-electron pseudopotential. However, the calculation of the total energy from the orbital energies necessitates some knowledge of the wave function (see Subsection 1). Furthermore, when the interaction between the two valence electrons is added in, there is no guarantee of an upper bound to the energy of the pair. The "derivation" of a pseudopotential formalism for a pair of electrons will now be considered. In Volume I, Chapter III, Section A, Eqs. (133) *et seq.*, we calculated the energy of a wave function built by antisymmetrizing the product of a core wave function, built from one-electron functions, and a valence pair wave function. By applying the variational principle, keeping the orbitals of the core fixed and maintaining the valence function strongly orthogonal to the core, we showed that the *valence* wave function should be an eigenfunction of

$$H_{12} = H^{\mathrm{HF}}(1) + H^{\mathrm{HF}}(2) + 1/r_{12} \,.$$

Here, the one-electron Hartree–Fock Hamiltonian $H^{\mathrm{HF}}$ includes operators for kinetic energy, electron–nucleus attraction, and electron–electron repulsion for all the spin orbitals in the core. The basic argument is due to Fock *et al.* [48], and was extended into a pseudopotential formalism by Szasz [49].

The core spin orbitals are eigenfunctions of $H^{\mathrm{HF}}$, the Hartree–Fock one-electron Hamiltonian. If we were willing to use an uncorrelated simple product,

$$\lambda_1(1)\lambda_2(2)$$

to approximate the valence pair function, we would simply choose $\lambda_1$ and $\lambda_2$ as the lowest-energy eigenfunctions of $H^{\mathrm{HF}}$ which are orthogonal to the core orbitals. According to the pseudopotential formalism, it is possible to replace the orthogonality constraint by addition of a repulsive potential $V'$ to $H^{\mathrm{HF}}$. Then we would seek the lowest eigenvalue of the equation

$$[H^{\mathrm{HF}}(1) + V'(1)]\lambda'(1) = E\lambda'(1)$$

without orthogonality constraints on $\lambda'$. Here,

$$V'f = \sum_{\mathrm{c}} \lambda_{\mathrm{c}}\langle\lambda_{\mathrm{c}}\,|\,f\rangle(E - E_{\mathrm{c}})$$

as usual. There is a formally identical equation for $\lambda_2$. Combining the two, we have

$$\sum_{i=1} [H^{\mathrm{HF}}(i) + V'(i)]\lambda_1'(1)\lambda_2'(2) = (E_1 + E_2)\lambda_1'(1)\lambda_2'(2)$$

If we are considering a closed-shell core and paired valence electrons ($E_1 = E_2 = E$ and $\lambda_1'$ and $\lambda_2'$ differing only in spin), we may carry out the spin integrations in $H^{\mathrm{HF}}$ and $V'$ to derive equations for the spatial parts. Then we have that the product of orbitals (spatial parts) itself satisfies

$$\sum_{i=1}^{2} [H^{\mathrm{HF}}(i) + V'(i)]\phi_1'(1)\phi_2'(2) = 2E\phi_1'(1)\phi_2'(2)$$

Here, $\phi_i'$ is the spatial part of $\lambda_i'$; we denote by $\phi_i$ the spatial part of the true eigenfunction $\lambda_i$. Since we have integrated over spins, $H^{\mathrm{HF}}$ and $V'$ now involve sums over orbitals instead of spin orbitals. Analogously, instead of seeking the eigenfunction of $H_{12}$ which is strongly orthogonal to the core orbitals, one should seek that of the "model potential"

$$H_{12}^{\mathrm{M}} = \sum_{i=1}^{2} (H^{\mathrm{HF}}(1) + V'(i)) + 1/r_{12} . \tag{37}$$

Now this is not rigorous, but merely reasonable when we invoke the physical interpretation of the pseudopotential $V'$. Szász [49] gave a different derivation of (37); for further discussion, see Weeks et al. [42]. There is no rigorous way of deriving such an equation. If $\phi(1, 2)$ is an eigenfunction of $H_{12}$ with eigenvalue $E$, and strongly orthogonal to the core orbitals, it will also be an eigenfunction of $H_{12}^{\mathrm{M}}$:

$$H_{12}^{\mathrm{M}}\phi(1, 2) = E\phi(1, 2) + \left[\sum_{i=1}^{2} V'(i)\right]\phi(1, 2)$$

and the second term vanishes. But suppose $\phi_0(1, 2)$ differs from $\phi(1, 2)$ by an admixture of some $\lambda_c$:

$$\phi_0(1, 2) = \phi(1, 2) + f(1)\lambda_c(1) + f(2)\lambda_c(1) \tag{38}$$

[we have made $\phi_0(1, 2)$ symmetric, like $\phi(1, 2)$, to an exchange of electrons]. In analogy to (34) et seq., we expect $\phi_0$ to be an eigenfunction of $H_{12}^{\mathrm{M}}$, but

$$\begin{aligned} H_{12}^{\mathrm{M}}\phi_0(1, 2) = {} & E\phi(1, 2) + (H^{\mathrm{HF}}(1) + V'(1) + E_c + E - E_c)f(1)\lambda_c(2) \\ & + (H^{\mathrm{HF}}(2) + V'(2) + E_c + E - E_c)f(2)\lambda_c(1) \\ & + 1/r_{12}(f(1)\lambda_c(2) + f(2)\lambda_c(1)) \\ = {} & E\phi_0(1, 2) + [H^{\mathrm{HF}}(1) + V'(1)]f(1)\lambda_c(2) \\ & + [H^{\mathrm{HF}}(2) + V'(2)]f(2)\lambda_c(1) + 1/r_{12}(f(1)\lambda_c(2) + f(2)\lambda_c(1)) \end{aligned}$$

so $\phi_0$ is not an eigenfunction of $H_{12}^{M}$. Then there is no rigorous justification for obtaining approximations to the best pair function by ignoring orthogonality constraints and minimizing the expectation value of $H_{12}^{M}$.

Nevertheless, good results were obtained [50, 51] for atoms, by approximating the eigenvalues of $H_{12}^{M}$ using the variational principle and ignoring the requirement of orthogonality to the core. This suggested the same be tried for the molecule. The Hartree–Fock operator for the molecule would have Coulomb and exchange operators corresponding to the filled inner-shell orbitals. To the extent that the $N$ molecular orbitals of the core are simply linear combinations of $N$ atomic orbitals, as in the simplest description of the diatom formed from closed-shell atomic systems, the Coulomb and exchange operators can be replaced by a sum of the corresponding operators for the core atomic orbitals on both centers. This is true even if the atomic core orbitals overlap, provided that the localized orbitals that one can derive from the MO's are not too different from the atomic orbitals (no distortion or polarization). Then the sum in $V'$ is likewise a sum over atomic orbitals, and $V'$ is a sum of atomic pseudopotentials.

Making these assumptions, Szász and McGinn wrote [51]

$$H_{12}^{M} = \sum_{i} [H_{A}^{HF}(i) + V_{A}'(1) + H_{B}^{HF}(i) + V_{B}'(i)] + 1/r_{12}, \qquad (39)$$

A and B referring to contributions of the atomic cores A and B. Thus, the pseudopotential for the valence electrons of $Na_2{}^{+}$ was the superposition of the pseudopotentials due to two $Na^{+}$ cores. For the molecule near $R_e$ the atomic cores should be essentially those of the Na atoms. For atom A, Szász and McGinn [51] wrote

$$H^{HF}(i) + V'(i) = T(i) + V_{H}(i)$$

and similarly for atom B, where $T$ is the kinetic energy operator and the potential $V_H$ was taken to be of the Hellmann form [Eq. (36)]. For the value of $a$, they used the core charge. The parameters $b$ and $c$ were chosen so that ground and excited energy levels for the atom, calculated by treating the one-electron problem variationally in the effective (Hellmann) potential, were in agreement with experiment. The molecular energy was calculated as the energy of the valence electrons moving in the field of the cores, plus the Coulombic repulsion energy between the cores. The wavefunction for the valence electrons was a Heitler–London function, with covalent and ionic terms, written in terms of hydrogen 1s orbitals of varying effective charge. Therefore it dissociated correctly to atomic orbitals for large $R$, and the energy approached the energy of two Na atoms as $R$ approached infinity. Since the problem was two-electron, the wave function for the valence electrons could be improved to include explicit correlation factors.

While the energy as a function of $R$ led to a minimum at 6.18 $a_0$ (experimental value, 5.80 $a_0$), the dissociation energy, computed as the difference between the energy of the valence electrons and twice the ionization potential of Na, was only 30% of the experimental value. This disappointing result might have come from any of the approximations and assumptions made, such as the use of the Hellmann form for the pseudopotential, or the insufficiently elaborate trial function.

Subsequently, these sources of error were removed. The exact Phillips–Kleinman pseudopotentials (see p. 318) were derived [52] for a number of atoms and ions with one valence electron. Numerical tables were presented from which the pseudopotentials could be obtained in a form convenient for molecular calculations. Then these atomic pseudopotentials were used [53] instead of the Hellmann pseudopotentials. A local potential was included to represent the exchange operator, and an additional effective potential representing the effect of core polarization was added to the Phillips–Kleinman pseudopotential. The core polarization term was derived from atomic calculations, and turned out to be necessary for reasonable results, being especially important for the molecule $K_2$. The terms occurring in the effective potentials for the atoms were fit to an analytical form, with a pure Coulomb potential for the H atom. For each of the molecules ($Li_2$, $Na_2$, $K_2$, LiH, NaH, and KH), the atomic pseudopotentials for the two cores were superposed. For the homonuclear cases, a Heitler–London covalent wave function was used to describe the valence electrons, addition of ionic terms having given little improvement in previous work. Introduction of a correlation factor, $1 + cr_{12}$, into the wave function, did not improve the energy for homonuclear molecules. Ionic terms in the wave function were of importance for the hydrides, as was the correlation factor. In all calculations, screening parameters in the valence electron wave function were varied to minimize the energy at several values of $R$. The dissociation energy was computed, as in the previous work, as the energy of the valence electrons minus the sum of ionization potentials for the atoms, while vibrational frequencies were derived by fitting the valence-electron energy (including core repulsion) to a Morse curve.

Some of the results are shown in Table III (c.p. stands for the core polarization term.) The results, overall, are not very good. In general, dissociation energies, with the best wave function, are much too small and equilibrium internuclear distances too large (except for $K_2$ and KH).

**Table III**

*Pseudopotential Calculations$^a$ with Core Polarization*

| Mole-cule | $R_e$ $(a_0)$ | | | $D_e$ (eV) | | | $\omega_e$ $(cm^{-1})$ | | |
|---|---|---|---|---|---|---|---|---|---|
| | no c.p. | with c.p. | expt. | no c.p. | with c.p. | expt. | no c.p. | with c.p. | expt. |
| $Li_2$ | 6.30 | 6.13 | 5.05 | 0.218 | 0.237 | 1.05 | 200 | 213 | 351 |
| $Na_2$ | 6.85 | 6.37 | 5.80 | 0.196 | 0.246 | 0.74 | 96 | 110 | 159 |
| $K_2$ | 9.20 | 6.62 | 7.39 | 0.083 | 0.240 | 0.52 | 45 | 79 | 93 |
| LiH | 3.66 | 3.60 | 3.01 | 1.438 | 1.463 | 2.516 | 1306 | 1333 | 1406 |
| NaH | 3.94 | 3.78 | 3.57 | 1.384 | 1.463 | 2.27 | 1148 | 1215 | 1172 |
| KH | 5.01 | 4.21 | 4.24 | 1.077 | 1.230 | 1.92 | 888 | 1105 | 985 |

$^a$ See Szász and McGinn [51–53].

Vibrational frequencies are of the right order of magnitude. It turned out that the use of the Hellmann pseudopotential was not a serious source of error; the Hellmann pseudopotential describes the interaction between core and valence electrons accurately. It was noted [53] that the results for $D_e$ were comparable to those obtained from all-electron calculations which described the valence electrons by a wave function similar to that used for the pseudopotential calculations.

In order to be able to work with a greater variety of variational functions for the valence electrons, Simons and Mazziotti [54] recently investigated the pseudopotential method for several molecules with two valence electrons, using Gaussian basis functions. For the constituent atoms they chose parameters in the Hellmann potential and also in the potential $c_1 r^2 + c_2$. The one-electron Schrödinger equation with the latter potential has a single Gaussian as eigenfunction. The constants were chosen so that the two lowest eigenenergies agreed with the two lowest valence state energies, derived from experiment. For the diatomic molecules, the effective potentials of the atoms were added. A Coulomb form was used for the effective potential of H. The trial functions for the valence electrons were of the Heitler–London form, including ionic terms. Simons and Mazziotti calculated $R_e$, $\omega_e$, and $D_e$, where

$$D_e = \mid E - E_A - E_B \mid$$

with $E_A$ and $E_B$ calculated energies for the atomic systems with one valence electron to which the molecule dissociates. The molecules considered included dialkalis and alkali hydrides.

In general, $R_e$-values were reliable for both Hellmann and Gaussian pseudopotentials, while dissociation energies were poor with Hellmann and fair with Gaussian for the dialkalis, and poor with either for the alkali hydrides. Vibrational frequencies were also better with Gaussian for dialkalis, and better with Hellmann for alkali hydrides, neither being very reliable. The inadequacies were explained [54] in terms of the errors associated with using a sum of atomic pseudopotentials for the molecular system. The Hellmann pseudopotential was preferred over the Gaussian for molecules, since it is Coulombic for large distances from a nucleus, such as at the position of another nucleus. A problem with both pseudopotentials is that they are radial functions; the exact Phillips–Kleinman atomic pseudopotential acts differently on functions of different angular momentum because of its nonlocal nature.

The problem of choosing parameters appears, whatever pseudopotential is used. If parameters are chosen mainly on the basis of successful results, pseudopotential schemes will become convenient and accurate algorithms for generating molecular properties, rather than approximations to *a priori* calculations, as may be said to be the case for certain extended Hückel calculations.

Kahn and Goddard [54a] tested the pseudopotential idea for molecular calculations by using potentials derived from the G1 method [Chapter II, Eq. (66) and discussion), which yields valence orbitals which are smooth, like the pseudo wave functions, in the core region. The pseudo potential was a super position of pseudo potentials derived for the $2^2S$, $2^2P$, and $3^2D$ states of Li. Different pseudo potentials corresponded to different angular momenta, as in Eq. (45). Wave functions and energies were calculated for $Li_2$ and LiH, and compared with all-electron G1 and Hartree–Fock results, and computed properties were compared to experiment. Results were good.

Baylis [55] has used pseudopotentials in a different way than the above authors, for calculation of the interaction between an alkali atom and a noble gas atom, a system of one valence electron. Low-lying excited states of the diatomic system, important for interpretation of scattering measurements, were considered as well as the ground state. Pseudopotentials were used to simulate two different overlap effects. One was the effect of overlap between the valence electron, which should be mainly localized on the alkali atom, with the noble gas core. The other was for the overlap between the cores of the alkali and noble gas atoms, which were considered to be unchanged from the free atoms.

The valence electron was described by a wave function built from the ground and low-lying excited alkali atom orbitals. Thus the alkali atom was represented as a superposition of the ground state term ($^2S_{1/2}$) and terms arising from the first excited configuration ($^2P_{1/2}$ and $^2P_{3/2}$). In the molecule, the $^2P_{1/2}$ term splits into two according to the two possible values of $M$, $\frac{1}{2}$ and $\frac{3}{2}$. The $8 \times 8$ secular equation breaks down into two identical $1 \times 1$ equations ($M = \pm\frac{3}{2}$) and two identical $3 \times 3$ equations ($M = \pm\frac{1}{2}$). Since overlap or antisymmetrization effects were not considered explicitly, the molecular wave function was written as a sum of products of alkali and noble-gas wave functions, without antisymmetrization to exchanges of electrons between the two. The noble gas was assumed always to be in its ground state. Correspondingly, the Hamiltonian was written as a sum of atomic Hamiltonians plus an interaction Hamiltonian. The interaction Hamiltonian consisted of the two pseudopotential terms (valence electron–noble gas core and core–core) plus an electrostatic term. For the electrostatic term, Baylis [55] used $-\frac{1}{2}\alpha F^2$, where $F$ was the instantaneous electric field due to the alkali atom and $\alpha$ the calculated polarizability of the noble gas atom. The term, $-\frac{1}{2}\alpha F^2$, had to be averaged over the alkali atom wave function. This represents the interaction of a polarizable sphere (the noble gas) with an electric field. Clearly, this is invalid if the valence electron penetrates the noble-gas electron cloud, so a cutoff parameter $r_0$ was introduced [55]. When the valence electron was closer than $r_0$ to the noble-gas nucleus, the energy was taken to be a constant, its value at $r_0$. While this bears some resemblance to the $r_< - r_>$ formulas in the multipole expansion (see Chapter I, Section A), the value of $r_0$ depended on the alkali atom as well as the noble-gas atom. Thus its physical interpretation is vague, and it can compensate for many errors.

The pseudopotential for repulsion of the valence electron from the noble-gas core was derived from quantum statistical theories (see Volume I, Chapter III, Section F and Gombás [41]). The effect of the exclusion principle for a homogeneous electron gas is that all states are filled up to the Fermi energy

$$E_F = (3\pi^2\hbar^3\varrho)^{2/3}/2m \tag{40}$$

where $\varrho$ is the density and $m$ the electronic mass. An incoming electron sees an effective repulsive potential, since it must go to energies higher than $E_F$. Thus Baylis [55] used the pseudopotential $(3\pi^2\hbar^3\varrho_B)^{2/3}/2m$ where $\varrho_B$ is the electronic density of the core, a function of distance from the noble-gas nucleus B. Quantum statistical formulas were also used for the core–core interaction. This term in the energy, which depends only on the internuclear separation, was approximated as

$$(\hbar^2/2m)(3\pi^2)^{2/3} \int d\tau \varrho_A \varrho_B^{2/3}$$

where $\varrho_A$ is the core density of the alkali. This is identical to the interaction for the valence electron, the pseudopotential simply being averaged over the total alkali core electron density. Another heuristic derivation of the formula is given by Baylis: It requires $\varrho_A$ small compared to $\varrho_B$. In all the pseudopotential formulas, the electron densities used were approximations to the SCF densities.

Having chosen $r_0$ to match well depths for the ground state interaction, Baylis [55] predicted values for $R_e$ for the four lowest states, as well as well depths and curvatures of $U(R)$ for excited states. Agreement seemed generally satisfactory, although there is uncertainty about the correct potentials, which are obtained from scattering experiments. For LiHe, the adiabatic $U(R)$ was compared with an *ab initio* SCF calculation; agreement was fair. A more recent multiconfiguration SCF calculation [56] yielded $R_e = 11.6$ $a_0$ and $D_0 \sim 1.62$ cm$^{-1}$ for LiHe. It should be added that $U(R)$ is extremely flat near $R_e$. The approximations used are expected to lead to inaccuracies for small $R$, say less than a few $a_0$, where the interactions are sharply repulsive. The predicted $R_e$ and $\varkappa$ for some ground states are given in Table IV, where

$$\varkappa = (R_e^2/D_0)(d^2U/dR^2)R_e .$$

Where two values are given, they correspond to two choices of $r_0$.

The exchange potential derived from the quantum statistical theories was used in an extensive series of Hartree–Fock calculations on atoms by Herman and Skillman [39a], and continues to attract interest for atomic and solid state calculations. The method, often referred to as the $X\alpha$ method or the Hartree–Fock–Slater method, has recently been reviewed by Slater and Wood [57]. It involves performing Hartree–Fock calculations with replacement of the exchange part of the Fock operator by a local potential depending on the occupied orbitals. Then the Fock operator becomes [41, 57]

$$F_\alpha(1) = \sum_j \int |\phi_j(2)|^2 r_{12}^{-1} d\tau_2 - 3\alpha \left[ (3/8\pi) \sum_j |\phi_j(1)|^2 \right]^{1/3} \qquad (41)$$

where $\phi_j$ is the spatial part of the $j$th spin orbital, so $\Sigma(j)|\phi_j|^2$ is the electron density. The value of the parameter $\alpha$, which was 1 in the original theory, is chosen to give best results. Little work has been done using the

Table IV

| | $D_0$ (cm$^{-1}$) | $\varkappa$ | $R_e$ ($a_0$) | |
|---|---|---|---|---|
| | | | Calc. | Expt. |
| LiHe | 2.9 | 69 | 10.6 | |
| | 3.4 | 69 | 10.2 | |
| LiNe | 2.1 | 82 | 12.5 | |
| | 2.2 | 87 | 12.4 | |
| LiAr | 42.8 | 49 | 9.4 | 9.4 |
| LiKr | 68.9 | 42 | 9.2 | 9.2, 8.8 |
| LiXe | 106 | 38 | 9.2 | 9.3 |
| NaHe | 2.6 | 69 | 10.9 | |
| | 3.3 | 67 | 10.5 | |
| NaNe | 1.9 | 86 | 12.9 | |
| | 2.0 | 87 | 12.8 | |
| NaAr | 44.4 | 44 | 9.4 | 9.5, 9.1 |
| NaKr | 70.0 | 37 | 9.2 | 9.4, 8.9 |
| NaXe | 105 | 33 | 9.2 | 9.6 |
| CsHe | 2.5 | 60 | 12.0 | |
| | 3.7 | 51 | 11.1 | |
| CsNe | 1.0 | 87 | 15.6 | |
| | 1.1 | 88 | 15.4 | |
| CsAr | 45.5 | 29 | 9.9 | 10.4 |
| CsKr | 70.1 | 23 | 9.6 | 10.3 |
| CsXe | 110 | 23 | 9.3 | 10.3 |

$X\alpha$ method for molecules, because it does not avoid the difficult integrals over basis functions, but Schwartz and Connolly [58] have proposed an approximate Hartree–Fock method which employs the $X\alpha$ method as well as the "muffin tin" approximation of solid state physics.

The muffin tin approximation, designed to avoid multicenter integrals, puts $F_\alpha(1)$ equal to its spherical average within spheres centered at the nuclei, and takes $F_\alpha(1)$ as constant between the spheres. The radii of the spheres are parameters to be chosen. With these approximations, the Hartree–Fock equations can be solved in closed form. The accuracy of the method is claimed [58] to be better than CNDO, and the time needed hardly greater.

As we indicated above [Eq. (40)], the quantum statistical theories provide a repulsive pseudo potential in the kinetic energy expression, since the $\varrho^{5/3}$-dependence makes too much buildup of electron density at any point unfavorable (which represents the effect of the Pauli principle). Gordon and Kim [58a] used this kinetic energy expression, together with the quantum statistical (electron gas) expressions for exchange and correlation energy [Volume I, Chapter III, Eqs. (418), (425), and p. 263] as functionals of the electron density, to calculate the interaction energy between closed-shell systems, including rare-gas atoms and positive and negative ions. The electron density was not calculated from the Thomas–Fermi–Dirac theory, which gives a density seriously in error in some regions of space (Volume I, Chapter III, Section F). Instead, a superposition of quantum mechanical atomic densities was used. The electrostatic interactions were of course correctly calculated. Results for potential curves were remarkably good [58a].

Other applications of pseudopotentials to calculation of $U(R)$ are found in the work of Dalgarno *et al.* [59] and Bardsley [60], who set up model potentials for the valence electrons of some simple molecules. Dalgarno *et al.* [59] suggested that, particularly for Rydberg states, the potential in which the valence electrons move could be written as a sum of spherically symmetric atomic core potentials plus a potential representing the core–core interaction:

$$V = V_A(r_A) + V_B(r_B) + V_C(\mathbf{R}, \mathbf{r}_A, \mathbf{r}_B). \tag{42}$$

If, for large $R$, the electron goes into an atomic orbital $\phi_A$, its energy relative to its energy at infinite $R$ would be approximately

$$U(R) = \langle \phi_A \mid V_B(r_B) + V_C(\mathbf{R}, \mathbf{r}_A, \mathbf{r}_B) \mid \phi_A \rangle. \tag{43}$$

$V_B(r_B)$ was taken as $-Z_B r_B^{-1} - \frac{1}{2}\alpha_B r_B^{-4}$, corresponding to a polarizable ion model for the core of atom B, with net charge $Z_B$ and polarizability $\alpha_B$. The expectation value of $V_B$ over $\phi_A$ is $-Z_B R^{-1} - \frac{1}{2}\alpha_B R^{-4}$. The van der Waals interaction of the electron with the core of B, since the core is less polarizable than the electron [higher excitation energies–see Chapter II, Eq. (47)], was approximated by $-\alpha_B R^{-6}\langle \phi_A \mid r_A^2 \mid \phi_A \rangle$. For LiLi+, this gives $-3.39$ for the van der Waals coefficent, compared to the correct value of $-3.32$. The core potential includes the Coulomb repulsion, $Z_A Z_B R^{-1}$, and terms corresponding to polarization of B by the core of A, $-\frac{1}{2}Z_A{}^2\alpha_B R^{-4}$ $+Z_A\alpha_B R^{-4}$. For $Li_2{}^+$, $U(R)$ calculated from these formulas gave good agreement with an *ab initio* calculation: $D_e$ and $R_e$ were 1.30 eV and 6.0 $a_0$, compared to 1.22 eV and 5.8 $a_0$ *ab initio*.

Bardsley [60] treated $Li_2^+$ and $Li_2$ by setting up an effective potential for the valence electrons. For the molecule $AB^+$, the effective potential was taken as $V_{eff}^A(r_A) + V_{eff}^{(B)}(r_B) + V_C(R)$, the last term being a Coulombic core-core repulsion and the first two terms atomic potentials. For an alkali atom, he assumed

$$V_{eff}^A(r_A) = -\frac{1}{r_A} - \frac{\alpha}{2(r_A^2 + d^2)^2} - \frac{\beta}{2(r_A^2 + d^2)^2} + V_{SR}(r_A) \qquad (44)$$

The first three terms are point charge attraction, charge-induced dipole, and charge-induced quadrupole terms; the parameter $d$ (core radius) is chosen to give best results. The short-range repulsion, $V_{SR}$, was written as

$$V_{SR}(r) = \sum_l (A_l r^p e^{-z l r^q}) \, | \, l \rangle \langle l \, | \qquad (45)$$

the last factor ensuring that a different potential is used for states of different angular momentum, as should be the case [see Eq. (33)]. Then linear variation was used to find $U(R)$ as the electronic energy of the molecule, using basis functions chosen from calculations on the atomic pseudopotential problem. For $Li_2^+$, Bardsley got $D_e = 1.23$ eV, $R_e = 5.9 \, a_0$. For $Li_2$, he used a 13-configuration CI for the valence electrons in the $Li_2^+$ potential, plus the interelectronic repulsion term. Values for $D_e$ and $R_e$ were quite close to the experimental values. Calculations [61] using a core pseudopotential in correction with Rittner's ionic model will be discussed in Section B.

Calculations on larger systems will have to be done to show whether pseudopotential calculations can simplify calculations of $U(R)$. For the simpler systems, the accurate *ab initio* calculations which one can now perform limit the contribution of pseudopotential approaches to providing simple physical interpretations of the effects determining the potential $U(R)$.

### 3. Atoms in Molecules

The atoms-in-molecules theories attempt to incorporate information about the constituent atoms into the molecular calculation in order to remove inaccuracies common to atomic and molecular calculations. Atomic results are also used in SCF theories, as a guide in the choice of basis functions, and in semiempirical calculations, to parametrize matrix elements of the Hamiltonian. In atoms-in-molecules methods, the inclusion of atomic results is made more explicit.

The origin of these methods is often traced to a paper of Moffitt [62]. Parr [25, Sect. 17] and Arai [63] have given reviews of the atoms-in-molecules methods. Moffitt [62] noted that a good part of the energy error in a molecular calculation was also present in the calculations for the constituent atoms. His suggestion was to analyze the molecular wave function into contributions of atomic states, and, by comparing exact and calculated atomic energies for such states, remove such errors. Another reason for an atoms-in-molecule approach is that the energy of a molecule differs from the energy of the constituent atoms by a relatively small quantity, although it is in this quantity and its variation with internuclear separation that all our interest lies. Calculation of a small quantity as a difference of much larger ones, evaluated approximately, is dangerous, unless one can ensure that at least some of the errors in the larger ones cancel. Moffitt [62] suggested that this could be done if the same atomic data were incorporated into both atomic and molecular calculations.

For infinite separation of the nuclei, the molecular wave function is exactly a product of atomic wave functions, each antisymmetric in some subset of the electronic coordinates. Since differential overlaps vanish, there is no need to antisymmetrize the overall product wave function to exchanges of electrons between different atoms. For finite nuclear separation, we must introduce a residual antisymmetrizer (see Chapter I, Section A). Since the set of atomic eigenfunctions on each atom is complete, the effect of the perturbation of one atom by the other can be considered to be the mixing in of other atomic states. Then the molecular wave function can be written

$$\Psi = \sum_{i,j} [\mathscr{A}'\{\Phi_i{}^{A}(1, \ldots, n_A)\Phi_j{}^{B}(n_A + 1, \ldots, n_A + n_B)\}]c_{ij} \qquad (46)$$

where $\mathscr{A}'$ is the residual antisymmetrizer, $\Phi_i{}^{A}$ is the $i$th state of atom A, containing $n_A$ electrons, $\Phi_j{}^{B}$ is the $j$th state of atom B, containing $n_B$ electrons, and the $\{c_{ij}\}$ are coefficients to be determined. This expansion is discussed in Chapter I, Sections A and B [Eq. (204)]. The overcompleteness problem associated with such an expansion is irrelevant here, since we will attempt to approximate the molecular wave function using only a few states in the sum of (46), including the atomic states to which the molecule dissociates. Since at small distances it may be important to use ions in the description (corresponding to ionic states in valence bond theory), the sum of (46) should be augmented by terms like

$$\Psi_{M} = \mathscr{A}''\{\Phi_i{}^{A^+}(1, \ldots, n_A - 1)\Phi_j{}^{B^-}(n_A, \ldots, n_A + n_B)\}$$

(see Chapter I, Section B.3). Here, $\Phi_i^{A^+}$ is the $i$th state of the ion $A^+$ (antisymmetric in $n_A - 1$ electrons) and $\Phi_j^{B^-}$ is the $j$th state of the ion $B^-$ (antisymmetric in $n_B + 1$ electrons); $\mathscr{A}''$ is not quite the same as $\mathscr{A}'$. We write any of the composite functions $\mathscr{A}'\{\Phi_i^A\Phi_j^B\}$, $\mathscr{A}''\{\Phi_i^{A^+}\Phi_j^{B^-}\}$, etc. as $\Psi_K$, $K$ now indicating the pair of states of the atoms or ions to which $\Psi_K$ corresponds at large $R$.

The expansion of the molecular wave function is now written

$$\Psi = \sum_K \Gamma_K \Psi_K \tag{47}$$

where each $\Psi_K$ is an eigenfunction of $H$ for large $R$. For homopolar molecules, the symmetry of the molecular wave function on inversion puts a condition on some of the $\Gamma_K$, as does the requirement that the molecular wave function be an eigenfunction of $S^2$, $S_z$, and $M_z$ (see Chapter I, Section B.1). If the sum were over a complete set we could, in principle, find coefficients $\Gamma_K(R)$ such that $\Psi$ is an eigenfunction of $H$ for internuclear distance $R$. With a finite set of $\Psi_K$, we seek an approximate eigenfunction of $H$ using the linear variation technique. This leads to the secular determinant,

$$| H_{KL} - ES_{KL} | = 0. \tag{48}$$

$H_{KL}$ and $S_{KL}$ are Hamiltonian and overlap matrix elements between $\Psi_K$ and $\Psi_L$. The formulas for the matrix elements were given in Chapter I, Section B [Eqs. (207) *et seq.*]. For the atoms-in-molecules methods, a derivation in terms of the representation matrices of the symmetric group can be given [63, 64]. In comparing the present calculation to that of Chapter I, Section B.4, we note that we now use a finite instead of an infinite set of basis functions, that the coefficients are calculated from the secular equation instead of by perturbation theory, and that both atomic and ionic functions are present in the basis.

Consider the integral of $H$ between

$$\Psi_K = \mathscr{A}'\{\Phi_i^A(1, \ldots, n_A)\Phi_j^B(n_A + 1, \ldots, n_A + n_B)\}$$

and the ionic function $\Psi_M$, above. In calculating $H_{MK}$ we would partition $H$ as in Eq. (206) of Chapter I, Section B to obtain:

$$H_{MK} = \int \Psi_M{}^* \mathscr{A}'\{(H^{OA} + H^{OB} + V)(\Phi_i^A\Phi_j^B)\} \, d\tau$$
$$= (W_i^A + W_j^B) \int \Psi_M{}^* \mathscr{A}'\{\Phi_i^A\Phi_j^B\} \, d\tau + \int \Psi_M{}^* \mathscr{A}'\{V\Phi_i^A\Phi_j^B\} \, d\tau. \tag{49}$$

Denoting the sum of atomic energies $W_i^A + W_j^B$ as $W_k$ (which we may consider as an element of a diagonal matrix $\mathbf{W}$) we have

$$H_{MK} = S_{MK}W_K + V_{MK}$$

the quantities being defined in (49). Now $H_{MK} = H_{KM}^*$ since $H$ is Hermitian, but $H_{KM}$ would be calculated as

$$H_{KM} = \int \Psi_K^* \mathscr{A}'' \{(H^{OA^+} + H^{OB^-} + V')\Phi_i^{A^+}\Phi_j^{B^-}\} \, d\tau$$

$$= (W_i^{A^+} + W_j^{B^-})S_{KM} + \int \Psi_K^* \mathscr{A}'' \{V'\Phi_i^{A^+}\Phi_j^{B^-}\} \, d\tau \tag{50}$$

The partitioning of $H$ would be different, to take advantage of the fact that $\Phi_i^{A^+}$ and $\Phi_j^{B^-}$ are eigenfunctions of Hamiltonians differing from $H^{OA}$ and $H^{OB}$ by, respectively, addition and subtraction of one electron. Thus $V'$ differs from $V$. If we write the last member of (50) as $W_M S_{KM} + V_{KM}$, we must remember that $\mathbf{V}$ is not a Hermitian matrix. $\mathbf{SW}$ is obviously not Hermitian (although $\mathbf{S}$ is), so $\mathbf{V}$ cannot be Hermitian if $\mathbf{H}$ itself is to be Hermitian. To simplify the problem, it is better to write

$$\mathbf{H} = \tfrac{1}{2}(\mathbf{SW} + \mathbf{WS}) + \tfrac{1}{2}(\mathbf{V} + \mathbf{V}^\dagger)$$

where both parts are now Hermitian, before solving the secular equation.

While exact energies of atomic states (the elements of $\mathbf{W}$) are available, the wavefunctions are not, and they are needed to compute $\mathbf{S}$ and $\mathbf{V}$. At this point, we have recourse to approximate wave functions. Following Moffitt [62], the tilde is here used to denote quantities calculated from approximate wave functions. These will generally be orbital approximations to make the calculation tractable. The atomic *energies* used are still the true, experimental energies. Let $H_{KL}$ be approximated as

$$H_{KL} \cong \tfrac{1}{2}(\tilde{S}_{KL}W_L + W_K\tilde{S}_{KL}) + \tfrac{1}{2}(\tilde{V}_{KL} + \tilde{V}_{KL}^\dagger). \tag{51}$$

The integrals $\tilde{V}_{KL}$ will be much more difficult to calculate than the overlaps $\tilde{S}_{KL}$. To eliminate then, Moffitt considered the Hamiltonian matrix $\mathbf{H}$ calculated from the approximate atomic wave functions.

$$\tilde{H}_{KL} \cong \tfrac{1}{2}(\tilde{S}_{KL}\tilde{W}_L + \tilde{W}_K\tilde{S}_{KL}) + \tfrac{1}{2}(\tilde{V}_{KL} + \tilde{V}_{KL}^\dagger) \tag{52}$$

Here, $\tilde{W}_K$ is the energy obtained from the approximate function when the Hamiltonian appropriate for $R = \infty$ is employed. Combining the expressions (51) and (52), we have for the matrix elements of $H$:

$$H_{KL} \cong \tilde{H}_{KL} + \tfrac{1}{2}(\tilde{S}_{KL}(W_L - \tilde{W}_L) + (W_K - \tilde{W}_K)\tilde{S}_{KL}). \qquad (53)$$

Equation (53) is not exact because $\tilde{S}_{KL}$ and $\tilde{V}_{KL}$ were used for $S_{KL}$ and $V_{KL}$ in (51). However, the major contribution to $H_{KL}$, the atomic energies $W_K$ and $W_L$, are being put in exactly. The hope is that errors due to use of approximate wave functions will not be as serious when these wavefunctions are used to calculate $\mathbf{S}$ and $\mathbf{V}$ as when they are used to calculate the entire energy, i.e., $\mathbf{H}$ It is understood that $\tilde{S}_{KL}$ is used for $S_{KL}$ in the overlap part of the secular equation. In approximate valence bond configuration interaction calculations using the same approximate atomic functions, one would simply be using $\tilde{H}_{KL}$ for $H_{KL}$. The correction terms in Eq. (53) are the basic improvements given by the atoms in molecules theories. They incorporate *atomic* spectroscopic data, since $W_L - \tilde{W}_L$ is the difference between observed and calculated energies for the atomic states denoted by $L$, the calculated energy $\tilde{W}_L$ being a sum of two expectation values for the appropriate atomic Hamiltonians over the approximate atomic wave functions.

If the atomic functions are good approximations to the exact functions, so that they represent the atomic states well, the correction terms will be unimportant. Also, if all the differences $W_K - \tilde{W}_K$ are equal, the correction has no effect on the secular equation except to change the roots by $W_K - \tilde{W}_K$. The molecular energies and the atomic energies will be changed by the same amount, and $U(R)$ will be unchanged. Note that, for large $R$, where a single term of Eq. (47) suffices, the corrected matrix elements automatically lead to the exact energy.

The approximations to the atomic wave functions cannot be poor ones for this method to be meaningful. In particular, the exponential parameters for the atomic orbitals in the approximate functions must have values appropriate to the atomic state [64, 65]. This complicates the calculation, since a large number of integrals over different orbitals, corresponding to different states, are needed for the $\tilde{S}_{KL}$. In addition, we know that the exponential parameters required to describe a molecular state in terms of atomic orbitals are generally larger than those optimized for the atomic states (charge contraction). (For $H_2$, the parameter should be increased by 20–30% at the equilibrium internuclear distance.)

This prompted Arai [64] to suggest that, for good convergence of the expansion (47), one should not use the actual atomic states to construct the molecular states, but rather contracted or otherwise deformed atomic states. He imagined atomic functions constructed from atomic orbitals corresponding to appropriate effective charges for the molecule, rather than those for the atom. This meant, in particular, an increase in the effective charges for the valence electrons, or perhaps an increase for all orbitals in the region of space close to the nucleus of the other atom. In the latter case, one would have to introduce $r$-dependent exponential parameters into the Slater orbitals. In Arai's theory, it is necessary to correct the experimental atomic energies for the deformation, since the $W_K$ in (53) would refer to the deformed atoms rather than the physical atoms. Since correlation energies in atoms are fairly insensitive to changes in effective charges, changes of correlation energies on deformation were neglected. Therefore the change in atomic energies due to deformation was set equal to the difference between energies calculated from two atomic wave functions of the Hartree–Fock type, one using the best atomic exponential parameters and the other the best molecular exponential parameters, in the basis functions. One may calculate directly this energy difference when the deformation is properly defined, using simple model wave functions for this purpose [66, 67]. The method was illustrated for Li$_2$ by Arai and co-workers [66, 67], both ground and excited states being considered. Several sets of orbital exponents were considered, and dissociation energies of 0.56, 0.83, and 0.96 eV were obtained. The experimental value is 1.05 eV, and 0.28 eV was obtained with no atoms-in-molecules correction.

Hurley [65] simultaneously developed a closely related scheme from examination of the results of calculations by Moffitt's method for H$_2$. Again, the exponential parameters in the atomic functions were allowed to vary with $R$. For each internuclear distance, one uses, in $\tilde{H}_{KL}$ and $\tilde{S}_{KL}$, the appropriate values of these parameters. Correlation energies being relatively invariant to change in nuclear charge, and a change in exponential parameter being equivalent to a change in nuclear charge, the difference between the exact and approximate energies of the hypothetical deformed atom should be essentially the difference between the exact and approximate energies of the undeformed atom. (It is understood in Arai's and Hurley's theories that the deformed and undeformed atomic wave functions differ only in a specific change in exponential parameters.) This means that, in Eq. (53), one uses $\tilde{H}_{KL}$ and $\tilde{S}_{KL}$ computed using exponential parameters appropriate to the molecule at internuclear distance $R$, but, for $W_L - \tilde{W}_L$, one may use energy differences for the undeformed atoms. Then $W_L$ is an experimental energy and $\tilde{W}_L$ is obtained from a calculation for $R = \infty$, with parameters optimized for this calculation. This scheme was referred to as the intraatomic correlation correction (ICC).

A new analysis of the problem was given by Hurley [68]. Again, it was supposed that we have a set of approximate wave functions, $\{\tilde{\Psi}_K^{(0)}\}$, for the separated atom states, such that, on changing parameters in these functions, they become a suitable basis for expansion of the exact wave function. Let the change of parameters in $\tilde{\Psi}_L^{(0)}$ produce $\tilde{\Psi}_L$, so that the expansion of the molecular wave function is

$$\Psi = \sum_K G_K \tilde{\Psi}_K \tag{54}$$

Since the $\tilde{\Psi}_K$ refer to a finite internuclear distance $R$, they have nonvanishing overlaps:

$$\tilde{S}_{KL} = \langle \tilde{\Psi}_K \mid \tilde{\Psi}_L \rangle$$

Let $V$ (for "valence") denote that state of the separated atoms which resembles as much as possible the state of the electrons in the molecule and let its wave function $\Psi_V^{(0)}$ be expanded in the $\tilde{\Psi}_K^{(0)}$:

$$\Psi_V^{(0)} = \sum_K c_K \tilde{\Psi}_K^{(0)} \tag{55}$$

Since the $\tilde{\Psi}_K^{(0)}$ are orthonormal, $\sum_K c_K{}^2 = 1$. The $\{c_K{}^2\}$ may be interpreted as occupation numbers. The analogous quantities for the set $\tilde{\Psi}_K$ are [68]

$$C_K{}^2 = \sum_L G_K \tilde{S}_{KL} G_L = G_K{}^2 + \sum_{L \neq K} G_K \tilde{S}_{KL} G_L \tag{56}$$

(all coefficients are assumed real here). This definition of occupation numbers takes into account the fact that, by virtue of the nonvanishing of the $\tilde{S}_{KL}$, there is some $\tilde{\Psi}_K$ in $\tilde{\Psi}_L$ and vice versa. Since $\Psi$ is normalized,

$$\sum_K C_K{}^2 = \sum_{K,L} G_K \tilde{S}_{KL} G_L = 1 \tag{57}$$

The valence state may be defined in a nonarbitrary way by taking $\mid c_K \mid$ $= \mid C_K \mid$ in Eq. (54). It may be expected that the error in the calculated molecular energy would be the same as the error in the energy calculated for the valence state $V$. The latter energy is

$$\sum_K c_K{}^2 \tilde{W}_K^{(0)}$$

and the corresponding experimental energy is

$$\sum_K c_K{}^2 W_K$$

where $\tilde{W}_K^{(0)}$ is the energy computed from $\tilde{\Psi}_K^{(0)}$ and $W_K$ the experimental energy for this state. Combining these hypotheses, we have for the corrected molecular energy

$$\langle H \rangle_\Psi = \sum_{K,L} G_K \tilde{H}_{KL} G_L + \sum_K c_K{}^2 (W_K - \tilde{W}_K^{(0)})$$

$$= \sum_{K,L} [G_K \tilde{H}_{KL} G_L + G_K \tilde{S}_{KL} G_L (W_K - \tilde{W}_K^{(0)})]$$

Since $(W_K - \tilde{W}_K^{(0)})$ may be replaced by $\frac{1}{2}(W_K - \tilde{W}_K + W_L - \tilde{W}_L^{(0)})$ without affecting the value of the sum, this corresponds to the correction of Eq. (53). We should minimize $\langle H \rangle_\Psi$, as corrected, with respect to the $\{G_K\}$, which leads to the secular equation (48), with $\tilde{S}_{KL}$ substituted for $S_{KL}$ and (53) for $H_{KL}$.

The predicted binding energy of the molecule is

$$W_A - \sum_{K,L} G_K H_{KL} G_L$$

where $W_A$ is the experimental energy of the dissociation products, $H_{KL}$ is computed using (53), and the $\{G_K\}$ are obtained from the secular equation. The diagram below shows the relations between some of the energies involved in a calculation by this method. The theory assumes that energy errors A and B are equal. It calculates the binding energy as the difference between $W_A$, and $\sum_{K,L} G_K \tilde{H}_{KL} G_L$ minus energy error B.

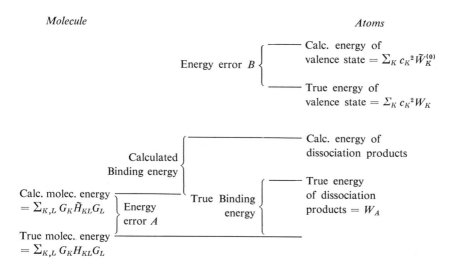

Hurley obtained [65, 68–70] remarkably good binding energies for a series of diatomic molecules using his ICC method, as shown in Table V. The corrected dissociation energies represent lower limits to the true energies, insofar as the theory is correct, since improvements in the molecular wave function would make $D_e$ larger. The results of the calculations in CO [70] and $N_2$ [69] actually gave a basis for choosing among several proposed experimental values for the binding energy (i.e., the largest value should be taken in each case). Extension of the method to excited states is possible.

**Table V**

*Binding Energies Using Interatomic Correlation Correction[a,b]*

|        | Uncorrected $D_e$ | With ICC corr. | Experiment          |
| ------ | ----------------- | -------------- | ------------------- |
| $H_2$  | 4.26              | 4.72           | 4.74                |
| $HeH^+$| 0.79              | 1.77           | —                   |
| LiH    | 1.67              | 2.09           | 3.15                |
| BH     | 2.10              | 2.72           | 3.74                |
| CH     | 1.60              | 2.94           | 3.65                |
| NH     | 1.01              | 3.21           | 3.9                 |
| OH     | 1.14              | 4.00           | 4.58                |
| FH     | 1.73              | 5.59           | 6.08                |
| $N_2$  | 3.29              | 9.18           | 9.91, 7.52          |
| CO     | 7.45              | 11.00          | 11.24, 9.74, 9.28   |

[a] See the literature [65–70].
[b] All values are in electron Volts.

A recent calculation using the methods of atoms-in-molecules is that of Balint-Kurti and Karplus [71] on LiF, $F_2$, and $F_2^-$. Illustrative calculations on $H_2$ were also performed. The methods used are:

(1) Direct *ab initio* calculation using the molecular basis functions built up from orbital approximations to atomic wavefunctions;
(2) atoms-in-molecules calculation with the same basis functions but using $H_{KL}$ of Eq. (53) instead of $\tilde{H}_{KL}$;
(3) Hurley's method of intraatomic correlation correction;
(4) A new modification of Moffitt's method, referred to as "orthogonalized Moffitt."

Method (4) was designed to cope with the objection that, since the anti-symmetrized products of atomic wavefunctions $\tilde{\Psi}_K$ are not orthogonal for finite $R$, we should not identify particular atomic corrections with particular states purely on the basis of their asymptotic form. To do this with more confidence, one should obtain an orthogonalized set of functions from the approximate functions $\tilde{\Psi}_K$. Balint-Kurti and Karplus [71] arranged the states in order of increasing atomic corrections $| W_K - \tilde{W}_K |$ and performed a Schmidt orthogonalization. Then the atoms-in-molecules correction was applied to the orthogonal states. The matrix elements for these states are just

$$H_{KL}^{or} = \tilde{H}_{KL}^{or} + (W_K - \tilde{W}_K)\delta_{KL}$$

$\tilde{H}_{KL}^{or}$ being computed over the orthogonalized set. The Hamiltonian matrix is automatically Hermitian. For a calculation on $H_2$ using atomic orbitals with exponential parameter 1 at $R = 0$ (an extreme case, where the covalent and ionic states are identical), the corrected Hamiltonian matrix elements will be identical for the covalent and ionic states. In AIM or ICC these identical states would have different matrix elements because they would have different correction terms.

For LiF the structures used [71] corresponded to Li + F and Li+ + F−. Including the different excited states deriving from the ground state configurations of the atoms (using only one of each symmetry for each atom) led to 13 possibilities of $^1\Sigma^+$ molecular symmetry. To use more than one state of a particular atomic symmetry would have demanded an intra-atomic configuration interaction calculation prior to the molecular calculation. The *ab initio* $U(R)$ was high by 0.1 a.u. compared to experiment near $R_e$, while AIM was too low by about the same amount. Orthogonalized Moffitt (OM) did well and ICC slightly less well, but ICC became inadequate at large $R$. Orthogonalized Moffitt and AIM energies exhibited "reasonable" behavior, but *ab initio* and ICC energies rose too sharply with $R$ compared to the expected true $U(R)$ (experimental data are unavailable in this region). For $F_2$, 18 basis functions (each an antisymmetrized product of atomic states) were used. Intra-atomic correlation correction and OM again were best near $R_e$ with the former better; AIM was much too low here. All four go to the proper limit at large $R$. Intraatomic corrections were much less important for $F_2$ than LiF. The dissociation energies were, in atomic units, 0.0510, 0.0747, 0.1385, and 0.0675 from *ab initio*, OM, AIM, and ICC, respectively; the experimental value is 0.06 a.u. $R_e$, for which the experimental value is 2.683 $a_0$, was 2.83, 2.80, 2.65, and 2.76 with the four methods. For $F_2^-$, using four many-electron functions, very similar results were obtained from *ab initio*, OM, and AIM, but ICC was ∼0.015 a.u. high for all $R$. The other methods predicted [71] $D_e = 0.087$ a.u., which is larger by ∼0.027 a.u. than for $F_2$, and $R_e = 3.3\ a_0$.

Arai's review article [63] gives a general analysis of atoms-in- molecules methods, which includes the methods we have discussed as special cases. Note that atoms-in-molecules methods are used to correct an *ab initio* valence bond calculation. It is possible to go a step further, and to perform a valence bond calculation, substituting the empirical parametrization of integrals for their direct calculation. Although the concepts of the valence bond theory are readily combined with chemical intuition, relatively few semiempirical valence bond schemes have been proposed.

Companion and Ellison [72], using a definition of the valence state closely related to that of Hurley [68], gave prescriptions for evaluating valence state energies of atoms in a molecule. These were used to obtain the intra-atomic parts of the matrix elements for valence bond calculations. Part of the interatomic contributions were calculated using the Mulliken approximation: for $\phi_a$ and $\phi_b$ atomic orbitals on different centers and $S_{ab}$ their overlap, they put

$$\phi_a(1)\phi_b(1) \cong \tfrac{1}{2}S_{ab}[\phi_a(1)\phi_a(1) + \phi_b(1)\phi_b(1)]. \tag{58}$$

This approximation reduces exchange integrals to Coulomb integrals. The remaining two-center integrals were approximated [72] by formulas parameterized so that the calculations for CH and OH gave good agreement with the experimental dissociation energies, while certain combinations of one-center integrals were expressed in terms of atomic properties, for which experimental values were introduced. Using the chosen parameters, calculations were made for NH.

In subsequent work [73], LiH, BeH, BH, and $H_2$ were treated. The inner-shell electrons were not treated explicitly. The atomic states used in setting up the valence bond function were taken as scaled, i.e., all atomic coordinates, relative to the nucleus, were multiplied by some scale factor. The matrix elements of the Hamiltonian between these states were written, as in the formalisms already discussed,

$$H_{KL} = \tfrac{1}{2}(S_{KL}(W_K + W_L) + V_{KL} + V_{LK}),$$

except that $W_K$ is the exact energy of the scaled atoms, with scaling factor appropriate to the states denoted by $K$. The scaling was imagined to affect only the electron with principal quantum number 2, not the $(1s)^2$ core, which should be the same for atom and molecule. Following Hurley, the difference between the energy of the hypothetical scaled atom and the experimentally observed energy for the atomic state was set equal to the difference between the atomic energies calculated using the approximate function with exponential parameters appropriate to the molecule and to the atom. The elements $V_{KL}$ were expressed in terms of the Coulomb and exchange integrals of valence bond theory (Chapter I, Section B), with the Mulliken approximation [Eq. (58)] invoked to express exchange integrals in terms of Coulomb integrals. Certain groupings of Coulomb integrals could be associated to energy differences of various atomic species, for which empirical values (corrected for scaling) were introduced. Some small terms in the energy expression were neglected. The only integrals that were calculated directly were the two-center overlaps. Fair dissociation and excitation energies were obtained, as well as other properties. Thus, $D_e$ was calculated as 2.39, 1.84, and 4.64 eV for LiH, BeH, and BH, respectively, with the experimental results being 2.52, 2.33, and 3.55 eV.

Ellison [74] recalculated the dissociation energies and other properties of these molecules using a "modified atoms-in-molecules model," in which the scaling of the atomic functions from which the composite molecular functions were formed was made more explicit. It was assumed that the energy of an atomic state could be written as a sum of "core" (inner shell) and "peel" (valence shell) contributions, and that the wave functions, correspondingly, could be written as a product of "core" and "peel" parts. The scaling needed for molecule formation applies only to the latter. Then the effect of scaling the peel was considered to be multiplication of the peel energy by $2s - s^2$, where it is the scaling factor. This may be justified as follows (see Volume I, Chapter III, Section C.1): The effect of the scaling is to multiply the potential energy $\langle V \rangle$ by $s$ and the kinetic energy $\langle T \rangle$ by $s^2$. Assume that the original atomic wave function is an eigenfunction of some atomic Hamiltonian with eigenvalue $E$, $E = \langle V \rangle + \langle T \rangle$, and that the scaled atomic wave function is an eigenfunction of some scaled Hamiltonian with eigenvalue $E_s$. The virial theorem being satisfied in both cases,

$$E_s = \langle V \rangle_s + \langle T \rangle_s = s\langle V \rangle + s^2\langle T \rangle = s(2E) + s^2(-E) \qquad (59)$$

or $E_s = (2s - s^2)E$. Ellison supposed that scaled atomic functions were employed [74, 75] in setting up the valence bond or atoms-in-molecules calculation. In the calculation of the intra-atomic part of a matrix element, such as

$$I = \int \Psi_M{}^* \mathscr{A}' \{ H^{OA}\Phi_i{}^A\Phi_j{}^B \} \, d\tau$$

[see Eq. (49)], we used the fact that $\Phi_i{}^A$ is an eigenfunction of $H^{OA}$ with eigenvalue $E^A$ to write $I$ as $E^A \int \Psi_M^* \mathscr{A}' \{\Phi_i{}^A \Phi_j{}^B\} \, d\tau$, the integral being an overlap. Ellison assumed that, when the scaled functions were used, $I$ could be approximated by $E^A(2s^A - (s^A)^2)$ times the corresponding overlap factor. Here, $s^A$ is the scaling factor appropriate to $\Phi_i{}^A$. Furthermore, it was assumed that, with core–peel separability, $E^A(2s^A - (s^A)^2)$ could be replaced by the sum of the core energy and the scaled peel energy for atom A. The scaled peel energy was $s^A(2 - s^A)E_p{}^A$, where $E_p{}^A$ is the unscaled peel energy.

Then Ellison [74] put, in the secular equation,

$$H_{KL} \rightarrow \tfrac{1}{2}\tilde{S}_{KL}(\tilde{W}_K{}^s + \tilde{W}_L{}^s) + \tfrac{1}{2}(\tilde{V}_{KL} + \tilde{V}_{LK}) \tag{60}$$

In this equation, the overlaps $\tilde{S}_{KL}$ and the interaction integrals $\tilde{V}_{KL}$ are calculated using orbital approximations to the unscaled atomic functions, but the $\tilde{W}_K{}^s$ are sums of energies for atoms which differ from the real atoms by a scaling factor in the peel (they are the distorted atoms of Arai). In fact, several approximations, including the Mulliken approximation, were used in computing $\tilde{S}_{KL}$ and $\tilde{V}_{KL}$, so that only simple integrals needed to be evaluated. One- and two-center nuclear attraction integrals and Coulombic electron repulsion integrals were obtained from semiempirical formulas in some earlier work, but evaluated in later work. The peel scaling factors were chosen to minimize the energy. For the ground states of LiH, BeH, and BH, the results were those shown in Table VI. Excited states were treated as well as ground states.

Theoretical justification for much of this theory was given by Ellison and Wu [75, 76], who produced a general formulation and applied it to $H_2$ and $He_2^+$. The intra-atomic contributions to the matrix element $H_{KM}$

Table VI[a]

*Semiempirical Valence Bond Calculations[b]*

|      | $\omega_e$ (cm$^{-1}$) | $R_e$ (Å)     | $D_0{}^0$ (eV) |
|------|------------------------|---------------|----------------|
| LiH  | 1429 (1406)            | 1.555 (1.595) | 2.39 (2.43)    |
| BeH  | 1639 (2059)            | 1.380 (1.343) | 2.01 (2.2)     |
| BH   | 2124 (2367)            | 1.315 (1.236) | 3.398          |

[a] Experimental values are given within the parentheses.
[b] See Ellison [74].

were evaluated exactly from experimental atomic properties, while the interatomic contributions $V_{KL}$ and $S_{KL}$ were evaluated using orbital approximations to these functions. A "calibrated Mulliken approximation" for a product of atomic orbitals on two centers was introduced:

$$\phi_A(1)\phi_B(1) = \tfrac{1}{2}S_{AB}K[\phi_A(1)\phi_A(1) + \phi_B(1)\phi_B(1)]$$

with the calibration parameter $K$ chosen to give correct energy for the ground state of $H_2$. Properties of excited state $U(R)$ curves were then much improved. Another value of $K$ was used for the states of $He_2^{2+}$. These calibration parameters served to give good results for several states of $HeH^+$, $He_2^+$, $H_2^-$, and HeH.

A semiempirical valence bond theory used by Vanderslice *et al.* [77] obtains the exchange integrals which appear in valence bond energies from measured molecular energies. It is assumed that all the molecular states which derive from particular atomic configurations have been calculated in the simplest valence bond approximation, so that all their energies are expressed in terms of the same exchange integrals. Then measurements of the $U(R)$ for several states suffice to give values for all the exchange integrals, which may then be used to derive energies for the other states. Futher detail will be found in the next section.

## REFERENCES

1. K. Jug, *Theor. Chim. Acta* **14**, 91 (1969).
2. H. H. Jaffe, *Accounts Chem. Res.* **2**, 136 (1969).
3. J. A. Pople, *Accounts Chem. Res.* **3**, 217 (1970).
4. G. Klopman and B. O'Leary, *Fortschr. Chem. Forsch.* **15**, 445 (1970).
5. J. A. Pople and D. L. Beveridge, "Approximate Molecular Orbital Theory." McGraw-Hill, New York, 1970.
6. N. D. Sokolov, *Uspekhi Khim.* **32**, 967 (1963).
6a. E. Hückel, *Z. Phys.* **70**, 204 (1931); **76**, 628 (1932).
7. R. G. Pearson, *J. Chem. Phys.* **17**, 969 (1949).
8. R. G. Pearson and H. B. Gray, *Inorg. Chem.* **2**, 358 (1963).
9. R. Hoffmann and W. N. Lipscomb, *J. Chem. Phys.* **36**, 2179 (1962); R. Hoffmann, *J. Chem. Phys.* **39**, 1397 (1963).
10. G. I. Kagan, G. M. Kagan, and I. N. Fundyler, *Theor. Exp. Chem. USSR* **3**, 254 (1967); *Teor. Eksp. Khim.* **3**, 444 (1967).
11. A. D. Walsh, *J. Chem. Soc.* pp. 2260, 2266, 2288, 2296, 2301, 2306, 2318, 2321, 2325, 2330 (1953).
11a. L. C. Allen, *Theor. Chim. Acta* **24**, 117 (1972).
12. L. C. Allen, *in* "Sigma Molecular Orbital Theory," (O. Sinanoğlu and K. B. Wiberg, eds.), p. 227. Yale Univ. Press, New Haven, 1970. S. D. Peyerimhoff, R. J. Buenker, and J. L. Whitten, *J. Chem. Phys.* **46**, 1707 (1967).

12a. F. P. Boer, M. D. Newton, and W. N. Lipscomb, *Proc. Nat. Acad. Sci. U. S.* **52**, 890 (1964).

13. J. Goodisman, *J. Amer. Chem. Soc.* **91**, 6552 (1969).

14. M. Wolfsberg and L. Helmholtz, *J. Chem. Phys.* **20**, 837 (1952).

15. C. K. Jorgenson, S. M. Horner, W. A. Hatfield, and S. Y. Tyree, Jr., *Int. J. Quantum Chem.* **1**, 191 (1967).

16. J. A. Pople, D. P. Santry, and G. A. Segal, *J. Chem. Phys.* **43**, S129 (1965).

17. K. Jug, *Theor. Chim. Acta* **16**, 95 (1970).

18. M. J. S. Dewar and G. Klopman, *J. Amer. Chem. Soc.* **89**, 3089 (1967).

19. L. C. Cusachs and J. W. Reynolds, *J. Chem. Phys.* **43**, S160 (1965).

20. R. Rein, N. Fukuda, H. Win, G. A. Clarke, and F. E. Harris, *J. Chem. Phys.* **45**, 4743 (1966).

21. F. E. Harris, *J. Chem. Phys.* **48**, 4027 (1968).

22. G. Klopman, *J. Amer. Chem. Soc.* **86**, 1463, 4550 (1964).

23. G. Klopman, *J. Amer. Chem. Soc.* **87**, 3300 (1965); M. J. S. Dewar and G. Klopman, *J. Amer. Chem. Soc.* **89**, 3089 (1967).

24. H. A. Pohl, R. Rein, and K. Appel, *J. Chem. Phys.* **41**, 3385 (1964).

25. R. G. Parr, "Quantum Theory of Molecular Structure," Chapter III. Benjamin, New York, 1963.

26. R. S. Mulliken, C. A. Rieke, D. Orloff, and H. Orloff, *J. Chem. Phys.* **17**, 1248 (1949).

27. F. E. Harris and H. A. Pohl, *J. Chem. Phys.* **42**, 3648 (1965).

28. R. Ferreira and J. K. Bates, *Theor. Chim. Acta* **16**, 111 (1970).

29. R. Ferreira, *J. Phys. Chem.* **68**, 2240 (1964).

30. G. H. Cheesman, A. J. T. Finney, and I. K. Snook, *Theor. Chim. Acta* **16**, 33 (1970).

31. J. A. Pople and G. A. Segal, *J. Chem. Phys.* **44**, 3289 (1966).

32. G. A. Segal, *J. Chem. Phys.* **47**, 1876 (1967).

33. J. A. Pople and G. A. Segal, *J. Chem. Phys.* **43**, S136 (1965).

34. J. M. Sichel and M. A. Whitehead, *Theor. Chim. Acta* **11**, 239 (1968).

35. H. Fischer and H. Kollmar, *Theor. Chim. Acta* **13**, 213 (1969).

36. J. A. Pople, D. L. Beveridge, and P. A. Dobosh, *J. Chem. Phys.* **47**, 2026 (1967); R. N. Dixon, *Mol. Phys.* **12**, 83 (1967); R. Sustmann, J. E. Williams, M. J. S. Dewar, L. C. Allen, and P. von R. Schleyer, *J. Amer. Chem. Soc.* **91**, 5350 (1969).

37. R. C. Baetzold, *J. Chem. Phys.* **55**, 4355 (1971).

38. K. Jug, *Theor. Chim. Acta* **23**, 183 (1971).

39. J. C. Slater, *Phys. Rev.* **81**, 385 (1951).

39a. F. Herman and S. Skillman, "Atomic Structure Calculations." Prentice-Hall, Englewood Cliffs, New Jersey, 1963.

40. P. Gombás and D. Kisdi, *Theor. Chim. Acta* **5**, 127 (1966).

41. P. Gombás, "Pseudopotentiale." Springer-Verlag, Berlin and New York, 1967.

42. J. D. Weeks, A. Hazi, and S. A. Rice, *Advan. Chem. Phys.* **16**, 283 (1969).

43. J. C. Phillips and L. Kleinman, *Phys. Rev.* **116**, 287 (1959).

44. H. Hellmann, *J. Chem. Phys.* **3**, 61 (1935); H. Hellmann and W. Kassatotschkin, *Ibid.* **4**, 324 (1936); *Acta Physicochim. URSS* **23**, 5 (1936).

45. A. U. Hazi and S. A. Rice, *J. Chem. Phys.* **45**, 3004 (1966); **47**, 1125 (1967); **48**, 495 (1968); J. D. Weeks and S. A. Rice, *Ibid.* **49**, 2741 (1968).

46. W. L. McMillan, *Phys. Rev. A* **4**, 69 (1971).

47. B. Cadioli, U. Pincelli, and G. Del Re, *Theor. Chim. Acta* **10**, 393 (1968); **14**, 253 (1969).

48. V. Fock, M. Wesselow, and M. Petrashen, *Zh. Eksp. Teor. Fiz.* **10**, 723 (1940).
49. L. Szász, *Z. Naturforsch.* **14a**, 1014 (1959).
50. L. Szász and G. McGinn, *J. Chem. Phys.* **42**, 2363 (1965).
51. L. Szász and G. McGinn, *J. Chem. Phys.* **45**, 2898 (1966).
52. L. Szász and G. McGinn, *J. Chem. Phys.* **47**, 3495 (1967).
53. L. Szász and G. McGinn, *J. Chem. Phys.* **48**, 2997 (1968).
54. G. Simons and A. Mazziotti, *J. Chem. Phys.* **52**, 2449 (1970).
54a. L. R. Kahn and W. A. Goddard III, *Chem. Phys. Lett.* **2**, 667 (1968). C. F. Melius, W. A. Goddard III, and L. R. Kahn, *J. Chem. Phys.* **56**, 3392 (1972).
55. W. E. Baylis, *J. Chem. Phys.* **51**, 2665 (1969).
56. G. Das and A. C. Wahl, *Phys. Rev. A* **4**, 825 (1971).
57. J. C. Slater and J. H. Wood, *Int. J. Quantum Chem. Symp.* **4**, 3 (1971).
58. K. Schwarz and J. W. D. Connolly, *J. Chem. Phys.* **55**, 4710 (1971).
58a. R. G. Gordon and Y. S. Kim, *J. Chem. Phys.* **56**, 3122 (1972).
59. A. Dalgarno, C. Bottcher, and G. A. Victor, *Chem. Phys. Lett.* **7**, 265 (1970); C. Bottcher, A. C. Allison, and A. Dalgarno, *Ibid.* **11**, 307 (1971).
60. J. N. Bardsley, *Chem. Phys. Lett.* **7**, 517 (1970).
61. A. C. Roach and M. S. Child, *Mol. Phys.* **14**, 1 (1968).
62. W. Moffitt, *Proc. Roy. Soc. Ser. A* **210**, 245 (1951).
63. T. Arai, *Rev. Mod. Phys.* **32**, 370 (1960).
64. T. Arai, *J. Chem. Phys.* **26**, 435 (1957).
65. A. C. Hurley, *Proc. Phys. Soc. London Sect. A* **68**, 149 (1955); **69**, 49 (1956).
66. T. Arai and M. Sakamoto, *J. Chem. Phys.* **28**, 32 (1958).
67. I. Mannari and T. Arai, *J. Chem. Phys.* **28**, 28 (1958).
68. A. C. Hurley, *J. Chem. Phys.* **28**, 532 (1958).
69. A. C. Hurley, *Proc. Roy. Soc. Ser. A* **248**, 119 (1958); **249**, 402 (1959); *Proc. Phys. Soc. London Sect. A* **69**, 301, 767, 868 (1956).
70. A. C. Hurley, *Rev. Mod. Phys.* **32**, 400 (1960).
71. G. G. Balint-Kurti and M. Karplus, *J. Chem. Phys.* **50**, 478 (1969).
72. A. L. Companion and F. O. Ellison, *J. Chem. Phys.* **28**, 1 (1958); **32**, 1132 (1960).
73. F. O. Ellison, *J. Phys. Chem.* **66**, 2294 (1962).
74. F. O. Ellison, *J. Chem. Phys.* **43**, 3654 (1965).
75. F. O. Ellison and A. A. Wu, *J. Chem. Phys.* **47**, 4408 (1967).
76. A. A. Wu and F. O. Ellison, *J. Chem. Phys.* **48**, 1103 (1968).
77. J. T. Vanderslice, E. A. Mason, and W. G. Maisch, *J. Chem. Phys.* **31**, 738 (1959); **32**, 515 (1960).

## B. Simple Models and Relations between Potential Constants

The semiempirical methods of Section A all derive from *ab initio* calculations of the electronic energy (in principle, at least). They introduce empirical data to reduce the amount of computation or to improve results. The "simple models" of this section generally contain a larger empirical element and are not derived from the Schrödinger equation for the molecule, but rather from ideas about how the wave function or electron density should behave. On the basis of simple physical or other assumptions, properties of the system are derived; the hope is not to get accurate results, but, rather, rough agreement with experiment with a minimum of calculation.

The "floating spherical gaussian orbital model" of Frost and co-workers, which we discuss first, differs from the others because it is a true *ab initio* calculation. It appears here because its aim, i.e., to obtain rapidly reasonably good results for a large number of systems, is more akin to the spirit of the other approaches, because it is intentionally based on a simple, almost naive, description of the electrons in atoms and molecules, and because it is characterized as a model by Frost.

The simple pictures on which the models are based, as well as the simple formulas given by many of them for $U(R)$ and its properties, lead to relations between $U(R)$ for different systems, or between different properties of $U(R)$ for the same system. In some cases, such relations were noticed in the results of experiment, and theoretical justification or model-building followed. The relations make it possible to obtain properties of $U(R)$ for a diatomic system when $U(R)$ for other, related, systems are known, or to obtain one property of the interaction from others. They are discussed in the second subsection of this section.

## 1. Simple Models

Frost and co-workers [1] have shown that one can obtain quite reasonable results for equilibrium bond lengths and other properties for a wide variety of atoms and molecules by a very simple calculation. The wave function for a system composed of $n$ electron pairs is represented by a single determinant of doubly filled orbitals, with each orbital written as a normalized spherical Gaussian

$$\chi_i = (2/\pi\varrho_i)^{3/4}e^{-(r/\varrho_i)^2}. \tag{61}$$

Both the exponential parameter $\varrho_i$ and the position of the center, from which the distance $r$ is measured, are taken as variational parameters. The localized orbitals may be inner shell, bonding, or lone pair orbitals. The ease with which integrals over Gaussian functions can be evaluated [2] makes it possible to carry out the large number of calculations needed to optimize the energy with respect to the $4n$ nonlinear parameters (one exponential parameter and three coordinates to define the center). Systematic ways of finding the best values for all parameters, taking advantage of the symmetries of the molecules, were discussed. The minimization of the energy with respect to nuclear positions as well led to predictions of molecular geometry, such as $R_e$ values for stable diatomic systems. The calculated energies, necessarily upper bounds to the Hartree–Fock energies, were quite far above the Hartree–Fock energies (by 0.038 a.u. for $H_2$ and by 15 a.u., out of 100 a.u., for HF, for example) but the agreement between calculated and observed bond lengths was good, considering the simplicity of the calculation. The accuracy was generally a few percent. For example, calculated $R_e$ [1, 3] for $HeH^+$, LiH, $BeH^+$, and HF were 1.287 $a_0$, 3.235 $a_0$, 2.625 $a_0$, and 1.48 $a_0$; the experimental values are 1.440, 3.014, 2.479, and 1.733 $a_0$.

For LiH, the forces on the nuclei, as well as other properties, were computed [1]. The forces were close to zero when the internuclear distance was close to $R_e$. Since all parameters in the trial function are minimized for each $R$, the Hellmann–Feynman theorem is obeyed (see Volume I, Chapter III, Section D). The virial theorem should also be obeyed because a scaling parameter is implicit in the parameters varied. This was checked as well. The dissociation energy to $Li^+ + H^-$ (both paired-electron systems) was 0.27 a.u., close to the experimental value of 0.26 a.u. Thus the energy error (about $1\frac{1}{2}$ a.u.) is roughly the same for the molecule as for the separated ions. The good value for $R_e$ shows the error does not vary very strongly with $R$. In addition the location and sizes of the orbitals coincided with what simple pictures of the molecule would lead one to expect. There was a tightly bound pair centered almost on Li, and a pair centered near H, with exponential parameter $\varrho_i$ somewhat smaller than that appropriate to $H^-$, corresponding to an orbital contraction. For the He–He system, interaction energies (relative to separated atoms) were 1.2619, 0.1141, and 0.0059 a.u. at $R = 1$, 2, and 3 $a_0$; the results of very accurate calculations were 0.929, 0.099, and 0.011 a.u.

In later calculations [4], the single Gaussians were replaced by double Gaussians. Of course, these more flexible functions produced a great improvement in the energy, but did not change predicted geometries much. Thus, the recent work of Chu and Frost [5] used a single Gaussian for each localized orbital. The inner-shell orbitals were assumed to be localized on the nucleus of their atoms, since allowing them to float off the nuclei had given little change in geometries. For NaH and $MgH^+$, the $L$-shell orbitals were arranged at the corners of a tetrahedron of fixed size, and with one vertex on the molecular axis on the opposite side of Na or Mg from the proton. The bond orbital was placed between Na (Mg) and H. This reduced the number of parameters, but also served the purpose of preventing the distance of the $L$ orbitals from Na (Mg) from becoming too small, as would occur if the distance were determined to minimize the energy. For the same reasons, some of the coordinates of the lone pair orbitals for other molecules were not allowed to vary from predetermined values. For $Be_2$ and $C_2$, the bond orbital was represented by two equal orbitals, symmetrically placed with respect to the internuclear axis; $C_2$ had in addition two lone pair orbitals on the axis. For the triple bond of $N_2$, Chu and Frost [5] used three orbitals forming an equilateral triangle in a plane perpendicular to the internuclear axis.

The energies were about 85% of the SCF energies for all molecules. The predicted internuclear distances are shown in Table VII. The average percentage deviation between the first and second column is less than 6%,

**Table VII**

*FSGO Results for Diatomic Molecules Composed of First- and Second-Row Atoms[a]*

| | $R_e$ (calc.) $(a_0)$ | $R_e$ (expt.)[b] $(a_0)$ |
|---|---|---|
| NaH | 3.758 | 3.566 |
| MgH$^+$ | 3.179 | (3.116) |
| AlH | 3.230 | 3.114 |
| HCl | 2.273 | 2.409 |
| Li$_2$ | 5.304 | 5.501 |
| Be$_2$ | 3.428 | (3.780) |
| B$_2$ | 3.613 | (3.005) |
| C$_2$ | 2.614 | (2.348) |
| N$_2$ | 2.304 | 2.075 |
| O$_2$ | 2.031 | (2.318) |
| F$_2$ | 2.810 | 2.725 |
| CO | 2.126 | 2.132 |
| SiO | 2.830 | (2.750) |
| CS | 2.996 | (2.900) |
| PN | 2.672 | (2.670) |
| P$_2$ | 3.242 | (3.580) |
| LiF | 2.718 | (2.888) |
| NaF | 3.481 | (3.629) |
| LiCl | 4.230 | (3.825) |
| NaCl | 4.420 | (4.461) |
| NaLi | 5.505 | (5.500) |
| Na$_2$ | 5.680 | (5.818) |

[a] See the literature [1–5].
[b] Parentheses indicate calculated value from accurate SCF.

with the two worst errors arising for molecules ($B_2$ and $O_2$) whose ground states are not singlets.

Simple molecular orbital and valence bond calculations have often suggested formulas for approximate computation. For instance, in the MO treatments of $H_2^+$ and $H_2$, the dissociation energies are roughly equal to the contribution of the "resonance" term in the Hamiltonian, which involves the integral of the product of the atomic orbitals on the different atoms. Mulliken [6] suggested that this term could be approximated as the product of the overlap $S$ between the atomic orbitals and the energy (ionization potential) $I$ of the valence orbital. (For homonuclear molecules, the atomic orbitals are the same on the two atoms.) This is closely related to the Mulliken approximation for integrals (see Chapter I, Section A). For several bonding orbitals, the bond energy would be a sum of terms of the form $AIS/(1 + S)$, $A$ being a semiempirical scaling constant, while antibonding orbitals would contribute terms of the form $AIS/(1 - S)$. The denominators arise from the normalization of the molecular orbitals. In the valence bond theory, similar-looking terms (except with denominators of the form $1 \pm S^2$) play the dominant role in determining bond energies.

Mulliken [6] therefore suggested that, at least near the equilibrium internuclear distance, the bond energy be represented by a sum of such terms. In line with the ideas of the VB theory, however, additional contributions to the energy, associated with preparation of the atoms for bonding, should also be included. Mulliken [6] gave expressions ("magic formulas") to be used for bond energies of many-electron, polyatomic molecules. The dissociation energy of a diatomic was written as a sum of terms of the form $AIS/(1 + S)$ for each bonding pair, plus contributions for each nonbonded pair, and an additive term representing promotion and resonance energy contributions. The parameters involved were obtained by fitting results for certain molecules.

The perfect pairing formula of the simple valence bond theory has been used to relate potential curves for different states of a molecule. The interaction energy aside from a relatively unimportant Coulomb term is written as a sum of (negative) exchange integrals between pairs of atomic orbitals on the two atoms: The exchange integral appears with a positive sign if the spins of the orbitals are paired, with a negative sign if the spins are parallel, and with a factor of $-\frac{1}{2}$ if the spins are unpaired. Vanderslice et al. [7] used this formula to derive approximate $U(R)$ for some states of $N_2$, $O_2$, and NO from known $U(R)$ for other states correlating with the same atomic configurations. Further discussion is found at the end of the next section [Eq. (114)].

We now turn to calculations invoking a simple physical picture. Recently, the treatment of the electrons in a bond as free electrons or as electrons in a square-well ("box") potential has come in for renewed interest, due to the contributions of Parr and co-workers (see page 354). This idea has a long history in quantum mechanics, the particle in the box being one of the few problems capable of simple solution in closed form.

A simple model of this kind was developed by Arnold [8] for bonds within polyatomic molecules as well as for diatomic molecules themselves. For a single bond between atoms A and B, the two valence electrons were imagined to move in an infinite square-well potential of size $2l_{AB}$. Their energy is thus

$$E_{AB} = 2(h^2/32ml_{AB}^2) \qquad (62)$$

where $m$ is the electronic mass. The effective distance $l_{AB}$ was taken as the bond length $R$ plus the extensions of the atoms on the sides of the nuclei away from the bond. The extensions were supposed to be characteristic of the atoms A and B. Then

$$2l_{AB} = l_A + l_B + R, \qquad (63)$$

where the diameters of atoms A and B are $2l_A$ and $2l_B$. Applying the same model to the valence electrons on A and B, their energy would be

$$E_A + E_B = (h^2/32ml_A^2) + (h^2/32ml_B^2). \qquad (64)$$

The bond energy was calculated [8] as $E_A + E_B - E_{AB}$, augmented by an electronegativity correction, 23.06 kcal mole$^{-1}$ times the square of the electronegativity difference. For double or triple bonds, the bond energies were multiplied by two or three. A similar discussion could be given [8] for the model in which the square wells were replaced by harmonic oscillators. In addition to the electronegativities (which are empirically defined quantities, but whose values may be considered as given for these calculations), Arnold required, for either model, the half-diameters of the atoms. He took them as proportional to nonbonding radii, the constant of proportionality (different for the harmonic oscillator and square well models) being a parameter to be chosen empirically. Except for diatomics involving N, O, and F, predicted bond energies were good to about 10%.

A more sophisticated free-electron model was presented by Müller and Dunken [9], to whose article we also refer for further bibliography on the subject. Motion in three dimensions was considered. Thus, for the homonuclear diatomic $A_2$, the electrons were supposed to be confined to a cylindrical box of radius equal to the radius of atom A, and length equal to twice this plus the internuclear distance $R$. An effective mass of $2.43m$ was used instead of the electronic mass $m$. The empirical parameters were thus this factor and the radii of the atoms. The energy levels in this potential having been calculated, they were filled in order by the valence electrons. One hundred twenty homonuclear diatomic molecules and molecular–ions were treated, and dissociation energies (molecular versus atomic energies) calculated. The agreement with experiment was surprisingly good—a few percent in many cases. Matlow's three-dimensional free-electron model [10] used cubes for atoms and rectangular parallelipipeds for homonuclear molecules. The dimension of an atomic box was a van der Waals diameter, which was also used as the transverse dimension of a molecular box. The length of a molecular box was the van der Waals diameter plus the inter-nuclear distance. The effective mass was chosen to give agreement with experiment for dissociation energies, and did not vary much for different elements.

Nuclear potentials in the form of delta-functions have been used [11] to correct free-electron–gas models, although not for diatomic molecules. Frost has also proposed a model [11] in which the entire nuclear attraction is represented by a delta function. In the box models of the preceding paragraphs, the confinement of the electrons walls of the box is supposedly an effect of the nuclear attraction.

The particle-in-a-box picture arises in another context. Borkman and Parr [12] found that the assumption that, near $R_e$,

$$Q = -R^{-1}(d/dR)[R^2\langle T\rangle] = \text{const} \tag{65}$$

$[\langle T\rangle =$ electronic kinetic energy: see Volume I, Chapter III, Eqs. (312) et seq.] led to some simple relations between the derivatives of $U$ near $R_e$. These relations proved to be verified in practice, and the problem of in-terpreting Eq. (65) then arose. The authors noted that if one expands $U(R)$ near $R_e$ as

$$U(R) = \sum_{i=0} W_i R^{-i} \tag{66}$$

(Kratzer potential), neither $W_1$ nor $W_2$ enters $Q$ of Eq. (65). This means that $Q$ is likely to be much closer to constant than $U$ itself. By including the next nonvanishing term in $Q$ (in $R^{-3}$), one can derive [12] more accurate correlations between the first three force constants than those given below, but Eq. (65) already gives results of accuracy comparable to what one can get from Hartree–Fock calculations.

The implication of (65) is

$$\langle T \rangle = T_0 + T_2 R^{-2} \qquad (67)$$

near $R_e$. The $R$-independent part, $T_0$, was associated with the core electrons, which are nonbonding and essentially unaffected by changing $R$. The second term should refer to the valence or bonding electrons. The simple physical picture of particles in a one-dimensional box is suggested because of the dependence of the (kinetic) energy on the inverse square of the length of the box [see Eq. (62)]. With this picture, the quadratic force constant could be expressed in terms of the kinetic energies of the atomic orbitals in the various bonds and the populations of these orbitals, and qualitative agreement with experiment was obtained.

The picture also helps to explain [12] the rule (see Subsection 2) that $R_e^2 \omega_e$ is roughly constant for many electronic states of the same molecule and for many molecules formed from atoms in the same rows of the periodic table. It follows from the virial theorem (Volume I, Chapter III, Section C) that

$$R_e (d^2 U / dR^2)_e = -(d\langle T \rangle / dR)_e$$

According to (67), $(d\langle T \rangle / dR)_e = -2T_2 R^{-3}$. Then, from the formula

$$\omega_e = (2\pi c)^{-1}[(d^2 U / dR^2)_e / \mu]^{1/2}$$

we obtain

$$R_e^2 \omega_e = [T_2 / 2\pi^2 \mu c^2]^{1/2}. \qquad (68)$$

According to the one-dimensional box model, the kinetic energy is $qh^2/8mL^2$ where $L$ is the box length and $q$ is the number of electrons involved or the effective charge on the particle in the box. We may write $T_2 = qh^2/8mv^2$, where $v$ is the ratio of $L$ to $R$ near $R_e$ ($v$ is of size unity), and the near constancy of $R_e^2 \omega_e$ is explained by the expected constancy of $q$ and $v$ for related molecules.

Borkman et al. [13] considered homonuclear diatomic molecules to derive the values of $q$ and $v$ that are required to reproduce the experimental values of $R_e$ and $k_e$ (force constant). For this purpose, the square-well model could be extended to predict $W_1$ of Eq. (66). Here $W_0$ is irrelevant for $R_e$ and $k_e$, whereas $W_2$ is the kinetic energy term $T_2$. This is seen by inserting a truncated expansion (66) in Eq. (299b) of Volume I, Chapter III:

$$T_0 + T_2 R^{-2} = -W_0 + 2W_2 R^{-2} - W_2 R^{-2} \qquad (69)$$

Similarly, from (299a) of Volume I, Chapter III, $W_1$ represents the potential energy of the valence electrons. (Note that the atomic virial theorem, $T_0 = -W_0 = -\frac{1}{2}V_0$, holds, in accord with the identification of this term with the atomic cores.) Borkman et al. [13] supposed that a charge $Z = \frac{1}{2}q$ was located at each atomic center and a charge of $-q$ at the center of the bond, giving a neutral system. The charge at the bond center represents the bond electrons; if the units are taken as electronic charge, $q$ can be identified with the number of valence electrons. The overall potential energy of the charges is $e^2R^{-1}(-7q^2/4)$. Then the truncated expansion (66) is:

$$U(R) = W_0 + W_1R^{-1} + W_2R^{-2}$$
$$= W_0 + (-7q^2e^2/4)R^{-1} + (h^2q/8mv^2)R^{-2} \tag{70}$$

[This does not conform to the usual convention that $U(\infty) = 0$; to obtain this one puts $W_0 = 0$]. Equation (70) is differentiated with respect to $R$ to give $q$ and $v$ in terms of $R_e$ and $k_e$. Setting $dU/dR$ equal to zero,

$$R_e = -2W_2/W_1 \tag{71}$$

while

$$k_e = 2W_1R_e^{-3} + 6W_2R_e^{-4} = W_1^4/8W_2^3. \tag{72}$$

For a series of molecules for which $v$ may be taken as constant, it is easy to see that $qR_e$ and $k_eR_e^5$ should be constant.

For heteronuclear diatomics a homopolar model was first tried, giving reasonable results [13]. It was recognized that different effective charges should be used for the two cores and that the center of electronic charge should not necessarily be at the midpoint. Then two additional parameters must be determined. Several schemes to determine them were discussed and implemented; remarkably, the values of $q$ were fairly independent of the model. Furthermore, these values again correlate with our ideas of bond orders. Similarly, the values of $v$ did not change much from those obtained from the homonuclear model. Finally, relations were predicted between values of $v$ for the A–B molecule and the values for the A–A and B–B molecules. The implications of the values of $q$ for bond charges, bond orders, and charge moments (which were poorly predicted) were discussed, as were the significance of the length parameters $v$ and their relation to atomic core radii. It was concluded that, pending further investigation, the parameters $q$ and $v$ should be considered simply as numbers whose values reproduce $R_e$ and $k_e$ in this model. Predictions of anharmonicities of $U(R)$ (see next subsection) from this formalism were successful, but give no significance to the values of the parameters.

According to this model, the electronic energy of a diatomic for $R$ near $R_e$ is given by

$$U(R) = W_0 + W_1 R^{-1} + W_2 R^{-2}$$

with $W_2 R^{-2}$ representing kinetic energy of the electrons in the bond and $W_1 R^{-1}$ the electrostatic potential energy for the fixed separated charges. Politzer [14] has discussed the interpretation of $W_0$, which should be associated with the localized charges on the nuclei. He considered that it should include the energy of two atoms at infinite separation, $U(\infty)$, plus the energy needed to remove charge from the nuclei to the bond region. The latter should be proportional to $q^2$. Using $D_e = U(\infty) - U(R_e)$ and Eqs. (70) and (71), one can derive an expression for the constant of proportionality $B$.

$$B = -D_e q^{-2} + 7e^2/8R_e \qquad (73)$$

For a series of related molecules, $B$ calculated in this way varies linearly with the mean of the ionization potentials of the constituent atoms. If $Bq^2$ equals the energy to remove a charge of $-\frac{1}{2}q$ from the highest filled orbital of each atom, we would have

$$Bq^2 = \tfrac{1}{4}q^2(\langle r_A^{-1}\rangle + \langle r_B^{-1}\rangle)$$

where $\langle r_A^{-1}\rangle$ and $\langle r_B^{-1}\rangle$ are atomic expectation values, permitting the calculation of $B$. The resulting values agree reasonably well with those derived from Eq. (73) [14].

Simons and Parr [15] have investigated this "bond charge" model further, especially the choice of parameters needed to describe diatomic molecules. They were able to predict the first and second anharmonicity constants to good accuracy and explain empirically observed correlations between different systems.

It may be noted that this model includes both quantum mechanical (particle in a box) and electrostatic (point charge) elements. The description of the electron density used for the kinetic energy is not the same as that used for the potential energy. According to the Hellmann–Feynman theorem, all forces on nuclei can be rigorously calculated from the charge distribution, provided that the charge distribution is derived from the exact wave function. If we do this, only Coulombic interactions between the charge densities and the nuclear point charges will appear. To get energies, we must integrate the force expression over nuclear displacements.

The resulting expression may not have a simple electrostatic interpretation, since the parameters of the charge distribution change with $R$. Some "electrostatic models" neglect this dependence, or take it into account by introducing additional terms into the energy expression. Usually, the energy is written as a combination of electrostatic interactions (in the sense of Chapter I, Section A) and other terms which may derive from quantum mechanical calculations.

An ionic model for the alkali halides, which was surprisingly successful in view of the approximations made, was that of Rittner [16]. In this model, the molecule was considered to be composed of polarizable and non-overlapping ions, characterized by charges $+e$ and $-e$ and polarizabilities $\alpha_1$ and $\alpha_2$. The field $E_1$ at the center of ion 1, due to the charge and induced dipole of ion 2, induces a dipole on 1 equal to $\alpha_1 E_1$. Similarly, the field at the center of ion 2, multiplied by $\alpha_2$, yields the induced dipole on ion 2. Solving the simultaneous equations gives the induced dipole, and hence the total dipole moment of the molecule, in terms of $\alpha_1$, $\alpha_2$, and the internuclear distance. Rittner showed that there was some justification for using the formulas even when the fact that the electric fields were not homogeneous over the ions was taken into account. The binding energy of the molecule was written as the electrostatic interactions between the charges and induced dipoles, plus the energy stored in the induced dipoles. The leading terms in the energy were

$$\phi = (-e^2/R) + (-e^2(\alpha_1 + \alpha_2)/2R^4) + (-2e^2\alpha_1\alpha_2/R^7) \qquad (74)$$

corresponding to charge–charge, charge–induced dipole, and induced dipole–induced dipole interactions. To $\phi$ Rittner added an exponential (inner shell) repulsion $Ae^{-R/\varrho}$, a van der Waals attraction $-CR^{-6}$, the thermal energy of vibration, and the difference between rotational and translational energies for the molecule and separated ions $(-\tfrac{1}{2}kT)$. Rittner used experimental values for $R_e$ and force constant $k$ for each molecule, and estimated values for $C$ in terms of the ionization potentials and electron affinities of the atoms [16].

For molecules for which $R_e$ and $k$ were known, values for the repulsion constants $\varrho$ and $A$ could be derived. For the nine cases considered, $\varrho$ was $0.31 \pm 0.01$ $A$. When $R_e$ was known, the binding energy could be computed. Binding energies agreed to a few percent [16, 17] with experimental values for the energy difference between the molecule and separated ions, i.e., the experimental binding energy plus the ionization potential of the alkali minus the electron affinity of the halogen. There was little contribution to the binding energy from the van der Waals attraction, thermal vibration energy, or $R^{-7}$ term in $\phi$. Internuclear distances $R_e$ were calculated in some cases and agreed well with experiment. Furthermore, assuming $\varrho = 0.31A$, Rittner calculated vibration frequencies from known or estimated values of $R_e$. Agreement with experiment was excellent—to 10 cm$^{-1}$ or so. The one exception was LiI, where it was thought the ionic picture was not applicable. Predicted molecular dipole moments were also good, supporting the applicability of the polarizable-ion picture to these molecules.

Varshni [18] considered the substitution of an inverse power repulsion $bR^{-n}$ for the exponential form $Ae^{-R/\varrho}$, and also derived values for the Dunham constants $\alpha_e$ and $\omega_e x_e$. The former depends on $d^3U/dR^3$ at $R_e$, the latter on the third and fourth derivatives. The data used for the predictions were $\alpha_1$, $\alpha_2$, $I_{\text{alk}}$, $A_{\text{hal}}$, $\omega_e$, and $R_e$. The agreement with experiment for $\alpha_e$ was good using the exponential repulsion, but not for the inverse power repulsion, which Varshni concluded was inappropriate. He also suggested, on the basis of good agreement of predicted results for $D_e$ and $\alpha_e$, that the experimental $\omega_e x_e$ were too low, and that $\omega_e$ for LiI was incorrect. Subsequently, Pearson and Gray [19] tested a number of variations of Rittner's model for the alkali chlorides, with an aim toward evaluating the ionicity of the compounds for correlation with simple quantum mechanical calculations. A large number of calculations using ionic models had been performed for these systems since Rittner's work [19]. Pearson and Gray first considered the hard sphere ion model, wherein

$$U(R) = -fR^{-1} - CR^{-6} + Ae^{-R/\varrho}.$$

For the diatomic molecule, $f = e^2$; the other constants were derived from virial coefficient data. Putting $R$ equal to the experimental equilibrium separation, the dissociation energies to ions were computed. The results were 6.57, 5.53, 4.63, 4.46, and 4.23 eV for LiCl, NaCl, KCl, RbCl, and CsCl; the experimental values were 6.70, 5.54, 4.88, 4.72, and 4.51 eV. The allowance for polarization of the halide plus substitution of $BR^{-9}$ for $Ae^{-R/\varrho}$, did not improve the overall agreement. Nor did a model with more quantum mechanical character, involving a molecular orbital calculation for the bond electrons with semiempirical evaluation of integrals. (This model did lead to improved bonding energies for polyatomic chlorides.)

The Rittner model has been tested [17, 20] for alkali hydrides, giving very satisfactory predictions for dissociation energies; it works much less well for alkaline earth oxides [20, 21] and hydrogen halides [20]. For the work on the alkali hydrides, it was impossible to obtain good results with neglect of the polarization and repulsion contributions to the energy, as was the case for the alkali halides [21]. The model also seemed applicable to the thallous halides [22]. Some explanations for the failures of the model for alkali hydrides and alkaline earth oxides were suggested [20, 21]. It has been noted [23] that, in order to get correct dipole moments from the Rittner model, values for halogen ion polarizabilities must been taken about $\frac{2}{3}$ as large as are needed to get good potential curves.

Frost and Woodson [24] treated some ionic molecules by combining the Coulombic attraction with a semi-empirical potential function previously used [25] for covalent molecules (see page 366). For the alkali halides, a term $e^2/R$ was added to $U(R)$ derived for the pair of inert gas atoms corresponding to each alkali halide molecule. The parameters in $U(R)$ for homonuclear pairs of inert gas atoms were taken from experiment, while $U(R)$ for heteronuclear pairs was obtained by combination rules (see page 370) from $U(R)$ for homonuclear pairs. This means that $U(R)$ is the same for two alkali halide molecules, like KBr and RbCl, corresponding to the same rare gas atoms. The potential curves for alkali halides resemble those from the Rittner model closely. Predicted values for $R_e$ and $D_e$ are compared with experiment in Table VIII; agreement is satisfactory.

Rittner's model for interionic interactions has been used, in conjunction with pseudopotentials, for semiempirical calculations on diatomics built from alkali atoms [26]. Only the valence electrons were treated explicitly, the interaction of the cores being calculated according to Rittner's formulas.

**Table VIII**

*Frost–Woodson Results for Alkali Halides[a]*

| Molecule | $R_e$ (calc.) $(a_0)$ | $R_e$ (expt.) $(a_0)$ | $D_e$ (calc.) (a.u.) | $D_e$ (expt.) (a.u.) |
|---|---|---|---|---|
| KCl  | 5.075 | 5.0392 | 0.1808 | 0.1814 |
| KBr  | 5.300 | 5.3304 | 0.1739 | 0.1735 |
| RbCl | 5.300 | 5.2664 | 0.1739 | 0.1763 |
| RbBr | 5.531 | 5.5649 | 0.1672 | 0.1681 |
| KI   | 5.555 | 5.7596 | 0.1633 | 0.1622 |
| CsCl | 5.555 | 5.4920 | 0.1633 | 0.1683 |
| RbI  | 5.846 | 6.0035 | 0.1556 | 0.1568 |
| CsBr | 5.846 | 5.8053 | 0.1556 | 0.1616 |
| CsI  | 6.354 | 6.2645 | 0.1402 | 0.1561 |

[a] From Frost and Woodson [24].

A pseudopotential was used to take into account the required orthogonality of the valence electrons to the cores (see Section A.2). Roach *et al.* [26] wrote the Hamiltonian as

$$H = H^{\text{val}} + V^{\text{core}}$$

where $H^{\text{val}}$ included the kinetic energy and mutual repulsion of the valence electrons, plus their attraction to the ionic cores. The core–core interaction was written as $R^{-1} + Ae^{-R/\varrho}$, sometimes with the addition of a polarization term. By expanding the valence electron wave functions in valence shell atomic functions of atom A, only the lack of orthogonality to core B had to be handled by the pseudopotential. Experimental information was used to parametrize the interactions and to evaluate some integrals approximately. Good results were obtained for $Na_2^+$, $K_2^+$, $K_2$, $Na_2$, NaK, NaK$^+$, and other molecules: $D_e$ was correct to 0.1 eV, $R_e$ to 0.15 Å, $\omega_e$ and $\omega_e x_e$ to 30%.

Another recent application of Rittner's model is to the halides of K and Cs. Maltz [27] studied the vibrational energy levels in $U(R)$ as given by Rittner's formulas, as well as the variation of dipole moment and rotational constant with vibrational excitation, using the vibrational wavefunctions.

Platt's model for diatomic hydrides, based on the united atom (see Chapter I, Section C), gives a picture of the molecule somewhat different from the ionic models. It assumes that the separation of the proton from the united atom nucleus should give only a small change in the electron density. A review of this work is given in the very interesting article of Platt in the *Encyclopedia of Physics* [28]. If the density of the hydride is considered to be the density of the united atom, irrespective of $R$, classical electrostatics allows the calculation of $R_e$ and $k_e$. The equilibrium internuclear distance is the value of $R$ for which the effective charge acting on the proton is unity, i.e., the radius beyond which lies one electron in the united atom. The force constant is just $4\pi e^2$ times the density at this point. For further details, see Chapter I, Section C. Despite an inconsistency with the virial theorem, $R_e$ values are predicted to 9% and force constants, using the density at the experimental $R_e$ rather than the computed one, to 7%. The contribution of d orbitals to the force constant is large, so the united atom wavefunction used must reproduce these accurately: Hartree rather than Slater functions must be used.

The alkali-ion–rare-gas systems have been treated by Menendez *et al.* [29], using a potential that combines electrostatic interactions with pseudo-potentials for repulsive and attractive "exchange forces." Letting A be the alkali ion and B the rare gas, the following interactions were taken into account: (a) Charge-induced dipole, going as $R^{-4}$; (b) Charge-induced quadrupole, going as $R^{-6}$; (c) Dispersion, going as $R^{-6}$; (d) Closed-shell repulsion, going as $R^{-4}$. The resulting potential,

$$U(R) = A/R^n - B/R^4 - C/R^6,$$

was parametrized and proved inadequate. Therefore, a "charge-exchange" or "resonance" term, corresponding to the fact that the wavefunction could contain a term corresponding to $AB^+$ as well as one corresponding to $A^+B$, was added. The form of the term in $U(R)$ was derived from a valence bond calculation, using a number of approximations in evaluating the integrals. A modified Thomas–Fermi–Dirac expression was used for (d), relying "principally on physical intuition." It was possible to choose parameters that led to reasonable $R_e$-values and scattering cross sections.

We close this subsection with mention of a simple method proposed for molecular calculations. This "cellular method," like the muffin-tin method mentioned previously, was first used for solid-state calculations. However, Fricker and Anderson [30] found that convergence problems made it necessary to make modifications in the method in order to use it for molecules. In this method, space is divided into cells, in each of which the potential is approximated in a simple way, and the wave function is constructed by joining solutions from the different cells. For diatomics, two cells were used, in each of which the potential seen by an electron was that of a single ionic core. The Schrödinger equation could be solved exactly within each cell. For diatomic alkali molecules, the calculations gave $R_e$ good to 10% and binding energies to 30%.

### 2. Relations between Potential Constants

In this subsection, we discuss rules and generalizations which relate properties of $U(R)$ for different systems, and different properties of $U(R)$ for the same system. Some of these rules are theoretically derived, while others are empirical discoveries.

Into the class of empirical rules falls one of the earliest and most frequently cited generalizations, Badger's rule [31], which relates the force constant $k_e$ and the equilibrium internuclear distance $R_e$. If $k_e^{-3}$ is plotted against $R_e$ for ground and excited states of diatomic molecules, the points fall into groups, according to the rows of the periodic table to which the constituent atoms belong. The points in any one group fall closely on a straight line, and the lines are roughly parallel. Analytically, we may represent the situation as

$$k_e(R_e - d_{ij})^3 = 1.86 \times 10^5 \qquad (75)$$

where $k_e$ is the force constant for molecule AB in dynes per centimeter, and $R_e$ and $d_{ij}$ are in Angstroms. The value of $d_{ij}$ depends on the rows $i$ and $j$ of the periodic table in which atoms A and B are found. An obvious interpretation of these constants is that they represent the sizes of the cores of the atoms. Rules of slightly different functional form, but similar to (75), seemed to be less successful [31].

By allowing the constant on the right side of (75) to depend on $i$ and $j$, so that one has

$$k_e(R_e - d_{ij})^3 = C_{ij}, \qquad (76)$$

one obtains even better agreement with experimental results. The values of $d_{ij}$ and $C_{ij}$ for various $i$ and $j$ were given by Badger [31], who also applied (76) to bonds in polyatomic molecules. For diatomics corresponding to a particular set of $i$ and $j$, (76) implies that an increase in $k_e$ and a decrease in $R_e$ go together, which is usually, but not always, the case. Both are generally associated with stronger bonding. Herschbach and Laurie [32] reviewed Badger's rule in 1961, to test it in the light of the new spectroscopic information available. They found it still worked as well as when originally proposed.

By perturbation theory, Murrell [33] derived a theoretical formula showing a reciprocal relation between $k_e$ and $R_e^{-3}$, mentioned in Volume I, Chapter III, Section B. According to Eq. (248) of Volume I, Chapter III, Section B,

$$\frac{k_e}{2\bar{Z}_A \bar{Z}_B e^2} = \frac{1}{R_e^3} - \Delta^{-1} \left\langle 0 \left| \sum_{i,j} \frac{\cos\theta_{Ai} \cos\theta_{Bj}}{r_{Ai}^2 r_{Bj}^2} \right| 0 \right\rangle$$

where $\Delta$ is an average energy denominator, $\bar{Z}_A$ and $\bar{Z}_B$ are effective nuclear charges, and the fact that $E^{(2)}$ of Chapter I is $\frac{1}{2}k_e$ has been used. Plotting $k_e/2Z_A Z_B e^2$ against $R_e^{-3}$ for a number of molecules, Murrell found that points for related molecules fell on smooth curves, while isoelectronic species ($N_2$, CO and BF; BCl, CS, SiO, and PN) gave straight lines. Some interpretations of these results and their relation to Badger's rule were given [33]. For strongly ionic metallic hydrides, the equation

$$k_e = 2/R_e^3$$

holds [34]. This corresponds to $\bar{Z}_A = \bar{Z}_B = 1$ and no electronic relaxation term. For one-electron diatomic molecules (see Volume I, Chapter III, Section C, p. 206), the relation $k_e R_e^3 = (Z_A + Z_B)S$ was derived [35], $S$ being an arbitrary function of $(Z_A + Z_B)R_e$.

Another proposed relation, based on observation of experimental data, was [36]

$$k_e R_e^6 = \bar{C}_{ij} \tag{77}$$

where the $\bar{C}_{ij}$ form a new set of constants depending on the rows of the periodic table for the atoms of the molecule. This differs only slightly from the constancy of $k_e R_e^5$ predicted by the free-electron model of Parr *et al.* [see Eq. (72)]; the constancy of $R_e^4 k_e$, predicted by a simpler model [Eq. (68)], is closer to Badger's rule.

An argument for the constancy of $R_e^4 k_e$ can be based on the virial theorem [28, Sect. 64]. Let $r$ be a scale parameter on the coordinates of the electrons and nuclei in a molecule. Then the potential energy goes as $-ar^{-1}$ and the kinetic energy as $br^{-2}$. The equilibrium value of $r$ is that which minimizes the total energy: $r_e = 2b/a$. As discussed in greater length in Volume I, Chapter III, Section C, the virial theorem may be obtained by arguments of this sort. Differentiating the energy twice with respect to $r$ gives a "force constant" for change of scale:

$$k_r = d^2(-ar^{-1} + br^{-2})/dr^2 = -2ar^{-3} + 6br^{-4}$$

At equilibrium $k_r r_e^3 = a$ and $k_r r_e^4 = b$. If $r$ is identified with the internuclear distance, $r_e$ is $R_e$ and $k_r = k_e$ for this value of $r$. For similar (particularly isoelectronic) molecules, the potential energy is more likely to vary than the kinetic energy. The constancy of $b$ implies $R_e^4 k_e$ is constant. Platt [28, Sect. 64] applied this to polyatomic isoelectronic hydrides.

Sutherland [36] pointed out that bond length–force constant relations and similar rules imply a similarity between the potential curves for different systems. If we assume a particular algebraic form for $U(R)$, certain relations between its derivatives follow, and these determine $R_e$, $k_e$, and other potential constants. Sutherland also suggested that, by taking the empirically determined rules as given, we could deduce information about the correct form for $U(R)$. [The possibility of a universal potential, from which $U(R)$ for any molecule can be obtained by simple transformations, will be mentioned later—the idea does not seem to hold to very high accuracy.] Sutherland took the proposed function

$$U(R) = -\alpha R^{-m} + \beta R^{-n}$$

and derived relations among $R_e$, $k_e$, and $D_e$. By comparison with empirically found relations, he drew conclusions about the values of the parameters in $U(R)$. In particular, it was suggested that

$$U(R) = -\alpha R^{-1} + \beta R^{-4} \tag{78}$$

was correct for the diatomic hydrides, with $\alpha$ fixed and $\beta$ varying along the series. Equation (78) is similar to what we obtain from an ionic model (see the previous section).

Linnett [37, 38] also considered the relations between $k_e$ and $R_e$ that followed from the assumption of different forms for $U(R)$, comparing them with (76) and other empirical relations, in hopes of providing a theoretical basis for them. For example, he used

$$U(R) = A(R - d)^{-4} - B(R - d)^{-n} \tag{79}$$

and determined $R_e$ by setting $dU/dR$ equal to zero and $k_e$ from the value of $d^2U/dR^2$ for $R = R_e$. With some algebraic manipulation, one can show that (79) leads to

$$k_e(R_e - d)^{n+2} = \frac{n^2 D_e(R_e - d)^n}{1 - B^{-1}D_e(R_e - d)^n} \tag{80}$$

where $D_e$ is the dissociation energy. It was suggested [37, 38] that $d$ be taken as Badger's $d_{ij}$. Then the validity of Badger's rule implies that $n = 1$ and that the right side of (80) is constant from molecule to molecule, as suggested by Sutherland [36]. Linnett argued that $D(R_e - d)$ and $B$ cannot separately be constant, since $B$ determines the strength of the attraction, and it is unreasonable for this to be the same in all cases. If $D(R_e - d)$ and $B$ each varied, the constancy of the right side of (80) would be fortuitous and hence unlikely. Thus he judged (79) unsatisfactory from this viewpoint.

The function

$$U(R) = A R^{-m} - B e^{-nR} \tag{81}$$

leads to

$$k_e R_e^{m+2} = mA + \frac{m^2 D_e R_e^m}{1 + A^{-1}D_e R_e^m} \tag{82}$$

With $m = 3$ and other parameters chosen to fit available experimental data, it was found [37, 38] that the first term on the right of (82) dominated the second, while $D_e R_e^3$ was relatively slowly varying. Then one has $k_e R_e^5$ = const, a relation which fits experimental results, but not as well as

Badger's. The relation

$$k_e R_e{}^5 = 3A(1 + c_{ij}k_e^{1/2}) \tag{83}$$

was derived [37, 38] using an empirical relation between $n$ and $k$, $c_{ij}$ being a constant depending on the rows of the periodic table of the atoms in the molecule. Furthermore,

$$D_e = \frac{A}{R_e{}^m} \frac{c k_e^{1/2}}{m - c k_e^{1/2}} \tag{84}$$

The error in (83) for 75 electronic states of 28 diatomic molecules was about 1.2%, which was better than for any other rule proposed at that time. The accuracy of (84) was, in general, considerably less good.

Wu and Yang [39] gave a similar discussion to Linnett's, using exponential repulsion and inverse-power attraction (Born–Mayer):

$$U(R) = Ae^{-R/\varrho} - BR^{-m} \tag{85}$$

They derived several relations between $k_e$ and $R_e$, one of which proved useful:

$$k_e R_e{}^{m+1} = (-Bm(m + 1)/R_e) + (Bm/\varrho). \tag{86}$$

In fact $k_e R_e^{m+1}$ was found, for a series of molecules, to be linear in $R_e^{-1}$. The value of $m$ was 4 or 6, depending on the periods of the atoms. The other constants, $B$ and $\varrho$, also depend on the periods of the atoms. It was found that (86) was as successful as the Badger rule. Further discussion of work relating empirical rules to the form of $U(R)$ is found in Volume I, Chapter II, Section D.

Lippincott [40, 41] proposed the use of the potential energy function

$$U(R) = D_e(1 - e^{-ns})(1 + af(R)) - D_e \tag{87a}$$

where

$$s = (R - R_e)^2/2R \tag{87b}$$

and the form of $f(R)$ is relatively unimportant except that $f(0)$ is infinite and $f \to 0$ as $R \to \infty$. With $a = 0$ in (87), it is easy to derive

$$D_e = k_e R_e/n \tag{88}$$

so that the three experimental quantities are related once one has a recipe for getting $n$. An empirical formula, in terms of the ionization potentials of the atoms A and B and their positions in the periodic table, was given

[40, 41]. The dissociation energies $D_e$ and bond dissociation energies ($D_e$ minus zero-point vibration), computed from (88), with $n$ calculated according to the recipe and $k_e$ and $R_e$ experimentally determined, were correct to a few percent for a large number of diatomics. Furthermore, the anharmonicity, derived from the cubic and quartic terms of the expansion of (87) with $a = 0$, was predicted to an average of six percent. In terms of the spectroscopic constants, the relation was

$$D_e = \frac{hc}{2} \frac{3\omega_e^2}{8\omega_e x_e - 3Be} .$$

Inclusion of the term, $af(R)$, in $U$ yields a potential function which behaves correctly for $R \to 0$ and gives an improved correlation between $k_e$, $D_e$, $\omega_e x_e$, $R_e$, and $\alpha_e$ (the vibration–rotation coupling constant) [40, 41]. The problem then becomes providing a theoretical justification for the values of $n$ used and a theory or model of chemical bonding which leads to an interatomic interaction of the form of Eq. (87) [42].

Frost and co-workers [24, 25] investigated the semiempirical function

$$U(R) = e^{-aR}(cR^{-1} - b)$$

which was first proposed for $H_2^+$ and $H_2$. The three parameters may be determined when $D_e$, $R_e$, and $k_e$ are known. Then values for $\alpha_e$ (vibration–rotation interaction constant) and $\omega_e x_e$ (anharmonicity) may be predicted with some success. The values of $c$ found in many cases were roughly products of effective charges for the atoms. However, modification was necessary if ionic molecules were to be treated, since $U$ should approach $R^{-1}$ as $R \to \infty$. Also, the London $R^{-6}$ interaction for large $R$ was to be incorporated [24]. But the addition of an attractive $R^{-6}$ term would make $U$ go to $-\infty$ instead of $+\infty$ for small $R$, so further modification was necessary. The resulting function [24] contained a large number of parameters, and was only used for some special cases, where one could simplify it on physical grounds.

The Morse potential also relates [40, 41] $k_e$, the anharmonicity, $D_e$, and $\omega_e$. The relationship

$$D_e = \omega_e^2/4\omega_e x_e$$

was derived by Morse [43]. The values thus predicted for $D_e$ are not in good accord with experiment [41]. In terms of $\bar{\varrho} = R/R_e$, the Morse function may be written:

$$U/D_e = e^{-2\bar{a}(\bar{\varrho}-1)} - 2e^{-\bar{a}(\bar{\varrho}-1)} \tag{89}$$

Since the potential constants are derivatives of $U$ with respect to $R$, evaluated at $\bar{\varrho} = 1$, they are simply proportional to powers of $\bar{a}$. Representing the second, third, and fourth derivatives of $U$ by $k_e$, $l_e$, and $m_e$, we have $k_e = 2\bar{a}^2 D_e/R_e^2$, $l_e = -6\bar{a}^3 D_e/R_e^3$, and $m_e = 14\bar{a}^4 D_e/R_e^4$. Then, for the dimensionless quotient of these,

$$(k_e m_e/l_e^2)^{1/2} = (7/9)^{1/2} = 0.88 \tag{90}$$

which holds reasonably well for most molecules.

Herschbach and Laurie [32] considered extending Badger's rule to correlate cubic and quartic force constants with $R_e$. Badger's rule appears when the logarithm of the quadratic force constant is plotted against $R_e$ for a number of molecules and states, the points tending to fall on straight lines (it should be noted that $R_e$ varies by only about 10% for a group of related molecules). Similar behavior was noted by Herschbach and Laurie for $-l_e$ and for $m_e$. Thus, we may write

$$\log F = -(R_e - a_{ij})/b_{ij} \tag{91}$$

where $F$ is $k_e$, $-l_e$, or $m_e$; $a_{ij}$ and $b_{ij}$ depend on the particular force constant being considered as well as on $i$ and $j$. Agreement for the higher force constants was found [32] to be about as good as Badger's rule for $k_e$. Exceptions were $Li_2$ and some excited states of LiH and $H_2$. It is interesting that $a_{ij}$ differs by only a few percent for the different force constants, as might be expected from its interpretation in terms of the sizes of the cores. In going from one family of molecules to another, the change in $a_{ij}$ could be interpreted in terms of the nature of the cores and how well they screen the nuclei [32]. A qualitative argument was given on the variation with $R$ of the electronic contribution to the force and its derivatives with respect to $R$.

The free-electron model of Borkman and Parr [12] also gives relations between derivatives of $U(R)$ at $R_e$. Equation (70) may be rewritten, using (71), as

$$R_e(U(R) - W_0)/W_1 = (R/R_e)^{-1} - \tfrac{1}{2}(R/R_e)^{-2} \tag{92}$$

According to (92), all derivatives of $U$ with respect to the reduced distance $\bar{\varrho} = R/R_e$ are proportional. The potential constants are proportional to the derivatives with respect to $\bar{\varrho}$. Thus $k_e = -W_1/R_e^3$, $l_e = 6W_1/R_e^4$, and $m_e = -36W_1/R_e^5$ (note that $W_1$ is negative and in units of energy times distance). Combining these, we have

$$(k_e m_e/l_e^2)^{1/2} = 1 \tag{93}$$

for comparison with Eq. (90). The accuracy of (90) and (93) was found to be quite comparable [12]. Borkman and Parr [12] also considered the calculation of the individual anharmonic force constants from $k_e$, using $l_e = -6k_e/R_e$, $m_e = 36k_e/R_e^2$, $n_e = -240k_e/R_e^3$, $o_e = 1800k_e/R_e^4$. For seven molecules for which all these constants were available from experiment, the average percentage errors for $l_e$, $m_e$, $n_e$, and $o_e$ were 22, 39, 47, and 53. According to the Morse potential, Eq. (89), $l_e = -3\bar{a}^4 k_e/R_e$, $m_e = 7\bar{a}k_e/R_e$, $n_e = -15\bar{a}^3 k_e/R_e^3$, and $o_e = 31\bar{a}^4 k_e/R_e^4$. The percentage errors for these, for the same molecules, are [12] 13, 38, 40, and 65, when a value for $\bar{a}$ derived from experiment is used. The rules from the free-electron potential and the Morse potential for $l_e$ and $m_e$ were tested on a number of other molecules with results comparable to those just mentioned. A discussion of the rule that $R_e^2 \omega_e$ should be constant was also given [12].

The proportionality between force constants is a consequence of the fact that, in Eq. (92), the "reduced energy" (interaction energy in dimensionless units) depends only on the "reduced distance" $\bar{\varrho}$. The only difference from molecule to molecule is in the "unit of energy" $W_1/R_e$ and the "unit of distance" $R_e$. Thus Eq. (92) defines a family of two-parameter functions for $U(R)$. In Eq. (89), the reduced energy depends as well on the parameter $\bar{a}$, which varies from molecule to molecule so that (89) defines a three-parameter form for $U(R)$. Equation (92) is a universal potential function, such as will be discussed at the end of this chapter. If a term in $R^{-3}$ is added to Eq. (70), $U(R)$, defined so that $U(\infty) = 0$, is three-parameter and ratios of force constants are no longer simply proportional to powers of $R_e$. Writing

$$U(R)R_e/W_1 = \bar{\varrho}^{-1} + A\bar{\varrho}^{-2} + B\bar{\varrho}^{-3}$$

with $A = W_2/W_1 R_e$ and $B = W_3/W_1 R_e^2$, we have

$$1 + 2A + 3B = 0$$

from the vanishing of $dU/dR$ at $R = R_e$. The resulting relations between force constants may be regarded as improvements on those leading to Eq. (33):

$$l_e = (-6k_e - 3d_e)/R_e; \qquad m_e = (36k_e + 36d_e)/R_e^2$$

where $d_e = -2W_1 B/R_e^3$. Borkman and Parr [12] treated $d_e$ as an adjustable parameter and found that increased accuracy could be obtained from these relations.

Recently, Calder and Ruedenberg [44] considered the correlations which exist between potential constants for 160 different diatomic molecules. They considered the constants in the Dunham expansion

$$U(R) = a_0\varrho^2(1 + a_1\varrho + a_2\varrho^2 + \cdots) - D_e \tag{94}$$

where $\varrho$ is the reduced displacement,

$$\varrho = (R - R_e)/R_e. \tag{95}$$

Note that, while the zero of energy in (94) corresponds to separated atoms, the expansion is valid only for $R < 2R_e$. (The same remark holds for the expansions in $\bar{\varrho}^{-1}$ just considered.) When the centrifugal term $B_e J(J + 1)/(1 + \varrho)^2$ is also expanded as a power series in $\varrho$, the computed energy levels are

$$E(v, J) = \sum_{p,q=0}^{\infty} Y_{pq}(v + \tfrac{1}{2})^p[J(J + 1)]^q$$

where $v$ and $J$ are the vibrational and rotational quantum numbers and the $Y_{pq}$ are expressed in terms of the $a_i$. Calder and Ruedenberg noted that, excluding molecules containing H or Li or formed from two alkali atoms, $a_1$ and $a_2$ were roughly constants. The values were $-3.22 \pm 0.09$ for $a_1$ and $6.44 \pm 0.75$ for $a_2$. For diatomics whose constituent atoms come from specific columns of the periodic table, the constancy of $a_1$ and $a_2$ was even more marked. It was also noted by Calder and Ruedenberg that $a_0$ is roughly constant for certain groups of molecules even though $\omega_e$ (which depends on $B_e$ as well as $a_0$) varies markedly.

Since $k_e$ is proportional to $a_0$, $l_e$ to $a_0 a_1$, $m_e$ to $a_0 a_2$, etc., $a_1 = $ const means $l_e/k_e = $ const, $a_2 = $ const means $m_e/k_e = $ const, etc. This is consistent with the previous discussion. In terms of spectroscopic constants, the empirical relations of Calder and Ruedenberg [44] are: $x_e\omega_e = 9.5B_e$, $\alpha_e\omega_e = 13.2B_e^2$, and $\alpha_e = 1.39x_eB_e$ [the last relation may be written, $4(1 - a_1) = 1.39(a_2 - 5a_1^2/4)$]. To these may be added the approximate proportionality of $\omega_e$ to $B_e$, corresponding to $\omega_e R_e^2 = $ constant [12]. For molecules whose constituent atoms come from specific columns of the periodic table, it was actually found [44] that $\omega_e$ is monotonic with $B_e$, but the slope is quite small for singly bonded molecules. Larger slopes were associated with multiple bonds, particularly molecules, both of whose atoms were from column 5A (N, P, As, Sb, Bi). After testing various functional forms, they found the relation

$$B_e/\omega_e = L_{ij} = K_{ij}\omega_e^{1/2}$$

expressed the correlation well. The values $L_{ij}$ and $K_{ij}$ depend on the columns in the periodic table where the atoms are found [see Eq. (83)].

The correlations mentioned may be combined in various ways to give a variety of formulas. Calder and Ruedenberg tested a number of these. They showed that they could be useful in finding a value for a desired spectroscopic constant unavailable from experiment, and also in detecting errors in reported values for spectroscopic constants. In addition, where one has several sets of potential constants which fit experiment equally well, the correlations may be used to choose between them.

Earlier systematic examination of experimentally measured constants, to find relations between them, was carried out by Varshni [45]. By plotting data for a number of molecules, he derived two approximate relations between three dimensionless combinations of the eight parameters: $\alpha_e$, $\omega_e$, $B_e$, $\omega_e x_e$, $k_e$, $R_e$, $D_e$, and the reduced mass. Additional correlations between potential constants will be mentioned in Subsection 3.

While the preceding discussion emphasized rules which relate one property of $U(R)$ to another, it is clear that many of the formulas can be used to obtain values for properties of $U(R)$ from one molecule when properties of $U(R)$ for other molecules are known. We have, basically, interpolation formulas for series' of related molecules. We now turn to specific schemes for estimating properties of $U(R)$ for one system from their values for another.

Perhaps the simplest such rule is the isoelectronic principle, which states that corresponding states of isoelectronic systems should have the same properties. For example, the ground state of $O_2^+(NO)$ has $\omega_e = 1876$ cm$^{-1}$ (1904), $\omega_e x_e = 16.5$ cm$^{-1}$ (14.0), $B_e = 1.67$ cm$^{-1}$ (1.70), $\alpha_e = 0.0198$ cm$^{-1}$ (0.0178), $R_e = 1.12$ cm$^{-1}$ (1.15). Another pair of systems frequently cited as evidence of the accuracy of the isoelectronic principle is $N_2$–CO.

There are many recipes for computing properties of $U(R)$ for the system AB from those for the systems A–A and B–B ("combination rules"). For the van der Waals attraction, $-CR^{-6}$, a number of approximate formulas have been suggested for $C_{AB}$ in terms of $C_{AA}$ and $C_{BB}$. According to the London formula [Chapter I, Eq. (49)],

$$C_{AA} = \tfrac{3}{4}\alpha_A{}^2 \varDelta_A \tag{96a}$$

and

$$C_{AB} = \frac{3}{2} \frac{\varDelta_A \varDelta_B}{\varDelta_A + \varDelta_B} \alpha_A \alpha_B \tag{96b}$$

so that

$$C_{AB} \approx (C_{AA} C_{BB})^{1/2} \frac{2\varDelta_A^{1/2} \varDelta_B^{1/2}}{\varDelta_A + \varDelta_B}. \tag{97}$$

The quantities $\alpha_A$ and $\alpha_B$ are the atomic polarizabilities. The second factor involves the "mean energy denominators." Equation (97) is an approximation because $\Delta_A$ and $\Delta_B$, defined in terms of $C_{AA}$ and $C_{BB}$, are not really the same mean energies as appear in $C_{AB}$. The energy quotient in Eq. (97) may be approximated as $2I_A^{1/2}I_B^{1/2}/(I_A + I_B)$, assuming ionization potentials are proportional to the mean energies, or, assuming $\Delta_A$ and $\Delta_B$ are about the same size, as unity, giving

$$C_{AB} \approx (C_{AA}C_{BB})^{1/2}.$$

The factor $2(I_A I_B)^{1/2}/(I_A + I_B)$ is necessarily less than or equal to 1, as can easily be shown by putting $I_A = cI_B$ and maximizing the expression with respect to $c$. However, as pointed out by Sristava and Madan [46], it usually does not differ much from unity; for the rare-gas cases they treated, it was never below 0.986. One can also prove that $C_{AB}$ must be less than the geometric mean of $C_{AA}$ and $C_{BB}$ [see Chapter I, Section A, Eqs. (66) et seq.].

It has been noted [47] that Eq. (96b), as well as a number of other approximate formulas [Chapter I, Eqs. (52), (53), and discussion], may be written in the form

$$C_{AB} = \alpha_A \alpha_B/(\beta_A + \beta_B) \tag{98}$$

where $\beta_A$ and $\beta_B$ are constants characteristic of atoms A and B. If (98) is used, with the same value of $\beta_A$, to relate $\alpha_A$ and $C_{AA}$, we have $\beta_A = \frac{1}{2}\alpha_A^2/C_{AA}$. Using this and $\beta_B = \frac{1}{2}\alpha_B^2/C_{BB}$ to eliminate the polarizabilities in (98) gives Eq. (101) below;

$$C_{AB} \approx \frac{2\alpha_A\alpha_B}{(\alpha_A^2/C_{AA}) + (\alpha_B^2/C_{BB})}$$

which works fairly well. The interpretation of $\beta_A$ and $\beta_B$ depends on the derivation of (98), but most derivations [47] suggest that the value of $\beta_A$ to be used in (98) should not depend on the atom B. Other combination rules for $C_{AB}$ will be mentioned later.

Kramer and Herschbach [48] compared a number of combination rules for van der Waals constants and found the above equation to be far superior to the others tested. The average error, considering all known van der Waals constants, was about 3%, which was about the uncertainty in the experimental values. Kramer and Herschbach also pointed out relations between various rules, e.g., the Slater–Kirkwood formula puts $\Delta_A = (N_A/\alpha_A)^{1/2}$ in (96b), where $N_A$ = number of electrons in the outer shell of A, the Kirkwood–Müller formula puts $\Delta_A = \frac{2}{3}\langle r^2\rangle_A/\alpha_A$, and the geometric mean law follows from the above equation if $\alpha_A^2/C_{AA} = \alpha_B^2/C_{BB}$.

Among the combination rules for other constants entering potential curves, the geometric mean rule is often used for well depths and other energies, and the arithmetic mean rule for equilibrium internuclear distances and other distances. The latter rule,

$$R_{AB} = \tfrac{1}{2}(R_{AA} + R_{BB}) \qquad (99)$$

holds in the hard-sphere model, in which the repulsive interaction is characterized by a single parameter (the hard sphere radius $r_A$) for each atom, and $R_{AB} = r_A + r_B$. The geometric mean rule for energies is suggested by the van der Waals constant combination rule. In the Lennard-Jones potential,

$$U = 4\varepsilon[(\sigma/R)^{12} - (\sigma/R)^6]$$

the equilibrium internuclear distance is $2^{1/6}\sigma$ and the depth of the well is $\varepsilon$. Since $\sigma$ is the collision diameter, or distance of closest approach for particles of zero energy, we could expect $\sigma_{AB} = \tfrac{1}{2}(\sigma_{AA} + \sigma_{BB})$, which implies Eq. (99). The coefficient of $R^{-6}$, $4\varepsilon\sigma^6$, may be equated to the van der Waals coefficient, for which the approximation of Eq. (97) may be used, with $\Lambda_A$ replaced by $I_A$. Then

$$\varepsilon_{AB} = \frac{C_{AB}}{4\sigma_{AB}^6} = \frac{(C_{AA}C_{BB})^{1/2}}{2\sigma_{AB}^6} \frac{I_A^{1/2}I_B^{1/2}}{I_A + I_B}$$

$$= \frac{(\varepsilon_{AA}\varepsilon_{BB})^{1/2}\sigma_{AA}^3\sigma_{BB}^3}{\sigma_{AB}^6} \frac{2I_A^{1/2}I_B^{1/2}}{I_A + I_B} \qquad (100)$$

If ionization potentials are about the same size and the internuclear distances obey a *geometric* mean law, $\varepsilon_{AB} = (\varepsilon_{AA}\varepsilon_{BB})^{1/2}$. Usually, the geometric mean and the arithmetic mean are not very different, so that the use of the geometric mean for $\varepsilon_{AB}$ is not inconsistent with the arithmetic mean for $\sigma_{AB}$. Hudson and McCoubrey [49] suggested, however, that one make a correction for the difference between the two means by writing,

$$\varepsilon_{AB} \approx 2^6(\varepsilon_{AA}\varepsilon_{BB})^{1/2}\sigma_{AA}^3\sigma_{BB}^3/(\sigma_{AA} + \sigma_{BB})^6$$

where (99) has been used. The factor involving ionization potentials in (100) was not corrected.

Some justification for the geometric mean rule for energies is found in the early paper of Hirschfelder and Roseveare [50]. Guggenheim and Mc-Glashan [51] and Mason [52] have referred to this rule as the "most reasonable." In some early work, Pauling [53] used a geometric mean rule for the "covalent component" of single bond energies. Writing the energy of a heteronuclear bond as $W_cE_c + W_iE_i$ (covalent and ionic components with $W_c + W_i = 1$), he put $E_c$ for A–B equal to the geometric mean of the bond energies of A–A and B–B. It was assumed that there was no ionic contribution for the homonuclear bonds. Sanderson [54] has argued that the covalent component of the energy of A–B should be corrected by multiplication by $R_c/R_e$, where $R_c$ is the sum of covalent radii of A and B. This is analogous to going from a geometric mean rule for $C_{AB}$ to one for $\varepsilon_{AB}$.

Using the Lennard-Jones potential with a geometric mean rule for $\varepsilon$ and an arithmetic mean rule for $\sigma$, Johnston [55] extended Badger's rule to rare gas interactions. The force constant for the Lennard-Jones potential is $72\varepsilon/2^{1/3}\sigma^2$. When $\varepsilon_{AB}$ and $\sigma_{AB}$ were computed by the combination rules, the straight lines of plots of $k_e^{-3}$ versus $R_e$ also ran close to the points for the noble gases. For instance, the following systems fall on one line: CO, $N_2$, C=C, C≡C, C=N, C–N, C–F, NO, O=O, BN, O—O, $F_2$, C—F, LiF, BeO, NeNe (included are diatomics and bonds in polyatomics). The only exceptions were the dialkalis.

There are other ways [56] of deriving combination rules for the well depth of the Lennard-Jones potential, leading to slightly different formulas. Instead of eliminating the polarizabilities between the London formulas (96a) and (96b), which led to Eq. (100), we can eliminate the mean energies or ionization potentials. Then

or

$$C_{AB} \approx \frac{2C_{AA}C_{BB}\alpha_A\alpha_B}{\alpha_B{}^2C_{AA} + \alpha_A{}^2C_{BB}} \qquad (101)$$

$$\varepsilon_{AB} \approx \frac{2\varepsilon_{AA}\varepsilon_{BB}\alpha_A\alpha_B\sigma_{AA}^6\sigma_{BB}^6}{\varepsilon_{AA}\sigma_{AA}^6\alpha_B{}^2 + \varepsilon_{BB}\sigma_{BB}^6\alpha_A{}^2}\,\frac{1}{\sigma_{AB}^6}. \qquad (102)$$

Either the arithmetic or the geometric mean rule may then be used for $\sigma_{AB}$. It is reasonable to assume that $\alpha_A/\sigma_{AA}^3$ and $\alpha_B/\sigma_{BB}^3$ are about the same, since the polarizability is often approximated by the atomic volume. Then, with the geometric mean rule for $\sigma_{AB}$, Eq. (102) becomes

$$\varepsilon_{AB} \approx 2\varepsilon_{AA}\varepsilon_{BB}/(\varepsilon_{AA} + \varepsilon_{BB}) \qquad (103)$$

This equation, which states that the reciprocals of $\varepsilon_{AB}$ follow an arithmetic mean law, is often invoked [56]. Equation (101) may be rewritten:

$$\frac{C_{AB}}{\alpha_A\alpha_B} \approx \frac{2(C_{AA}/\alpha_A{}^2)(C_{BB}/\alpha_B{}^2)}{(C_{AA}/\alpha_A{}^2) + (C_{BB}/\alpha_B{}^2)}$$

which is of the form of (103), with $C_{AB}/\alpha_A\alpha_B$ inserted for $\varepsilon_{AB}$. This relation holds remarkably well [56]. The Kirkwood–Müller approximation for the Van der Waals coefficient (Chapter I, Eq. (52)], is

$$C_{AB} \approx \frac{6mc^2\alpha_A\alpha_B}{(\alpha_A/\chi_A) + (\alpha_B/\chi_B)}$$

with $\chi_A$ equal to the diamagnetic susceptibility of atom A. This may also be used [57] to obtain combination rules. Eliminating $\alpha_A$ and $\alpha_B$ from the expressions for $C_{AA}$, $C_{BB}$, and $C_{AB}$ yields

$$C_{AB} \approx \frac{2C_{AA}C_{BB}\chi_A^{-1}\chi_B^{-1}}{C_{AA}\chi_A^{-2} + C_{BB}\chi_B^{-2}} \tag{104}$$

In conjunction with combination rules for $\sigma_{AB}$, this yields combination rules [58] for $\varepsilon_{AB}$:

$$\varepsilon_{AB} = \frac{2\varepsilon_{AA}\varepsilon_{BB}\sigma_{AA}^6\sigma_{BB}^6\chi_A^{-1}\chi_B^{-1}}{\varepsilon_{AA}\sigma_{AA}^6\chi_A^{-2} + \varepsilon_{BB}\sigma_{BB}^6\chi_B^{-2}} \frac{1}{\sigma_{AB}^6} \tag{105}$$

If $\chi_A \approx \chi_B$ and $\sigma_{AA} \approx \sigma_{BB}$, this also leads to (103).

Good and Hope [59] have criticized the use of the arithmetic mean (39) and suggested that the geometric mean be used instead:

$$R_{AB} = (R_{AA}R_{BB})^{1/2}$$

Their argument was based on the combining rules derived, by arguments like those we have given, for the constants A and C in the potential

$$U = -AR^{-n} + CR^{-m}$$

of which the Lennard-Jones potential is a special case. In terms of the equilibrium internuclear distance $R_e$ and the well depth $\varepsilon$,

$$A = [m/(m-n)]R_e^n\varepsilon \quad \text{and} \quad C = [n/(m-n)]R_e^m\varepsilon$$

so that we may derive combining rules for A and C from those for $R_e$ and $\varepsilon$. According to these equations, if any two of the four constants $A$, $C$, $R_e$, and $\varepsilon$ follow a geometric mean law, so do the other two. Since $A$ and $C$ are both energy parameters and thus probably obey a geometric mean law, and since the approximations of $C_{AB}$ by $(C_{AA}C_{BB})^{1/2}$ and of $\varepsilon_{AB}$ by $(\varepsilon_{AA}\varepsilon_{BB})^{1/2}$ both seem to hold fairly well, Good and Hope [59] suggested one should use

$$R_{AB} = (R_{AA}R_{BB})^{1/2} \tag{106}$$

and showed that it generally works better than (99). For the exponential-six potential, Eq. (107), a geometric mean rule for $A$ and $C$ and the assumption that $p$ is constant implies [59] that other constants obey geometric mean rules. Other possible combining rules were briefly discussed [56, 59].

The combining rules for the exponential-six potential had been previously discussed, from a semiempirical point of view, by Mason [52]. He wrote Eq. (85) with $m = 6$, for the molecule AB, as

$$U_{AB}(R) = A_{AB}e^{-R/p_{AB}} - C_{AB}R^{-6} \qquad (107)$$

and similarly for $U_{AA}$ and $U_{BB}$. According to the previous paragraphs, $C_{AB}$ is approximated as the geometric mean of $C_{AA}$ and $C_{BB}$. By analogy, Mason [52] approximated $A_{AB}$ as the geometric mean of $A_{AA}$ and $A_{BB}$, and previous approximate work suggested

$$p_{AB}^{-1} = \tfrac{1}{2}(p_{AA}^{-1} + p_{BB}^{-1}).$$

The resulting formula,

$$U_{AB}(R) = (A_{AA}A_{AB})^{1/2}\exp[-R(p_{AA}^{-1} + p_{BB}^{-1})/2] - (C_{AA}C_{BB})^{1/2}R^{-6}$$

may be compared with

$$U(R) = D_e(1 - 6/\alpha)^{-1}[(6/\alpha)e^{\alpha(1-R/R_e)} - (R_e/R)^6]$$

to give combination rules for $D_e$, $R_e$, and $\alpha$. For mixtures of rare gases, these rules worked fairly well.

Mason [52] also suggested that it is reasonable to expect $R_e$ for a molecule AB to be the arithmetic mean of the values for AA and BB, and $D_e$ for AB to be the geometric mean of the values for the AA and BB molecules. For a two-parameter potential like the 6–12, these rules suffice to give $U_{AB}$ when $U_{AA}$ and $U_{BB}$ are known. The resulting potentials, for mixtures of rare gases, accounted for properties at ordinary temperatures only slightly less well than did the exponential-six potentials (three-parameter).

There have been a large number of attempts, not all of which we can mention, to test and compare the combining rules for the Lennard-Jones and exponential-six potentials, using values of the parameters derived from experiment. Sristava and Sristava [60] criticized the Mason–Rice combination rules for the exponential-six potential [52, 61, 61a]. In (107), Mason and Rice suggested a geometric mean rule for $C$, an arithmetic mean rule for $1/p$, and a geometric mean rule for $A$. Since their experimental data showed inadequacies in these rules, Sristava and Sristava attempted to replace the third by a geometric mean rule for $\varepsilon$ or an arithmetic mean rule for $\sigma$ in

$$U(R) = \frac{\varepsilon\alpha}{6-\alpha}\left[\frac{6}{\alpha}e^{\alpha(1-R/\sigma)} - \left(\frac{\sigma}{R}\right)^6\right]$$

They found the former worked better. Sristava and Madan [46] derived $\varepsilon_{AA}$, $\varepsilon_{BB}$, $\varepsilon_{AB}$, $\sigma_{AA}$, $\sigma_{BB}$, and $\sigma_{AB}$ for the Lennard-Jones potential by using it to interpret diffusion and thermal diffusion data for rare gases and other molecules. They compared the values of $\varepsilon_{AB}$, computed from Eq. (100),

$$\varepsilon_{AB}\sigma_{AB}^6 = 2(\varepsilon_{AA}\varepsilon_{BB}\sigma_{AA}^6\sigma_{BB}^6)^{1/2}(I_A I_B)^{1/2}/(I_A I_B)$$

with "measured" values, and concluded that it worked well. The "measured" values of $\sigma_{AA}$, $\sigma_{BB}$, and $\sigma_{AB}$ were used. Using the experimental $\varepsilon_{AB}$, they tested this equation as a combining rule for $\sigma_{AB}$, and found it worked better than the arithmetic mean law.

Weissman, Saxena, and Mason [62] gave a similar analysis, discussing the exponential-six potential as well as the Lennard-Jones. For the exponential-six potential in the second form used above, they found that an arithmetic mean rule for $\sigma$ and $\alpha$ worked as well as an arithmetic mean rule for $p^{-1}$ and a geometric mean rule for C in (107). They concluded that the combination rules worked as well as do the models themselves, i.e., the exponential-six could be made to fit all the data, even using the combination rules, while the Lennard-Jones could not [62]. This conclusion was reaffirmed subsequently by Mason et al. [58] who stated that no combination rule could work well if the Lennard-Jones potential was used to interpret their experimental data. Fender and Halsey [57] tested combining rules (100) (using ionization potentials), (103), (105), and the geometric mean rule, for $\varepsilon_{AB}$ of the Lennard-Jones potential, derived from second virial coefficients of Ar and Kr. The arithmetic mean rule was used for $\sigma_{AB}$ throughout. They found that (103) worked best. Saxena and Mathur [62a] pointed out that the experimental data used by previous workers could lead to erroneous conclusions about the combination rules. If values for the parameters chosen to fit experiment were used to test the rules, neither the Lennard-Jones nor the exponential-six results were very good. However, the values of the parameters are sensitive to small changes in experimental results. Saxena and Mathur argued that, in fact, parameters derived using combination rules for either potential worked as well as did experimentally derived parameters in correlating measurements of diffusion and other properties.

Recently, Lin and Robinson [63] considered combination rules for the energy and length parameters required to treat mixtures of rare gases by a reduced potential of other than Lennard-Jones form (see Subsection 3 for a discussion of reduced potentials). In

$$U(R) = \varepsilon V(R/\sigma), \tag{108}$$

Lin and Robinson used, for $V$, a numerical potential derived for Ar–Ar by Dymond and Alder [64]. Once $V$ is chosen, Eq. (108) is a two-parameter potential function. Virial coefficient data were fit [63] using the Dymond–Alder potential to derive $\varepsilon$ and $\sigma$ for various homonuclear and heteronuclear pairs of rare gas atoms. Then six different combination rules for $\varepsilon_{AB}$ were considered: the geometric mean law, Eq. (102), Eq. (100), Eq. (100) with the ionization potential factor set equal to unity, Eq. (103b), and

$$\varepsilon_{AB} = (\sigma_{AB})^{-3}(\varepsilon_{AA}\varepsilon_{BB}\sigma_{AA}^3\sigma_{BB}^3)^{1/2}.$$

For $\sigma_{AB}$, both "experimental" values and arithmetic means of $\sigma_{AA}$ and $\sigma_{BB}$ were used. None of the rules was wholly satisfactory; the best was the second, except for systems involving He. Values of the correction factor,

$$k_{AB} = 1 - (\varepsilon_{AB}^2/\varepsilon_{AA}\varepsilon_{BB})^{1/2}$$

were tabulated [63]. Good and Hope [56] compared the arithmetic and geometric mean formulas for $\sigma_{AB}$ or $R_{AB}$ in intermolecular interactions, and concluded that the geometric mean was superior. The potential used was the Lennard-Jones potential, with a variety of combining rules, like those used by Lin and Robinson [63], for $\varepsilon$.

Calvin and Reed [65] considered combination rules for

$$U(R) = BR^{-n} - CR^{-6} \tag{109}$$

of which the Lennard–Jones potential is a special case. For $C$ they used (97) with ionization potentials. They suggested that a geometric mean law should hold for the repulsive interaction, implying $n_{AB} = \frac{1}{2}(n_{AA} + n_{BB})$ and $B_{AB} = (B_{AA}B_{BB})^{1/2}$. These in turn led to combination rules for the well depth and the value of $R$ for which $U = 0$, which were tested on the rare gases and seemed to work well.

For the rare gas interactions, one can use the van der Waals coefficient from experiment or combination rules, combined with parameters for the repulsive interaction from Hartree–Fock calculations at small and intermediate $R$, to get $U(R)$. This approach was investigated by Nesbet [66] and Konowalow [66a], with mixed success.

For other systems, Gilbert [67] suggested combination rules could be used for the hard-core repulsion. He took $U(R)$ for the alkali halides as

$$U_{AB}(R) = -R^{-1} - A_{AB}R^{-4} + B_{AB}\exp(-R/p_{AB}) \qquad (110)$$

and determined the constants from spectroscopic data. He defined $R_{AB} = p_{AB}\log(B_{AB}/fp_{AB})$ with $f$ fixed ($f$ represents a standard force to compress $R$ to $R_{AB}$) and assumed $R_{AB}$ would be a sum of atomic contributions ("soft-sphere radii"), and that $p_{AB} = \frac{1}{2}(p_{AA} + p_{BB})$. The choice $f = 0.14$ a.u. minimized the average deviation of $R_{AB}$ from measured alkali halide distances in crystals. Then $R_{AA}$ and $p_{AA}$ were determined from SCF calculations and used to generate $R_{AB}$ and $p_{AB}$ via the combination rules. The results deviated from the measured values from (110) by only a few percent.

Despite the fact that the arguments of the preceding paragraphs build on expressions for the van der Waals constants and the Lennard-Jones 6–12 potential, the combination rules are often invoked for diatomic systems other than the rare gases. Even though the attractive and repulsive forces may be quite different, it is common to approximate the equilibrium internuclear distance of the diatomic molecule A–B by the average of the A–A and B–B distances, and the well depth or binding energy of A–B by the geometric mean of the A–A and B–B binding energies.

A modification of the exponential-six potential (107) suitable for alkali halide molecules is

$$U(R) = c'R^{-1} + be^{-aR} - dR^{-6}$$

the additional term representing the electrostatic attraction of ions ($c' < 0$). Frost and Woodson [24] assumed that $a$, $b$, and $d$ could be obtained from the exponential-six potentials for rare gases, since they arise basically from the atomic cores. They further assumed that the values of $a$, $b$, and $d$ for heteronuclear rare gas pairs could be calculated by Mason's combining rules from values for homonuclear pairs. The potential curves thus generated for the alkali halides predicted $R_e$ and $D_e$ quite well (see Subsection 1).

Redington [68] discussed combination rules for the valence repulsion term in $U(R)$ for ionic molecules. Writing this term as a sum of exponentials,

$$\sum_j A_j e^{-R/j\varrho}$$

he found correlations between the $A_j$ and parameters which represented the effective charges and sizes of the ions.

Recently, there has been some interest in the correlation of potential constants with electronegativities. Of course, one of the original definitions of electronegativity [53] was as a correlation between dissociation energies of heteronuclear molecules and dissociation energies of the corresponding homonuclear molecules. Pauling [69] suggested that the difference between $D_e$ for A–B and the mean of $D_e$'s for A–A and B–B could be expressed as $(X_A - X_B)^2$, where $X_A$ is the electronegativity of atom A in suitable units. (At least as recently as 1971, this kind of relation was found useful [70] for a new class of compounds: the rare-earth aurides.) Thus a correlation of $k_e$ with Pauling's electronegativity implies a correlation between $k_e$ and $D_e$, possibly involving several molecules.

There are other scales of electronegativity, such as Mulliken's, which makes the electronegativity of an atom the mean of its electron affinity and its ionization potential. The different scales are more or less in agreement. Gordy found [71] empirically that force constants correlated with Mulliken electronegativities according to

$$k_e = aN\{X_A X_B R_e^{-2}\}^{3/4} + b. \tag{111}$$

Here, $X_A$ is the Mulliken electronegativity of atom A, $N$ the bond order of the molecule AB, and $a$ and $b$ are empirical constants. Ferreira demonstrated [72] a linear relation between $k_e$ and $X/R_e^2$ for the hydrogen halides, $X$ referring to the halogen.

Much modern work on electronegativity has related to fitting the electronegativity concept into the molecular orbital and valence bond theories and *ab initio* calculations. Ferreira [73] has given a review of such work. The idea of electronegativity as a property of the isolated atom has given way to a more dynamic concept, in which electronegativity is a function of charge or occupation number within the molecule. Electron transfer occurs until electronegativities are equalized. Using the principle of electronegativity equalization, Ferreira [72] gave an interpretation of Pauling's rule that the difference between the bond energy of a heteropolar molecule and the mean of the corresponding homopolar bond energies is proportional to the square of the electronegativity difference. He derived

$$D_{AB} = (D_{AA} D_{BB})^{1/2} + \Phi + A_A(2q - q^2) - I_B q^2$$

and showed it was correct to 2% for 14 hydrogen halides and alkali halides. Here, $A$ is the electron affinity, $I$ the ionization potential, $q$ the partial charge transferred, and $\Phi$ the electrostatic energy between the ions $B^{+q}$ and $A^{-q}$.

Hussain [74] noted that, for homonuclear molecules formed from a particular group in the periodic table, one could write

$$\omega_e = qX\mu^{-p} \tag{112}$$

with $p$ and $q$ constants, $X$ the electronegativity, and $\mu$ the reduced mass. Previous authors [74] had given correlations of $\omega_e$ for homonuclear diatomic molecules with atomic weight, atomic number, and atomic ionization potential. Like (112), these relations include one or more constants characteristic of the locations of the atoms in the periodic table, so they actually relate $\omega_e$ values for molecules formed from chemically related atoms. The electronegativity formula seemed [74] to work best overall. When (112) was generalized [75] to heteronuclear diatomics by using the geometric mean of the electronegativities of the two atoms, good agreement with experimental values was obtained (often to 1% or better). This constitutes a relation between $\omega_e$ for AB and $\omega_e$ for the homonuclear pairs.

For the diatomic halides, a correlation between the rotational constant $B_e$ [proportional to $(\mu R_e^2)^{-1}$] and the electronegativity,

$$\log B_e = m \log(X/\mu^2) + C$$

has been put forward [76]. The constants $m$ and $C$ depend on the group of the nonhalide. Average deviations of 1.6% within a group were reported.

Van Hooydonk and Eeckhardt [77] have discussed various semiempirical relations of molecular properties to electronegativity. By studying the relation of electronegativity to Hückel-type calculations, they attempted to correct Pauling's scale. They derived

$$D_{AB} = \tfrac{1}{2}(X_A + X_B)(1 + i^2), \tag{113}$$

where $i$ is an ionicity parameter, equal to $b^2 - a^2$ when the valence electrons were represented by $a\phi_A + b\phi_B$. It was argued that $i = (X_B - X_A)/(X_A + X_B)$. Then, combination of the formula for $D_{AB}$ with Sutherland's relation [36], $D_{AB} = k_{AB}R_e^2$, and other formulas suggested that $k$ should be linear in $(X_A X_B)^{1/2}R_e^{-2}(1 + i^2)(1 - i^2)^{-1/2}$. Equation (113), applied to 29 diatomics, gave errors of about 10%, although it was admitted that there were problems with the derivation of the formula [78]. The agreement was as good as with a previously proposed [79] improvement on Pauling's formula, giving bond energies in terms of electronegativities.

Finally, we mention that the simple valence bond theory may be used to establish correlations for $U(R)$ for different states of the same molecule, as was suggested by Vanderslice *et al.* [7]. They invoked the perfect pairing formula for the interaction energy:

$$U(R) = \sum_{\text{paired}} J_{ij} - \tfrac{1}{2} \sum_{\text{nonpaired}} J_{ij} - \sum_{\text{parallel}} J_{ij}. \tag{114}$$

Here, $J_{ij}$ is the (negative) exchange integral between orbitals $i$ and $j$, the first sum is over pairs of orbitals whose spins are paired (antiparallel), the last over pairs of orbitals whose spins are parallel, and the other over pairs of orbitals whose spins are unpaired. The Coulomb integral, which is relatively small, has been omitted from (114). Considering the molecular states formed from a particular set of atomic configurations, $U(R)$ for each state will be a linear combination of the same exchange integrals, the coefficient of a particular $J_{ij}$ depending on the structure of the valence bond function for the state. Thus the $J_{ij}$ may be obtained from known $U(R)$ for a few states and used to calculate $U(R)$ for the others. Vanderslice *et al.* [7], considering the six states of NO which dissociate to ground state atoms, were able to express $U(R)$ for four of these in terms of $U(R)$ for the other two. Here, there were only two different nonvanishing exchange integrals. The same situation obtained for $O_2$. A discussion was given of how the necessary information regarding valence bond functions could be extracted from the molecular orbital descriptions of the states. The results for the repulsive states of $O_2$ were in agreement with experimental data. This treatment is valid only for large values of $R$ [7]. Recently, Stallcup [80] gave a similar treatment for the states of $N_2^+$ which correlate with ground state atoms.

### 3. *Reduced Potential Curves*

A general solution to the problem of deriving information on one potential function from those for other systems would be in hand if there existed a general reduced potential function for all chemically related diatomic systems. The existence of a reduced potential would mean that one could write

$$U(R) = \varepsilon V'(\varrho) \tag{115}$$

where $\varepsilon$ is an energy characteristic of a molecule, $\varrho$ a function of $R$, and $V'$ a universal function. The factor of $\varepsilon$ in (115) gives the scale of the ordinate in the plot of $U$ versus $R$, and the use of $\varrho$ (usually a linear function of $R$) gives the scale of the abscissa. The assumption is that, aside from these factors, the shape of $U(R)$ is the same for related systems. Of course, $U(R) \to \infty$ for $R \to 0$ and $U(R) \to 0$ for $R \to \infty$ for all molecules.

For the rare gases and spherical molecules, the notion of a universal or reduced potential has played an important role in theoretical and experimental developments. It is linked to the notion of corresponding states, although the principle of corresponding states arose historically from the van der Waals equation. Pitzer [61] gave the conditions necessary for the existence of a principle of corresponding states. It is required that, in (115), the argument of $V'$ be of the form

$$\varrho = R/R^* \tag{116}$$

with $R^*$ characteristic of the particular diatomic system. If equal numbers of molecules A and B are enclosed in containers with linear dimensions proportional to $R_A^*$ and $R_B^*$, the two collections of molecules correspond as far as spatial dimensions are concerned. Since kinetic energies are proportional to temperature, correspondence in energy is attained when the temperatures are in the ratio, $\varepsilon_A/\varepsilon_B$. Then the two collections of molecules should have similar behavior, i.e., be in corresponding states. For instance, one can show [51]

$$B/V^* = \phi(T/T^*)$$

where $B$ is the second virial coefficient, $\phi$ is a universal function, and $V^*$ and $T^*$ are characteristic of the diatomic systems. It may be shown that $T_{AA}^*/\varepsilon_{AA} = T_{BB}^*/\varepsilon_{BB}$, and $V_{AA}^*/(R_{AA}^*)^3 = V_{BB}^*/(R_{BB}^*)^3$. Pitzer [61] gave a mathematical derivation of the corresponding states principle from (115) and (116) which, like the preceding argument, also assumed that classical mechanics was valid.

Guggenheim and McGlashan [51] suggested that, for a mixture of A and B, we should assume Eqs. (115) and (116) and define the characteristic temperature and volume for the mixture by

$$T_{AB}^*/\varepsilon_{AB} = T_{AA}^*/\varepsilon_{AA}; \qquad V_{AB}^*/(R_{AB}^*)^3 = V_{AA}^*/(R_{AA}^*)^3.$$

Then it follows that

$$B_{AB}/V_{AB}^* = \phi(T/T_{AB}^*).$$

If an arithmetic mean law was used for $R_{AB}^*$ and a geometric mean law for $\varepsilon_{AB}$ (see the preceding section), the virial coefficients seemed to follow the law reasonably well.

The Lennard-Jones potential is clearly of the form of equations (115) and (116), and is the most commonly used for rare gases and other spheri-

cally symmetric molecules. Recently, Lin and Robinson [63] proposed
that an accurate numerical potential, derived from experiments on Ar–Ar
by Dymond and Alder [64], be used to describe the interactions of all pairs
of rare-gas atoms via the scaling represented by (115) and (116). They
considered the accuracy with which virial coefficient data could be fit,
and found that the two-parameter function represented by (115) and (116)
worked much better than the two-parameter Lennard-Jones function and
slightly better than the three-parameter Kihara potential:

$$U(R) = 4\varepsilon\left[\left(\frac{\sigma - 2a}{R - 2a}\right)^{12} - \left(\frac{\sigma - 2a}{R - 2a}\right)^{6}\right], \qquad R \geq 2a$$

$$U(R) = \omega, \qquad\qquad\qquad\qquad\qquad\qquad\qquad R \leq 2a$$

(This potential seems to work well for rare-gas systems [81].) The values
of $\varepsilon$ and $\sigma$ for various pairs of rare-gas atoms were derived and discussed
(see the previous subsection).

Other potentials of the form of Eqs. (115) and (116) are Eq. (92), the
Morse potential [Eq. (89) if $a$ is fixed], and Eq. (94) if $a_1$ and $a_2$ are universal
constants. These functions may describe systems with chemical bonds. The
use of universal reduced potentials for such systems is not as widespread
as for rare gases.

Frost and Musulin [82] considered the existence of universal functions
$V'$ with $\varrho$, referred to as a reduced internuclear distance, given by

$$\varrho = (R - R_{ij})/(R_e - R_{ij}). \qquad (117)$$

Here, $R_{ij}$ depends on the atoms composing the molecule, and was supposed
to take into account the inner-shell repulsion. Thus, it should be zero for
$H_2^+$ and $H_2$. The value of $\varrho$ is proportional to the distance between the
hard cores and is unity for $R = R_e$. While negative values of $\varrho$ are not
excluded, $V'$ should become very large when $\varrho$ goes below zero. The $V'$
curves for $H_2^+$ and $H_2$ could be made to coincide fairly well [82]. If $V'(\varrho)$
is universal, $U(R)$ for any molecule could be derived from the potential
curves for $H_2^+$ or $H_2$, except for the values of $R_e$, $R_{ij}$, and $D_e$.

Frost and Musulin defined

$$(d^2V'/d\varrho^2)_{\varrho=1} \equiv \varkappa \qquad (118)$$

$\varkappa$ being a dimensionless (and presumably universal) constant. Since $\varkappa = 1$
corresponds to $R = R_e$, $\varkappa$ is related to the force constant $k_e$. Assuming
$k_e$ and $D_e$ to be known, one can find $R_{ij}$ by inserting equations (115) and

(117) into (118). The result is

$$R_{ij} = R_e - (\varkappa D_e/k_e)^{1/2}. \tag{119}$$

If the theory is correct, it should be possible to choose a single value for $\varkappa$ such that, for different molecules, $R_{ij}$ depends only on the cores of the atoms, and is zero for $H_2^+$ and $H_2$. In the first test of the theory, $\varkappa$ was computed as $k_e R_e^2/D_e$ for $H_2^+$ and $H_2$ and essentially the same value was obtained for both (3.96 and 4.14). Taking $\varkappa = 4$, $R_{ij}$ was calculated for a number of molecules. The values, which agreed roughly with Badger's $d_{ij}$ [Eqs. (75) and (76)], indeed seemed to be characteristic of the rows of the periodic table where the constituent atoms of the molecule were found, i.e., of the cores of the atoms.

A second test of the universal potential function involved the higher terms in the expansion of $V'$ in a Taylor series in $\varrho$ about $\varrho = 1$. The force constant $k_e$ is proportional to the term in $(\varrho - 1)^2$, and $l_e$ and $m_e$ to the terms in $(\varrho - 1)^3$ and $(\varrho - 1)^4$. We have, in general

$$\left(\frac{d^n V'}{d\varrho^n}\right)_{\varrho=1} = \frac{(R_e - R_{ij})^n}{D_e} \left(\frac{d^n U}{dR^n}\right)_{R=R_e}$$

Using the $R_{ij}$ values they had obtained, Frost and Musulin [82] computed, for a number of molecules,

$$L = (d^3 V'/d\varrho^3)_{\varrho=1} \quad \text{and} \quad M = (d^4 V'/d\varrho^4)_{\varrho=1},$$

using experimental data on $l_e$ and $m_e$. The values $L$ and $M$ were constant to 13.2% and 42.0%, respectively (this has since been noted by several authors—preceding section). Since the derivatives themselves vary over about two orders of magnitude, this result is useful for obtaining qualitative estimates of $l_e$ and $m_e$. However, the results also imply that equation (115) does not represent reality to high accuracy.

In his review article [45], which discusses the merits of a large number of forms for $U(R)$ for diatomic molecules, Varshni considered the existence of a universal potential energy function. He plotted, for a number of molecules, the dimensionless quantities

$$F = \frac{\alpha_e \omega_e}{6B_e^2} = -\left[1 + \frac{1}{3} \frac{U'''(R_e)}{U''(R_e)} R_e\right]$$

and

$$G = \omega_e x_e R_e^2 \mu_A = c\left[\frac{5}{3}\left(\frac{U'''(R_e)}{U''(R_e)} R_e\right)^2 - \frac{U''''(R_e)}{U''(R_e)} R_e^2\right]$$

against

$$\Delta = k_e R_e^2 / 2 D_e .$$

Here, $c$ is a combination of physical constants, $\mu_A$ is the reduced mass in atomic weight units, and primes on $U$ represent derivatives with respect to $R$. If a universal potential function existed, it would yield universal relations between $F$ and $\Delta$ and between $G$ and $\Delta$. In the plots, however, the points for the different molecules scattered without falling on a smooth curve. Thus any general relation that exists must be extremely complicated, and a universal potential curve would seem to be ruled out, except as a rough approximation. It is possible to draw straight lines through the points: $F = 0.11\Delta + 0.36$, $G = 5\Delta + 9$, which may be of some use in estimation of unknown constants. There is the possibility that a single reduced potential function could describe a group of similar molecules, but the available data were insufficient to confirm or deny this.

Jenč [83] has taken up the possibility of using reduced potential curves to study diatomic molecules systematically. In Eq. (116), the definition of the reduced displacement was changed to

$$\varrho = \frac{R - \varrho_{AB}(1 - e^{-R/\varrho_{AB}})}{R_e - \varrho_{AB}(1 - e^{-R/\varrho_{AB}})} = 1 + \frac{R - R_e}{R_e - \varrho_{AB}(1 - e^{-R/\varrho_{AB}})} \tag{120}$$

where

$$\varrho_{AB} = \frac{R_e - (\varkappa D_e / k_e)^{1/2}}{1 - e^{-R_e/\varrho_{AB}}} \tag{121}$$

with $\varkappa = 3.96$. In (121), the spectroscopic constants have their usual meanings and $\varrho_{AB}$ is calculable and different for each molecule AB. Values of $\varrho_{AB}$ were tabulated by Jenč. The quantity $\varkappa$ is the universal reduced force constant of Eq. (118). The Frost–Musulin choice of $\varrho$, Eq. (117), was modified [84] so that $\varrho$ would always be positive and approach zero only for $R \to 0$, but still equal 1 for $R = R_e$. With the "natural" [84] assumption that $R_{ij}$ of (117) should decrease exponentially with $R$, we obtain (120), while (121) follows if we require $d^2 V' / d\varrho^2 = \varkappa$.

The reduced potential curves of the hydrides and of other groups of molecules composed of similar atoms (e.g. $O_2$, $N_2$, CO, and NO) coincided closely, the coincidence being much better than for the Frost–Musulin functions. Just as did the Frost–Musulin definition, Jenč's formulation leads to a relation between potential constants, but it is more complicated:

$$k_e / D_e [R_e - \varrho_{AB}(1 - e^{-R_e/\varrho_{AB}})]^2 = \varkappa \tag{122}$$

This may be used to derive values for potential constants, if other potential constants are known and if the reduced potential curve (RPC) has been constructed for related molecules, by assuming the RPC's coincide. The correct value of the unknown constant is the one that makes them do so. For example, in order to get a good coincidence between the RPC of CN and RPC's for other diatomics formed from C, O, and N, Jenč had to use 8.2 eV for the dissociation energy of CN, which was one of several possible values derived from experiment [85]. Furthermore, since the calculated RPC's coincide well for hydrides, one can derive a good approximation to the potential curve for any diatomic hydride (neutral molecule or positive ion) from a calculated or measured potential curve for, say, $H_2$.

In this connection, Jenč has discussed [85] how an error in an experimentally determined molecular constant, or an approximation made in the calculation of a theoretical curve, affects the derived RPC. For most of Jenč's work [85, 86], potential curves derived from spectroscopic data by the RKR method were used. SCF calculations were not reliable enough unless corrections, such as those for MECE and incorrect dissociation (see Chapter II, Section C), were made.

The trends observed [83] for the 22 molecules for which $R_e$, $D_e$, and $\omega_e$ were known with sufficient accuracy were: (a) RPC's for molecules, the atomic numbers of whose atoms are close, coincide closely. (b) For larger atomic numbers, the RPC's are broadened and turned to larger $R$. The possibility of proving these trends mathematically was considered [83]. Considering the ground states of halogen and interhalogen molecules [85, 86], Jenč found that $Cl_2$ did not fit trends [83] found for other series of molecules. This suggested that one of the experimentally determined parameters was in error. Since an error of the size required for the discrepancy seemed unlikely, Jenč [86] considered errors connected with construction of the RKR curves, and also the possibility of modifying the definition of the RPC.

In the absence of sufficiently good data to construct the RKR curves for a large number of molecules, Jenč considered the Hulburt–Hirschfelder functions, parametrized for 30 molecules. Five spectroscopic constants, $\omega_e$, $B_e$, $\omega_e x_e$, $D_e$ and $\alpha_e$, are needed to construct this function:

$$U(R) = D_e[(1 - e^{-x})^2 + cx^3 e^{-2x}(1 + bx)]$$

$$x = \tfrac{1}{2}\omega_e(B_e D_e)^{-1/2}(R - R_e)/R_e$$

$$c = 1 + a_1(D_e/a_0)^{1/2}$$

$$b = 2 + \left(\frac{7}{12} - \frac{D_e a_2}{a_0}\right)\Big/c$$

Here, $a_0$, $a_1$, and $a_2$ are the constants in the Dunham expansion. The RPC's derived from these curves behaved like those derived from RKR curves, with some exceptions (such as the absence of the discrepancy for $Cl_2$). Some of these were attributed to deficiencies of the Hulburt–Hirschfelder curve in describing highly ionic molecules. Further applications of the RPC method were suggested [83], such as the possibility of using RPC's to understand the nature of the bonding in different molecules, or to construct accurate potential curves for regions of $R$ not experimentally measured.

Stwalley [87] has recently tested the idea of reduced potential curves for chemically related systems by considering the constancy of ratios of potential parameters. Thus, if $P$ and $Q$ are two measured observables involving the potential energy curves for systems A–B and C–D, and A–B and C–D are chemically similar, we could expect

$$P(A–B)/P(C–D) \approx Q(A–B)/Q(C–D). \tag{123}$$

This implies similarities of potential curves in the region of $R$ to which the measurement is sensitive. Because $\varrho$ of Eq. (120) is not proportional to $R$, Jenč's RPC formalism does not imply, and is not implied by, relations of the form (123). The hypothesis expressed by Eq. (116) would imply (123), but not vice versa. Stwalley [87] considered measured values of the van der Waals interaction constant, the well-depth parameter $\varepsilon$, and $\varepsilon R_e$ for pairs of atoms, one being a rare gas or Hg, and the other He, H, or an alkali atom. The correlation expressed by (123) seemed to hold. Various combination rules (arithmetic mean, geometric mean, arithmetic mean for reciprocals) were tested.

## REFERENCES

1. A. A. Frost, B. H. Prentice, III, and R. A. Rouse, *J. Amer. Chem. Soc.* **89**, 3064 (1967); A. A. Frost, *J. Chem. Phys.* **47**, 3707 (1967).
2. I. Shavitt, *Methods Comp. Phys.* **2**, 1 (1963); H. Preuss, *Z. Naturforsch.* **20a**, 18, 21, 1290 (1965).
3. A. A. Frost, *J. Chem. Phys.* **47**, 3714 (1967); *J. Phys. Chem.* **72**, 1289 (1968).
4. R. A. Rouse and A. A. Frost, *J. Chem. Phys.* **50**, 1705 (1969).
5. S. Y. Chu and A. A. Frost, *J. Chem. Phys.* **54**, 760, 764 (1971).
6. R. S. Mulliken, *J. Phys. Chem.* **56**, 295 (1952).
7. J. T. Vanderslice, E. A. Mason, and W. G. Maisch, *J. Chem. Phys.* **31**, 738 (1959); **32**, 515 (1960).
8. J. R. Arnold, *J. Chem. Phys.* **24**, 181 (1956).
9. H. Müller and H. Dunken, *Theor. Chim. Acta* **3**, 97 (1965).

10. S. L. Matlow, *J. Chem. Phys.* **34**, 1187 (1961).
11. A. A. Frost, *J. Chem. Phys.* **22**, 1613 (1954).
12. R. F. Borkman and R. G. Parr, *J. Chem. Phys.* **48**, 1116 (1968).
13. R. G. Parr and R. F. Borkman, *J. Chem. Phys.* **49**, 1055 (1968); R. F. Borkman, G. Simons, and R. G. Parr, *J. Chem. Phys.* **50**, 58 (1969).
14. P. Politzer, *J. Chem. Phys.* **52**, 2157 (1970).
15. G. Simons and R. G. Parr, *J. Chem. Phys.* **55**, 4197 (1971).
16. E. S. Rittner, *J. Chem. Phys.* **19**, 1030 (1951).
17. J. L. Margrave, *J. Phys. Chem.* **58**, 258 (1954).
18. Y. P. Varshni, *Trans. Faraday Soc.* **53**, 132 (1957).
19. R. G. Pearson and H. B. Gray, *Inorg. Chem.* **2**, 358 (1963).
20. W. A. Klemperer and J. L. Margrave, *J. Chem. Phys.* **20**, 527 (1952).
21. L. Brewer and D. Mastick, *J. Amer. Chem. Soc.* **73**, 2045 (1951).
22. A. Altshuller, *J. Chem. Phys.* **21**, 2074 (1953).
23. H. W. DeWijn, *J. Chem. Phys.* **44**, 810 (1966).
24. A. A. Frost and J. H. Woodson, *J. Amer. Chem. Soc.* **80**, 2615 (1958).
25. A. A. Frost and B. Musulin, *J. Chem. Phys.* **22**, 1017 (1954); P. S. K. Chen, M. Geller, and A. A. Frost, *J. Phys. Chem.* **61**, 828 (1957); A. A. Frost and J. H. Woodson, *J. Amer. Chem. Soc.* **80**, 2615 (1958).
26. A. C. Roach and M. S. Child, *Mol. Phys.* **14**, 1 (1968); A. C. Roach and P. Baybutt, *Chem. Phys. Lett.* **7**, 7 (1970).
27. C. Maltz, *Chem. Phys. Lett.* **3**, 707 (1969).
28. J. R. Platt, *in* "Encyclopedia of Physics" (S. Flügge, ed.), Volume XXXVII/2, Sect. 57–58. Springer-Verlag, Berlin and New York, 1961.
29. M. G. Menendez, M. J. Redmon, and J. F. Aebischer, *Phys. Rev.* **180**, 69 (1969).
30. H. S. Fricker and P. W. Anderson, *J. Chem. Phys.* **55**, 5028 (1971); H. S. Fricker, *Ibid.* **55**, 5034 (1971).
31. R. M. Badger, *J. Chem. Phys.* **2**, 128 (1934); **3**, 710 (1935).
32. D. R. Herschbach and V. W. Laurie, *J. Chem. Phys.* **35**, 458 (1967).
33. J. N. Murrell, *J. Mol. Spectrosc.* **4**, 446 (1960).
34. L. Salem, *J. Chem. Phys.* **38**, 1227 (1963).
35. P. J. Laurenzi and A. F. Saturno, *J. Chem. Phys.* **53**, 579 (1970).
36. G. B. B. M. Sutherland, *J. Chem. Phys.* **8**, 161 (1940).
37. J. W. Linnett, *Trans. Faraday Soc.* **36**, 1123 (1940).
38. J. W. Linnett, *Trans. Faraday Soc.* **38**, 1 (1942).
39. C. K. Wu and C. T. Yang, *J. Phys. Chem.* **48**, 295 (1944).
40. E. R. Lippincott, *J. Chem. Phys.* **21**, 2070 (1953).
41. E. R. Lippincott and R. Schroeder, *J. Chem. Phys.* **23**, 1131 (1955).
42. E. R. Lippincott, *J. Chem. Phys.* **23**, 603 (1955).
43. P. M. Morse, *Phys. Rev.* **34**, 57 (1929).
44. G. V. Calder and K. Ruedenberg, *J. Chem. Phys.* **49**, 5399 (1968).
45. Y. P. Varshni, *Rev. Mod. Phys.* **29**, 664 (1957).
46. B. R. Sristava and M. P. Madan, *Proc. Phys. Soc. London Sect. A* **66**, 278 (1953).
47. J. N. Wilson, *J. Chem. Phys.* **43**, 2564 (1965); **49**, 3325 (1968); A. D. Crowell, *Ibid.* **49**, 3324 (1968).
48. H. L. Kramer and D. L. Herschbach, *J. Chem. Phys.* **53**, 2792 (1970).
49. G. H. Hudson and J. C. McCoubrey, *Trans. Faraday Soc.* **56**, 767 (1960).
50. J. O. Hirschfelder and W. E. Roseveare, *J. Phys. Chem.* **43**, 15 (1939).

51. E. A. Guggenheim and M. L. McGlashan, *Proc. Roy. Soc. Ser. A* **206**, 448 (1951).
52. E. A. Mason, *J. Chem. Phys.* **23**, 49 (1955); E. A. Mason and W. E. Rice, *Ibid.* **22**, 522 (1954).
53. L. Pauling, "The Nature of the Chemical Bond," 3rd ed. Cornell Univ. Press, Ithaca, New York, 1960.
54. R. T. Sanderson, *J. Inorg. Nucl. Chem.* **28**, 1553 (1966).
55. H. S. Johnston, *J. Amer. Chem. Soc.* **86**, 1643 (1964).
56. R. J. Good and C. J. Hope, *J. Chem. Phys.* **55**, 111 (1971).
57. B. E. F. Fender and G. D. Halsey, *J. Chem. Phys.* **36**, 1881 (1962).
58. E. A. Mason, M. Islam, and S. Weissman, *Phys. Fluids* **7**, 1011 (1964).
59. R. J. Good and C. J. Hope, *J. Chem. Phys.* **53**, 540 (1970).
60. B. N. Sristava and K. P. Sristava, *J. Chem. Phys.* **24**, 1275 (1956).
61. K. S. Pitzer, *J. Chem. Phys.* **7**, 583 (1939).
61a. E. A. Mason and W. E. Rice, *J. Chem. Phys.* **22**, 843 (1954); R. J. Munn, *Ibid.* **40**, 1439 (1964).
62. S. Weissman, S. C. Saxena, and E. A. Mason, *Phys. Fluids* **3**, 510 (1960).
62a. S. C. Saxena and B. P. Mathur, *Chem. Phys. Lett.* **1**, 224 (1967).
63. H.-M. Lin and R. L. Robinson, Jr., *J. Chem. Phys.* **54**, 52 (1971).
64. J. H. Dymond and B. J. Alder, *J. Chem. Phys.* **51**, 309 (1969).
65. D. W. Calvin and T. M. Reed, III, *J. Chem. Phys.* **54**, 3733 (1971).
66. R. K. Nesbet, *J. Chem. Phys.* **48**, 1419 (1968).
66a. D. D. Konowalow, *J. Chem. Phys.* **50**, 12 (1969).
67. T. L. Gilbert, *J. Chem. Phys.* **49**, 2640 (1968).
68. R. L. Redington, *J. Phys. Chem.* **74**, 181 (1970).
69. L. Pauling, *J. Amer. Chem. Soc.* **54**, 3570 (1932).
70. K. A. Gingerich and H. C. Finkbeiner, *Chem. Commun.* p. 901 (1969); *J. Chem. Phys.* **54**, 2621 (1971).
71. W. Gordy, *J. Chem. Phys.* **14**, 305 (1946).
72. R. Ferreira, *Trans. Faraday Soc.* **59**, 1064, 1075 (1963).
73. R. Ferreira, *Advan. Chem. Phys.* **13**, 55 (1967).
74. Z. Hussain, *Can. J. Phys.* **43**, 1690 (1965).
75. P. L. Goodfriend, *Can. J. Phys.* **45**, 3425 (1967).
76. S. P. Tandon, M. P. Bhutra, and P. C. Mehta, *Indian J. Pure Appl. Phys.* **8**, 18 (1970).
77. G. Van Hooydonk and Z. Eeckhaut, *Ber. Bunsenges. Phys. Chem.* **74**, 323, 327 (1970).
78. G. Van Hooydonk, *Theor. Chim. Acta* **22**, 157 (1971).
79. R. S. Evans and J. E. Huheey, *J. Inorg. Nucl. Chem.* **32**, 373, 777 (1970).
80. J. R. Stallcop, *J. Chem. Phys.* **54**, 2602 (1971).
81. J. S. Rowlinson, *Mol. Phys.* **9**, 197 (1965).
82. A. A. Frost and B. Musulin, *J. Amer. Chem. Soc.* **76**, 2045 (1954).
83. F. Jenč, *J. Chem. Phys.* **47**, 127 (1967).
84. F. Jenč and J. Pliva, *Collect. Czech. Chem. Commun.* **28**, 1449 (1963).
85. F. Jenč, *Collect. Czech. Chem. Commun.* **29**, 1507, 1521 (1964).
86. F. Jenč, *Collect. Czech. Chem. Commun.* **30**, 3772 (1965).
87. W. T. Stwalley, *J. Chem. Phys.* **55**, 170 (1971).

## SUPPLEMENTARY BIBLIOGRAPHY

Below are listed, alphabetically by author, articles relevant to the content of this chapter which came to our attention too late to be included in the manuscript. In addition to author, title, and journal reference, we have included, in some cases, additional information (in parentheses), and given the Section where related material is discussed.

A. B. Anderson and R. G. Parr, Diatomic Vibrational Potential Function from Integration of a Poisson Equation. *J. Chem. Phys.* **55**, 5490 (1971); **56**, 5204 (1972). (B1, B2)

A. B. Anderson and R. G. Parr, A Poisson Equation for Vibrational Potentials of Diatomic Molecules. *Theor. Chimica Acta* **26**, 301 (1972). (B1, B2)

A. B. Anderson and R. G. Parr, Universal Force Constant Relationships and a Definition of Atomic Radius. *Chem. Phys. Lett.* **10**, 293 (1971). (B1, B2)

T. Anno, On the Evaluation of the Energies of Hybridized Valence States (for atoms in molecules theories). *Theor. Chimica Acta* **25**, 248 (1972). (A3)

N. J. Brown and R. J. Munn, Inert Gas Potentials for Mixed Interactions. *J. Chem. Phys.* **57**, 2216 (1972). (B2)

C. Butterfield and E. H. Carlson, Ionic Soft Sphere Parameters from Hartree–Fock–Slater Calculations. *J. Chem. Phys.* **56**, 4907 (1972). (B1)

S. Y. Chu, SCF Calculations for $H_2^+$, $Li_2^+$, and $LiH^+$ with Atomic Basis Sets Enlarged by Bond Functions. *Theor. Chimica Acta* **25**, 200 (1972). (B1)

W. H. Fink, Approach to... the Li-He interaction Potential. *J. Chem. Phys.* **57**, 1822 (1972). (B1)

W. Jakubetz, H. Lischka, P. Rosmus and P. Schuster, On the Role of Configuration Interaction in Semiempirical Methods (effect on calculated potential curves). *Chem. Phys. Lett.* **11**, 38 (1971). (A1)

L. R. Kahn and W. A. Goddard III, Ab Initio Effective Potentials in Molecular Calculations. *J. Chem. Phys.* **56**, 2684 (1972). (A2)

T. Kihara and K. Yamazaki, Estimated van der Waals Potential Between Halogen and Inert Gas Atoms. *Chem. Phys. Lett.* **16**, 32 (1972). (B2)

G. Klopman and R. Polak, Semiempirical SCF Theory with "Scaled" Slater Orbitals. *Theor. Chimica Acta* **25**, 223 (1972). (A1)

D. D. Konowalow, P. Weinberger, J.-L. Calais, and J. W. D. Connolly, SCF-α Cluster Calculations for the Ground State $Ne_2$ Molecule. *Chem. Phys. Lett.* **16**, 81 (1972). (A2)

B. J. Laurenzi, Isoelectronic Molecules: First and Second-Order Properties. *Theor. Chim. Acta* **28**, 81 (1973). (B2)

J. N. Murrell and A. J. Harget, "Semiempirical SCF-MO Theory of Molecules." Interscience, New York (1971). (A1)

R. G. Parr, J. M. Finlan, and G. W. Schnuelle: Static Charge Distributions which Generate Exact Potential Curves for Diatomic Molecules. *Chem. Phys. Lett.* **14**, 72 (1972). (B1)

P. Politzer, R. K. Smith, and S. D. Kasten, Energy Calculations with the Extended Hückel method (and modifications of it). *Chem. Phys. Lett.* **15**, 226 (1972). (A1)

H. O. Pamuk, Semiempirical ... Correlation Energies of BH, CH, NH, OH, HF, $N_2$... *Theor. Chim. Acta* **28**, 85 (1972). (A, B)

A. I. M. Rae, A Theory for the Interactions Between Closed Shell Systems. *Chem. Phys Lett.* **18**, 574 (1973). (A2)

A. C. Roach, Theoretical Ground State and Excited State Potential Energy Curves for Alkali Diatomic Molecules (by a semiempirical pseudopotential method). *J. Mol. Spec.* **42**, 27 (1972). (A2, B1)

S. I. Sandler and J. K. Wheatley, Intermolecular Potential Parameter Combining Rules for the Lennard-Jones 6-12 Potential. *Chem. Phys. Lett.* **10**, 375 (1971). (B2)

A. B. Sannigrahi and S. N. Mohammed, Simplified SCF Calculations on the Diatomic Alkali Metal Molecules. (Semiempirical valence electron no). *Mol. Phys.* **24**, 905 (1972) (A1)

M. E. Schwartz and J. D. Switalski, Valence Electron Studies with Gaussian-Based Model Potentials . . . Applications to $Li_2^+$, $Na_2^+$, and $LiH^+$. *J. Chem. Phys.* **57**, 4132 (1972). (A2)

G. Simons, Pseudopotential Studies of the Water and Hydrogen Fluoride Molecules. *Chem. Phys. Lett.* **18**, 315 (1973). (A2)

F. T. Smith, Atomic Distortion and the Combining Rule for Repulsive Potentials. *Phys. Rev.* **A5**, 1708 (1972). (B1, B2)

K. G. Spears, Repulsive Potentials of Atomic Ions, Atoms, and Molecules (determined semiempirically). *J. Chem. Phys.* **57**, 1842 (1972). (A2, B1)

A. van Wijngaarden, B. Miremadi, and W. E. Baylis, The Repulsive Part of the Potential Between Hg Atoms and Light Atomic Projectiles. *Can. J. Phys.* **50**, 1938 (1972). A2

# General Bibliography

The discussions of the preceding chapters assume a knowledge of quantum mechanics such as would be obtained from a one-semester course. Material which is standard or elementary has not been included, or merely summarized. Instead, reference has been made to one or more quantum chemistry texts. These basic works are listed in Section 1. In the present volume, all sources in Section 1 of this Bibliography are referred to by the name of the first author, and pertinent page, in upper-case letters and *without* a reference number. In Section 2, we give some books closely related to the subject matter discussed here, to which we have probably not referred as often as we should have.

The coverage of the literature of diatomic molecule calculations emphasizes recent work. In particular, an attempt was made to mention all papers published in 1970 and 1971, since the most recent of the bibliographies listed below (Richards, Walker, and Hinkley) includes papers through 1969. The articles and books listed under Bibliographies of Calculations (Section 3) and Reviews of the Literature (Section 4) provide complete coverage of the published literature.

## 1. *Basic Texts in Quantum Chemistry*

H. Eyring, J. Walter, and G. E. Kimball, "Quantum Chemistry." Wiley, New York, 1944.
W. Kauzmann, "Quantum Chemistry." Academic Press, New York, 1957.

F. L. Pilar, "Basic Quantum Chemistry." McGraw-Hill, New York, 1968.

R. McWeeny and B. T. Sutcliffe, "Methods of Molecular Quantum Mechanics." Academic Press, New York, 1969.

I. N. Levine, "Quantum Chemistry," Vol. I. Allyn and Bacon, Boston, 1970.

## 2. Books Discussing the Calculation of $U(R)$

G. Herzberg, "Molecular Spectra and Molecular Structure, I. Spectra of Diatomic Molecules." Van Nostrand, New York, 1950.

J. O. Hirschfelder, C. F. Curtiss and R. Byron Bird, "Molecular Theory of Gases and Liquids." Wiley, New York, 1954.

J. O. Hirschfelder, ed., "Intermolecular Forces" (Advances in Chemical Physics, Vol. 12). Interscience, New York, 1967.

H. Margenau and N. R. Kestner, "Theory of Intermolecular Forces." Pergamon, Oxford, 1969 (2nd Ed. 1971).

J. C. Slater, "Quantum Theory of Molecules and Solids, Vol. I, Electronic Structure of Molecules." McGraw-Hill, New York, 1963.

I. M. Torrens, "Interatomic Potentials." Academic Press, New York, 1972.

## 3. Bibliographies of Calculations

H. Yoshizumi, *Advan. Chem. Phys.* **2**, 323 (1959). A bibliographical survey of methods for treating interelectronic correlation. Papers dealing with SCF and other calculations for simple molecules are listed.

J. C. Slater, "Quantum Theory of Molecules and Solids," Vol. I. McGraw-Hill, New York, 1963. A bibliography is given (p. 401 *et seq.*) of papers and books giving a "fairly complete account of the literature" on molecular electronic structure. Articles are listed by author, with cross-references, and titles are given.

R. G. Parr, "Quantum Theory of Molecular Electronic Structure." Benjamin, New York, 1963. In Section 7 is a summary of a variety of *ab initio* calculations on small molecules.

R. K. Nesbet, *Advan. Quantum Chem.* **3**, 1 (1967). Methods for approximate Hartree–Fock calculations on small molecules are discussed, together with the results of such calculations.

A. D. McLean and M. Yoshimine, *IBM J. Res. Devel.* **12**, 206 (1968). Methods and accomplishments in the computation of molecular properties and structure.

M. Krauss, Natl. Bur. Stds. Technical Note 438 (U.S. Dep't. of Commerce, Dec. 1967). Gives references to *ab initio* molecular electronic calculations from 1960 to 1967, including material unpublished at the time of the report. For each molecule, a list of calculations is given, with references, together with information on the method used, the properties calculated, and whether part of the potential curve was derived. In separate tables, the best calculated values for dissociation energies, spectroscopic constants, and other quantities are given.

G. Klopman and B. O'Leary, *Fortschritte der Chem. Forsch.* **15**, 445 (1970). A book-length discussion and critique of semi-empirical molecular orbital calculations, with a complete bibliography and tables of results.

R. G. Clark and E. T. Stewart, *Quart. Rev.* **24**, 95 (1970). Here are tabulated nonempirical LCAO-MO calculations for systems of 4 or more electrons, published between 1960 and 1968, indicating where calculations were performed for several internuclear distances.

W. G. Richards, T. E. H. Walker, and R. K. Hinkley, "A Bibliography of *ab initio* Molecular Wave Functions." Oxford Univ. Press, London, 1971. Includes all *ab initio* calculations published through the end of 1969. For each molecule, calculations are listed with references, the nature of each calculation being given. The energy obtained by each (a rough measure of accuracy) is indicated, together with indications of which properties were calculated in addition to the energy. Properties considered include potential curve, spectroscopic constants, and dissociation energies.

## 4. *Reviews of the Literature*

Reviews of the literature and surveys of recent work appear regularly as chapters in *Annual Reviews of Physical Chemistry*. These are generally complete (each covering the period since the previous review of the same subject), with the amount of discussion variable. Below are listed the chapters that have appeared, since 1958, dealing with molecular calculations.

M. Kotani, Y. Mizuno, K. Kayama, and H. Yoshizumi, Quantum Theory of Electronic Structure of Molecules. **9**, 245 (1958).

J. A. Pople, Quantum Theory. Theory of Molecular Structure, and Valence. **10**, 331 (1959).

P.-O. Löwdin, Quantum Theory of Electronic Structure of Molecules. **11**, 107 (1960).

O. Sinanoğlu and D. F.-T. Tuan, Quantum Theory of Atoms and Molecules. **15**, 251 (1964).

B. M. Gimarc and R. G. Parr, Quantum Theory of Valence. **16**, 451 (1965).

F. Prosser and H. Shull, Quantum Chemistry. **17**, 37 (1966).

A. Golebiewski and H. S. Taylor, Quantum Theory of Atoms and Molecules. **18**, 353 (1967).

C. Schlier, Intermolecular Forces. **20**, 191 (1969).

L. C. Allen, Quantum Theory of Structure and Dynamics. **20**, 315 (1969).

A. D. Buckingham and B. D. Utting, Intermolecular Forces. **21**, 287 (1970).

F. E. Harris, Quantum Chemistry. **23**, 415 (1972).

# Author Index

Numbers in parentheses are reference numbers and indicate that an author's work is referred to although his name is not cited in the text. Numbers in italics show the page on which the complete reference is listed.

## A

Adams, W. H., 145, 146(6), 176, 177, *181*
Aebischer, J. F., 360(29), *388*
Ahlrichs, R., 272, 273(117), 274(117), *289*
Albat, R., *290*
Alder, B. J., 377, 383, *389*
Alexander, M. H., 56, *63*, 108(51), *117*
Allavena, M., 131(18), *136*, 213(13), 215(13), *286*
Allen, L. C., 109, *117*, 151, 158(50, 59), *181*, *182*, *183*, 220(25), 233, 273, 274(118), *287*, *289*, 299(11a), 300(12), 314(36), *345*, *346*, *394*
Allison, A. C., 331(59), *347*
Alper, J. S., 159, *183*
Altshuller, A., 359(22), *388*
Amemiya, A., 155(24), *182*, 189(7), 190(7), 191(7), *206*
Amos, A. T., 76, 77, 89(15, 16), 90, *116*, 142(1, 2), 173, *181*
Anderson, A. B., *290*, *390*
Anderson, P. W., 361, *388*
Andreev, E. A., *136*
Anex, B. G., 252(82), 268, *288*
Anno, T., *390*
Appel, K., 305(24), 306(24), 309(24), *346*
Arai, T., 333, 334(63,64), 336(64), 337(66), 340(66, 67), 342, *347*
Arnau, C., 158(53), *182*

## A (cont.)

Arnold, J. R., 352, *387*
Arrighini, G. P., 58, *63*
Askar, A., 55(89), *63*

## B

Baber, W. G., 119, *135*
Bader, R. F. W., 132, *136*, 247(73), *288*
Badger, R. M., 361, 362(31), *388*
Baetzold, R. C., 315, *346*
Bagus, P. S., 150(14), *181*, 193(14), *206*, *291*
Balint-Kurti, G. G., 279, 281, *289*, 340, 341, 342, *347*
Bandrauk, A. D., 247(73), *288*
Banyard, K. E., 160, 161(71), *183*
Bardsley, J. N., *290*, 331, 332, *347*
Barker, J. A., 30, 42, *62*
Barnett, G. P., 265, *288*
Barnett, M. P., 155, 160, *182*
Basilevsky, M. V., *136*
Bates, J. K., 306(28), 307, 308(28), 309, *346*
Baybutt, P., 359(26), 360(26), *388*
Baylis, W. E., 327, 328, 329, *347*, *391*
Beach, Y., 50, *63*
Beckel, C. L., 208, *286*
Belford, R. L., 149(12), *181*, 236(59), 252(59), *288*
Bell, R. J., 30, 31(51), 40, *62*

395

# Subject Index

## A

AlH, 238–242

Alkali atom-rare gas interactions, 30, 31, 39, 259, 283, 327–330, 360

Alkali halides, 223, 259–263, 350, 357, 359, 378, 379

Alkali molecules, 30, 284, 325–327, 359, 361, 369

Anharmonicity, 217, 226, 255, 356, 366–369, *see also* Spectroscopic constants

Antisymmetrization, 3, 7, 8, 51, 68, 78, 81, 101, 171, 173, 190, 328, 333

$Ar_2$, 114, 234, 377, 383

$Ar_2^+$, 256

Ar Hg, 28

Asymptotic series, 12, 14

Atoms-in-molecules methods, 93, 113, 201, 279, 294, 332–345

Average energy, *see* Mean energy denominator

## B

$B_2$, 280

BC, 281

BF, 243, 247–248

BH, 127, 236, 238–242, 272–279

BN, 245, 248

Badger's rule, 361, 363–365, 367, 373, 384

Basis functions, 149–161, 223, 295, 300

Basis set, 225, 234, 256

BeF, 251

BeH, 238–242

$BeH^+$, 202, 236, 272

$BeH^-$, 252

BeO, 222, 224, 245, 247–250, 281–282

Binding energy, *see* Dissociation energy

Bond charge model, 354–357

Bond energy, *see* Dissociation energy

Brillouin Theorem, 147, 150, 167, 170, 172, 220, 226, 230

Brillouin-Wigner perturbation theory, 88–90, 93, 122

## C

$C_2$, 245–247, 350

$C_2^+$, 246

CF, 258–259

CH, 238–242

CN, 280, 386

CNDO, 311–314, 330

CO, 227, 247–248, 282, 283, 340

$CO^+$, 257

$Cl_2$, 246, 279, 386

$Cl_2^-$, 256

ClO, 251

Closed shell, 165, 166, 171, 175, 179, 194, 219, 323, 331

Combination rules, 370–378

Configuration, 173, 198, 215, 222, 227, 228, 233, 234, 250, 266, 284

Configuration interaction, 16, 58, 106, 108, 111, 114, 115, 134, 135, 184, 197, 198–206, 211, 213, 220, 236, 264–286, 336

Core polarization, 325, 360

Correlated wave function, 185, 325

Correlation, 37, 55, 134, 198, 200, 203, 204, 219, 256, 260–262, 272, 276, 278, 284

406

# N

$N_2$, 242–245, 340, 350
$N_2^+$, 243, 257–258, 381
$Na_2$, 324–325
NF, 246
NH, 238–242, 274–275
NO, 231, 281, 381
NaF, 284
NaH, 226, 238–242, 349
NaHe, 259
NaLi, 283–284, *see also* Alkali molecules
$NaLi^+$, 283–284
Natural orbital, 142, 198, 205, 216, 265, 272, 275, 277, *see also* Iterative natural orbital method
$Ne_2$, 114, 234
$Ne_2^+$, 256
$NeH^+$, 254
Noble gas, *see* Rare gas
Noncrossing rule, 72, 144

# O

$O_2$, 37, 246, 280–281, 381
OH, 230, 238–242
$OH^-$, 252–254
One-center expansion, 159–161, 199, 213, 237, 251, 252, 268, 269
Open shell, 19, 165, 167, 168, 172, 175, 184, 192–194, 219, 230, 231, 245, 259, 278, 280
Optimized valence configurations, 201, 203–205, 267, 276–278, 284, 286
Order of smallness, 8, 75, 103, 210
Oscillator strength, 23–25, 30, 41, 45, 47, 52
Oscillator strength distribution function, 23–25
Overlap, 92–95, 98, 324, 327, 338 *see also* Differential overlap

# P

PH, 238–242
Padè approximant, 40, 47, 61
Paired electrons, *see* Electron pairs
Particle in box, 352–356
Partitioning theory, 83–87, 197

Permutation group, *see* Symmetric group
Perturbation theory, 2, 7, 8, 12, 14 *see also* Brillouin-Wigner theory
Polarizability, 15, 16, 22, 24–28, 32, 33, 38, 39, 49, 52, 257, 259, 317, 328, 331, 357, 359, 373
frequency-dependent, 28, 32, 34, 36, 41, 46, 54, 55
Polarization function, 158, 179, 231, 237, 243, 248, 259, 260, 263, 279
Power series, 208–211
Projected Hartree-Fock, 174, 193
Projection operator, 3, 43, 81, 86, 87, 126, 172–174, 185, 189–192, 197, 319
Pseudoeigenvalue equation, 142, 194, 230, 297
Pseudokinetic energy correction, 218, 267
Pseudopotential, 294, 316–332, 359, 360

# Q

Quadrupole moment, 16, 17, 258, 332, *see also* Moment, permanent

# R

$R_e$, *see* Equilibrium internuclear distance
Rare gas interactions
dispersive, 29–32, 38, 40–42, 46, 56, 58, 371, 375, 377
repulsive, 114, 180, 233–235, 259, 264–267, 331, 375, 377
Reduced energy, 368
Reduced internuclear distance, 368, 383, 385
Reduced potential curve, *see* Universal potential curve
Relativistic energy, 20, 60, 222, 223, 238, 240, 241, 245, 248, 250, 260–262
Restricted Hartree-Fock, 144, 165, 167, 169–179, 200, 203, 207, 219–232, 251, 260, 265, 270, 278
Roothaan equations, 145, 149–150

# S

SF, 179, 263
SH, 238–242
$SH^-$, 252–254

# Physical Chemistry

## A Series of Monographs

**Ernest M. Loebl,** Editor

Department of Chemistry, Polytechnic Institute of

Brooklyn, Brooklyn, New York